Reproductive Biology of Crustaceans

Case Studies of Decapod Crustaceans

Reproductive Biology of Crustaceans

Case Studies of Decapod Crustaceans

Editor
Elena Mente
Department of Ichthyology and Aquatic Environment
University of Thessaly
Greece

and

Honorary Research Fellow
University of Aberdeen
UK

Science Publishers
Enfield (NH) Jersey Plymouth

Science Publishers
234 May Street
Post Office Box 699
Enfield, New Hampshire 03748
United States of America

www.scipub.net

General enquiries : *info@scipub.net*
Editorial enquiries : *editor@scipub.net*
Sales enquiries : *sales@scipub.net*

Published by Science Publishers, Enfield, NH, USA
An imprint of Edenbridge Ltd., British Channel Islands
Printed in India

ISBN 978-1-57808-529-3

Library of Congress Cataloging-in-Publication Data

Reproductive biology of crustaceans : case studies of decapod crustaceans/editor, Elena Mente.
p. cm.
Includes bibliographical references and index.
ISBN 978-1-57808-529-3 (hardcover)
1. Decapoda (Crustacea)—Reproduction. I. Mente, Elena.

QL444.M33R464 2008
595.3'8146—dc22

2008000561

To my husband
Ioannis

The review indicates that very little is known about the behaviour of crustacean parents, the resources they invest into individual offsprings, the benefits and costs of this reproductive behaviour and the evolution of social behaviour.

Knowledge of reproductive biology of females has come mainly from studies on sexual maturity by macroscopical investigations and by histological methods (Adiyodi and Subramonian, 1983). To date eyestalk ablation still represents the most commonly used manipulation to induce maturation and spawning in female crustaceans. This is an outcome of the success of this method in increasing reproductive output and spawning frequency and of its achievement in shortening the latency to first spawn (Racotta et al. 2003). Female maturity is more easily defined since attached eggs provide conclusive proof although there are other indicators such as relative abdomen width (in *Homarus*), ovigerous setae and cement gland development (Aiken and Waddy 1980). The oocytes and ovaries of *Nephrops* species develop from white through yellow and pale green to very dark green at maturity (Table 1). The investigation on oogenesis in many crustacean mature females allows the identification of growing stages and their comparison between species. The literature for many crustaceans distinguishes 4 to 5 growing stages (Adiyodi and Subramonian, 1983). However, more species-specific research using histological, histochemical and ultrastructural approaches is needed to investigate egg (oocyte) development and improve understanding of oogenesis and management of crustacean. Central to female reproduction is the process of vitelogenesis including the biosynthesis of yolk proteins and their transport and storage in the ovary for sustenance of the developing embryo. In many decapoda, there is evidence that yolk protein synthesis depends on extraoocytic sequestration of vitellogenin, and haemolymph protein fraction specific to vitellogenic females (Desantis et al. 2001). Chapter 6 of this book focuses mainly on research on the hormonal control of vitellogenesis and ovarian maturation. It illustrates recent advances in identifying the pathways of vitellogenin synthesis and describes its main regulators and advances in their isolation and mode of action. The research detailed in Chapter 6 has shown that the endocrine regulation of both reproduction and molting is controlled by a complex interaction of neuropeptides, juvenoids and steroids. Although studies have confirmed that female gonadal development could successfully be induced through injections or transplants of these hormones and endocrine glands, the effects are highly variable depending on factors such as species, developmental stage of the animal and concentrations of hormones and neurotransmitters given. For future applications of hormones in the manipulation of breeding stocks in

To my husband
Ioannis

Preface

Crustaceans adapt to a wide variety of habitats and way of life. They have a complex physiological structure particularly with regard to the processes of growth (molting), metabolic regulation, and reproduction. Crustaceans are ideal model organisms for the study of endocrine disruption and stress physiology in aquatic invertebrates. This book is an overview of the extensive research that has taken place over the recent years on issues of crustacean reproduction.

Acknowledgements

The editor is grateful to the referees who found time to provide constructive comments on drafts of individual chapters. She also gratefully acknowledges the help of Professor R. G. Hartnoll for his detailed guidance on the structure and content of the book, Dr. Katrin Schimdt Researcher of the British Antarctic Survey for helpful comments and suggestions on chapter six, and, she is particularly grateful to Maria Theodossiou who had the patience to read the entire earlier version of the manuscript and make editorial comments. Finally the editor would like to express her sincere gratitude to her family and colleagues for their support and contribution to this book.

Contents

List of Contributors

Arvanitoyannis Ioannis S.

Associate Professor, School of Agricultural Sciences, Department of Ichthyology and Aquatic Environment University of Thessaly, Fytoko Street, GR- 38446 Nea Ionia Magnesias, Greece.

Tel.: +30 24210 93104, Fax: +30 24210 93137

E-mail: parmenion@uth.gr

Barki Assaf

Researcher, Aquaculture Research Unit, Institute of Animal Science, Agricultural Research Organization, Volcani Center, P.O. Box 6, Bet Dagan 50250, Israel.

E-mail: barkia@volcani.agri.gov.il

Chang Erest S.

Professor of Animal Science and of Neurobiology, Physiology and Behavior, Bodega Marine Laboratory, University of California-Davis PO Box 247, 2099 Westshore Road Bodega Bay CA 94923, U.S.A.

Tel: 707-875-2061, Fax: 707-875-2091

E-mail: eschang@ucdavis.edu

Cuzon Gerard

Researcher, Ifremer/COP, BP 7004 Taravao, Tahiti, French Polynesia, France. Tel: 689546035, Fax: 689546099

E-mail: Gerard.Cuzon@ifremer.fr

Exadactylos Athanasios

Assistant Professor, School of Agricultural Sciences, Department of Ichthyology and Aquatic Environment, University of Thessaly, Fytoko Street GF- 38446, Nea lonia Magnesias, Greece. Tel: 3024210 93173, Fax: 30 24210 93157

Email: exadactuputh.gr

Gaxiola Gabriela

Professor, Unidad Multidisciplinaria de Docencia e Investigation, Faculdad de Ciencias UNAM, Porto de Abrigo s/n, Sisal, Yucatan, Mexico.

Tel: 5298891 20147-49, Fax: 5298891 20147-49

E-mail: mggc@hp.fciencias.unam.mx

Klaoudatos S. Spyros

Professor, School of Agriculture Sciences, Department of Ichthyology and Aquatic Environment, University of Thessaly, Fytoko Street, GR- 38446 Volos, Greece.

Tel.: +30 3 24210 93145, Fax: +30 3 210 8991738

E-mail: sklaoudat@uth.gr

Klaoudatos Dimitris S.

Research Fellow, School of Agriculture Sciences, Department of Ichthyology and Aquatic Environment, University of Thessaly Fytoko Street, GR- 38446 Volos, Greece.

E-mail: dklaoudatos@yahoo.com

Lizárraga-Cubedo Héctor A.

Research Fellow, School of Biological Sciences (Zoology), University of Aberdeen, Tillydrone Avenue, Aberdeen AB24 2TZ, UK.

E-mail: halizarraga@yahoo.com

Mente Elena

Assistant Professor, School of Agricultural Sciences, Department of Ichthyology and Aquatic Environment, University of Thessaly, Fytoko Street, GR- 38446 Nea Ionia Magnesias, Greece and

Honorary Research Fellow, University of Aberdeen, UK.

Tel.: +30 24210 93176, Fax: +30 24210 93134

E-mail: emente@uth.gr, e.mente@abdn.ac.uk

Neofitou Christos

Professor, School of Agricultural Sciences, Department of Ichthyology and Aquatic Environment, University of Thessaly, Fytoko Street, GR- 38446 Nea Ionia Magnesias, Greece.

Tel.: +30 24210 93064, Fax: +30 24210 93065

E-mail: chneofit@uth.gr

Papadopoulou Nadia K.

Senior Researcher, Hellenic Centre for Marine Research Crete (HCMR), Institute of Marine Biological Resources, P.O. Box 2214, Heraklion 71003, Crete, Greece.

Tel: +302810 337827, Fax: +302810 337822

E-mail: nadiapap@her.hcmr.gr

Parnes Shmulik

Department of Life Sciences, Ben-Gurion University of the Negev, P.O. Box 653, Beer-Sheva 84105, Israel.

E-mail: parnes@bgumail.bgu.ac.il

Pierce Graham J.

Reader, School of Biological Sciences (Zoology), University of Aberdeen, Tillydrone Avenue, Aberdeen AB24 2TZ, UK.

Tel: +44 1224 272459, Fax: +44 1224 272396

E-mail: g.j.pierce@abdn.ac.uk

Raviv Shaul

Department of Life Sciences, Ben-Gurion University of the Negev, P.O. Box 653, Beer-Sheva 84105, Israel.

E-mail: ravivsha@bgu.ac.il, shauli.raviv@gmail.com

Ritar Arthur J.

Tasmanian Aquaculture and Fisheries Institute, Marine Research Laboratories, University of Tasmania, Nubeena Crescent, Taroona Tasmania, 7053, Australia.

Tel: +61 03 6227 7294, Fax: +61 03 6227 8035

E-mail: Arthur.Ritar@utas.edu.au

Rosas Carlos

Professor, Unidad Multidisciplinaria de Docencia e Investigation, Faculdad de Ciencias UNAM, Porto de Abrigo s/n, Sisal, Yucatan, Mexico.

Tel: +529889120147-49, Fax: +529889120147-49

E-mail: crv@hp.fciencias.unam.mx

Sagi Amir

Professor, Department of Life Sciences and National Institute for Biotechnology in the Negev, Ben Gurion University, PO Box 653, Beer Sheva 84105, Israel

Tel: 972-(0)8-6461364, Fax: 972-(0)8-6472992

E-mail: sagia@bgu.ac.il

Santos M. Begoña

Researher Senior, Instituto Español de Oceanografía, Centro Oceanográfico de Vigo, P.O. Box 1552, 36200 Vigo, Spain.

Tel: +34 986492111, Fax: +34 986498626

E-mail: m.b.santos@vi.ieo.es

Smith Christopher J.

Research Director, Hellenic Center for Marine Research Crete (HCMR), Institute of Marine Biological Resources, P.O. Box 2214, Heraklion 71003, Crete, Greece.

Tel: +30 2810 337752, Fax: +30 2810 337822

E-mail: csmith@her.hcmr.gr

Smith Greg G.

Researcher, Australian Institute of Marine Science, Marine Biotechnology, PMB No. 3 Townsville MC, Queensland, 4810 Australia

Tel: +61 07 47534178, Fax: +61 07 47725852

E-mail: g.smith@aims.gov.au.

Stowasser Gabriela

Researcher, British Antarctic Survey, Natural Environment Research Council, High Cross, Madingley Road, Cambridge CB3 0ET, UK

Tel. +44 01223/221241, Fax: +44 01223/221259

E-mail: gsto@bas.ac.uk

1

An Overview

E. Mente[1] and C. Neofitou[2]

Assistant Professor, School of Agricultural Sciences, Department of Ichthyology and Aquatic Environment, University of Thessaly, GR- 38446 Volos, Greece. and Honorary Research Fellow, School of Biological Sciences, University of Aberdeen, Tillydrone Avenue, AB24 3TZ Aberdeen, UK. E-mails: emente@uth.gr; e.mente@abdn.ac.uk

Professor, School of Agricultural Sciences, Department of Ichthyology and Aquatic Environment, University of Thessaly, GR- 38446 Volos, Greece. E-mail: chneofit@uth.gr

Interest in the theoretical and the applied biology of crustaceans has been growing steadily over a long time, not only because of the commercial importance of the species but also because they are an excellent study species for physiological, biochemical and neurobiological research. This book is intended to provide a synthesis of the diverse of research that has been carried out on the reproductive biology of crustaceans and to emphasise as case studies current active research on lobsters and shrimps. In recent years the need to understand and control reproductive processes and mechanisms in crustaceans has been stimulated by increasing interest in the culture of shrimps, lobsters and crabs.

In the subphylum Crustacean there are 6 classes and the Malacostraca contain some 70% of all species (Chapter 2). However, within Malacostrace most of the species are contained within the three orders Decapoda, Amphipoda and Isopoda (Chapter 2). It seems that Crustacea have a multiplicity of groups but most of the species are concentrated in a few of these groups and the groups which are relatively poor in species are often important in terms of individuals (Warner, 1977). The decapod is the group that contains the most familiar crustaceans as, lobsters, shrimps and crabs. Convenient size, easy availability, important commercial fisheries species, complexity and diversity make the decapod model crustaceans to study.

The application of biotechnology to crustacean aquaculture is of an increasing in importance (Chapter 3). Early doubts that such approaches were not fruitful in view of the crude state of development of the aquaculture industry are replaced by the recognition that sophisticated molecular approaches are required to tackle key problems of disease recognition and control, and to achieve higher production through the development of domesticated strains. Progress in penaeid genetic and biotechnological research has been slow because of a lack of knowledge on fundamental aspects of their biology. However, data is beginning to emerge from research projects during the last decade, and this is likely to increase rapidly in the future. Highly variable markers are developed that allow the genetic variation in crustacean stocks to be assessed, even in highly inbred cultured lines. Greater attention is devoted to developing cell lines for crustaceans, and to the development of molecular probes for the identification of pathogens. Chapter 3 concludes that these newly emerged molecular techniques are transforming the research and practice of animal agriculture, and are instrumental in developing profitable, environmentally friendly agriculture and in meeting the demand of market worldwide. The complete genomic sequences and DNA microarrays featuring the whole genome of major farm animal species are expected to be available in the near future. Novel approaches to enhancing animal productivity and health are expected to be developed, and new biotherapeutics and new animal breeds with desired characteristics to be produced. In addition, transgenic farm animals appear to be a progressing avenue to much needed pharmaceuticals and immunologically suitable organs for transplantation.

The large variety of reproduction patterns in crustacean is a result of adaptive processes determined by evolutionary pressures to improve survival of the offspring (Hartnol and Gould, 1988). Knowledge of the reproductive biology of the species in relation to their growth patterns is important for sustainable future stocks assessments. Size at sexual maturity and fecundity are of great importance for understanding population dynamics and reproductive strategies (Kennelly and Watkins, 1994). Chapter 4 of this book explores some of the most important aspects of crustacean reproduction in relation to their exploitation by fisheries. It reviews relationships between size classes and fecundity, egg development and their spatial variability, catch-effort restrictions, minimum legal landing size, a ban on landings of berried females, closed seasons and protected areas. It discusses stock assessment methods and migration patterns. Environmental factors affect crustacean reproduction through modifications of size at maturity, delay in time of moulting and spawning and hatching periods. Temperature is an important factor affecting catch rates in the short term and this must be considered when using catch per unit effort as an abundance index. The presence of external and internal maturity indicators is more evident in females than in males and female maturity is more relevant for stock assessment. Therefore analysis of maturity in females is a more common practice in crustacean maturity studies. Chapter 4 presents a list of maturity indicators and their respective relationships to body size and maturity for different species of crustaceans around the world for both males and females. Knowledge on maturity and reproductive biology in relation to fishery practices is essential to achieve the sustainable management of the species. Future studies should focus on sustainable management in fishing resources in the ecosystem and trophic level context.

Chapter 5 reviews the different aspects of sexual selection and demonstrates the diversity of mating behaviour and mating systems in Crustacea. Numerous studies have investigated various factors influencing the mating behaviour, strategies and systems in crustaceans. Christy (1987) discussed the adaptive value of different mating strategies in male crabs. Hartnoll (2000) analysed the adaptive value of the two sexes to different mating strategies, considering mating strategy as a process of co-evolution in both sexes. He concluded that changes in mating strategy in crabs retaining indeterminate growth have been harder to account for especially in those retaining an aquatic habit at the same time. Further research is needed to clarify this uncertainty. Further research is also necessary to investigate the interplay of multiple ecological, reproductive and life history factors underlying the observed diversity in mating behaviour and mating systems. Crustaceans offer an excellent model for investigating the complex network of relationships caused by the interplay of multiple factors, and for studying current issues related to sexual selection such as sexual conflict, mate choice, male sperm limitation and sperm allocation. In-depth knowledge of the mating behaviour and an understanding of the interplay of multiple factors affecting the reproductive behavioural ecology of crustaceans would contribute to the conservation of these important macro-invertebrates and of the freshwater and marine ecosystems inhabited. Thiel, (2000) reviews parental care behaviour from different crustacean taxa and environments but no general pattern has been recognized. The high diversity of crustaceans with extended parental care suggests that this reproductive behaviour has evolved independently in a variety of crustacean taxa under a variety of environmental conditions (Thiel, 2000).

The review indicates that very little is known about the behaviour of crustacean parents, the resources they invest into individual offsprings, the benefits and costs of this reproductive behaviour and the evolution of social behaviour.

Knowledge of reproductive biology of females has come mainly from studies on sexual maturity by macroscopical investigations and by histological methods (Adiyodi and Subramonian, 1983). To date eyestalk ablation still represents the most commonly used manipulation to induce maturation and spawning in female crustaceans. This is an outcome of the success of this method in increasing reproductive output and spawning frequency and of its achievement in shortening the latency to first spawn (Racotta et al. 2003). Female maturity is more easily defined since attached eggs provide conclusive proof although there are other indicators such as relative abdomen width (in *Homarus*), ovigerous setae and cement gland development (Aiken and Waddy 1980). The oocytes and ovaries of *Nephrops* species develop from white through yellow and pale green to very dark green at maturity (Table 1). The investigation on oogenesis in many crustacean mature females allows the identification of growing stages and their comparison between species. The literature for many crustaceans distinguishes 4 to 5 growing stages (Adiyodi and Subramonian, 1983). However, more species-specific research using histological, histochemical and ultrastructural approaches is needed to investigate egg (oocyte) development and improve understanding of oogenesis and management of crustacean. Central to female reproduction is the process of vitelogenesis including the biosynthesis of yolk proteins and their transport and storage in the ovary for sustenance of the developing embryo. In many decapoda, there is evidence that yolk protein synthesis depends on extraoocytic sequestration of vitellogenin, and haemolymph protein fraction specific to vitellogenic females (Desantis et al. 2001). Chapter 6 of this book focuses mainly on research on the hormonal control of vitellogenesis and ovarian maturation. It illustrates recent advances in identifying the pathways of vitellogenin synthesis and describes its main regulators and advances in their isolation and mode of action. The research detailed in Chapter 6 has shown that the endocrine regulation of both reproduction and molting is controlled by a complex interaction of neuropeptides, juvenoids and steroids. Although studies have confirmed that female gonadal development could successfully be induced through injections or transplants of these hormones and endocrine glands, the effects are highly variable depending on factors such as species, developmental stage of the animal and concentrations of hormones and neurotransmitters given. For future applications of hormones in the manipulation of breeding stocks in

Table 1 Ovary staging based on ovary colour for *Nephrops*.

Studies	*Stage*	*Stage*	*Stage*	*Stage*	*Stage*	*Stage*
Thomas, 1964	0-I	II	II-III	III-IV	V	VI
Farmer, 1975	1	2	3	4		
Bailey, 1984	0	1	2	3	4	
	White ovary	Cream ovary	Pale green ovary Mature animals	Dark green ovary Mature animals	Dark green, swollen Mature animals	
Briggs, 1988	I	II	III	IV	V	
	Ovaries as thin white threads	Thicker than I with a tinge of pale green colour	Larger than II, and a dark green colour	Larger than III. Clearly visible through the carapace, Dark green/black	Pale green with dark green specks.	
Relini et al 1998	I	II	III	IV		
	White juvenile or under-developed adult	Cream at beginning of maturation	Pale green – intermediate stages of ovoverdins storage	Dark green – near to the extrusion on pleopods		
Mente et al.	1	2	3	4	5	6
(Unpublished data). External visual scale	White or cream ovar Immature	Pale green ovary	Dark green ovary and abdomen Mature	Berried females (carry green coloured eggs)	Berried females (carry brown coloured eggs)	Hatched eggs
Mente et al. (Unpublished data). Histology scale (Size of oocytes in the ovary in µm)	25-196	196-680 (392-490 mean)	588-882 (588-784 mean)	200-480 (200-400 mean)	Similar to previous no further oocyte development	Larger than stage 1 and smaller than stage 2

captivity, it is first necessary to fully understand the interactions and functions of these endocrine systems. Only then can research be directed at finding a hormonal treatment that has clear advantages over eyestalk ablation and that is also effective in terms of reproductive output. In species where close links exist between environmental factors and reproduction, the interruption of established metabolic pathways may have detrimental effects on the survival of the population and possibly also on their consumers. The resolution of connecting pathways between environmental stimuli and their hormonal responses would therefore be also beneficial for the conservation of populations in the wild.

Knowledge of male reproductive biology and physiology is less developed. Hence, further research on the hormonal control of male reproduction and factors affecting their reproduction activities is necessary. Research on crustacean endocrinology is still needed. Chapter 7 gives a brief historical overview of hormonal regulation of reproduction in male crustaceans, including the androgenic gland hormone (AGH), and some details of recent work on its isolation, characterization, and mode of action. It also summarises evidence for the role of other chemical mediators in the regulation of male reproduction (ecdysteroids, vertebrate-like steroids, and methyl farnesoate). The isolation and characterization of crustacean pheromones is important for both basic research and for the development of novel methods of harvesting commercially important fisheries species and for the control of nuisance and alien crustaceans. Whether testosterone has a physiological role in male reproduction in crustaceans is an issue where further research is needed. However, reports suggest that ecdysteroids are involved in male reproduction. Except for the ecdysteroid receptor, research on crustacean hormone receptors is limited. The identification of the AGH in decapod crustaceans enables a better understanding of the regulation of sexual differentiation and masculine reproductive physiology. If the AG active factor is found to be a member of the insulin-like super family, this would support the view that insulin may have evolved in the context of regulating sexual differentiation. Chapter 7 concludes that there are indications of other chemical mediators involved in male reproduction. These factors include neurotransmitters and peptides such as gonad-inhibiting hormone (GIH, also known as vitellogenesis-inhibiting hormone, VIH). Most of the work on these factors is focused on females. More research needs to be conducted to determine if these factors have a role in male reproduction. Due to the biological importance of Crustacea, it is likely that much more research would be devoted to the determination of the effects of exogenous chemicals upon various physiological processes, including male reproduction. Crustaceans are likely to be useful indicator species for environmental contamination.

Many factors play an important role on growth and maturation processes such as the irregular and relatively asynchronous molting, the extended spawning season, the number of spawnings, the prolonged period of recruitment and the duration of larval stages (Sarda and Demestre, 1989; Demestre and Fortuño, 1992; Kapiris and Thessalou-Legaki, 2001). Hartnoll's (1982) review on growth in crustaceans covers both absolute and relative growth. Hartnoll's (2001) review identifies four aspects of crustacean growth: the hormonal control of moulting, the effects of external factors, the patterns of growth and the determination of age. He concludes that despite the development of modelling approaches and an increased body of data our understanding of the underlying principle of the diversity of crustacean growth patterns remains speculative. The adaptive value of the great diversity of growth patterns in crustaceans is not adequately explained and further research is needed beyond the descriptive phase. The interrelationships between reproductive and molt cycles during the reproductive season are discussed in Chapter 9. The results of the chapter stress the necessity for further molecular research on the patterns of expression of genes under the control of eyestalk neuropeptides in both the molt and reproductive cycles.

Chapter 11 reviews the reproductive biology of shrimps and reports biological differences between wild and captive shrimps. The most enigmatic phenomenon is the female's "abstinence" from reproduction in captivity. Further research, including comparisons with observations in the wild, is needed to understand the reproduction of penaeid shrimps.

Harrison (1990; 1997) reviews the special nutritional requirements of crustacean broodstock. The effects of feeding level and diet quality on reproductive performance and in broodstock nutrition in captivity in penaeids shrimps are well known (Cuzon et al., 1995; 1998; 2004). Dietary deficiencies in essential nutrient levels in the diets of maturing fish are associated with reduced fecundity, egg size, and oocytes formation. Less clear is the relationship between nutrition and male reproduction. Rosas et al. (1993) showed that maturation of *L. setiferus* adult males increases with the age of shrimp and shrimp weighing 10 g are suitable as sperm donors for artificial insemination. Nutritional requirement, in relation to reproduction takes into account not only the nutrition of breeding animals but also the larvae development (Racotta et al., 2003). The importance of nutrition is clearly demonstrated through successive spawning under normal environmental conditions. Nevertheless, for most species, reproduction in captivity relies on eyestalk ablation and nutritional solutions (including additives). This enables the industry to maintain shrimp in captivity. However, more research is needed to induce ovary development without ablation in captivity induce ovary development without ablation in (Chapter 8). The growing industry of shrimp production requires an improvement of knowledge in the areas of maintenance and nutrition of breeding animals in order to manage several families and select for different charactersistics (Goyard et al., 2004). Maintaining animals in good health is a constant concern at a research, and of course, a farming level. Scientific progress in the understanding of the nutritional requirements of penaeids should benefit the culture of other species such as crabs, Scylla, and spiny lobsters (Chapter 8). The combined efforts of endocrinologists and nutritionists are required to unravel the relationship between nutrition and crustacean maturation.

While much is known about the reproductive cycles of male and female, producing large and predictable numbers of eggs for aquaculture is still a complex process. Chapter 10 reviews the reproductive biology and growth of clawed lobsters and rock lobsters. Lobsters are of considerable scientific interest as they are abundant, widely distributed, large, long-lived, and of great ecological consequence (Chapter 10). Understanding their biology is important. They also offer a unique model for fisheries because their abundance, lower mobility and hardiness make them more amenable to detailed research. Wild broodstocks remain under considerable pressure since they are perceived by many hatchery operators as producing a higher and more consistent quality of larvae than reared or artificially matured ones , even though scientific evidence for this seems equivocal (Wickins and Lee, 2002). The commercial value of *Nephrops norvegicus*, (Norway lobster) is particularly high (Smith et al., 2001; Rotlant et al., 2001) and recent trends in consumer demand justify the interest in the possibility of commercial aquaculture. It forms the basis of valuable fisheries in the northwest Atlantic and western Mediterranean. There is also scientific interest regarding the physiology, reproductive biology and fishery management of the species (Chapter 13). The several studies of the reproductive biology of Nephrops show that the duration of its reproductive cycle varies geographically and this may well be related to environmental factors such as temperature (Chapter 13).

Environmental or dietary treatments, sometimes in conjunction with eyestalk ablation, are used to rear several commercially important penaeid species though successive generations (Wickins and Lee, 2002; Chapter 12 in this book). Today, eyestalk ablation is the only commonly used technique that provides predictable maturation and spawning results (Chapter 12). The removal of the eyestalk effectively removes maturation inhibition. However, as the eyestalk is the source of numerous hormones including those required for moulting, sugar balance and metabolism, there are numerous adverse side effects of ablation (Browdy,

1992). Eyestalk ablation often results in spawner exhaustion, because ablated animals mature at a faster rate and have a reduced period between spawning events. This may affect their ability to sequester sufficient nutrients required for vitellogenesis and hence compromise fecundity and larval quality (Hansford and Marsden, 1995).

The procedure of unilateral eyestalk ablation (removal of one eye) is generally practiced and provides sufficient inhibition to allow maturation to proceed without maximizing the stress associated with bilateral ablation (Wickens and Lee, 2002). Eyestalk ablation is not practiced in males, as it is not been shown to produce additional reproductive benefits (Browdy, 1992). In the decades since eyestalk ablation became commonplace, research focuses on shrimp endocrinology (Huberman, 2000) but no viable alternatives are found.

Chapter 14 investigates the factors affecting both the reproductive cycle and growth of crustaceans and assesses the risk of all the parameters involved per stage by applying FMEA, Cause & Effect Diagram, HACCP and ISO 22000. Failure mode and effects analysis (FMEA) was first developed for systems engineering but it is useful for assessing potential failures in products or processes. Hazard Analysis and Critical Control Points (HACCP) is a systematic preventative approach to food safety that addresses physical, chemical and biological hazards as a means of prevention rather than as a finished product inspection. HACCP is used in the food industry to identify potential food safety hazards, so that key actions, known as Critical Control Points (CCP's) can be taken to reduce or eliminate the risk of the hazards being realised. The diagram shows that the low growth rate is attributed to inappropriate use of gonad hormones, improper larval rearing, an improper feeding schedule, inadequate control of pH, inadequate control of pellets and fresh feed, too rapid a dissolution of pellets and incomplete staff training.

A thorough understanding of how crustaceans achieve their biological success is an important scientific contribution. This book represents a revision and consolidation of the knowledge on the reproductive biology of crustaceans. In addition it offers a number of case studies on decapod crustaceans.

References

Adiyodi, R. G., Subramoniam, T. 1983. Arthropoda-Crustacea. In: Reproductive Biology of Invertebrates: Oogenesis, Oviposition, and Oosorption. Adiyodi, K. G., Adiyodi, R. G. (eds), Vol I, pp. 443-495. Wiley, Chichester, England

Aiken, D.E., and Waddy, S.L. 1980. Reproductive biology. In: The Biology and Management of Lobsters, Vol I. pp. 215-268. Cobb, J.S and Phillips, B., eds. Academic Press Inc, New York.

Arnold J. M. and Lois, D. Wiliams-Arnold. 1987. Cephalopoda: Decapoda. pp. 243-290. In: Reproduction of marine invertebrates. Volume IX. General Aspects: Seeking Unity in Diversity. Edited by Arthur C. Giese, John S. Pearse and Vicki B. Pearse. Blackwell Scientific Publications and The Boxwood Press, 1987.

Bailey, N. 1984. Some aspects of reproduction in Nephrops. ICES (Shellfish and Benthos Committee) 33: 1-15.

Briggs, R. P. 1988. Preliminary observations on the effects of a codend cover or lifterbag on catch composition in the Northern Ireland Nephrops fishery. ICES Fish Capture Comm., CM 1981/b:11 (mimeogr.).

Browdy, C. 1992. A review of the reproductive biology of Penaeus species: Perspectives on controlled shrimp maturation systems for high quality nauplii production. pp. 22-51. In: Wyban, J., (ed.) Proceedings of the Special Session on Shrimp Farming. World Aquaculture Society, Baton Rouge, LA, USA.

Charniaux-Cotton, H., Payen, G.G. and Ginsburger-Vogel, T. 1983. Arthropoda-Crustacea: Sexual Diferentation. Chapter 17. In: Adiyodi, R.G. and Adiyodi, K.G. (eds), Reproductive Biology of Invertebrates, Volume II. Spermatogenesis and sperm function. J. Wiley and Sons, Chichester: 443-495.

Cristy, J.H. 1987. Competitive mating, mate choice and mating associations of brachyuran crabs. Bull. Mar. Sci. 41:177-191.

Cuzon, G., Patrois J., Cahu C. and Aquacop. 1995. Some fundamentals of peneids breeders nutrition. In Proceedings III Congreso Ecuatoriano de Acuicultura, Guayaquil, Oct.-Nov.,1995.

Cuzon, G., Arena L., Goguenheim, J., Goyard, E. and Aquacop. 2004. Is it possible to raise, offspring of the 25th generation of *Litopenaeus.vannamei* (Boone) and 18th generation *Litopenaeus stylirostris* (Stimpson) in clear water to 40g? Aquaculture Research, 1-9.

Cuzon, G. and Aquacop. 1998. Nutritional review of *Penaeus stylirostris*. Reviews in Fisheries Science. 6:129-141.

Demestre and Furtuno, J.M. 1992. Reproduction of deep water shrimp *Aristeus antenattus* (Decapoda: Dendrobrachiata). Mar. Ecol. Prog. Ser. 84, 41-51.

Desatis, S., Labate., M., Maiorano, P., Tursi., Labate, G.M. and Ciccarelli, M. 2001. A histological and ultrastructural study of oogenesis in *Aristaeomorpha foliacea* (Risco, 1827). Hydrobiologia, 449, 253-259.

Farmer, A.S.D. 1975. Synopsis of biological data on the Norway lobster *Nephrops noruegicus* (Linnaeus, 1758). *FAO Fish. Synop.*, 112: 1-97.

Garcia-Rodriguez, M. and Esteban, A. 2000. A comparison between the biology and the exploitation level of two pink shrimp (*Aristeus antennatus*) stocks from two different areas in the Spanish Mediterranean. pp. 721-732. In: The Biodiversity Crisis and Crustacea. Proceedings of the fourth international crustacean congress, Amsterdam, Netherlands, 20-24 July 1998, Volume 2. Edited by: J. Carel Von Vaupel Klein and Frederick R. Schram. A.A. Balkema/ Rotterdam/ Brookfield/ 2000.

Goyard, E., Bedier, E., Patrois, J., Vonau, V., Pham, D., Cuzon, G., Chim, L., Penet, L., Dao, T., Boudry, P. 2004. Synthesis of the "shrimp" genetic programme in Tahiti: What profit for Caledonian shrimp industries? Styli 2003. Thirty years of shrimp farming in New Caledonia. Proceed. Symp., Noumea-Kone, 2-6 June 2003. Actes Colloq. Ifremer. no. 38:126-133.

Hansford, S.W. and G.E. Marsden 1995. Temporal variation in egg and larval productivity of eyestalk ablated spawners of the prawn *Penaeus monodon* from Cook Bay, Australia. J. World Aquac. Soc. 26, 396-405.

Harrison, K.E. 1990. The role of nutrition in maturation, reproduction and embryonic development of decapod crustaceans: a review. Journal of Shellfish Research 9(1), 1-28.

Harrison, K.E. 1997. Broodstock nutrition and maturation diets. In: *Crustacean Nutrition* (eds L. R. Abramo, D. E. Conklin and D. M. Akiyama), pp. 390-408. Advances in World Aquaculture Vol. 6, World Aquaculture Society, Baton Rouge, LA, USA.

Hartnoll, R.G. 2000. Evolution of brachyuran mating behavior: Relation to the female molting pattern. pp. 519-525. In: The Biodiversity Crisis and Crustacea. Proceedings of the fourth international crustacean congress, Amsterdam, Netherlands, 20-24 July 1998, Volume 2. Edited by: J. Carel Von Vaupel Klein and Frederick R. Schram. A.A. Balkema/ Rotterdam/ Brookfield/ 2000.

Hartnoll, R.G. and Gould, P. 1988. Brachyuran life history strategies and the optimization of egg production. In: A.A. Fincham and P.S. Rainbow 9 (eds), Aspects of Decapod Crustacean Biology: 1-9. Symposia of the Zoological Society of London.

Hartnoll, R. G. 1982. Growth. In: The Biology of Crustacean, vol 2, Embryology Morphology and Genetics, Bliss, D.E. and Abele, L. G. eds. Academic Press, New York, pp 111-196.

Hartnoll, R.G. 2001. Growth in Crustacea-twenty years on. Hydrobiologia, 449:11-122.

Hinsch, C. 1983. Arthropoda-Crustacea: Sexual behaviour and Receptivity. Chapter 18. In: Adiyodi, R.G. and Adiyodi, K.G. (eds), Reproductive Biology of Invertebrates, Volume II. Spermatogenesis and sperm function. J. Wiley and Sons, Chichester: 443-495.

Huberman, A. 2000. Shrimp endocrinology. A review. Aquaculture 191, 191-208.

Kapiris, K. and Thessalou-Legaki, M. 2001. Sex-related variability of rostrum morphometry of *Aristeus antennatus* (Decapoda: Aristeidae) from the Ionian Sea (Eastern Mediterranean, Greece). Hydrobiologia, 449, 123-130.

Kennely, S.T. and Watkins, D. 1994. Fecundity and reproduction and their relationship to catch rates of spanner crab, Ranina ranina, off the coast of Australia. J. Crust. Biol. 14:146-150.

Pochon-Masson, J. 1983. Arthropoda-Crustacea. Chapter 21. In: Adiyodi, R.G. and Adiyodi, K.G. (eds), Reproductive Biology of Invertebrates, Volume II. Spermatogenesis and sperm function. J. Wiley and Sons, Chichester: 443-495.

Racotta, I. S., E. Palacios, and A. M. Ibarra. 2003. Shrimp larval quality in relation to broodstock condition. Aquaculture 227: 107-130.

Relini, L.O., Zamboni, A. Fiorentino, F. Massi D. 1998. Reproductive patterns in Norway lobster Nephrops norvegicus in différent Méditerrannean areas. Sci. Mar. 62(1):25-41.

Rosas, C., Sanchez,A., Chimal, M.E., Saldaña, G., Ramos L. and Soto, L.A. 1993. Effect of electrical stimulation on spermatophore regeneration in white shrimp *Penaeus setiferus*. Aquatic Living Resources, 8, 139-144.

Rotlant, G., M. Charmantier -Daures, K. Anger, F. Sarda 2001. Effects of diet on *Nephrops norvagicus* (L.) larval and postlarval development, growth and elemental composition. *Jour. of Shellfish Research* 20:347-352.

Sarda, F. and Demestre, 1989. Shortening of the rostrum and rostral variability in *Aristeus antenattus* (Risso, 1816) (Decapoda: Aristeidae). J. Crust. Biol., 9:570-577.

Smith, C.J., Papadopoulou K.-N., Êallianiotis A., Vidoris P. Chapman C.J. & Vafides D. 2001. Growth and natural mortality of *Nephrops norvegicus* with an introduction and evaluation of creeling in Mediterranean waters. European Commission Final report, DGXIV Project N 96.013, 195 pp.

Thiel, M. 2000. Extended parental care behavior in crustaceans-A comparative overview. pp.211-226. In: The Biodiversity Crisis and Crustacea. Proceedings of the fourth international crustacean congress, Amsterdam, Netherlands, 20-24 July 1998, Volume 2. Edited by: J. Carel Von Vaupel Klein and Frederick R. Schram. A.A. Balkema/Rotterdam/Brookfield/ 2000.

Thomas, HJ., 1964. The spawning and fecundity of the Norway Lobsters (Nephrops norvegicus L.) around the Scottish coast, Journal du Conseil, 29: 221-229.

Ungaro, N., Marano, C.A., Marsan, R. and Pastorelli, A.M. 2000. On the reproduction of *Nephrops norvegicus* in the Southern Adriatic Sea: sex ratio, maturity lengths and potential fecundity. pp. 553-561. In: The Biodiversity Crisis and Crustacea. Proceedings of the fourth international crustacean congress, Amsterdam, Netherlands, 20-24 July 1998, Volume 2. Edited by: J. Carel Von Vaupel Klein and Frederick R. Schram. A.A. Balkema/ Rotterdam/Brookfield/2000.

Warner, G.F. 1977. The biology of Crabs. Elek Science London, London, 202 pp.

Wickens, J.F. and Lee, D.O'C 2002. Crustacean Farming: Ranching and Culture 2nd Edition. Blackwell Science, Oxford, 446p.s

2

Phylogeny Biology and Ecology of Crustaceans (Phylum Arthropoda; Subphylum Crustacea)

S.D. Klaoudatos and D.S. Klaoudatos

School of Agriculture Sciences, Department of Ichthyology and Aquatic Environment, University of Thessaly, Fytoko Street, 38446 New Ionia Magnisia, Greece.
E-mail: sklaoudat@uth.gr

OVERVIEW

Crustaceans constitute one of the largest groups of the animal kingdom and a general description of the reproductive biology and ecology of this group would be impossible to accomplish. The problem arises from the extent of the fundamental differences relating to reproductive biology and ecology within the group. For example, prawns of the genus *Penaeus* exhibit a wide range of behaviours in their reproductive biology and ecology. The genus has therefore been further subdivided into five more genera, and the only species left in the original genus is *Penaeus monodon* (Perez Farfante and Kensley, 1997). This division was imperative owing to the differences exhibited in the thelycum (the external reproductive organ of the female), which is open in some species and closed in others. In addition, some species are active during the day, while others are active during the night. Further differences within the genus are exhibited in burrowing activity and the relationship of the biological cycle to the salinity levels of the surrounding water.

In general there seems to be a diversification in the reproductive cycle within the subphylum. In several species, females release oocytes and sperm simultaneously directly into the aquatic environment (Penaeidae), where fertilization and embryogenesis occur. In other species, fertilization occurs during spawning of the oocytes when they pass through the spermatophores, stored inside the females during prior copulation earlier in the year. The fertilized oocytes are not released directly into the aquatic environment but are glued to the pleopods of the female until hatching (Palaemonidae, Astacidae, Nephropidae).

Pronounced differences are exhibited within the subphylum in relation to the reproductive ecology. Several species remain in shallow coastal waters throughout their lives, while others exhibit small-scale migratory movements to deeper waters during the winter, where they spawn before returning to shallow coastal waters during spring and summer. The species *Macrobranchium rosenbergii* of the family Palaemonidae incubate the fertilized oocytes attached on the pleopods by migrating from the fresh river water into estuarine water masses of higher salinity (12-14 ppm). After hatching the larvae return to the fresh river water, where they remain until they attain reproductive maturity.

In view of the diversity exhibited in the reproductive biology and ecology of the subphylum Crustacea, the authors of this chapter have described the reproductive biology and ecology of every taxonomic group separately while keeping the most accurate and up-to-date taxonomic classification (Martin and Davis, 2001).

Although the classification of crustaceans has been quite variable, the system used by Martin and Davis (2001) is the most authoritative, with some changes based on more recent morphological and molecular studies (Dixon et al., 2004; Porter et al., 2005), and is the system used in this chapter.

The current system maintains the earlier taxonomic distribution with an alteration in the groups of Reptantia and Natantia. Reptantia (meaning "those that walk"), a group of decapod crustaceans including lobsters and crabs, is a suborder alongside Natantia, with Reptantia containing the walking forms using the pereiopods and Natantia the forms swimming through the water with the pleopods (prawns, shrimps and boxer shrimp) (Scholtz and Richter, 1995). However, many reptants are able to propel themselves through the water, and many natants can walk. This classification is no longer valid, however, and Reptantia is now a clade within the suborder Pleocyemata, as the earlier suborders of Natantia and Reptantia were replaced (Burkenroad, 1963) with the monophyletic groups Dendrobranchiata (prawns) and Pleocyemata. Pleocyemata contains all the members of the Reptantia (which is still used, but at a lower rank), as well as the Stenopodidea (which contains the boxer shrimp or barber-pole shrimp) and Caridea, which contains all the true shrimps.

These taxa are united by a number of features, the most important of which is that the fertilized eggs are incubated by the female and remain stuck to the pleopods (swimming legs) until they are ready to hatch. It is this characteristic that gives the group its name.

Classification within the order Decapoda depends on the structure of the gills and legs and the way in which the larvae develop, giving rise to two suborders: Dendrobranchiata and Pleocyemata. Prawns (including many species colloquially referred to as "shrimp", such as the Atlantic white shrimp) make up the Dendrobranchiata. The remaining groups, including true shrimp, are the Pleocyemata.

Less formally, it can be stated that the most important groups of crustaceans are barnacles, branchiopods, copepods and Malacostraca (crabs, lobsters, shrimps and krill). There are around 1,320 barnacle species, 1,000 branchiopods, 14,000 copepods, and 30,000 malacostracans.

The most recent scientific classification of the Arthropods is the following:

Kingdom: Animalia

Subkingdom: Ecdysozoa

Phylum: Arthropoda

Subphylum: Trilobitomorpha

Class: Trilobita – Trilobites are a group of formerly numerous marine animals that died in the mass extinction at the end of the Permian.

Subphylum: Chelicerata – Chelicerates include spiders, sea spiders, mites, scorpions and related organisms. They are characterized by the presence of chelicerae.

Class: Arachnida – spiders, scorpions, etc.

Class: Merostomata – horseshoe crabs, etc.

Class: Pycnogonida – sea spiders

Subphylum: Myriapoda – Myriapods comprise millipedes and centipedes and their relatives and have many body segments, each bearing one or two pairs of legs. They are sometimes grouped with the hexapods.

Class: Chilopoda – centipedes

Class: Diplopoda – millipedes

Class: Pauropoda

Class: Symphyla

Subphylum: Hexapoda – Hexapods comprise insects and three small orders of insect-like animals with six thoracic legs. They are sometimes grouped with the myriapods, in a group called Uniramia.

Class: Insecta – insects

Order: Diplura

Order: Collembola – springtails

Order: Protura

Subphylum: Crustacea (55,000) – Crustaceans are primarily marine (a notable exception being woodlice) and are characterized by biramous appendages. They include lobsters, crabs, barnacles, and many others.

Class: Malacostraca (30,000), the largest class of crustaceans, represent two thirds of all crustacean species and contain all the larger forms (lobsters, crabs, shrimps, krill and woodlice).

Subclass: Eumalacostraca (22,000) (containing almost all living malacostracans)

Superorder: Eucarida

Order: Decapoda (10,000)

Suborder: Dendrobranchiata – prawns

Suborder: Pleocyemata

Infraorder: Caridea – true shrimp

Infraorder: Achelata – spiny lobsters, slipper lobsters

Infraorder: Astacidea – lobsters and crayfish

Infraorder: Thalassinidea – mud lobster, ghost shrimp

Infraorder: Brachyura (4,500) – true crabs

Infraorder: Anomura (1,400) – hermit crabs, squat lobsters

Infraorder: Stenopodidea – barber pole shrimp

Infraorder: Eryonoidea (34) – blind, benthic, lobster-like crustaceans

Order: Euphausiacea (86) – small, shrimp-like marine invertebrate animals, important organisms of the zooplankton (krill).

Superorder: Syncarida

Superorder: Peracarida

Order: Cumacea (1,400) (hooded shrimps).
Order: Isopoda (5,300) (pillbugs, sowbugs, woodlice)
Order: Amphipoda (7,000) – small, shrimp-like crustaceans
Order: Spelaeogriphacea (4) – small, shrimp-like crustaceans
Order: Thermosbaenacea (34)
Order: Mysida (1,000) (opossum shrimp)
Order: Lophogastrida
Order: Mictacea
Order: Tanaidacea (750)
Subclass: Phyllocarida (*Nebalia*)
Subclass: Hoplocarida (mantis shrimp)
Class: Ostracoda (50,000 many extinct) – small crustaceans laterally compressed, encased by two valves (mussel or seed shrimps)
Subclass: Myodocopa
Subclass: Podocopa
Class: Branchiopoda (1,000) – primitive and primarily freshwater crustaceans, mostly resembling shrimp
Subclass: Sarsostraca
Order: Anostraca (brine shrimp *Artemia*)
Subclass: Phyllopoda
Order: Notostraca (tadpole shrimp *Triops cancriformis*)
Order: Cladocera (*Daphnia*)
Class: Remipedia (12) – blind crustaceans found in deep caves
Order: Enantiopoda (extinct)
Order: Nectiopoda
Class: Cephalocarida (9) – shrimp-like benthic species (horseshoe shrimps)
Order: Brachypoda
Class: Maxillopoda (15,600) – barnacles, copepods, ostracods, fish lice, etc.
Subclass: Thecostraca (1,320) – barnacles
Infraclass: Facetotecta
Infraclass: Ascothoracida (100)
Infraclass: Cirripedia (1,220)

Subclass: Tantulocarida (30) – ectoparasites that infest deep-sea copepods, isopods, tanaids, and ostracods

Subclass: Branchiura (130) – ectoparasites on marine and freshwater fishes (carp lice or fish lice)

Subclass: Pentastomida (110) – obligate parasites of the respiratory tracts of vertebrates

Subclass: Mystacocarida (13) – interstitial crustaceans, part of the meiobenthos

Subclass: Copepoda (14,000) – group of small crustaceans found in the sea and nearly every freshwater habitat, many planktonic more benthic, some parasitic

Order: Calanoida (2,000)

Order: Cyclopoida

Order: Harpacticoida (3,000)

Numbers in parentheses are current species counts.

PHYLUM ARTHROPODA

Arthropods (Greek for "jointed feet") are the largest phylum of animals and include the insects, arachnids, and crustaceans. Approximately 80% of extant animal species are arthropods, with over a million modern species described and a fossil record reaching back to the early Cambrian. Arthropods are common throughout marine, freshwater, terrestrial, and even aerial environments and include various symbiotic and parasitic forms. They range in size from microscopic plankton (~0.25 mm) up to forms several metres long.

Arthropods are characterized by the possession of a segmented body (metamerism) with appendages on each segment (Koenemann and Jenner, 2005). All arthropods are covered by a hard exoskeleton made of chitin, a polysaccharide, and possess a muscular system that moves the animal by pulling on the exoskeleton. The skeleton of arthropods protects them against attack by predators and is impermeable to water, making them less prone to dehydration. The exoskeleton takes the form of plates called sclerites on the segments, plus rings on the appendages that divide them into segments separated by joints. This is in fact what gives arthropods their name (jointed feet) (Fortey and Thomas, 1997). The arthropod body is divided into a series of distinct segments, plus a presegmental acron that usually supports compound and simple eyes and a postsegmental telson. These are grouped into distinct, specialized body regions called tagmata. Each segment primitively supports a pair of appendages.

An arthropod in order to grow must shed its old exoskeleton and secrete a new one. This process of moulting (or molting in American English) is expensive in terms of energy consumption and during the moulting period an arthropod is vulnerable. Ecdysis is the moulting of the cuticula, which is shed during growth, and a new covering of larger dimensions is formed. In preparation for ecdysis, the arthropod will become inactive for a period of time and during this time will undergo apolysis. During apolysis, an arthropod becomes dormant for a period of time. Enzymes are secreted to digest the inner layers of the existing cuticle, detaching the animal from the outer cuticle. This allows the new cuticle to develop without being exposed to the environment. Directly after apolysis, ecdysis occurs. Ecdysis is the actual emergence of the arthropod into the environment (by crawling movements the old integument is pushed forward, splitting down the back and allowing the animal to emerge). The newly emerged animal then hardens and continues its life. Following the shedding of the old cuticula, a new layer is secreted during a second period of inactivity. Other reasons for moulting are damaged tissue and missing limbs. Over a series of moults, a missing limb can be regenerated, the stump being a little larger with each moult until it is of normal size again. Growth increments at moulting can be substantial; for example, the common shore crab *Carcinus maenas* increases its carapace width by an average of about 30% (Anderson, 2002).

Arthropods have an open circulatory system, a dorsal heart and a ventral nervous system. The circulatory fluid in a cavity called the haemocoel bathes the organs directly with no distinction between blood and interstitial fluid; this combined fluid is called haemolymph. Haemolymph, a copper-based blood analogue, is propelled by a series of hearts into the body cavity, where it comes in direct contact with the tissues. Arthropods are protostomes (Greek for first the mouth, mostly comprising a taxon of animals with bilateral symmetry and three germ layers). There is a coelom, but it is reduced to a tiny cavity around the reproductive and excretory organs, and the dominant body cavity is a haemocoel, filled with haemolymph, which bathes the organs directly.

SUBPHYLUM CRUSTACEA

The crustaceans are a large group of arthropods (55,000 species), usually treated as a subphylum. They include various familiar animals, such as lobsters, crabs, shrimps and barnacles. The majority of species are marine with many freshwater species and a few groups adapted to terrestrial life, such as terrestrial crabs, terrestrial hermit crabs and woodlice (Hayward and Ryland, 1998). Crustaceans occur in all marine habitats; most are motile, moving about independently, although a few taxa are parasitic and live attached to their hosts (including sea lice, fish lice, whale lice, tongue worms, and *Cymothoa exigua*, a parasitic crustacean), and adult barnacles live a sessile life attached head-first to the substrate. The scientific study of crustaceans is known as carcinology.

The crustacean body is segmented and organized into three distinct regions: head, thorax, and abdomen (or pleon), with the additional tail piece the telson, although the head and thorax may fuse to form a cephalothorax. The cephalothorax (prosoma) is derived from the fusion of the head (cephalon) and the trunk (thorax) and therefore includes all the mouthparts, antennae, and the thoracic appendages, such as the legs of a spider or lobster. The remainder of the body is the abdomen (opisthosoma), which may also bear lateral appendages as well as the tail, if present (Ruppert et al., 2003). The abdomen is variously developed throughout the phylum and is often much reduced.

In crustaceans, the carapace is a part of the exoskeleton that covers the cephalothorax and is particularly well developed in lobsters and crabs, functioning as a protective cover over the cephalothorax (the dorsal side of the body). Crustaceans have two pairs of antennae on the head, one pair of compound eyes and three pairs of mouthparts. Smaller crustaceans respire through their body surface by diffusion and larger crustaceans respire with gills and abdominal lungs, as shown by *Birgus latro* (the coconut crab, the largest terrestrial arthropod in the world).

In common with other arthropods, crustacean body is chitinous and usually reinforced by calcium carbonate forming a rigid exoskeleton that needs to be shed to allow the animal to grow (ecdysis). Various parts of the exoskeleton may be fused together and this is particularly noticeable in the carapace (thick dorsal shield seen on many crustaceans). Thoracic and abdominal appendages show a wide range of modification throughout the phylum. Crustacean appendages are typically biramous, meaning they are divided into two parts. Arthropod appendages are most commonly branched into a gill and leg with a common root at a body segment. Each leg/gill structure will be paired with a second biramous structure on the other side of the body, including the second pair of antennae (but not the first) with an inner (endopod) and an outer (exopod) ramus, the two rami often very different morphologically and serving different functions (Fish and Fish, 1996). Simple biramous limb structure is present in some primitive crustaceans, but in most groups the exopodite comprises the major functional unit of the limb, while the endopodite is reduced, lost, or adapted to serve a different function such as feeding, cleaning or respiration (Hayward and Ryland, 1998). At the end of the pleon is the tail fan, comprising a pair of biramous uropods and the telson, which bears the anus. Together, the uropods and the telson are used for steering while swimming, and in the caridoid escape reaction. In crabs and some other carcinized decapods, the abdomen is folded under the cephalothorax.

The external skeleton, which offers advantages regarding support and locomotion, poses a major problem to a growing animal. The solution evolved by crustaceans is the periodic shedding of the old exoskeleton and the formation of a new larger one, in a process called moulting or ecdysis. Ecdysis is a traumatic process frequently occurring at night and temporarily suspending normal behaviour for short periods before and after the event. Ecdysis constitutes an important part of the reproductive cycle, since this is the time when copulation takes place. There are generally four recognized stages in the moult cycle: premoult, ecdysis, postmoult and intermoult (Ruppert and Barnes, 1994). During ecdysis, several species of the Penaeidae family together with the exoskeleton expel the spermatophores stored in the closed thelycum during copulation if spawning did not take place earlier in the year. The spermatophores are replaced during the next copulation when they are transferred from the males through the closed thelycum to the female (Klaoudatos, 1984).

The intermoult period after sexual maturity is different between the sexes, with females having longer intermoults mainly due to the energy expenditure for reproduction and the associated attendance of the offspring (egg bearing). Two alternative strategies of size limitation occur in brachyuran crabs; the first is a steady decrease of the moult increment resulting in a limit to growth, as in *Cancer pagurus* (Pearson, 1908; Olmstead and Baumberger, 1923; Bennett, 1974), whereas the second involves a terminal anecdysis at a certain age, as with *Carcinus maenas* (Carlisle, 1957; Crothers, 1967). A size difference between the sexes in Crustacea with males being larger than females of the same age has been well documented and results from the differences in growth rates between sexually mature males and females (Hartnoll, 1978).

SEXES: REPRODUCTIVE ORGANS

Most crustaceans have separate sexes and are distinguished by appendages on the abdomen (pleopods). The exception is sessile Cirripedia (class Maxillopoda, infraclass Cirripedia), in which hermaphroditism is generally the rule (Hayward and Ryland, 1998). Pleopods are primarily swimming legs but are also used for egg brooding

Eumalacostraca, including many familiar groups, such as crabs, lobsters, prawns and shrimps. The Order Decapoda is the largest order of the crustaceans, comprising approximately 10,000 species (Ruppert and Barnes, 1994). As their name implies, all decapods have ten legs; these are the last five of the eight pairs of thoracic appendages characteristic of crustaceans.

The front three pairs function as mouthparts and are generally referred to as maxillipeds; the remainder are primarily walking legs and are also used for gathering food and are referred to as pereiopods. In many decapods, however, one pair of legs has enlarged pincers; the claws are called chelae, so those legs may be called chelipeds. Further appendages are found on the abdomen, with each segment capable of carrying a pair of biramous pleopods also known as swimmerets, and, as the name implies, are in some groups used in locomotion. In prawns, shrimps and lobsters, the uropods form a tail fan with the telson, which bears the anus, while in the true crabs the abdomen is reduced in size and folded under the thorax. The head and thorax are protected by a well-developed carapace that extends laterally to protect the gills (Wallace and Taylor, 2002). The wide range of body designs among decapods is easiest to understand in terms of adaptations for locomotion and habitat.

Decapods have separate sexes generally easily distinguishable (male prawns have specially modified limbs or claspers to grasp the female during release of sperm, female crabs have a broader abdominal flap than males that is used to brood eggs). In Penaeidae Family the sexes are distinguished by the petasma found in males. Petasma comprise two chitinous apophysis, which are connected, found in the endopodite of the first pleopods. Spermatophores are transferred through the petasma to the female. The thelycum is located between the bases of the fourth and fifth walking legs and, with a few exceptions (*Litopenaeus vannamei, Litopenaeus stylirostris, Litopenaeus schmitti*), is closed in most species of the Penaeidae family. Most decapods copulate, mating couples apparently finding each other with the help of released chemicals such as pheromones (Bliss, 1982). Elaborate courtship behaviour often takes place. In many cases the female undergoes a moult, the precopulatory moult just before mating. In prawns, shrimps and lobsters, fertilization is external, the eggs being fertilized as they leave the female, but in the true crabs it is internal, sperm being transferred to the spermathecae of the female (Wallace and Taylor, 2002). In many decapods, the eggs are retained by the females (a female brooding eggs is said to be "in berry", and should not be collected in this condition in order to sustain the reproductive capacity of the species) and attached by sticky secretions to fine hairs on the pleopods, where they may be held for several months.

The eggs hatch into free-swimming larvae. The zoea is the free-swimming larval stage of crustaceans and goes through several changes before turning into miniature adults. It follows the nauplius stage and precedes the post-larva stage. Zoea larvae swim with their thoracic appendages (nauplii use cephalic appendages, and megalopa use abdominal appendages for swimming). In many decapods, because of their accelerated development, the zoea is the first larval stage (Weldon, 1889), whereas for others the nauplius is the first larval stage. Zoea consists of a head and a telson, the thorax and abdomen not having developed yet. It has three pairs of appendages with which it swims; these become, in the adult, the antennules, the antennae, and the mandibles. Metanauplius is an early larval stage of some crustaceans such as krill following the nauplius stage (Gurney, 1942).

Shrimps and Prawns

The order Decapoda is subdivided into two suborders (Dendrobranchiata and Pleocyemata), depending on the structure of the gills and legs and the mode of reproduction. Prawns (including many species colloquially referred to as "shrimp", such as the Atlantic white shrimp) make up the Dendrobranchiata. The remaining groups, including true shrimp, crabs, and lobsters, are the Pleocyemata. The words "prawn" and "shrimp" lack precise definition but popular usage has led to the larger species being referred to as "prawns" and the smaller as "shrimps" (Fish and Fish, 1996). Prawns and shrimps are distinguished from other decapods by their swimming habit using the pleopods as organs for propulsion. The exoskeleton is thin and light and the abdomen is extended. The carapace is often extended between the eyes to form a spine or rostrum, particularly in those animals referred to as "prawns".

Prawns are edible, shrimp-like crustaceans belonging to the suborder Dendrobranchiata. They are distinguished from the superficially similar shrimp by the gill structure, which is branching in prawns (hence the name, *dendro* = tree; *branchia* = gill) and lamellar in shrimps. True shrimps are small, swimming decapod crustaceans belonging to the suborder Pleocyemata in the infraorder Caridea and found around the world in both fresh and salt water. A number of more or less unrelated crustaceans also have the word "shrimp" in their common name. Examples are the mantis shrimp and the opossum or mysid shrimp, both of which belong to the same class (Malacostraca) as the true shrimp but constitute two different orders within it, the Stomatopoda and the Mysidacea. *Triops longicaudatus* or *Triops cancriformis* are also popular animals in freshwater aquaria and are often called shrimp, although they belong to the Notostraca, a quite unrelated group.

Most prawns and shrimps have separate sexes. Reproduction is primitive in prawns with eggs being released into the water and hatching as nauplii. Shrimps and all other pleocyemates brood their eggs (cemented to the pleopods) for varied periods of time until they hatch and larvae usually hatch at a stage more developed than the nauplius. Prawns can be recognized by the small chelae on the first three pairs of walking legs, whereas shrimps can have chelae on various pairs of walking legs, but never on each of the first three.

There is, however, much confusion between the two, especially among non-specialists, and many shrimps are called "prawns" and many prawns are called "shrimps". In commercial farming and fishery, the terms are generally used interchangeably. In European countries, particularly the United Kingdom, the word "prawns" is more commonly on menus than the term "shrimp", which is used more often in the United States. In Australia the word "prawn" is used almost exclusively.

Penaeidae is a family of prawns, although they are often referred to as "penaeid shrimp". It contains many species of economic importance, such as the tiger prawn *Penaeus monodon*. Penaeoid shrimps (superfamily Penaeoidea) are the most commercially important shrimp species worldwide and the most often encountered members of the suborder Dendrobranchiata in shallow waters (Perez Farfante and Kensley, 1997).

The infraorder Caridea (suborder Pleocyemata) includes many larger temperate-water species, and the large freshwater tropical species of *Macrobrachium*. In the tropical marine environment many small species have evolved specialized lifestyles, living commensally on various species of sponges, corals, and echinoderms. Carideans are easily distinguished from penaeids by the pleura (side plate) of the second abdominal segment, which overlaps the pleura of both the preceding and following segments in the carideans but only the following segment in the penaeids (and all other decapods). The second walking leg is chelate in the carideans and may or may not occur as a massive cheliped.

Lobsters

The name "lobster" is reserved for members of two very different decapod infraorders, Astacidea and Achelata. Clawed lobsters should not be confused with spiny lobsters, which have no claws (chelae) and are not closely related. The closest relatives of clawed lobsters are the reef lobster *Enoplometopus* and the three families of freshwater crayfish (Tschudy, 2003).

The infraorder Astacidea (order Decapoda, suborder Pleocyemata) comprises **four** superfamilies: one of true lobsters, the clawed lobsters that have the first pair of legs modified as massive chelipeds (*Homarus*, *Nephrops*); two crayfish (*Astacus*, *Pacifastacus*, freshwater crustaceans resembling small lobsters, to which they are closely related); and one of reef lobsters (*Enoplometopus*, small lobsters that live on tropical coral reefs). The infraorder Achelata comprises members that lack the claws that are found on almost all other decapods (from the Greek *a* = not, *chela* = claw) and are further united by the great enlargement of the first antennae and by the special phyllosoma form of the larva. It includes the spiny lobsters, also known as rock lobsters, crayfish, sea crayfish, or crawfish (*Palinurus*, *Panulirus*, about 45 species of achelate crustaceans), slipper lobsters (Scyllaridae) and furry lobsters (Synaxidae).

Spiny lobsters are found in almost all warm seas, including the Caribbean and the Mediterranean, but are particularly common in Australasia, where they are referred to commonly as crayfish (*Jasus novaehollandiae* and *Jasus edwardsii*), and South Africa (*Jasus lalandii*). Spiny lobsters tend to live in crevices of rocks and coral reefs, only occasionally venturing out at night to seek snails, clams, crabs, sea urchins or carrion to eat. Sometimes, they migrate in groups, forming long files of lobsters across the sea floor. Potential predators may be deterred from eating spiny lobsters by a loud screech made by the antennae of the spiny lobsters rubbing against a smooth part of the exoskeleton.

Rock lobster, Australia's commercial crayfish (*Jasus edwardsii*, infraorder Achelata), spends 18 months as phyllosoma larva, then another year as a prawn-like puerulus larva. After 2 years the tiny juveniles develop over 5-6 years to reach maturity and mate. Information like this is essential in devising regulations to protect commercial species from overexploitation and in developing aquaculture farms for suitable species. For many varieties, such information on life histories is unknown.

Lobsters live on rocky, sandy, or muddy bottoms from the shoreline to beyond the edge of the continental shelf. They generally live singly in crevices or in burrows under rocks. Lobsters primarily feed on live fish, dig for clams and sea urchins, and feed on algae and eel-grass. An average adult lobster is about 230 mm long and weighs 700 to 900 g. Lobsters grow throughout their lives and are long-lived, surviving to 15+ years, although often fishing pressure ensures that few lobsters reach old age. The largest lobster was caught in Nova Scotia, Canada, and weighed 20.14 kg. Because a lobster lives in a murky environment at the bottom of the ocean, its vision is poor and it mostly uses its antennae as sensors. In general, lobsters move slowly by walking on the bottom of the sea floor.

However, when they are in danger and need to flee, they swim backwards quickly by curling and uncurling their abdomen. A speed of 5 m per second has been recorded.

Like all arthropods, lobsters are largely bilaterally symmetrical; clawed lobsters, such as the king crab, often possess unequal, specialized claws. The anatomy of the lobster includes the cephalothorax, which is the head fused with the thorax, both covered by the carapace, and the abdomen. The lobster's head consists of antennae, antennules, mandibles, the first and second maxillae, and the first, second, and third maxillipeds. Its abdomen includes pleopods and its tail is composed of uropods and the telson. Its colouring while still alive is often brownish-bluish, which changes to red when cooked because the most stable form of the pigmentation is red.

Lobsters as a form of identification mark their territory by urine. Individual recognition in the lobster *Homarus americanus* is based on detection of urine pheromones via chemoreceptors of the lateral antennular flagellum (Johnson and Atema, 2005). Individual recognition is a key element in the social life of many invertebrates and crustaceans have the potential for relatively high-order knowledge about conspecifies (Gherardi et al., 2005). Only the alpha male (the male in the community whom the others follow and defer to) of each group procreates with female lobsters in a region. Lobsters mature at 5 or 6 years of age. The sexes are separate and, after mating in summer, the eggs remain attached to the pleopods of the female for 9 to 11 months until they hatch into free-swimming larvae. The larvae swim for about a year, moulting between 14 and 17 times before settling to the bottom and starting to acquire adult characteristics. Spawning takes place once a year, declining to about once every two years in later life.

Lobsters crawl briskly over the ocean floor and swim backward with great speed by scooping motions of the muscular abdomen and tail, but they are clumsy on land. They are scavengers but also prey on shellfish and may even attack live fish and large gastropods. Over a period of 5 years they grow to an average weight of 1.4 kg. The common American lobster, *Homarus americanus,* is found inshore in summer and in deeper waters in winter from Labrador to North Carolina, but especially along the New England coast, where the chief lobster fisheries are located. Lobsters are caught in slatted wooden traps, or "pots," baited with dead fish. In Europe a species of *Homarus* similar to the American is found, but the smaller Norway lobster (*Nephrops norvegicus*) is the chief seafood variety and is frequently eaten, often under the name "scampi". Lobsters are an economically important seafood and are considered a delicacy around the world. They constitute the basis of a global industry netting $1.8 billion in trade annually.

INFRAORDER THALASSINIDEA

Thalassinidea is an infraorder of decapod crustaceans belonging to the suborder Pleocyemata, within the class of Malacostraca, that live in burrows in muddy bottoms of the world's oceans (mud lobster, ghost shrimp). Recent molecular analyses have shown this group to be most closely related to Brachyura (crabs) and Anomura (hermit crabs). There are believed to be 516 extant species of thalassinideans, with the greatest diversity in the tropics, although with some species reaching latitudes above 60° north (Dworschak, 2000). About 95% of species live in shallow water, with only three taxa living below 2,000 m.

True Crabs

The term crab is often applied to several different groups of short (nose to tail) decapod crustaceans with thick exoskeletons, but only members of the infraorder Brachyura (Greek for "short tail"), which include more than 4,500 species making up 28 families, are true crabs. Brachyurans represent the most diverse and successful group from the highly successful decapods (Whittington and Rolfe, 1963; Ruppert and Barnes, 1994). Other taxa, such as hermit crabs, porcelain crabs and king crabs, are, despite superficial similarities, not crabs at all; rather, they belong to the infraorder Anomura, which includes over 1,400 species in 12 different families. In the infraorder Anomura, the last pair of pereiopods is hidden inside the carapace, so only four pairs are visible (counting the claws), whereas uninjured true crabs (infraorder Brachyura) always have five visible pairs (Ingle, 1980).

True crabs are crustaceans in the infraorder Brachyura, in the order Decapoda within the class of Malacostraca. Brachyurans are the most advanced of the decapods in that they have the body most modified from the primitive shrimp-like decapod ancestor. The crab body is short, wide, and flat. The abdomen, once a muscular organ used for swimming, is now simply a flap used to cover reproductive appendages and hold eggs (with the exception of Raninoida, a superfamily of crabs) and is tightly folded under the carapace. The uropods, which along with the telson form the tailfan in other decapods, are totally absent in the crabs. In all but a few small groups all five pairs of walking legs are large. The first pair is modified as chelipeds. The antennae and antennules are greatly reduced and originate before the eyestalks (unlike anomurans such as porcelain crabs). The evolution of abdominal reduction and flexion in brachyurans was probably a locomotory adaptation, shifting the centre of gravity forward to a point beneath the locomotory appendages (Ruppert and Barnes, 1994). Crabs are a very diverse group, mostly found in salt water, but with some groups living in freshwater or on land. Although famed for their tendency to walk sideways, crabs are in fact able to walk in any direction.

The carapace, often well calcified throughout, extends ventrally inwards to the limb bases. Various degrees of calcification of the exoskeleton are associated with a defensive lifestyle. *Cancer* spp. are relatively slow-moving, ponderous animals that rely on a heavy shell and mechanically strong chelae for protection (Jeffries, 1966; Warner and Jones, 1976). The less heavily calcified portunids, in contrast, are extremely agile and very fast swimmers. They will normally use their chelipeds in agonistic display, but when the encounter demands it, they will defend themselves with their chelae (Teytaud, 1971; Jachowski, 1974), which have a lower mechanical advantage but faster muscle fibres than those in *Cancer* (Warner and Jones, 1976). Heterochelic and heterodontic crabs, such as portunids, also have the advantage of possessing a stronger, major chela (crusher) on one side and a faster, minor chela (cutter) on the opposite site (Stevcic, 1971; Williams, 1974; Vermeij, 1977). Distinctive spines and ridges, which provide some protection, are also common among brachyurans.

Reproduction is controlled by a number of physical environmental factors such as temperature, day-length, and food availability, and also by biotic factors such as social interactions. Generally reproduction occurs within the narrowest environmental range of conditions, compared to those suitable for survival and growth of a species. Temperature, day-length (intensity and periodicity), and food availability (quality and quantity) and, for those species inhabiting coastal and estuarine regions, salinity are the major exogenous determinants of the reproductive cycle (Sastry, 1983), while chemical factors and biotic factors, such as parasitism, may also influence reproduction. Many environmental parameters act as cues and induce responses via the endocrine organs influencing growth and reproduction.

The sexes are separate and the abdomen of the female is usually broader than that of the male. Only the first two pairs of pleopods are present in males and are modified for the transference of sperm during copulation. Sperm are stored in the spermathecae of the female and fertilization is internal (Schram, 1986). The eggs are carried on the pleopods of the female for a varying length of time before hatching as planktonic larvae known as zoeae. These undergo several moults, the number of which depends on the species, before developing into a second larva type, the megalopa. The megalopa is planktonic and moults into a benthic, juvenile crab (Ingle, 1992). Soft crabs (known as "peelers") are occasionally found on the shore and the soft exoskeleton indicates that the crab has recently moulted. The skeleton requires a few days to harden and during that time the crab seeks shelter under rocks and stones to avoid predators.

Brachyurans are in general fast-growing but slow-breeding arthropods and the continuous body growth by periodic moulting is not deterred by reproduction except when the eggs are brooded on the pleopods, which prevents the onset of the next moulting period (Subramoniam, 2000). Hence, the integration between moulting and reproduction is a physiological necessity in female crustaceans. In brachyurans reproduction and growth are programmed as antagonistic events (Subramoniam, 2000). The intermoult period is long enough to accommodate one or more reproductive cycles; while the animals enter the comparatively short premoult only after oviposition and liberation of the brood.

There are seasonal differences in timing of moulting and reproductive development between males and females (Carlisle, 1953; Cheung, 1969) resulting in a seasonal diversion of energy for somatic growth or reproduction. Pillay and Nair (1971) reported that the amplitude of gonadal index for females is higher than that of males, since females appear to divert a greater amount of energy for production of ova compared to the energy spent by males for production of sperm. The antagonism between somatic and reproductive growth in female decapods resulting in smaller females compared with males of the same age has been established for a number of Brachyurans (Drach, 1955; Bliss, 1966; Cheung, 1969; Adiyodi and Adiyodi, 1970; Bennett, 1995), including the shore crab (Demeusy, 1965).

In brachyuran crabs, moulting and copulation follow a lengthy courtship controlled by chemical and tactile stimuli (Hartnoll, 1973). In some brachyuran families, such as the Cancridae (*Cancer*) and Portunidae (*Necora, Carcinus*), there is a premoult attendance of the female by the male in which the male carries the female about beneath him, her carapace beneath his sternum. He releases her so that she can moult. Copulation occurs shortly afterwards (Klaoudatos, 2003). During copulation the female lies beneath the male or in the reverse position, with ventral surfaces opposing each other (Hartnoll, 1969; Ruppert and Barnes, 1994). In the brachyuran crabs, as in most invertebrates, the sperm cells (packaged in spermatophores) are introduced into the females during mating (Subramoniam, 1993), where they are stored in paired seminal receptacles or spermathecae for long periods of time pending fertilization. Sperm is transmitted by the anterior two pairs of copulatory pleopods of the male to spermathecae of the female, which receives spermatophores by way of spermathecal openings (Hill et al., 2004). The oviducts are usually unmodified and open via gonopores on to the coxae or near the coxae of the third pair of legs (Ruppert and Barnes, 1994).

A part of the sexually mature brachyuran female population undergoes ecdysis immediately after egg hatching and copulates with a usually larger male, whereas another part will repeat the cycle without the need for moulting and copulation. The sperm in the spermatheca of the shore crab (*Carcinus maenas*) has remained viable (Naylor, 1962) and is able to fertilize more than one batch of eggs during one intermoult period (Chiba and Honma, 1972; Adiyodi and Subramoniam, 1983; Sastry, 1983). Edwards (1989) suggested that spawning of the female edible crab (*Cancer pagurus*) can occur the same year that moulting and mating takes place or may be delayed until the following year. In addition, one mating may be used for multiple spawnings since stored sperm may fertilize eggs for successive clutches during one intermoult cycle (Hartnoll, 1963; Ryan, 1967; Cheung, 1969; Pillay and Ono, 1978; Edwards, 1989). Sequential spawning also occurs in the velvet crab (*Necora puber*), according to Norman and Jones (1993), with up to three batches of eggs being produced during the period of one intermoult (Gonzalez-Gurriaran, 1981).

In certain brachyuran crabs, moulting and reproduction alternate with one another with the sequence occurring almost all through the year for sexually mature females (Subramoniam, 2000). The moult cycles, where ecdysis and copulation take place, are timed to occur after hatching of the offspring thereby minimizing metabolic competition. The concept of antagonism between moulting and reproduction has been suggested from several authors on work done with a number of decapods (Passano, 1960; Bliss, 1966; Cheung, 1969; Adiyodi and Adiyodi, 1970; Sochasky, 1973; Kleinholz, 1976; Aiken and Waddy, 1976; Sastry, 1983; Gonzalez-Gurriaran, 1985; Heasman et al., 1995; Mori and Zunino, 1987; Mantelatto and Fransozo, 1999).

The brachyuran reproductive cycle includes a series of events for the sexually mature individuals of each population. Each of these events takes a set length of time for each individual of a population in a given environment. The time between the events of the reproductive cycle and the duration of each event is variable for different brachyuran species, resulting in the production of different temporal reproductive patterns between species. The temporal pattern of the reproductive cycle of a population in a given environment has been selected to ensure successful reproduction of the species in this particular environment. The interaction between the reproductively capable individuals and the environment is expressed by the timing of migrations to suitable locations for mating or breeding, by behaviours associated with mating and copulation, and by the occurrence of berried females (Klaoudatos, 2003).

In the Brachyura, the ovary undergoes gametogenesis during intermoult when levels of moult-inhibiting hormone (Lachaise and Hoffmann, 1977) and gonad-stimulating hormone (Otsu, 1963; Gomez and Nayar, 1965) are high in the haemolymph and those of ecdysteroids and gonad-inhibiting hormone are low (Panouse, 1943, 1947; Charniaux-Cotton and Kleinholz, 1964; Rangneker and Deshmukh, 1968). The ovary is a paired organ and in the brachyurans is H-shaped, connected by a central bridge of ovarian tissue, and is restricted to the cephalothorax (Fig. 1) (Ryan, 1967; Payen, 1974).

Fig. 1 Dorsally dissected edible crab, *Cancer pagurus*, showing the H-shaped, orange ovary (Klaoudatos, D. 2003).

In the fully mature state the contours of the ovarian limbs are difficult to distinguish as the ovary completely fills the dorsal region of the body cavity, located either dorsal or dorsolateral to the gut. The oviducts are generally lateral extensions of the ovary. They are short tubes connecting the ovary with the gonopores. The position of the opening of the oviduct may vary among different species. A spermatheca with glandular epithelium is found in proximal regions of the oviduct in species with internal fertilization. The spermatheca of crabs not only serves as a depository of spermatophores but also shows high secretory activity, which may be related to sperm maintenance (Adiyodi and Subramoniam,

1983). In brachyuran crabs, with the ovary confined to the much-broadened convex cephalothorax, the germinal zone is centrally placed in both limbs of the ovary and is active continually throughout the reproductive life of the female (Adiyodi and Subramoniam, 1983).

Classification within the crabs is traditionally based on the position of the gonopores (genital pore): whether they are found on the legs or on the thorax. In the two "primitive" sections (sometimes called collectively the Podotremata), the gonopores are found on the legs (as in all other decapods); in the Heterotremata, the male gonopores are on the legs, and the female gonopores are on the sternum; in the Thoracotremata, the gonopores are on the sternum in both males and females (Ingle, 1983). Most crabs show clear sexual dimorphism and so can be easily sexed. The abdomen, which is held recurved under the thorax, is narrow in males. In females, however, the abdomen is considerably wider and retains a greater number of pleopods. This relates to the carrying of the fertilized eggs by the female crabs (as seen in all pleocyemates). In species in which no such dimorphism is found, the position of the gonopores must be used instead. In females, these are on the third pereiopod, or nearby on the sternum in higher crabs; in males, the gonopores are at the base of the fifth pereiopods or, in higher crabs, on the sternum nearby.

The North Atlantic shore crab *Carcinus maenas* (infraorder Brachyura) is one of the commonest crabs on the shores of north-west Europe and is found on all types of shore from high water to depths of about 60 m, first reported in the western Atlantic in 1817 (Cohen et al., 1995). It is an extremely eurythermal and euryhaline species (Taylor and Taylor, 1992) ubiquitous on all British coasts, from northern Norway to Iceland to West Africa, in north-eastern America, the Indo-West Pacific and Japan (Hayward and Ryland, 1995). Because of its potential for extensive ecosystem alteration, it is today considered a pest species in a large body of literature (Legeay and Massabuau, 2000).

Carcinus maenas reaches maturity after the first year at a carapace width of 25-30 mm (males) and 15-31 mm (females) (Fish and Fish, 1996). The abdomen of the female is broader than that of the male and has seven segments compared with five in the male. The male seeks out the female about to moult and carries her under his body for a few days. When the female moults, copulation takes place. The female then burrows in sand, forms a large cavity beneath her and lays the eggs, which are attached to the pleopods. Eggs are held for several months, with as many as 185,000 eggs carried at one time. After a larval life of two to three months in the plankton, young crabs settle on the bottom.

The Christmas Island red crab (*Gecarcoidea natalis* infraorder Brachyura) is a species of terrestrial crab endemic to both Christmas

Island and the Cocos (Keeling) Islands in the Indian Ocean. It is estimated that up to 120 million red crabs may live on Christmas Island, making it the most abundant of the 14 terrestrial crab species. Christmas Island red crabs eat mostly fallen leaves and flowers but will occasionally be cannibalistic. Male crabs are generally larger than the females, while adult females have a much broader abdomen and usually have smaller claws. The broader abdomen of the female Christmas Island red crab becomes apparent only in the third year of growth. They live in burrows in order to shelter from the sun. Since they still breathe through gills, the possibility of drying out is a great danger. Christmas Island red crabs mate seasonally and are famous for their annual migration to the sea in order to lay their eggs in the ocean. During the migration, the crabs cover the routes to the coast so densely that they can be seen from the air.

Members of the family Gecarcinidae (infraorder Brachyura) possess inflated, highly vascular branchial cavities that act as lungs. Eyes are set close together on short stalks, and mature males of some species have a disproportionately large major chela. They are not fully terrestrial, in that they need to return to the sea to release their larvae. Three species of Gecarcinids are common in the British Virgin Islands. *Cardisoma guanhumi*, the giant land crab, is light in colour (blue-gray to lavender) and is found in low-lying wooded areas, where it lives in deep burrows up to 20 cm in diameter. Young of this species live just behind the beach and resemble ghost crabs. The other two species, *Gecarcinus quadratus* and *G. ruricola*, are darker purple or red, living inland and often seen roaming after rain.

Crabs make up 20% of all marine crustaceans caught and farmed worldwide, with over 1.5 million t consumed annually. Global capture production 1950-2004 FAO. Retrieved on August 26, 2006.

INFRAORDER ANOMURA

Anomura is an infraorder of decapod crustaceans belonging to the suborder Pleocyemata, within the class of Malacostraca (also called Anomala), and include hermit crabs (superfamily Paguroidea), mole crabs (superfamily Hippoidea), squat lobsters (families Galatheidae and Chirostylidae), porcelain crabs (family Porcellanidae) and others, superficially crablike, holding the abdomen beneath the body and most often possessing large chelae on the first legs. The infraorder Anomura is divided into four superfamilies:

Superfamily Lomisoidea contains a single family, which contains a single species *Lomis hirta*, the hairy stone crab. It is a crab-like crustacean that lives in the littoral zone of southern Australia. It is slow-moving and covered in brown hair, which camouflages it against the rocks it lives upon (Davey, 2006).

Superfamily Hippoidea contains the two families of mole crabs, Hippidae and Albuneidae. Mole crabs, sand fleas or sand crabs live under sand in shallow water near the shore and live two to three years. They are the colour of rippled sand at the water's edge and live mostly buried in the sand, with their antennae reaching into the water. These antennae filter plankton and organic debris from the water. Their camouflage protects them from their predators, chiefly fish and birds. Females grow to about 35 mm long and carry their bright orange-coloured eggs under their telson during the summer months until they are ready to hatch. Males are smaller, reaching only 20 mm, making the sexes easy to tell apart when fully grown.

Superfamily Galatheoidea contains squat lobsters and related animals in five families: Galatheidae, Chirostylidae (squat lobsters), Porcellanidae (porcelain crabs), Aeglidae (a small group of South American freshwater anomurans with no common name in English), and Kiwaidae, containing the newly discovered *Kiwa hirsuta* discovered at a vent site 2,300 m underwater, 1,500 km south of Easter Island.

Squat lobsters are decapod crustaceans of the families Galatheidae and Chirostylidae, including the common genera *Galathea* and *Munida* (Macpherson et al., 2006). They are not lobsters, although they share a number of characteristics, and are, in fact, more closely related to porcelain crabs, hermit crabs and, more distantly, true crabs. Flesh from this animal is often commercially sold in restaurants as "langostino" (a Spanish word meaning prawn, not to be confused with langoustes, spiny lobsters) or sometimes called merely "lobster" when incorporated in seafood dishes. The body of a squat lobster is usually flattened, the abdomen is folded under itself, and the first pereiopods are greatly elongated and armed with long chelae. The fifth pair of pereiopods is usually hidden within the gill chamber, under the carapace, giving squat lobsters the appearance of having only eight pereiopods.

Porcelain crabs are decapod crustaceans in the family Porcellanidae, which superficially resemble true crabs. They live in all the world's oceans, except the Arctic and the Antarctic. Porcelain crabs are small, usually with a body width of 1 to 2 cm. Their bodies are more compact and flattened than those of squat lobsters. This is an adaptation for living and hiding under rocks, as well as squeezing into little nooks and crannies. They are common under rocks and can often be found and observed on rocky beaches and shorelines. Porcelain crabs can be distinguished from the true crabs by the apparent number of walking legs (three instead of four pairs, the fourth pair being hidden under the carapace), the apparent lack of a wrist (carpal) segment on the chelipeds, and long antennae originating in front of the eyestalks. The abdomen of the porcelain crab is long and folded underneath it. When alarmed, the porcelain crab might swim by flapping its abdomen. Porcelain crabs are quite fragile animals, hence their name, and will often shed their limbs in order to escape if a limb is grabbed by a predator or caught between rocks. The lost appendage can grow back over several moults. Porcelain crabs have large chelae used for territorial struggles, but not for catching food. Feeding is accomplished by filtering plankton and other organic particles from the water column using long setae on the mouthparts. Feeding also involves scavenging on the sea floor for detritus.

Kiwa hirsuta is an Anomuran crab belonging to the superfamily Galatheoidea discovered in 2005 in the South Pacific Ocean. This decapod, approximately 15 cm long, is notable for the quantity of silky blond setae (resembling fur) covering its pereiopods. Its discoverers dubbed it the "yeti lobster" or "yeti crab". It was found 1,500 km south of Easter Island in the South Pacific, at a depth of 2,200 m, living on hydrothermal vents along the Pacific-Antarctic Ridge. The "hairy" pincers contain filamentous bacteria, which may be used to detoxify poisonous minerals from the water emitted by the hydrothermal vents. Alternatively, it may feed on the bacteria, although it is thought to be a general carnivore. Its diet also consists of green algae and small shrimp.

Superfamily Paguroidea includes hermit crabs and their relatives: Coenobitidae, Diogenidae, Paguridae, Parapaguridae, Pylochelidae, and the king crabs or stone crabs in the Lithodidae. Hermit crabs are decapod crustaceans of the superfamily Paguroidea and are not closely related to true crabs. Most hermit crabs salvage empty seashells to shelter and protect their soft abdomens, from which they derive the name "hermit". Only the anterior part of the carapace is calcified (often referred to as "the hard carapace") (McLaughlin, 2003). There are about 500 known species of hermit crabs in the world, most of which are aquatic, living at a range of depths from shallow coral reefs and shorelines to deep bottoms, although some species are terrestrial. Hermit crabs live in the wild in colonies of 100 or more and do not thrive in smaller numbers. As hermit crabs grow, they must exchange their shell for a larger one that has been thoroughly tested for weight and size. Since intact gastropod shells are not an unlimited resource, there is frequently strong competition for the available shells, with hermit crabs fighting over shells (Tricarico and Gherardi, 2006). The availability of empty shells depends on the abundance of the gastropods and hermit crabs and on the frequency of organisms that prey on gastropods but leave the shells intact. The abdomen is soft, asymmetrical and twisted, allowing it to fit into the coiled shell. The crab is held securely by the specially adapted left uropod, which is hook-like and grips the inside of the shell, while the fourth and fifth pereiopods are held against the wall of the shell to give further purchase. The pleopods have been lost in males but are retained on the left side of the body in females and are used to carry the eggs. The first pereiopods are chelate, the right being larger than the left in most species. The fourth and fifth pereiopods are generally poorly developed (Hayward and Ryland, 1998).

There are approximately 15 terrestrial species in the world. Species range in size from the Pacific hermit crab (*Coenobita compressus*), which rarely grows larger than a peach, to the Caribbean hermit crab (*Coenobita brevamanus*), which can approach the size of a coconut. Terrestrial hermit crabs can be found several kilometres from the sea, returning to the shore periodically only to release their eggs. At that time they may find new gastropod shells to carry back inland with them. Terrestrial hermit crabs begin their lives in the sea but, through a series of moults, develop the ability to breathe air. Reduced gills and highly vascular branchial cavities act as lungs, freeing these crabs from the marine environment (Farrelly and Greenaway, 2005). After the last developmental moult, the young hermit crab will drown if left in water for an indefinite period of time. Their link with the sea is never entirely broken, however, as hermit crabs carry a small amount of water in their shells at all times to keep their abdomen moist and their modified gills hydrated. It is believed that *C. brevamanus* is the species of *Coenobita* best adapted to life on land (aside, of course, from *Birgus latro*) (Grubb, 1971).

The coconut crab (*Birgus latro*) is the largest terrestrial arthropod in the world. It is a derived hermit crab that is known for its ability to crack coconuts with its strong pincers in order to eat the contents (Altevogt and Davis, 1975). It is sometimes called the robber crab or palm thief (in German, Palmendieb), because some coconut crabs are rumoured to steal shiny items such as pots and silverware from houses and tents. Another name is the terrestrial hermit crab, due to the use of shells by the young animals (although terrestrial hermit crab also applies to a number of other hermit crabs such as the Australian land hermit crab *Coenobita variabilis*). *Birgus latro* reaches a weight of up to 4 kg, a body length of up to 400 mm, and a leg span of 1 m, with males generally being larger than females. Some reports claim weights up to 17 kg and a body length of 1 m. They can reach an age of up to 60 years (references vary). The body of the coconut crab is, as in all decapods, divided into a front section (cephalothorax), which has 10 legs, and an abdomen. The foremost legs have massive claws used to open coconuts, and these claws (chelae) can lift objects up to 29 kg in weight. The next three pairs have smaller tweezer-like chelae at the end and are used as walking limbs. In addition, these specially adapted limbs enable the coconut crab to climb vertically up trees (often coconut palms) up to 6 m high. The last pair of legs is very small and serves only to clean the breathing organs. These legs are usually held inside the carapace, in the cavity containing the breathing organs (Stensmyr et al., 2005).

Although *Birgus latro* is a derived type of hermit crab, only the juveniles use salvaged snail shells to protect their soft abdomens, and adolescents sometimes use broken coconut shells to protect their abdomens. Unlike other hermit crabs, the adult coconut crabs do not carry shells but instead harden their abdominal armour by depositing chitin and calcium. They also bend their tails underneath their bodies for protection, as do most true crabs. The hardened abdomen protects the coconut crab and reduces water loss on land but has to be moulted at periodic intervals. Moulting takes about 30 days, during which the animal's body is soft and vulnerable, and it stays hidden for protection (Held, 1963). Coconut crabs cannot swim and will drown in water within a few hours or minutes. They use a special organ in the rear of the cephalothorax called a branchiostegal lung to breathe. The branchiostegal lung is a developmental stage between gills and lungs and is one of the most significant adaptations of the coconut crab to its habitat. In addition, the coconut crab has a rudimentary set of gills probably used to breathe under water in the evolutionary history of the species, but these no longer provide sufficient oxygen.

Coconut crabs mate frequently and quickly on dry land in the period from May to September, especially in July and August. The male and the female fight with each other, and the male turns the female on her back to mate. The whole mating procedure takes about 15 minutes. Shortly thereafter, the female lays her eggs and glues them to the underside of her abdomen, carrying the fertilized eggs underneath her body for a few months (Barnett et al., 1999). At the time of hatching, usually October or November, the female coconut crab releases the eggs into the ocean at high tide. It is reported that all coconut crabs do this on the same night, with many females on the beach at the same time. The zoea larvae float in the ocean for 28 days, during which a large number of them are eaten by predators. Afterwards, they live on the ocean floor and on the shore as hermit crabs, using discarded shells for protection for another 28 days. At that time, they sometimes visit dry land. As with all hermit crabs, they change their shells as they grow. After these 28 days, they leave the ocean permanently and lose the ability to breathe in water. Young coconut crabs that cannot find a seashell of the right size also often use broken coconut pieces. When they outgrow even coconut shells, they develop a hardened abdomen. About 4 to 8 years after hatching, the coconut crab matures and can reproduce. This is an unusually long development period for a crustacean.

The diet of coconut crabs consists primarily of fruit, including coconuts and figs. However, they will eat nearly anything organic, including leaves, rotten fruit, tortoise eggs, dead animals, and the shells of other animals, which are believed to provide calcium. They may also eat live animals that are too slow to escape, such as freshly hatched sea turtles. Coconut crabs live alone in underground burrows and rock crevices, depending on the local terrain. They dig their own burrows in sand or loose soil. During the day, the animals stay hidden in order to protect themselves from predators and reduce heat-induced water loss. These crabs live almost exclusively on land, and some have been found up to 6 km from the ocean.

A number of species, most notably king crabs, have abandoned seashells for a free-living life; these species have forms similar to true crabs and are known as carcinized hermit crabs. King crabs, also known as stone crabs, are a family of crab-like decapod crustaceans chiefly found in cold seas. Their large size means that many species are widely caught and sold as food. King crabs are generally believed to be derived from hermit crab ancestors, which may explain the asymmetry still found in the adult forms. Although some doubt still exists about this theory, king crabs are the most widely quoted example of carcinization among the order Decapoda (McLaughlin and Lemaître, 1997; McLaughlin et al., 2004).

INFRAORDER STENOPODIDEA

The Stenopodidea is a small group of decapod crustaceans (infraorder: Stenopodidea), often confused with shrimps or prawns but belonging to a group closer to the reptant decapods, such as lobsters and crabs. They are easily recognized by the enlargement of the third pereiopod, which is chelate and stouter than the first pair. Many species are particularly attractive, with striped red and white bodies, giving them the alternative name "barber-pole shrimp". Stenopodideans, referred to as cleaner shrimps, are often found in the open on coral reefs, where they clean ectoparasites off fish (Lowry, 2006).

INFRAORDER ERYNOIDEA

The infraorder Eryonoidea of the Order Decapoda contains 34 species (family Polychelidae) of blind, benthic lobster-like crustaceans. They are found throughout the world's tropical, subtropical and temperate oceans, including the Mediterranean Sea and the Irish Sea. The family Polychelidae is notable for the number of chelate limbs, with either four or all five pairs of pereiopods bearing claws. This gives rise to the scientific names Polycheles (many-clawed) and Pentacheles (five-clawed) (Galil, 2000). The first pair of pereiopods is greatly elongated but often breaks off while specimens are being brought to the surface. The rostrum is very short or absent and, although eyestalks are present, the eyes are absent. This family can be seen as evidence of the transition from shrimp-like animals to lobster-like animals, since they possess a number of primitive characters such as the pointed telson, in contrast to the rounded telson in lobsters. Although apparently widespread, they were first discovered only in the late 19th century. The reason that polychelids remained unknown for so long is that they often live at great depths (the family as a whole has a depth range from less than 100 m to over 5,000 m) (Ahyong and Brown, 2002). This also accounts for the lack of eyesight, since almost none of the sun's light penetrates to such abyssal depths. The larvae of polychelids are distinctive and were first described under the name "Eyoneicus". Over 40 different larval forms are known, although few can be ascribed to known adult species.

ORDER EUPHASIACEA

The order Euphausiacea (superorder Eucarida) is divided into two families. The family Bentheuphausiidae has only one species (*Bentheuphausia amblyops*), a bathypelagic krill living in deep waters below 1,000 m and considered to be the most primitive living species of all krill. The other family, Euphausiidae, contains 10 different genera with a total of 85 species. Of these, the genus *Euphausia* is the largest, with 31 species. It is subject to commercial krill fishery including the Antarctic krill (*Euphausia superba*), the Pacific krill (*Euphausia pacifica*) and the Northern krill (*Meganyctiphanes norvegica*) (Brinton et al., 2000). Commercial fishing of krill is done in the Southern Ocean and in the waters around Japan. The total global production amounts to 150,000-200,000 t annually. A peak in krill harvest was reached in 1983 with more than 528,000 t in the Southern Ocean alone (of which the Soviet Union produced 93%) (Nicol, 1997; Nicol and Endo, 1997). Most krill is used for aquaculture and aquarium feed, as bait in sport fishing, or in the pharmaceutical industry. In Japan and Russia, krill is also used for human consumption and known as "okiami" in Japan. For mass-consumption and commercially prepared products, krill must be peeled because the fluorides contained in their exoskeletons are toxic in high concentrations.

Krill are shrimp-like marine invertebrates forming the basis of the trophic chain in several ecosystems (Antarctica), particularly as food for baleen whales, mantas, whale sharks, crab eater seals (*Lobodon carcinophagus*) and other seals, and a few seabird species that feed almost exclusively on krill. Their scientific name is Euphausiids, after their taxonomic order Euphausiacea. The name krill comes from the Norwegian word krill meaning "young fry of fish". Krill can be easily distinguished from true shrimps by their externally visible gills. Most krill are 1 to 2 cm long as adults, with a few species growing to sizes of the order of 6 to 15 cm. The largest krill species is the mesopelagic *Thysanopoda* (Mauchline and Fisher, 1969).

Krill occur worldwide with transoceanic distribution and several species have endemic or neritic restricted distribution (Brinton, 1962). Krill are considered a keystone species near the bottom of the food chain because they feed on phytoplankton and to a lesser extent zooplankton, converting these into a form suitable for many larger animals for which krill makes up the largest part of the diet (Bamstedt and Karlson, 1998). In the Southern Ocean the Antarctic krill (*Euphausia superba*) makes up a biomass of hundreds of millions of tonnes, similar to the entire human consumption of animal protein (Ross and Quetin, 1986).

Many krill are filter-feeders; their thoracopods form very fine combs with which they can filter out their food from the water. These filters can be very fine in species that feed primarily on phytoplankton (*Euphausia* spp.), in particular on diatoms. However, it is believed that all krill species are mostly omnivorous and some few species are carnivorous, preying on small zooplankton and fish larvae (Kils and Marshall, 1995).

Krill are bioluminescent animals (except for the *Bentheuphausia amblyops*), having photophores able to emit light. The light is generated by an enzyme-catalysed chemoluminescence reaction, wherein a luciferin (a kind of pigment) is activated by a luciferase enzyme (Shimomura, 1995; Herring and Widder, 2001; Johnsen, 2005). Krill probably do not produce this substance themselves but acquire it as part of their diet, which contains dinoflagellates. Krill photophores are complex organs with lenses and focusing abilities and they can be rotated by muscles. The precise function of these organs is as yet unknown; they might have a purpose in mating, social interaction or in orientation (Dunlap et al., 1980; Lindsay and Latz, 1999). Some researchers have proposed that krill use the light as a form of counter-illumination camouflage to compensate their shadow against the ambient light from above to make themselves more difficult for predators to see from below.

Most krill are swarming animals; the size and density of the swarms vary greatly depending on the species and the region. *Euphausia superba* swarms have been reported to contain between 10,000 and 30,000 individuals per cubic metre. Swarming is a defensive mechanism, confusing smaller predators that would prefer to pick out single individuals. When disturbed, a swarm scatters, and some individuals have even been observed to moult instantaneously, leaving the exuvia behind as a decoy (Brierley et al., 2002).

When krill hatch from the eggs, they go through several larval stages, the nauplius, pseudometanauplius, metanauplius, calyptopsis, and furcilia stages, each of which is subdivided into several sub-stages. The pseudometanauplius stage is exclusive of species that lay their eggs within an ovigerous sac (sac spawners). The larvae grow and moult multiple times during this process, shedding their rigid exoskeleton and growing a new one whenever it becomes too small. Smaller animals moult more frequently than larger ones. Until and including the metanauplius stage, the larvae are nourished by yolk reserves within their body. Only by the calyptopsis stage has differentiation progressed far enough for them to develop a mouth and a digestive tract, when they begin to feed upon phytoplankton. By that time, the larvae have reached the photic zone and their yolk reserves are exhausted. During the furcilia stages, segments with pairs of pleopods are added, beginning at the foremost segments. Each new pair becomes functional only at the next moult. The number of segments added during any one of the furcilia stages may vary even within one species, depending on environmental conditions (Knight, 1984).

After the final furcilia stage, the krill emerges in a shape similar to an adult, but it is still immature. During the mating season, which varies with the species and the climate, the male deposits a spermatophore at the genital opening (thelycum) of the female. The females can carry several thousand eggs in the ovary, which may then account for as much as one third of the animal's body mass. Krill can have multiple broods in one season, with interbrood periods in the order of days.

There are two types of spawning mechanisms. The 57 species of the genera *Bentheuphausia*, *Euphausia*, *Meganyctiphanes*, *Thysanoessa*, and *Thysanopoda* are "broadcast spawners": the female eventually just releases the fertilized eggs into the water, where they usually sink into deeper waters, disperse, and are on their own. These species generally hatch in the nauplius one stage but recently have been discovered to hatch sometimes as metanauplius or even as calyptopis stages (Gómez-Gutiérrez, 2002). The remaining 29 species of the other genera are sac spawners, where the female carries the eggs with her attached to the hindmost pairs of thoracopods until they hatch as metanauplii, although some species such as *Nematoscelis difficilis* may hatch as nauplius or pseudometanauplius (Gómez-Gutiérrez, 2003).

Some high-latitude species of krill such as *Euphausia superba* can live in excess of 6 years; others, such as the mid-latitude species *Euphausia pacifica*, live only for 2 years. Longevity decreases in subtropical or tropical species: *Nyctiphanes simplex* usually lives for only 6 to 8 months.

Moulting occurs whenever the animal outgrows its rigid exoskeleton. Young animals grow faster and therefore moult more often than older and larger ones. The frequency of moulting varies widely from species to species and is, even within one species, subject to many external factors such as latitude, water temperature, or the availability of food (Marinovic and Mangel, 1999). The subtropical species *Nyctiphanes simplex*, for instance, has an overall intermoult period in the range of 2 to 7 days: larvae moult on the average every 3 days, while juveniles and adults do so on the average every 5 days. For *E. superba* in the Antarctic sea, intermoult periods range between 9 and 28 days (at temperatures between –1°C and 4°C), and for *Meganyctiphanes norvegica* in the North Sea the intermoult periods range also from 9 to 28 days (at temperatures between 2.5°C and 15°C) (Buchholz, 2003).

Disturbances of an ecosystem resulting in a decline of the krill population can have far-reaching effects. During a coccolithophore bloom in the Bering Sea in 1998, for instance, the diatom concentration dropped in the affected area. However, krill cannot feed on the smaller coccolithophores and, consequently, the krill population (mainly *E. pacifica*) in that region declined sharply (Hyoung-Chul and Nicol, 2002). This in turn affected other species: the shearwater population dropped, and the incident was even thought to have been a reason for salmon not returning to the rivers of western Alaska in that season. Other factors besides predators and food availability also can influence the mortality rate in krill populations. There are several single-celled endoparasitic ciliates of the genus *Collinia* that can infect different species of krill and cause mass dying in affected populations (Gómez-Gutiérrez, 2003). Such diseases have been reported in *Thysanoessa inermis* in the Bering Sea and in *E. pacifica, Thysanoessa spinifera,* and *T. gregaria* off the North American Pacific coast.

SUPERORDER SYNCARIDA

The superorder Syncarida within the class of Malacostraca are shrimp-like crustaceans bearing no carapace or oostegites and having exopodites on all thoracic limbs. The body is divided into a five-segmented head, a thorax with seven to eight segments, and a six-segmented abdomen. The antennules and antennae are moderately long. Some or all thoracic appendages are biramous. Unlike in the other malacostracan orders Decapoda, Amphipoda and Isopoda, the foremost thoracic legs are never modified as gnathopods. The abdomen may or may not carry leg-like appendages as long as those on the thorax. There are two extant orders, Anaspidacea and Bathynellacea. Anaspidacea comprises two families

(Anaspididae, Koonungidae) with moderate-sized benthic or hypogean species and two families (Psammaspididae, Stygocarididae) of mainly smaller, hypogean to interstitial species. The geographic range of this order is restricted to south-eastern Australia, South America, and New Zealand. In Australia, the family Anaspididae occurs in Tasmania and family Koonungidae in Victoria and Tasmania (De Deckker, 1980). The order Bathynellacea (Parabathynellidae, Bathynellidae), which is scarce but cosmopolitan, is entirely composed of small, blind crustaceans that live underground.

SUPERORDER PERACARIDA

The superorder Peracarida within the class of Malacostraca is a large group of crustaceans with nine orders, having members in marine, freshwater, and terrestrial habitats. Peracaridans have the first thoracic segment united with the head, the cephalothorax usually larger than the abdomen, and some thoracic segments free from the carapace.

Peracaridans have been found in hot springs living at temperatures up to 40°C, and in the deepest parts of the ocean. The common pill bug, the isopod *Amradillidium vulgaris*, has been transported all around the world in the soil of house plants, while its deep-sea relative, *Bathynomus gigas*, is the largest peracarid, growing to a length of 44 cm. The group can be recognized by several shared characteristics. In females, flaps (endites) termed oostegites, from the first segment (coxa) of the thoracic appendages, form a brood pouch (marsupium) (except in Thermosbaenaceans, which brood in the carapace). Development is direct in that the young hatch out at a manca stage, which resembles the adult but lacks the last pair of thoracic appendages. The carapace is generally reduced and, if developed, is not fused to the posterior thoracic segments.

ORDER CUMACEA

Cumacea is an order of small marine crustaceans belonging to the superorder of Peracarida, within the class of Malacostraca, occasionally called hooded shrimps. Cumaceans are characterized by a well-developed carapace that is fused dorsally with at least the first three thoracic somites and overhangs the sides (Bacescu, 1988). Cumaceans are found on all continents and are mainly marine crustaceans with some species surviving in low salinity water (e.g., estuaries). In the Caspian Sea they even reach some rivers that flow into it. Few species live in the intertidal zone. Their unique appearance and uniform body plan makes them easy to distinguish from other crustaceans. Cumaceans have a strongly enlarged carapace and pereon (breast shield), a slim abdomen and a forked tail (Day, 1980). The length of most species varies between 1 and 10 mm. The carapace of a typical cumacean is composed of several fused dorsal head parts and the first three somites of the thorax. This carapace encloses the appendages that serve for respiration and feeding. In most species, there are two eyes at the front side of the head shield, often merged into a single eye lobe. Cumaceans are epimorph, which means that the number of body segments does not change during the different developmental stages. This is a form of incomplete metamorphosis (Jones, 1963).

Most species live only one year or less and reproduce twice in their lifetime. Cumaceans exhibit sexual dimorphism: males and females differ significantly in their appearance. Both sexes have different ornaments (setation, knobs, and ridges) on their carapace. Other differences are the length of the second antenna, the existence of pleopods in males, and the development of a marsupium in females. There are generally more females than males, and females are also larger than their male counterparts. Females carry the embryos in their marsupium for some time. The larvae leave the marsupium during the so-called manca-stadium, in which they are almost fully grown but do not have the last pair of pereiopods (Bacescu, 1992).

Deep-sea species have a slower metabolism and presumably live much longer. Many shallow water species show a diurnal cycle, with males emerging from the sediment at night and swarming to the surface. Cumaceans are an important food source for many fishes and an important part of the marine food chain. They feed mainly on microorganisms and organic material from the sediment. Species that live in the mud filter their food, while species that live in sand browse individual grains of sand (Jones, 1976). In the genus *Campylaspis* and a few related genera, the mandibles are transformed into piercing organs, which can be used for predation on foraminifera and small crustaceans.

ORDER ISOPODA

Isopods belong to the superorder of Peracarida, within the class of Malacostraca. Isopods are one of the most diverse orders of crustaceans, with many species living in all environments, and common in shallow marine waters. Unlike most crustaceans, isopods are successful on land (suborder Oniscidea, woodlice), although their greatest diversity remains in the deep sea (suborder Asellota) (Wilson, 1989). The isopods are an ancient group. Fossils from the Carboniferous (suborder Phreatoicidea, family Paleophreatoicidae) differ slightly from modern southern hemisphere freshwater phreatoicideans.

Marine isopods occur from the intertidal to the deep sea. The body is usually dorsoventrally flattened and the head and first thoracic segment are fused, forming a cephalothorax and seven visible thoracic segments. There is no carapace and the eyes are not stalked. There are usually seven pairs of pereiopods of similar appearance. Isopods exhibit large diversity in feeding and include herbivores, carnivores and scavengers.

The sexes are separate and sperm are transferred to the female in spermatophores. The eggs are held on the underside of the female in a brood chamber formed from overlapping oostegites. The embryos develop in the brood chamber and emerge as juveniles closely resembling the adult but in some species having only six pairs of pereiopods. In a few species, embryonic development takes place in internal pouches leading from the brood chamber. There is no pelagic larva. Many isopods are parasitic on fish, while others parasitize crustaceans (Shields and Gómez-Gutiérrez, 1996).

ORDER AMPHIPODA

Amphipods belong to the superorder Peracarida, within the class of Malacostraca. Most amphipods are marine, with a few freshwater and terrestrial shrimp-like crustaceans. Amphipods have many characteristics in common with isopods, such as lack of a carapace. The body of the amphipod is laterally compressed and divided into three regions: head, pereon (thorax) and pleon (abdomen), with a terminal telson attached dorsally to the last pleon segment. The pleon consists of six segments with the first three segments typically bearing a pair of flattened pleopods, while the last three bear narrow, backwardly directed uropods (Fish and Fish, 1996; Hayward and Ryland, 1998). There are seven visible thoracic segments; the appendages of the first two are known as gnathopods and are used in feeding and grasping. The remaining five segments of the thorax bear pleopods. The range of feeding habits exhibited by amphipods includes suspension feeding, scavenging, grazing and detritus feeding. Amphipods are one of the few animal groups frequently seen when submarines venture to the deepest parts of the oceans. Other benthic amphipods are the primary food of gray whales.

Marine amphipods are generally benthic in habit and may burrow into sand and mud. Amphipods swim using the pleopods and some are caught in high densities in the intertidal plankton especially at night. Most species of pelagic amphipods are parasitic and some are mutualistic, living in association with jellyfish and salps. *Phronima* is a relatively common genus of pelagic amphipod that kills and cleans out the barrel-shaped body of a salp to live inside and raise its young.

Of the relatively few species of free-living, planktonic amphipods, the most abundant is *Themisto gaudichaudii* in the Southern Ocean, congregating in dense swarms, a voracious predator of copepods and other small members of the zooplankton. *Themisto* is the most abundant member of the mesozooplankton in the Southern Ocean after copepods, krill and salps. In the Southern Ocean, amphipods are the most abundant benthic crustaceans.

The sexes are generally separate and some species show precopula behaviour, during which the male holds the female using the gnathopods and carries her for some days before mating. Sperm are deposited in the brood chamber of the female, where external fertilization takes place, and where the eggs develop to the juvenile stage. The brood chamber is formed of brood plates (oostegites) arising from the bases of some thoracic appendages (Fish and Fish, 1996).

ORDER SPELAEOGRIPHACEA

The order Spelaeogriphacea of the superorder Peracarida belonging to the class Malacostraca is a group of small, cave-dwelling, freshwater crustaceans that grow to no more than 10 mm, with a short carapace, two pairs of elongate and biramous antennae, eye lobes without eyes, and appendages on all abdominal somites. Only four species, all subterranean, have been described. Spelaeogriphaceans are known only from caves in South Africa, Brazil, and Western Australia (Poore and Humphreys, 1998). They feed on plant detritus by sweeping their mouth appendages over the substrate, picking up small particles. Females carry 10-12 eggs in the brood pouch.

ORDER THERMOSBAENACEA

Thermosbaenacea is an order of the superorder Peracarida belonging to the class Malacostraca. It is a group of blind crustaceans that live in thermal springs and in fresh and brackish water. The order comprises seven genera arranged in four families with a total of 34 known species. All extant species in this order are known from ground waters. Thermosbaenaceans have a small, elongate body with a short head possessing a short carapace that extends backwards over the first few thoracic somites. The abdomen consists of six somites and a telson. Eyes are reduced or absent. Thermosbaenaceans move primarily by walking but can also swim by using their thoracic limbs for propulsion. They have been found principally in thermal water but occasionally in cool, fresh or brackish water, lakes, springs, interstitial coastal areas, and cave pools

Thermosbaenaceans living in hot springs feed on blue-green algae, diatoms, and other microalgae lining the rocks. One species is known to feed on plant detritus. The sexes are separate. The eggs are incubated in a dorsal brood pouch formed by the swollen carapace of the mature female and are bathed in water passing the respiratory epipod. The developing embryos are carried by the adult under its carapace until hatching. The young resemble miniature adults when they hatch (Wagner, 1994).

ORDER MYSIDA

The order Mysida belongs to the superorder of Peracarida, within the class of Malacostraca. Mysidacea are a group of small, often transparent, shrimp-like animals with a flexible carapace enveloping the thoracic region along the sides, stalked eyes, and a well-developed tail fan. The group includes the species *Neomysis americana* (opossum shrimp). The order Mysida comprises four families and as a whole includes slightly over 1,000 described species. They are sometimes known as "opossum shrimps" because of the marsupium, or external pouch, formed by specially developed plates on the inner sides of the thoracic limbs of adult females, though that name is also used for individual species (Kallmeyer and Carpenter, 1996). Despite their name, and their superficial resemblance to shrimp, they are distantly related to the true shrimps, which are classified in the order Decapoda. Mysidacea are usually found in shallow coastal waters but some species live in greater depths. Many species of deep-sea mysids are found on or just above the ocean floor at depths of 5,700-7,200 m. Some are abundant in inshore waters in rock pools and among seaweed. Some extend into estuaries often in dense swarms, several miles long and three feet or or more in diameter, and are important in estuarine food webs (Morgan, 1982).

Most mysids are fairly small, between 10 and 30 mm long, and have a well-developed carapace that covers most of the thorax, but it is never fused with more than four of the thoracic segments. The pereiopods are biramous, except sometimes the last pair, which may be reduced. The abdomen has six segments, the first five of which usually have small appendages, the pleopods, while the last segment has a pair of uropods. The pleopods are reduced and modified in males. They have usually a statocyst in the endopods of their uropods and possess stalked eyes. When feeding, the beating of the thoracic appendages creates a current of water from which suspended matter is filtered. Most mysids are filter-feeders, removing fine detritus, rotifers, mollusc larvae, diatoms, and other planktonic organisms out of the water while swimming just above the bottom and creating a suspension feeding current. All filter-feeding mysids may also feed raptorially. Members of some mysid genera, including *Neomysis* and *Siriella*, have been seen to catch small live crustaceans (copepods, cladocerans, amphipods) as well as small molluscs.

Male mysids do not actively search for females during reproduction. After shedding a previous brood, the female soon moults and is ready to breed again. At that time she produces a pheromone stimulating the nearby males. Mating is very quick and takes place at night. The male either lies under the female head-to-tail and belly-to-belly, or doubles up and grasps the anterior part of the female's abdomen with his antennae. The sperm are either injected into the female's brood pouch or shed between the mating individuals and swept by currents produced by the thoracic appendages into the marsupium. The copulating pair soon separate, and within half an hour the female's eggs are extruded into her brood chamber and fertilized there. The incubation period and frequency of mating depend on the species and the water temperature and can range from a few weeks to several months. The young are shed as juveniles with complete sets of appendages. Released mysids need about a month to reach their adult stage at a water temperature of about 20°C.

ORDER LOPHOGASTRIDA

The order Lophogastrida of the superorder Peracarida belonging to the class Malacostraca is a group of small, shrimp-like marine crustaceans with elongate carapaces covering, but not fused to, all thoracic somites, stalked eyes, and strongly developed swimming legs on the first five abdominal somites.

The body of a lophogastrid is long and shrimp-like, with the head and thorax covered by a loosely fitting carapace. The carapace is extended in front as a rostrum. The rear portion of the carapace may cover from five to seven thoracic somites. It extends over the sides of the animal to the bases of the thoracic legs. Both pairs of lophogastrid antennae are biramous. The thoracic limbs of lophogastrids are modified in various ways; however, all have well-developed endopods and exopods. The first pair is always modified as a maxilliped because the first thoracic somite is incorporated into the head. Most lophogastrid pereiopods bear gills at their base. Oostegites, or brood plates, are present on all seven pairs of pereiopods. All lophogastrids, both males and females, have well-developed biramous propulsive pleopods on the first five abdominal somites, and broad flattened uropods on the sixth abdominal somite. The uropods do not possess statocysts. A flattened and somewhat pointed telson terminates the body.

Lophogastrids occur in the pelagic oceanic zone in all oceans except the Arctic. They are generally bathypelagic, found below 1,000 m, but some species may be found in waters as shallow as 50 m and others as deep as 4,000 m. Most of the known species are found primarily in the Pacific and Indian Oceans, but a few are common in the Atlantic. The lophogastrids investigated to date are almost all predators. Only *Gnathophausia* seems to use its mouthparts to filter large particles from sea water. Lophogastrids spend almost all their time swimming, using the pleopods for propulsion and the exopods to maintain a flow of water around the gills for respiration.

The sexes are separate; eggs are extruded into the female's ventral brood pouch, where they are kept until hatching. When the young hatch, they possess a full set of appendages on the thorax. In males of the genus *Gnathophausia*, the endopod of the second pair of pleopods is slightly modified for sperm transfer.

ORDER MICTACEA

Mictacea is a proposed order of the superorder Peracarida (class: Malacostraca) established for two small crustacean species, *Hirsutia bathyalis* and *Mictocaris halope*. The two species share many features common to other peracaridans but differ sufficiently to justify their assignment to a distinct order with two monotypic families. Common peracaridan features include a brood pouch formed by basal lamellae of the pereiopods (oostegites) in the female, a small movable process (lacinia mobilis) on the mandible, free thoracic somites not fused to a carapace shield, a single maxilliped of typical peracarid form, and partially immobile pereiopodal basal segments. *Mictocaris* is a cave-dwelling species, whereas *Hirsutia* has been found only in soft muddy sediment of the deep sea and is thought to be carnivorous. In contrast, the feeding appendages in *Mictocaris* resemble those of thermosbaenaceans, which scrape food particles from the substrate. Mictacea appear to be most closely related to the Thermosbaenacea, Spelaeogriphacea, and Mysidacea (McGraw-Hill, 2006).

ORDER TANAIDACEA

The order Tanaidacea of the superorder Peracarida (class Malacostraca) is a group of mostly marine-dwelling crustaceans that are among the most diverse and abundant creatures in some marine environments with more than 750 species. The Tanaidacea are distributed in all oceans of the world, from tropical to temperate regions, and even in polar waters. Most species are benthic, although some species have been found in the plankton. Although almost exclusively marine dwellers, some species have penetrated freshwater habitats. They are raptorial feeders, eating detritus and its associated microorganisms (Holdich and Jones, 1983). The larger food particles are manipulated by the chelipeds and maxillipeds and transferred toward the mouth. Some species are filter or suspension feeders. In this feeding mechanism, the simultaneous action of the thoracic limbs creates a suspension-feeding current by alternate movements of adjacent limb pairs. Some tanaids are predators, catching the prey with chelate pereiopods or even directly with the mouth appendages. Then, with various mouthparts, mainly the mandibles, they shear, tear, or grind the prey.

There is a carapace covering the cephalothorax, which is formed from the head fused with the first two thoracic segments; the cephalothorax bears one pair of antennules and one pair of antennae. Compound eyes can be absent or present. The pleopods can be present or absent; when present, they are used in swimming and to produce a ventilatory current in tube-dwelling species. Most species of Tanaidacea are small, ranging from 1 mm to 2 cm in length. Many tanaids can produce cement-like material from tegumental glands in the pereonal cavity; this cement emerges from the tips of the dactyli of the anterior pereiopods and is used to construct tubes with particles of sand and detritus. Generally, the tubes are open at both ends, and tanaids use the pleopods to create water currents. During the incubatory period, ovigerous females may close the ends of their tubes. The tubes built by tanaids stabilize the sediment, favouring the colonization of other sedentary species.

As in all other peracarid crustaceans (superorder Peracarida), tanaid offspring are held within a marsupium on the ventral face of an ovigerous female during their early development and generally are released only when they have developed most of their appendages. The fertilized eggs develop into embryos, which gradually develop into the first larval stage (manca I) and then into a second, more mobile larval stage (manca II). In this second stage, the manca has partially formed the sixth pereiopods; this is when the tanaid generally leaves the marsupium. Juveniles will develop either into preparatory males or preparatory females. However, hermaphroditism can occur in tanaids in many ways. For example, in *Apseudes spectabilis*, some specimens possess both mature eggs and male gonads filled with sperm. Hence, several researchers believe that self-fertilization is possible. In the case of conventional reproduction of tube-dwelling species, a copulatory male enters the tube of a female, and the courtship takes place for a long period of time. Male and female lie with their ventral sides adjacent to one another, and sperm are deposited in the marsupium. After closing the marsupium, the female releases the eggs. Then, the female expels the male from the tube, closes the ends of the tube, and incubates the brood.

SUBCLASS PHYLLOCARIDA

The subclass Phyllocarida of the class Malacostraca contains the single order Leptostraca, comprising less than 40 described species worldwide. These animals all are marine, and most are epibenthic. They extend from the intertidal zone to a depth of 400 m and most feed by stirring up bottom sediments. The order Leptostraca is a small, poorly known group of crustaceans comprising only three extant families, Nebaliidae, Nebaliopsididae, and Paranebaliidae. The head consists of five somites, the thorax of eight, and the abdomen of seven excluding the telson (Walker-Smith and Poore, 2001). The carapace is bivalved, laterally compressed, with an adductor muscle, lacking the hinge or hinge line, and encloses the thorax but not the cephalon (Rolfe, 1969). These families together encompass only seven genera and 38 species.

SUBCLASS HOPLOCARIDA

The subclass Hoplocarida of the class Malacostraca is represented by a single order, the Stomatopoda, and includes the "mantis-shrimps" (Ahyong and Harling, 2000). Mantis (from the Greek *mantes* = prophet or fortune teller) shrimps are neither shrimps nor mantids, but receive their name purely from the physical resemblance to both the terrestrial praying mantis (the common colloquial name for an insect of the order Mantodea named for the typical "prayer-like" stance) and the shrimp (Mauchline, 1984). They may grow to a length of 20–30 cm, although most species are considerably smaller. The carapace covers only the rear part of the head and the first three segments of the thorax. Mantis shrimps appear in a variety of colours, from dull browns to stunning neon. Called sea locusts, they are sometimes also referred to as "thumb splitters" because of the relative ease with which the creature is able to mutilate small appendages. Mantis shrimps sport powerful claws, formed like jackknives and used to attack and kill prey by spearing, stunning or dismembering it.

Mantis shrimps are aggressive and typically solitary, spending most of their time hiding in rock formations or burrowing in the sea bed. They rarely leave their homes except to feed or relocate and can be diurnal, nocturnal or crepuscular, depending on the species. Most species live in tropical and subtropical seas (Indian and Pacific Oceans between eastern

Africa and Hawaii), although some live in temperate seas. Mantis shrimps are the only animals with hyperspectral colour vision and are considered to have the most complex eyes in the animal kingdom.

In a lifetime, they can have as many as 20 or 30 breeding episodes. Depending on the species, the eggs can be laid and kept in a burrow or carried around under the female's tail until they hatch (Caldwell, 1991). Also depending on the species, male and female only come together to mate or bond in monogamous long-term relationships (Caldwell, 1992). In the monogamous species, the mantis shrimp remains with the same partner for up to 20 years. They share the same burrow, and there are reasons to suspect that these pairs can coordinate their activities (Shuster and Caldwell, 1989). Both sexes share the care of the eggs (biparental care). In *Pullosquilla* and some species of *Nannosquilla*, the female will lay two clutches of eggs, one that the male tends and one that the female tends. In other species, the female will look after the eggs while the male hunts for both of them (Montgomery and Caldwell, 1984). Once the eggs hatch, the offspring may spend up to three months as plankton.

CLASS OSTRACODA

Ostracoda is a class of the subphylum Crustacea, known as mussel or seed shrimps because of their appearance. The word *ostracod* is derived from the Greek word *ostrakon*, a shell; this shell or carapace has numerous morphological characteristics that allow taxonomic and phylogenetic studies to be made on living and fossil specimens. Ostracods are small crustaceans, typically around 1 mm, but varying between 0.2 and 30 mm, laterally compressed and protected by a bivalve-like, chitinous or calcareous valve or "shell". The hinge of the two valves is in the upper, dorsal region of the body.

The body of an ostracod is enclosed between the two valves and consists of two main parts, the head (or cephalon) and the thorax, separated by a slight constriction, but segmentation is unclear. The abdomen is regressed and fused with the thorax or absent. The adult gonads are relatively large. There are five to eight pairs of appendages marking the existence of ancestral segmentation of the head and thorax. The branchial plates are responsible for oxygenation. The body is entirely enveloped in a cuticle secreted by the epidermis. The carapace develops from two lateral folds or duplicatures in the epidermis, each with an outer lamella and an inner lamella. The space between these lamellae is a continuation of the body cavity and may house certain reproductive and digestive organs.

Ostracods are part of the zooplankton and the benthos, living on or inside the upper layer of the sea floor. Many ostracods are also found in freshwater but some are also found in humid continental forest soils (Athersuch et al., 1989). Some ostracods are filter-feeders; others are scavengers, detritivores, herbivores or predaceous carnivores. Some are parasitic or commensal on other crustaceans, polychaete worms, echinoderms and sharks. Some 50,000 extinct and extant species have been identified, grouped into two subclasses Myodocopa and Podocopa.

CLASS BRANCHIOPODA

Branchiopoda is a class of the subphylum Crustacea with primitive and primarily freshwater crustaceans, mostly resembling shrimp. Their name arises from the feet of branchiopods, which are supposed to perform the function of gills. The branchiopods are characterized by their possession of broad, leaf-like appendages fringed with bristles used for filter-feeding and propulsion. There are over 900 known species worldwide, with the well-known representatives of *Artemia* (brine shrimp or Sea-Monkeys), and *Daphnia* both being used as aquarium food and in aquaculture (Dumont and Negrea, 2002). Branchiopods should not be confused with Brachiopods (Phylum Brachiopoda, known as lamp shells, sessile two-shelled marine animals with an external morphology resembling clams of the Phylum Mollusca, to which they are unrelated). The thorax and abdomen are fused or indistinguishable in most species except those in the order Anostraca. Their appendages are generally phyllopodus (leaf-like), although some groups have more appendages than others. They have a combination of paired compound eyes and/or a single simple median eye. Many are capable of parthenogenesis (growing from unfertilized eggs) but several use other reproductive strategies, ranging from releasing eggs simultaneously with the adult moults to encapsulating the eggs in a modified moult (Tasch, 1969). As artemia is an inhabitant of biotopes characterized by unusual environmental conditions, its survival during periods of extreme conditions is ensured by the production of dormant embryos (diapause).

ORDER ANOSTRACA

The order Anostraca (Greek for "without shell") of the subclass Sarcostraca is divided into seven families, which include the fairy shrimp and the brine shrimp (*Artemia*). Fairy shrimps are widely considered the most primitive living crustaceans. The main habitat of fairy shrimps is rain pools. The pools may be filled by periodic and predictable rains, or only erratically at long intervals. A number of anostracans, however, have adapted to other water bodies: high freshwater mountain lakes and Arctic ponds and saline lakes. The life cycle of fairy shrimps in Arctic or Antarctic ponds is not regulated by alternation between wetting and drying, but by alternation between freezing and thawing. Anostracans are well adapted to living in arid areas where water is present for only part of the year. Their eggs will survive drought (an extended period of below-normal rainfall) for several years and hatch about 30 hours after rains fill the pools where they live. Some eggs may not hatch until going through several wet/dry cycles, ensuring the animals' survival through times that the pools do not last long enough for the shrimp to reproduce (Belk and Brtek, 1995).

One peculiar feature of all anostracan species is that they swim upside down. Some are largely translucent and hard to spot in the water; others, however, may develop bands or zones of bright colour. Fairy shrimps, like zooplankton, migrate vertically over a 24 hour period if they live in sufficiently deep pools; they usually come to the surface at night. Females tend to live below the males in the water column. The vast majority of fairy shrimps are filter-feeders. They use specialized endites on their legs to collect food in the ventral food groove. The food is then pushed towards the mouth. Filter feeding allows them to collect particles as small as bacteria and as large as algal cells. Fairy shrimps can consume even rotifers, nauplii (crustacean larvae), and nauplii of their own species (Erikson and Belk, 1999).

The brine shrimps (*Artemia*) are found worldwide in salt water, though not in oceans. *Artemia* is a well-known genus as one variety, the *Artemia nyos*, a hybrid of *Artemia salina*, is sold as a novelty gift, most commonly under the trademark Sea-Monkeys. Views are mixed as to whether all brine shrimps are part of one species or whether the varieties that have been identified are properly classified as separate species.

The sexes are separate, except in some strains of *Artemia*, which may be parthenogenetic. Males have modified second antennae with which to grasp females. The antennae may demonstrate structural complexity. In addition to the clasping organ, male anostracans have two penes. Copulation may take place so rapidly as to be hardly visible to the unaided eye, as in most streptocephalids, or last for many hours as the couple swims around in tandem formation, as in *Artemia*. Following copulation and internal fertilization, the eggs are deposited in an external brood pouch of variable shape, depending on the genus and species. Each batch, or clutch, may contain several hundred eggs, and a female may produce up to forty clutches in her lifetime; thus, total fertility may reach 4,000 eggs per female. The eggs are usually shed freely into the water. They may either sink to the bottom, as is usually the case in freshwater species, or float on the surface, to be deposited eventually along the lake shore. The eggs or cysts of anostracans are noteworthy because they are surrounded by a thick wall that allows them to resist drought and high temperatures. They develop into a gastrula containing about 4,000 cells and then stop developing in order to survive adverse conditions. This stage of latency may continue for long periods of time, possibly more than a century in some strains of *Artemia*. This characteristic is called cryptobiosis (Greek for "hidden life") or diapause. On the other hand, *Artemia* is the only known genus in which viviparity may occur. Once placed in water, the cyst-like eggs hatch within a few hours. The nauplii, or larvae, of brine shrimp are less than 500 μm when they first hatch and will grow to a mature length of around 1 cm on average. They eat microalgae but will also eat yeast, wheat flour, soybean powder, or egg yolk. Brine shrimps have a biological life cycle of one year. The nutritional properties of newly hatched brine shrimps make them particularly suitable to be sold as aquarium food and in aquaculture as they are high in lipids and unsaturated fatty acids. These nutritional benefits are likely to be one reason that brine shrimps are found only in highly salinated waters, areas uninhabitable for potential predators (Persoone et al., 1980).

The genus *Artemia* is of considerable economic importance. The cysts of this species are harvested, cleaned, dried, packed and sold as fish food in the aquarium business. The cysts are also used in industrial aquaculture to feed fish larvae. The Libyan Fezzan desert contains several spring-fed dune lakes that have turned saline with time. Small communities living around these lakes use *Artemia* as their main source of animal protein.

ORDER NOTOSTRACA

The order Notostraca of the subclass Phyllopoda comprises small crustaceans in the class Branchiopoda. Notostracans (*Triops*, tadpole shrimp or shield shrimp) have two internal compound eyes (visual organ found in certain arthropods) and one naupliar eye in between and a flattened carapace covering the head and leg-bearing segments of the body. The order contains a single family, with only two extant genera. Their external morphology has apparently not changed since the Triassic appearance of *Triops cancriformis* around 220 million years ago (Kelber, 1999).

Tadpole shrimps usually live near the ground of astatic pools, where they move with their ventral side down. However, lack of oxygen can force them to swim upside down with their gill-like legs close under the surface of the water. Notostraca are omnivorous, looking not only for plankton but also for larger prey such as worms, chironomid larvae and even weak tadpoles.

Triops survives in temporary pools all over the world and is short-lived. *Triops longicaudatus* has a maximum lifespan of about 50 days and *T. cancriformis* a maximum lifespan of about 90 days, with some individuals beginning to die off as soon as two weeks after hatching. These ponds usually dry up during certain times of the year when there is no rainfall. Although the adult *Triops* die during such droughts, the embryos remain in a state of diapause until the rains return and fill up the temporary pools once again, allowing them to hatch.

In northern and central Europe, with few exceptions, all tadpole shrimps are female, whereas the sexes in southern and western Europe as well as in northern Africa are nearly equal in number. These "females" possess hermaphroditic glands and parthenogenesis (Greek *parthenos* = virgin, and *genesis* = birth) takes place. Modified appendages of the 11th pair of limbs bear the ovisacs with the mature eggs.

ORDER CLADOCERA

The order Cladocera of the class Branchiopoda are small crustaceans forming a monophyletic group currently divided into four suborders, 11 families, 80 genera, and about 400 species. The most commonly known genus is *Daphnia* (freshwater water fleas), which is the most researched in this group. *Daphnia* are small planktonic crustaceans, between 0.2 and 5 mm in length. *Daphnia* are one of the several small aquatic crustaceans commonly called water fleas (characterized by their jumping or jerky mode of swimming). They live in various aquatic environments ranging from acidic swamps to freshwater lakes, ponds, streams and rivers (Pirow et al., 1999).

The body is divided into segments; the head is fused and is generally bent down towards the body with a notch separating the two. In most species the rest of the body is covered by a carapace, with a ventral gap with five or six pairs of legs. The most prominent features are the compound eyes, the second antennae, and a pair of abdominal setae. In many species, the carapace is translucent, making them excellent subjects for the microscope.

A few *Daphnia* prey on tiny crustaceans and rotifers, but most are filter-feeders, ingesting mainly unicellular algae and various sorts of organic detritus including protists and bacteria. *Daphnia* can be kept easily on a diet of yeast. Beating of the legs produces a constant current through the carapace that brings suspended material into the digestive tract. The trapped food particles are formed into a food bolus, which then moves down the digestive tract. The first and second pairs of legs are used in the organisms' filter feeding, ensuring large unabsorbable particles are kept out, while the other sets of legs create the stream of water rushing into the organism. Swimming is powered mainly by the second set of antennae, which are larger than the first set. The action of this second set of antennae is responsible for the jumping motion.

Daphnia reproduce parthenogenetically usually in the spring until the end of the summer. One or more juvenile animals are nurtured in the brood pouch inside the carapace. The newly hatched *Daphnia* moult several times before they become a fully grown adult usually after about two weeks. The young are small copies of the adult; there are no true nymphal or instar stages. The fully mature females are able to produce a new brood of young about every 10 days under ideal conditions. The reproduction process continues while the environmental conditions continue to support their growth. Winter or drought conditions bring an end to production of new female generations, changing the reproductive method. Parthenogenic males are produced, followed by mating and fertilization of the eggs. Fertilized eggs are termed winter eggs and are provided with an extra shell layer called ephippium. The extra layer preserves and protects the egg inside from harsh environmental conditions until the more favourable times, such as spring, when the reproductive cycle is able to take place once again.

Males are found only at times of harsh environmental conditions, during periods of scarce resources due to population overgrowth or winter conditions, and even then may make up considerably less than half the population, in some species being unknown entirely. They are much smaller than the female and they typically possess a specialized abdominal appendage used to grasp the female from behind and insert a spermatheca. The male appearance is for the creation of resting or winter eggs, allowing for the survival of the population through harsh conditions. It has been proposed that this switch to sexual reproduction allows greater offspring variation (through genetic recombination), which is useful in varied or unpredictable conditions (the lottery model) (Chesson and Warner, 1981).

The lifespan of a *Daphnia* does not exceed one year and is largely temperature-dependent. However, during the winter, when harsh conditions limit the population, females living for over 6 months have been recorded. These females generally grow at slower rate but in the end are larger than ones living under normal conditions. *Daphnia* are commonly used to test the toxicity of chemicals in solution or for water pollution. They are also an important food source for many larger aquatic organisms including various fish species and are used as food for aquarium fish.

CLASS REMIPEDIA

Remipedia is a class of colourless, slow-moving, blind crustaceans found in deep caves connected to salt water, in Australia and the Caribbean Sea. The first described remipede was the fossil *Tesnusocaris goldichi* but, since 1979, about a dozen living species have been found. These species have been assigned to the order Nectiopoda and to the two families of Godzilliidae and Speleonectidae.

The body comprises a head and up to 32 similar body segments forming an elongated trunk. The swimming appendages are lateral on each segment, and the animals swim on their backs. They have fangs connected to secretory glands; it is still unknown whether these glands secrete digestive juices or poisonous venom, or whether remipedes feed primarily on detritus or on living organisms. They have a generally primitive body plan in crustacean terms and have been thought to be a basal, ancestral crustacean group. It has been suggested that Remipedia might be the sister taxon to Malacostraca, regarded as the most advanced of the crustaceans (Fanenbruck and Harzsch, 2005).

CLASS CEPHALOCARIDA

Cephalocarida is a class in the subphylum Crustacea that comprises only about nine shrimp-like benthic species. Cephalocarids are commonly referred to as horseshoe shrimps and are generally considered to belong in two families: Hutchinsoniellidae and Lightiellidae. A total of nine species in four genera are currently known (Schram, 1986). The best-known species is *Hutchinsoniella macracantha,* from the east coast of North and South America, about 3-4 mm long, with a shovel-shaped head and a slender, very flexible body that is not covered by a carapace. Cephalocarids move through the upper few millimetres of the sediment by moving the thoracic limbs. They appear to be able to swim or burrow. Even though there is no fossil record of cephalocarids, most specialists believe them to be primitive among crustaceans. Their anatomy is relatively simple in comparison to that of other crustaceans (Sanders, 1955). The body is small (2-3.7 mm long) and comprises a head, an eight-segmented thorax with biramous appendages, and an 11-segmented abdomen that bears a telson but no other appendages. The second pair of maxillae closely resembles the appendages of the thorax. The eyes are very small and buried in the exoskeleton, which gives the creature the appearance of being eyeless. Small organic particles appear to be the primary food source of cephalocarids. Cephalocarids are found from the intertidal zone down to a depth of 1,600 m, in all kinds of sediments feeding on marine detritus (Hessler and Sanders, 1973). Most cephalocarids have commonly been found in flocculent surface muds with high organic content. A few, however, are known from coral rubble and substrates with fine sediment particles. The nine species are known from the eastern and western United States, the Caribbean Islands, Brazil, Peru, southwest Africa, New Caledonia, New Zealand, and Japan. To bring in food particles, they generate currents with the thoracic appendages like the branchiopods and the malacostracans. Food particles are then passed anteriorly along a ventral groove leading to the mouthparts (Myers, 2001). Cross-fertilization is likely since cephalocarids are functional hermaphrodites. Eggs are carried on the reduced appendages of the ninth post-cephalic somite and young hatch as a metanauplius.

CLASS MAXILLOPODA

Maxillopoda is a class of the subphylum Crustacea, characterized by a reduction of the abdomen and its appendages. The Maxillopods are divided into six subclasses with the most species-rich groups being the barnacles (subclass Thecostraca) and copepods (subclass Copepoda). Most species are small. Most feed by means of their maxillae (rather than filter feeding using thoracic appendages to move water); barnacles, however, are an exception. Maxillopods have a basic plan of 5 head and 10 trunk segments (6 thoracic and usually 4 abdominal), followed by a terminal telson. The abdominal segments typically lack appendages; appendages elsewhere on the body are usually biramous.

INFRACLASS CIRRIPEDIA

The most important infraclass of the class Maxillopoda is the infraclass Cirripedia, constituting about 1,220 known species including the barnacles. Barnacles are perhaps the best-known intertidal organisms and are so common and widespread in their distribution that they have been described as ubiquitous. Around 1,220 barnacle species are currently known, all exclusively marine (Escobar-Briones and Alvarez, 2002). Cirripedia are highly modified crustaceans (Hayward and Ryland, 1998). The name "Cirripedia" means "curl-footed". Barnacles were first fully studied and classified by Charles Darwin during his development of the theory of evolution and natural selection. For centuries, they were thought to be molluscs because of their apparent possession of a shell. However, scientists in the 19th century discovered that they are actually crustaceans.

Common barnacles are loosely divided into goose or stalked barnacles and acorn or sessile barnacles characterized by a low conical shape and a roughly circular base. Goose barnacles are the less common of the two varieties and are to be found suspended in the sea from floating objects. They have a fleshy protruding stalk at the end of which is a body or "head" (also known as the capitulum) bearing a number of calcareous plates. The name "goose barnacle" stems from the legend that barnacle geese hatched from barnacles growing on trees at the water's edge (Fish and Fish, 1996). *Lepas*, the common goose barnacle, has five shell plates. It lives attached to hard surfaces of rocks and flotsam in the ocean intertidal zone and has very few predators, hence the reduced number of plates. A relative, *Lithotrya*, commonly bores into limestone as it grows on tropical shores consisting of raised coral reefs, as in East Africa. Living burrowed into the rock enables this barnacle to develop a long stalk but live on the shore where its burrow protects it from danger. *Conchoderma* is another variation on the theme. It still possesses shell plates but has further reduced their size because its habitat, whale skin, is safe. Stalked barnacles are occasionally washed ashore on floating debris and are believed to be among the most primitive members of the infraclass (Hayward and Ryland, 1998).

Acorn barnacles do not posses a stalk and the wall is made up of a number of calcareous plates that completely cover the animal inside and are cemented to a firm surface. The calcareous plates are mostly white but can be pink or deep purple. The number, size and shape of the plates vary from species to species, in some species the original number of eight has been reduced by loss and/or fusion. The number and arrangement of these plates are important in species identification. During low tide the plates fit firmly together, but when submerged the top plates can open slightly, allowing the feeding appendages to emerge. Barnacles develop four, six or eight calcareous plates to surround and protect their bodies, forming a volcano-like cover. The top entrance is covered by two more plates. When feeding, these two top plates open and basket-like cirri (feeding "baskets") limbs wave into the oncoming current of water and direct food into the mouth. Barnacles typically have six pairs of biramous thoracic appendages used to filter suspended food particles from sea water. The barnacle exoskeleton is unique in not being completely shed at each moult and the calcareous plates, like the shells of molluscs, exhibit concentric growth lines (Hayward and Ryland, 1998).

British shores are inhabited by the common British barnacle, *Semibalanus balanoides*, together with two other species of *Chthamalus* that are subtly different from each other. The latter two species tend to live higher up the shoreline than *Semibalanus balanoides*. It is thought that this most primitive of acorn barnacles originally had eight shell plates that evolved from a ring of shell plates around the top of the stalk. Reduction in the number of shell plates reduces water loss and subsequent desiccation and predation from molluscs. *Elminius modestus*, an acorn barnacle with four shell plates, was introduced to Britain during the Second World War from New Zealand via Australia in 1943. This barnacle has grown well on British shores, including estuaries, and has become prolific, particularly in sheltered environments. It has also spread to Europe, from Norway to France.

The common acorn barnacles are hermaphroditic and alternate male and female roles over time. They do not release their gametes into the sea but are able to fertilize one another, and copulation and cross-fertilization take place between neighbouring individuals. When ready to reproduce, an adult barnacle uncoils its long, tubular penis and extends it out through the operculum (a covering or lid that closes an aperture) to search for a nearby receptive neighbour. When the sperm is transferred, the fertilized eggs are brooded within the shell of the receiver adult until they develop into a nauplius larva (which first measures about a tenth of a millimetre long) after about 4 months, often coinciding with periods when plankton is plentiful. The nauplius larva looks nothing like an adult barnacle. It has a shield-shaped body, one eye, a spiny tail and three pairs of limbs with which it swims jerkily. A single adult barnacle may release over 10,000 larvae.

Barnacles have two larval stages. In the first stage (nauplius) the animal spends its time as part of the plankton, drifting according to the wind, waves, currents, and tides, while eating and moulting. The planktonic stage lasts for about 2 weeks until the second stage is reached. At this point the nauplius metamorphoses into a non-feeding, more strongly swimming cyprid larva. The barnacle cyprids are the final, lecithotrophic, larval stage of barnacles. Metamorphosis into a cyprid usually follows five or six planktotrophic nauplius stages with the nauplius eventually approaching a millimetre in length and moulting into the cypris larva. The free-swimming cyprid larva has six pairs of thoracic swimming appendages, and a pair of first antennae used for selecting and attaching to a substrate (Hayward and Ryland, 1998). The time that the cyprid spends in the plankton can range from a few days (*Balanus amphitrite*) to weeks (*Semibalanus balanoides*). The role of the cyprid is to successfully locate and attach to an area in which environmental cues indicate a safe and productive environment for adult growth and survival. It explores immersed surfaces using modified antennular structures that can recognize adults of its own species and suitable rocky environments. When an appropriate place is found, the cyprid larva pours a secretion from the ducts of the first antennae and permanently cements itself headfirst to the surface using a proteinaceous adhesive (Forest and Von Vaupel Klein, 2006). The cyprid then moults and rotates its body so that the appendages now face upward, modified to form long and feathery cirripeds that sweep through the water for planktonic food. Metamorphosis into a juvenile barnacle then proceeds quickly, generally within 24 hours. If the cyprid larva fails to settle down it will eventually die.

Barnacles live firmly attached to one spot on the rock, unable to move about and dependent upon the high tides for food. They are usually found in the intertidal zone (also known as the littoral zone). Interestingly, many species of barnacle live quite high on the shore and may be covered with water for only a few hours each day, enduring extreme weather conditions. Barnacles are usually found in groups, which can sometimes consist of thousands of animals. Settling near other barnacles of the same species ensures that a barnacle can reproduce, as the barnacle penis can extend only about ten times its body height to fertilize another barnacle's eggs.

Barnacles are often found attached to mobile animals such as whales, turtles, or other invertebrates that can provide a suitable surface. In addition, some barnacle species bore into the shells of molluscs by chemical and mechanical means, using teeth-like structures on the mantle. The adult barnacles are a few millimetres in length and their presence is revealed by tiny slit-like holes in the shell (Saxena, 2005). Barnacles often attach themselves to man-made structures, sometimes to the structure's detriment. They are often perceived to be major fouling agents, because they will grow abundantly on ships and in doing so disrupt the flow of water over the hull and slow down the ship. Barnacles will also clog the pipes of power installations on the coast, and anti-fouling paints and other anti-fouling devices have been developed in the attempt to prevent barnacle growth. The additional cost to the shipping and power industries every year is huge, and barnacles are the most prolific fouling animals of the sea. However, scientists are now making use of their remarkable capacity to accumulate concentrations of poisonous metals to help manage environmental pollution. Barnacles have been used to locate sources of high metal availability, as in the Hong Kong Harbour and Junk Bay areas, which have significant pollution problems. Scientists have found copper concentrations in barnacles from the Chai Wan Kok area to be 3,460 ppm, an astronomical level of accumulated copper. A normal barnacle from a clean site would contain only around 60 ppm.

Members of the superorder Rhizocephala (Greek for "root-heads") of the infraclass Cirripedia are derived from barnacles. This relationship is apparent from the larval stages of the cirripede crustacean, but not from the adult forms of the animals, which are extensively modified (Fish and Fish, 1996). As adults the animals lack appendages, segmentation, and all internal organs except gonads and the remains of the nervous system. Other than the naupliar stages, the only distinguishable portion of a rhizocephalan body is the externa or reproductive portion. The adult form consists of a network of threads penetrating the body of the host that resemble the root of a plant penetrating the soil.

Members of the genus *Sacculina* are parasitic on crabs. A female nauplius settles on a host and metamorphoses as it penetrates the internal portion of the animal. It then ramifies, or grows in a similar manner to a root system, through the host, centring on the digestive system. Once mature, the female produces a sac-like externa on the abdomen of the host. The externa is immature until a male nauplius settles on it and fuses with it. The externa then produces two types of eggs: small ones that become females and large ones that become males. When a female *Sacculina* is implanted in a male crab it will interfere with the crab's hormonal balance. This sterilizes it and changes the bodily layout of the crab to resemble that of a female crab by widening and flattening its abdomen, among other things. Because the externa is located in the same place as the host's egg sac would be, the host treats it as if it were its own egg sac and never moults again (female crustaceans do not moult until they release their eggs or young from the brood pouch). This behaviour extends even to male hosts, which would never have carried eggs or young in a brood pouch but care for the externa in the same way as females. The female *Sacculina* has even been known to cause the male crabs to perform mating gestures typical of female crabs. *Sacculina carcini* (Thompson), widely distributed in northwest Europe, is found on the underside of the abdomen of *Carcinus maenas* and other crabs of the family Portunidae, as a conspicuous, yellow-brown mass that bears a superficial resemblance to the crab's eggs (Begon et al., 2005).

SUBCLASS TANTULOCARIDA

Tantulocarida is a small subclass of the class Maxillopoda (subphylum Crustacea) comprising four families, 22 genera, and about 30 species. They are typically ectoparasites that infest deep-sea copepods, isopods, tanaids, and ostracods. Members of this subclass are less than 0.3 mm in length and have a reduction in body form compared to other crustaceans, with an unsegmented sac-like thorax and a much reduced abdomen. As a result of their parasitic mode of life, adult females have lost all resemblance to crustaceans; males are free-living but non-feeding.

Tantulocaridans have a double life cycle, involving a sexual phase and an asexual phase. The asexual phase is encountered much more frequently than the sexual phase. The sac-like asexual female releases fully formed tantulus larvae, which are capable of infecting a new host and developing directly into another asexual female, without mating and without even moulting (Boxshall and Lincoln, 1987). The tantulus larva is minute, ranging from 85 μm to about 180 μm in length. It consists of a head, which has an oral disc but lacks limbs, a trunk of six leg-bearing segments and a maximum of two limbless abdominal segments. The swimming legs are biramous and have reduced endites. Infection of the host occurs at the tantulus larval stage, which is believed to occur immediately after the larva is released. Tantulocaridans exhibit varying degrees of host specificity. After successfully infecting a host, the tantulus larva develops into the asexual female, and the post-cephalic trunk of the larva is shed, so the female remains attached to the host by the adhesive oral disc of the preceding larval stage. The host surface is punctured by a stylet, which protrudes through a minute pore in the centre of this disc. Nutrients are obtained via the puncture into the host. The trunk of the female expands to accommodate the growing larvae until they are released (Huys et al., 1993).

In the sexual phase, the cycle again begins with the infective tantulus larva attaching to its host by the oral disc. A sac-like expansion begins to grow near the back of the trunk, within which either an adult male or an adult female then develops. Both are supplied with nutrients from the host, transported via an umbilical cord originating in the still-attached larval head. Fully formed adults develop within the sac, which remains attached to the host by the oral disc of the larva. On reaching maturity these sexual adults are released by rupturing of the sac wall (Boxshall and Lincoln, 1987). These sexual stages have never been observed alive, but it is assumed that the male, which has well-developed swimming legs and a large cluster of antennulary chemosensors, actively searches out and locates the receptive female. The male carries a large penis and presumably inseminates the female by the single ventral copulatory pore. The fertilized eggs are presumed to develop within the expandable cephalothorax of the female until ready to hatch, as a fully formed tantulii.

SUBCLASS BRANCHIURA

The subclass Branchiura of the class Maxillopoda are a group of parasitic crustaceans, commonly called carp lice or fish lice. Branchiurans attach to the skin of their fish host and feed on its blood and external tissues. The subclass contains four genera in a single family with about 175 species, almost all of which are ectoparasites on fish mainly living in freshwater habitats. The body is flattened and comprises a head of five limb-bearing segments and a short trunk divided into a thoracic region, carrying four pairs of strong swimming legs and a short, unsegmented abdomen. The head has well-developed carapace lobes (Overstreet et al., 1992).

Branchiurans have a tubular sucking mouth equipped with rasping mandibles located at the tip of the mouth tube (Gresty et al., 1993). The four pairs of thoracic swimming legs are biramous and directed laterally. The first and second legs commonly carry an additional process, the flagellum, originating near the base of the exopod. The third and fourth legs are usually modified in the male and are used for transferring sperm to the female during mating. The abdomen contains the paired testes in the male and, in the female, the paired seminal receptacles, where sperm are stored until needed to fertilize eggs. The sexes are separate and, in most branchiurans, males transfer sperm directly to the females using a variety of structures on the third and fourth thoracic legs. In *Dolops*, sperm are transferred in spermatophores.

The mature female *Argulus* will leave its host after taking a meal and lay eggs in rows cemented on a hard submerged surface. These eggs hatch into free-swimming larvae functioning as a dispersal phase and will moult into the second stage, in which strong claws have replaced the setae on the antenna and the setose palp of the mandible is lost. Branchiurans are parasitic from the second stage onwards but appear to leave the host and then find a new host at intervals throughout development. Changes during the larval phase are gradual, mainly involving the development of the thoracic legs and reproductive organs, except for the maxillule, which undergoes a metamorphosis around stage five, changing from a long limb bearing a powerful distal claw to a powerful circular sucker.

SUBCLASS PENTASTOMIDA

Pentastomida (Greek for "five openings") are a subclass of parasitic invertebrates (class Maxillopoda) commonly known as tongue worms due to the resemblance of some species to a vertebrate tongue. Pentastomida was once classified as a phylum, a fact reflected in most modern textbooks of parasitology. A relationship between tongue worms and branchiuran crustaceans (class Maxillopoda, subclass Branchiura) is supported by results of studies of sperm of other species and by comparison of ribosomal 18S recombinant RNA sequences in the two groups (Abele et al., 1989). There are about 110 extant species of pentastomids distributed between two orders: the primitive Cephalobaenida and the advanced Porocephalida. Pentastomids are obligate endoparasites of the respiratory tracts of tetrapods with correspondingly degenerate anatomy. They have five anterior appendages. One is the mouth; the others are two pairs of hooks, which they use to attach to the host (Riley, 1986).

Tongue worms inhabit the respiratory systems of terrestrial vertebrates and all aspects of their structure and function are accordingly adapted. With the exception of species belonging to the genus *Linguatula*, which browse on cells and mucus lining the nasal sinuses, all tongue worms feed on blood. They possess an elongated, worm-like, mostly cylindrical body, rounded both anteriorly and posteriorly. The body is differentiated into an anterior head region, bearing a small ventral mouth and a long posterior trunk carrying numerous raised annuli that are not true segments. Sexual dimorphism is pronounced; females are invariably larger than males but are as small as 1.5 × 0.3 mm (*Rileyiella*) to 12 × 1 cm (*Armillifer*). Males are much shorter and proportionately more slender. Human infection occurs frequently in Africa and the Orient, where humans are an accidental intermediate host of the nymphal form. The liver is a common site of infection, and large numbers of larvae may produce serious and even fatal effects.

The dorsal ovary leads into a paired oviduct, which passes around the gut. Close to the junction between the oviduct and the uterus are paired spermathecae, responsible for long-term storage of sperm. The lower reproductive tract of males consists of paired, elongate penises (thin, coiled tubes of chitin) called dilators. Dilators may be extruded through the anterior genital pore by muscles and thereby carry the tip of the penis either to the entrance of the spermathecal ducts in cephalobaenids or into the vagina/uterus in porocephalids. The testis is dorsal. Copulation is a lengthy and complex process, entailing docking of the paired penises in the narrow spermathecal duct and the transfer of millions of filiform sperm. The vagina is equipped with a sieve-like mechanism that retains small, undeveloped eggs but allows mature eggs to escape. Thus, continuous egg production is usual (Riley, 1994).

SUBCLASS MYSTACOCARIDA

Mystacocarida is a subclass of the class Maxillopoda and comprises primitive interstitial crustaceans that are part of the meiobenthos, are about 5 mm long and have a worm-like body (Pennak and Zinn, 1943). There are 13 described species divided between two genera, *Derocheilocaris* and *Ctenocheilocharis* (Lombardi and Ruppert, 1982). Mystacocarids graze on microalgae and bacteria living on the surfaces of sand grains. The sexes are separate. Fertilized eggs are shed freely into the interstitial habitat. Development occurs through a series of moult stages in which body somites and appendages are added in a gradual manner (Hessler, 1988; Zinn et al., 1982).

SUBCLASS COPEPODA

Copepods are a subclass of small crustaceans belonging to the class Maxillopoda, found in the sea and nearly every freshwater habitat, with many planktonic, benthic and continental species. Many species live underground in marine and freshwater caves, sinkholes, or stream beds. The Copepoda is the second largest class of the Crustacea (after the Malacostraca) (Hayward and Ryland, 1998) and the group contains 10 orders with some 14,000 described species. Copepod structure, although similar in basic pattern, varies greatly, with three patterns being displayed in free-living forms. Calanoid (*Calanus finmarchicus*), cyclopoid (*Cyclops*, also called water fleas), and harpacticoid (*Harpacticus*).

Unlike most crustaceans, copepods have no carapace. They have a teardrop-shaped body with an armoured exoskeleton composed externally of a series of small chitinous cylinders, or segments, some of which may be more or less fused, except in some parasitic genera. In most species the body is almost totally transparent. The head bears two pairs of large antennae, often used in locomotion or prehension, and a single eye, usually bright red, in the centre. The thorax, consisting of six segments, the first usually fused into the head, bears the swimming feet. The last thoracic segment bears the genital apertures and is therefore often called the genital segment. The abdomen consists of one to five segments and bears no appendages. The last abdominal segment bears two furcal rami, which are usually setose. The copepod body plan consists of two regions, the anterior prosome and the posterior urosome. Behind the cephalosome are the free prosomal segments with swimming leg pairs, either four or five of them. In the gymnoplean body plan, as found in Platycopioida and Calanoida, there are five leg pairs, whereas in the podoplean body plan, as found in all the other eight orders, there are only four pairs. The boundary between the prosome and the urosome is less distinct in the Harpacticoida, many of which have slender cylindrical bodies. The most characteristic feature of copepods is the form of the swimming legs: members of each leg pair are fused to a median intercoxal sclerite, which ensures that left and right legs always beat together. The legs are two-branched (biramous) and each branch (endopod and exopod) consists of a maximum of three segments (Boxshall, 1992).

Copepods exhibit diverse feeding behaviours. In the plankton, many feed on small particles such as unicellular phytoplankton, protists, and other microorganisms. They capture these particles by generating water currents, using the slow swimming movements of antennae and maxillipeds. Other oceanic copepods are scavengers and specialist detritus feeders, which have acute chemosensory systems associated with their feeding systems. Benthic copepods feed on organic matter of all kinds, both living and dead. About half of all copepod species are symbionts living in association with multicellular animal hosts, including sponges, polychaetes, echinoderms, molluscs, crustaceans, and chordates, particularly tunicates and fishes. Their lifestyles are often poorly known, but many are parasitic, living inside or on the outer surface of their hosts (Huys and Boxshall, 1991; Razouls et al., 2005).

The sexes are separate and sexual dimorphism is frequently evidenced. Very often the female has one abdominal segment less than the male of the same species, due to a fusing of the first and second segments (Hayward and Ryland, 1998). Mating behaviour typically involves chemical signalling between the sexes. Females release pheromones, which provide information concerning species identity as well as the state of sexual receptivity. Males are attracted by these pheromonal trails and copulation results in the transfer of a single chitinous packet or a pair of packets of sperm (spermatophores) on to the female genital region, and these usually discharge their contents into storage organs (seminal receptacles) inside the female. The female typically carries the eggs in paired egg sacs, although many marine calanoids broadcast their eggs instead. The basic life cycle comprises six naupliar and five copepodid stages preceding the adult. Many parasites, however, have large yolk-filled eggs and an abbreviated developmental pattern. The commonest abbreviation is the reduction or even the complete loss of the naupliar phase. Sea lice, for example, have only two non-feeding naupliar stages, instead of the usual six. The nauplius phase is typically a dispersal phase.

Most species are very small, often less than 0.5 mm, but most sizes range from 0.5 mm to about 2 mm in length. The largest species, a parasite of the fin whale, grows to a length of about 32 cm. Most of the smaller copepods feed directly on phytoplankton, catching single cells, but a few of the larger species are predators on other copepods, fish larvae, and protozoa. Herbivorous copepods, particularly those in rich cold seas, store up energy from their food as oil droplets, taking up half of the body volume, during the spring and summer plankton blooms.

According to some scientists, copepods form the largest animal biomass on earth, competing for this title with the Antarctic krill (*Euphausia superba*). Because of their smaller size and relatively faster growth rates, however, and because of their more even distribution throughout the world's oceans, copepods almost certainly contribute far more to the secondary productivity of the world's oceans than krill do. Commonly copepods constitute about 70% of the zooplankton, and at times their numbers are so great as to impart a pinkish colour to the sea. Planktonic copepods are important food organisms for small fish, whales, seabirds and other crustaceans such as krill (order Euphausiacea) in the ocean and in freshwater. Copepods play a vital role in the economy of the oceans, forming the middle link in food chains leading from phytoplankton up to commercially important fish species. Consideration of global carbon flux suggests that copepods play an equally important role in the global carbon cycle, since a major proportion of fixed carbon dioxide passes through the oceanic food web. Many planktonic copepods feed near the surface at night, sinking into deeper water during the day. Their moulted exoskeletons, faecal pellets and respiration at depth all bring enough carbon to the deep sea to have an impact on the carbon cycle, in view of their abundance. Some freshwater copepods act as vectors of human parasites such as guinea worm, while others are important predators of mosquito larvae and are actively used in biological control of mosquitoes in malarial areas. Sea lice and other fish parasitic copepods are major pests in aquaculture of both fishes and molluscs (Boxshall and Defaye, 1993).

Three orders, Calanoida, Cyclopoida, and Harpacticoida, are wholly or primarily free-living and because of overwhelming numbers are of greatest general interest. The remaining orders (Poecilostomatoida, Siphonostomatoida, Monstrilloida, Misophrioida, Gelyelloida, Platycopioida and Mormonilloida) in the adult state are parasitic for at least part of their lives.

ORDER CALANOIDA

The order Calanoida (subclass Copepoda) includes the larger and more abundant of the pelagic species with nearly all planktonic representatives. Calanoid copepods include 43 families with about 2,000 species of both marine and freshwater representatives (Boltovskoy et al., 1999). The anterior part of the body is cylindrical with five or six segments, and much broader than the posterior part. The first antennae are not for locomotion but are stabilizers and sinking retarders, and they also have an olfactory function. The second antennae and mandibular palps are biramous and create water currents for feeding and slow movement. The five pairs of swimming legs are biramous, but the last pair is sometimes reduced or absent in the female, and the male's right fifth leg may be modified for grasping the female (Mauchline, 1998).

ORDER CYCLOPOIDA

The order Cyclopoida (subclass Copepoda) comprise abundant freshwater and salt-water species that form an important link in the food chains of aquatic life, consuming tiny plants, animals, and detritus and, in turn, comprising the main source of food for small fish, some large fish, and other organisms. Some species are important intermediate hosts for human parasites. The front part of the body is oval and sharply separated from the tubular hind end, which bears two caudal rami with distinctly unequal setae. Usually 10 body segments are present in the male, and nine in the female because of fusion of two to form a genital segment. Cyclopoids are intermediate hosts of the parasitic guinea worm (*Drucunculus medinensis*), and sometimes the fish tapeworm (*Dibothriocephalus latus*). Most freshwater species live in shallow water, swimming from plant to plant, but salt-water species are generally water-treaders. Food is not secured by filtration but is seized and eaten with the biting mouthparts. Many species have a worldwide distribution.

ORDER HARPACTICOIDA

The order Harpacticoida (subclass Copepoda) comprises 463 genera and about 3,000 species of mainly benthic copepods, with worldwide distribution mostly in the marine environment and in freshwater. In the sea they range from the shore to abyssal depths. In general, they are free-living and benthonic; some species are pelagic, parasitic, or commensal. Harpacticoids represent the second-largest meiofaunal group in marine sediment milieu, after the nematodes (Greek for "thread-like"). Their size varies in length from 0.4 to 3 mm. As a rule, the first thoracic segment is incorporated with the cephalothorax, and the last thoracic segment is included in the abdomen.

References

Abele, L., W. Kim and B.E. Felgenhauer. 1989. Molecular evidence for inclusion of the phylum Pentastomida in the Crustacea. Mol. Biol. Evol., 6: 685-691.

Adiyodi, K.G., and R.G. Adiyodi. 1970. Endocrine control of reproduction in Decapod Crustacea. Biol. Rev. Cambridge Phil. Soc., 45: 121-165.

Adiyodi, R.G., and T. Subramoniam. 1983. Arthropoda-Crustacea. *In:* Reproductive Biology of Invertebrates: Oogenesis, Oviposition, and Oosorption, Vol. I. K.G. Adiyodi and R.G. Adiyodi (eds.). Wiley, Chichester, England, pp. 443-495.

Ahyong, S.T., and D.E. Brown. 2002. New species and new records of Polychelidae from Australia (Crustacea: Decapoda). The Raffles Bull. Zool., 50(1): 53-79.

Ahyong, S.T., and C. Harling. 2000. The phylogeny of the Stomatopod Crustacea. Austral. J. Zool., 48(6): 607-642.

Aiken, D.E., and S.L. Waddy. 1976. Controlling growth and reproduction in the American lobster. Proceeding of Annual Meeting of the World Mariculture Society, 7: 415-430.

Altevogt, R., and T.A. Davis. 1975. *Birgus latro*: India's monstrous crab. A study and an appeal. Bulletin of the Department of Marine Sciences, University of Cochin.

Anderson, D.T. 2002. Invertebrate Zoology, 2nd ed. Oxford University Press. USA.

Athersuch, J., D.J. Horne and J.E. Whittaker. 1989. Marine and brackish water ostracods. Synopses of the British Fauna (New Series) No. 43, 343 pp, 8 pls, E.J. Brill. ISBN 90-04-09079-7.

Bacescu, M. 1988. Cumacea I. Crustaceorum Catalogus, 7: 1-173.

Bacescu, M. 1992. Cumacea II. Crustaceorum Catalogus, 8: 175-468.

Begon, M., R.T. Colin and J.L. Harper. 2005. Ecology: From Individuals to Ecosystems, 4th ed. Blackwell Publishing Professional. Oxford, UK.

Bamstedt, U., and K. Karlson. 1998. Euphausiid predation on Copepods in coastal waters of the Northeast Atlantic. Mar. Ecol. Progr. Ser. 172: 149-168.

Barnett, L.K., C. Emms and D. Clarke. 1999. The coconut or robber crab (*Birgus latro*) in the Chagos Archipelago and its captive culture at London Zoo. *In:* Ecology of the Chagos Archipelago. C.R.C. Sheppard and M.R.D. Seaward (eds.). Linnean Society Occasional Publications, 2. Westbury Publishing, pp. 273-284.

Belk, D., and J. Brtek. 1995. Checklist of the Anostraca. Hydrobiologia, 298: 315-353.

Bennett, D.B. 1974. Growth of the edible crab (*Cancer pagurus* L.) off south-west England. J. Mar. Biol. Assoc. U.K., 54: 803-823.

Bennett, D.B. 1995. Factors in the life history of the edible crab (*Cancer pagurus* L.) that influence modelling and management. ICES Mar. Sci. Symp., 199: 89-98.

Bliss, D.E. 1966. Relation between reproduction and growth in decapod crustaceans. Am. Zool., 6: 231-233.

Bliss, D.E. 1982. Shrimps, Lobsters, and Crabs: Their Fascinating Life Story. New Century Publishers, Inc., New Jersey.

Boltovskoy, D., M.J. Gibbons, L. Hutchings and D. Binet. 1999. General biological features of the South Atlantic. *In:* Zooplankton of the South Atlantic. D. Boltovskoy (ed.). Backhuys Publishers, Leiden. 2 vols, i-xvi+1706 pp.

Boxshall, G.A. 1992. Copepoda. Microscopic Anatomy of Invertebrates, Vol. IX. Crustacea. F.W. Harrison (ed.). Chichester, U.K., John Wiley and Sons.

Boxshall, G.A. and D. Defaye. 1993. Pathogens of Wild and Farmed Fish: Sea Lice. Ellis Horwood, Chichester, England.

Boxshall, G.A., and R.J. Lincoln. 1987. The life cycle of the Tantulocarida (Crustacea). Phil. Trans. Roy. Soc. London, B315: 267-303.

Brierley, A.S., P.G. Fernandes, M.A. Brandon, F. Armstrong, N.W. Millard, S.D. McPhail, P. Stevenson, M. Pebody, J. Perrett, M. Squires, D.G. Bone and G. Griffiths. 2002. Antarctic krill under sea ice: elevated abundance in a narrow band just south of ice edge. Science, 295 (5561): 1890–1892.

Brinton, E. 1962. The distribution of Pacific euphausiids., Bull. Scripps Inst. Oceanogr., 8(2): 51-270.

Brinton, E., M.D. Ohman, A.W. Townsend, M.D. Knight and A.L. Bridgeman. 2000. Euphausiids of the World Ocean, World Biodiversity Database CD-ROM Series. Springer Verlag. ISBN 3-540-14673-3.

Buchholz, F. 2003. Experiments on the physiology of Southern and Northern krill, *Euphausia superba* and *Meganyctiphanes norvegica*, with emphasis on moult and growth – a review. Mar. Freshwater Behav. Physiol., 36(4): 229-247.

Burkenroad, M.D. 1963. The evolution of the Eucarida (Crustacea, Eumalacostraca), in relation to the fossil record. Tulane Stud. Geol., (1): 1-17.

Caldwell, R.L. 1991. Variation in reproductive behavior in stomatopod crustacea. *In:* Crustacean Sexual Biology. R. Bauer and J. Martin (eds.). Columbia University Press, New York, pp. 67-90.

Caldwell, R.L. 1992. Recognition, signalling and reduced aggression between former mates in a stomatopod. Anim. Behav., 44: 11-19.

Carlisle, D.B. 1953. Studies on *Lysamata seticaudata* Risso (Crustacea: Decapoda). V. The ovarian inhibiting hormone and the hormonal inhibition of sex-reversal. Pub. Stn. Zool. Napoli, 24: 355-372.

Carlisle, D.B. 1957. On the hormonal inhibition of moulting in decapod Crustacea. 2. The terminal anecdysis in crabs. J. Mar. Biol. Assoc. U.K., 36: 291-307.

Charniaux-Cotton, H., and L.H. Kleinholz. 1964. Hormones in invertebrates other than insects. *In:* The Hormones. Vol. 4. G. Pincus, K.V. Thimann and E.B. Astwood (eds.). Academic Press, New York, pp. 135-198.

Chesson, P.L., and R.R. Warner. 1981. Environmental variability promotes coexistence in lottery competitive systems. Am. Nat., 117: 923-943.

Cheung, T.S. 1969. The environment and hormonal control of growth and reproduction in the adult female stone crab, *Menippe mercenaria* (Say). Biol. Bull. (Woods Hole, Mass.), 136: 327-346.

Chiba, A., and Y. Honma. 1972. Studies on gonad maturity in some marine invertebrates. Seasonal changes in the ovary of the lined shore crab. Bull. Jap. Soc. Sci. Fish., 38: 323-327.

Cohen, A.N., J.T. Carlton and M.C. Fountain. 1995. Introduction, dispersal and potential impacts of the green crab *Carcinus maenas* in San Francisco Bay, California. Mar. Biol., 122: 225-237.

Crothers, J.H. 1967. The biology of the shore crab, *Carcinus maenas* (L.). 1. The background – anatomy, growth and life history. Field Stud., 2: 407-434.

Davey, K. Species bank: *Lomis hirta*. Department of the Environment and Heritage. Retrieved on 2006-08-15.

Day, J. 1980. Southern African Cumacea. Pt. 4. Families Gynodiastylidae and Diastylidae. Ann. South African Museum, 82: 187-292.

De Deckker, P. 1980. New records of *Koonunga cursor* Sayce, 1908 (Syncarida, Anaspidacea). Trans. Roy. Soc. South Austral., 104(2): 21-25.

Demeusy, N. 1965. New results concerning the relations between somatic growth and reproductive function in the Brachyuran decapod *Carcinus maenas* L. Case of females followed during winter. C. R. Hebd. Seances Acad. Sci., 260: 323-326.

Dixon, C.J., F.R. Schram and S.T. Ahyong. 2004. A new hypothesis of decapod phylogeny. Crustaceana, 76(8): 935-975.

Drach, P. 1955. The eyestalk endocrine system, intermoult duration and vitellogenesis in *Leander serratus* (Pennant) Crustacea, Decapoda. C. R. Seances Soc. Biol. Ses Fil., 149: 2079-2083.

Dumont, H.J., and S.V. Negrea. 2002. Introduction to the Class Branchiopoda. Backhuys Publishers, Leiden, The Netherlands.

Dunlap J.C., J.W. Hastings and O. Shimomura. 1980. Crossreactivity between the light-emitting systems of distantly related organisms: novel type of light-emitting compound. Proc. Natl. Acad. Sci. USA, 77(3): 1394-1397.

Dworschak, P.C. 2000. Global diversity in the Thalassinidea (Decapoda). J. Crustacean Biol., 20: 238-243.

Edwards, E. 1989. Crab fisheries and their management in the British Isles. *In:* Marine Invertebrate Fisheries: Their Assessment and Management. J.F. Caddy (ed.). New York.

Escobar-Briones, E., and F. Alvarez. 2002. Modern Approaches to the Study of Crustacea. Springer, New York.

Erikson, C., and D. Belk. 1999. Fairy Shrimp of California's Puddles, Pools and Playas. Mad River Press, Eureka, Calif.

Fanenbruck, M., and S. Harzsch. 2005. A brain atlas of *Godzilliognomus frondosus* Yager, 1989 (Remipedia, Godzilliidae) and comparison with the brain of *Speleonectes tulumensis* Yager, 1987 (Remipedia, Speleonectidae): implications for arthropod relationships. Arthropod Struct. Dev., 34(3): 343-378.

Farrelly, C.A., and P. Greenaway. 2005. The morphology and vasculature of the respiratory organs of terrestrial hermit crabs (*Coenobita* and *Birgus*): gills, branchiostegal lungs and abdominal lungs. Arthropod Struct. Dev., 34(1): 63-87.

Fish, J.D., and S. Fish. 1996. A Student's Guide to the Seashore, 2nd ed. Cambridge University Press, Cambridge, 564 pp.

Forest, J., and J.C. Von Vaupel Klein. 2006. The Crustacea (Treatise on Zoology-Anatomy, Taxonomy, Biology). Brill Academic Publishers, Netherlands.

Fortey, R.A., and R.H. Thomas. 1997. Arthropod Relationships (The Systematics Association Special Volume Series). Springer, London.

Galil, B. 2000. Crustacea Decapoda: Review of the genera and species of the family Polychelidae Wood-Mason, 1874. *In:* Résultats des Campagnes MUSORSTOM, Vol. 21. A. Crosnier (ed.), Mémoires du Muséum national d'Histoire naturelle. 184: 285-387.

Gherardi, F., E. Tricarico and J. Atema. 2005. Unraveling the nature of individual recognition by odor in hermit crabs. J. Chem. Ecol., (12): 2877-2896.

Gomez, R., and K.K. Nayar. 1965. Certain endocrine influences in the reproduction of the crab *Paratelphusa hydrodromous*. Zool. Jahrb., Abt. Allg. Zool. Physiol. Tiere, 71: 694-701.

Gómez-Gutiérrez, J. 2002. Hatching mechanism and delayed hatching of the eggs of three broadcast spawning euphausiid species under laboratory conditions. J. Plankton Res., 24(12): 1265-1276.

Gómez-Gutiérrez, J. 2003. Hatching mechanism and accelerated hatching of the eggs of a sac-spawning euphausiid *Nematoscelis difficilis*. J. Plankton Res., 25(11): 1397-1411.

Gómez-Gutiérrez, J., W.T. Peterson, A. De Robertis and R.D. Brodeur. 2003. Mass mortality of krill caused by parasitoid ciliates. Science, 301(5631): 339.

Gonzalez-Gurriaran, E. 1981. Preliminary data on the population dynamics of the velvet swimming crab (*Macropipus puber* L.) in the Ria de Arousa (Galicia, NW Spain). ICES CM 1981/K.

Gonzalez-Gurriaran, E. 1985. Reproducion de la *Necora Macropipus puber* (L.) (Decapoda, Brachyura), y ciclo reproductivo el la Ria de Arousa (Galicia, N. W. Espana). Bol. Inst. Espanol Oceanogr., 2(1): 10-32.

Gresty, K.A., G.A. Boxshall and K. Nagasawa. 1993. The fine structure and function of the cephalic appendages of the branchiuran parasite, *Argulus japonicus* Thiele. Phil. Trans. Roy. Soc. London, B339: 119-135.

Grubb, P. 1971. Ecology of terrestrial decapod crustaceans on Aldabra. Phil. Trans. Roy. Soc.: Biol. Sci., 260: 411-416.

Gurney, R. 1942. Larvae of decapod crustacea. Royal Society Publ. 129, London.

Hartnoll, R.G. 1963. The biology of Manx spider crabs. Proc. Zool. Soc. London, 141: 423-496.

Hartnoll, R.G. 1969. Mating in the brachyura. Crustaceana, 16: 161-181.

Hartnoll, R.G. 1973. Factors affecting the distribution and behaviour of the crab *Dotilla fenestrata* on east African shores. Est. Coas. Mar. Sci., 1: 137-152.

Hartnoll, R.G. 1978. The determination of relative growth in Crustacea. Crustaceana, 34(3): 281-293.

Hayward, P.J., and J.S. Ryland. 1998. Handbook of the Marine Fauna of North-West Europe. Oxford University Press, New York.

Heasman, M.P., D.R. Fielder and R.K. Shepherd. 1995. Mating and spawning in the mudcrab, *Scylla serrata* (Forskal) (Decapoda: Portunidae), in Moreton Bay, Queensland. Austral. J. Mar. Freshwater Res., 36: 773-783.

Held, E. 1963. Moulting behaviour of *Birgus latro*. Nature, 200: 799-800.

Herring, P.J., and E.A. Widder. 2001. Bioluminescence in plankton and nekton; *In:* Encyclopedia of Ocean Science, Vol. 1. J.H. Steele, S.A. Thorpe and K.K. Turekian (eds.). Academic Press, San Diego, pp. 308-317.

Hessler, R.R. 1988. Mystacocarida. *In:* Introduction to the Study of Meiofauna. R.P. Higgins and H. Thiel (eds.). Smithsonian Institution Press,. ISBN 0-87474-488-1.

Hessler, R.R., and H.L. Sanders. 1973. Two new species of Sandersiella, including one from the deep sea. Crustaceana, 13: 181-196.

Hill, R.W., G.A. Wyse and M. Anderson. 2004. Animal Physiology. Sinauer Associates Pub., New York.

Holdich, D.M., and D.A. Jones. 1983. Tanaids: Keys and Notes for the Identification of the Species of England. Synopses of the British Fauna, no. 27. Cambridge University Press, Cambridge.

Huys, R., and G.A. Boxshall. 1991. Copepod Evolution. The Ray Society, London.

Huys, R., G.A. Boxshall and R.J. Lincoln. 1993. The tantulocaridan life cycle: the circle closed? J. Crustacean Biol., 13: 432-442.

Hyoung-Chul, S., and S. Nicol. 2002. Using the relationship between eye diameter and body length to detect the effects of long-term starvation on Antarctic krill *Euphausia superba*. Mar. Ecol. Progr. Ser. (MEPS), 239:157-167.

Ingle, R.W. 1980. British Crabs. British Museum and Oxford University Press. Oxford.

Ingle, R.W. 1983. Shallow-water crabs: keys and notes of the identification of the species. *In:* Synopses of the British Fauna No. 25. D.M. Kermack and R.S.K. Barnes (eds.). Linnean Society of London and The Estuarine and Brackish-Water Science Association, Cambridge University Press, U.K.

Ingle, R.W. 1992. Larval stages of northeastern Atlantic crabs: An illustrated key. Chapman and Hall Identification Guide No. 1. Natural History Museum Publications and Chapman and Hall, London.

Jachowski, R.L. 1974. Agonistic behaviour of the blue crab, *Callinectes sapidus* Rathbun. Behaviour, 50: 232-253.

Jeffries, H.P. 1966. Partitioning of the estuarine environment by two species of *Cancer*. Ecology, 47: 477-481.

Johnsen, S. 2005. The red and the black: bioluminescence and the color of animals in the deep sea. Integr. Comp. Biol., 45: 234-246.

Johnson, M.E., and J. Atema. 2005. The olfactory pathway for individual recognition in the American lobster *Homarus americanus*. J. Exp. Biol., 208: 2865-2872.

Jones, N.S. 1963. The marine fauna of New Zealand: Crustaceans of the Order Cumacea. New Zealand Oceanographic Institute, Memoir 23: 1-80.

Jones, N.S. 1976. British Cumaceans. Synopses of the British Fauna No. 7, Academic Press. New York.

Kallmeyer, D.E., and J.H. Carpenter. 1996. *Stygiomysis cokei*, New Species, a troglobitic mysid from Quintana Roo, Mexico (Mysidacea: Stygiomysidae). J. Crustacean Biol., 16(2): 418-427.

Kelber, K.P. 1999. *Triops cancriformis* (Crustacea, Notostraca): Ein bemerkenswertes Fossil aus der Trias Mitteleuropas. *In:* Trias – Eine ganz andere Welt, III.16: 383-394. N. Hauschke and V. Wilde (eds.). Verl. Dr. F. Pfeil, Munich.

Kils, U., and P. Marshall. 1995. Der Krill, wie er schwimmt und frisst – neue Einsichten mit neuen Methoden (The antarctic krill, feeding and swimming performances – new insights with new methods). *In:* Biologie der

Polarmeere - Erlebnisse und Ergebnisse (Biology of the Polar Oceans). I. Hempel and G. Hempel (eds.). Fischer, ISBN 3-334-60950-2, pp. 201-210.

Klaoudatos, D. 2003. Reproductive ecology, population genetics and population dynamics of selected Decapod Crustaceans. PhD Thesis, University of Wales Swansea.

Klaoudatos, S. 1984. Contribution to the biology and under controlled conditions reproduction and breeding of the prawn *Penaeus monodon* (Forskal 1775). PhD Thesis, University of Patras, Greece.

Kleinholz, L.H. 1976. Crustacean neurosecretory hormones and physiological specificity. Am. Zool., 16: 151-166.

Knight, M.D. 1984. Variation in larval morphogenesis within the Southern California Bight population of Euphausia pacifica from winter through summer, 1977-1978. CalCOFI Report Vol. XXV.

Koenemann, S., and R.A. Jenner. 2005. Crustacea and Arthropod Relationships. CRC Press. New York.

Lachaise, F., and J.A. Hoffmann. 1977. Ecdysone and ovarian development in a decapod, *Carcinus maenas*. C. R. Hebd. Seances Acad. Sci., Ser. D, 285: 701-704.

Legeay, A., and J.C. Massabuau. 2000. The ability to feed in hypoxia follows a seasonally dependent pattern in shore crab *Carcinus maenas*. J. Mar. Biol. Ecol., 247: 113-129.

Lindsay, S.M., and M.I. Latz. 1999. Experimental Evidence for Luminescent Countershading by some Euphausiid Crustaceans. Poster presented at the American Society of Limnology and Oceanography (ASLO) Aquatic Sciences Meeting, Santa Fe.

Lombardi, J., and E.E. Ruppert. 1982. Functional morphology of locomotion in *Derocheilocaris typica* (Crustacea: Mystacocarida). Zoomorphology, 100: 1-10.

Lowry, J.K. 2006. Crustacea, the Higher Taxa: Description, Identification, and Information Retrieval. Version: 2 October 1999. http://crustacea.net/.

McGraw-Hill. 2006. Mictacea. Answers.com. McGraw-Hill Encyclopedia of Science and Technology. The McGraw-Hill Companies, Inc., 2005. http://www.answers.com/topic/mictacea, accessed September 21, 2006.

McLaughlin, P.A. 2003. Illustrated keys to families and genera of the superfamily Paguroidea (Crustacea: Decapoda: Anomura), with diagnoses of genera of *Paguridae*. Mem. Mus. Vic., 60(1): 111-144.

McLaughlin, P.A., and R. Lemaître. 1997. Carcinization in the Anomura - fact or fiction. 1. Evidence from adult morphology. Contr. Zool., 67: 79-123.

McLaughlin, P.A., R. Lemaître and C.C. Tudge. 2004. Carcinization in the Anomura - fact or fiction. 2. Evidence from larval, megalopal and early juvenile morphology. Contr. Zool., 73(3).

Macpherson, E., W. Jones and M. Segonzac. 2006. A new squat lobster family of Galatheoidea (Crustacea, Decapoda, Anomura) from the hydrothermal vents of the Pacific-Antarctic Ridge. Zoosystema, 27(4): 709-723.

Mantelatto, F.L.M., and A. Fransozo. 1999. Reproductive biology and moulting cycle of the crab *Callinectes ornatus* (Decapoda, Portunidae) from the Ubatuba region, Sao Paulo, Brazil. Crustaceana, 72: 63-76.

Marinovic, B., and M. Mangel. 1999. Krill can shrink as an ecological adaptation to temporarily unfavourable environments. Ecol. Lett., 2: 338-343.

Martin, J.W., and G.E. Davis. 2001. An Updated Classification of the Recent Crustacea, Science Series, Natural History Museum of Los Angeles County Science Series (39): 1-124.

Mauchline, J. 1984. Euphausiid, stomatopod, and leptostracan crustaceans. Keys and notes for the identification of species. Synopses of the British Fauna, no. 30. The Linnean Society of London and the Estuarine and Brackish-Water Sciences Association.

Mauchline, J. 1998. The Biology of Calanoid Copepods. Advances in Marine Biology. Academic Press, New York and London.

Mauchline, J., and L.R. Fisher. 1969. The biology of euphausiids. Adv. Mar. Biol., 7.

Montgomery, E.L., and R.L. Caldwell. 1984. Aggressive brood defense by females in the stomatopod *Gonodactylus bredini.* Behav. Ecol. Sociobiol., 14: 247-251.

Morgan, M.D. 1982. Ecology of Mysidacea (Developments in Hydrobiology, Vol. 10). W. Junk Publishers, The Hague, The Netherlands.

Mori, M., and P. Zunino. 1987. Aspects of the biology of *Liocarcinus depurator* (L.) in the Ligurian Sea. Invert. Pesq., 51: 135-145.

Myers, P. 2001. "Cephalocarida" (On-line), Animal Diversity Web. Accessed September 22, 2006 at http://animaldiversity.ummz.umich.edu/site/accounts/information/Cephalocarida.html.

Naylor, E. 1962. Seasonal changes in a population of *Carcinus maenas* L. in the littoral zone. J. Anim. Ecol., 31: 601-609.

Nicol, S. 1997. Krill Fisheries of the World, Technical Paper 167. Rome: Food and Agriculture Organization of the United Nations.

Nicol, S., and Y. Endo. 1997. Krill Fisheries of the World, FAO Fisheries Technical Paper 367.

Norman, C.P., and M.B. Jones. 1993. Reproductive ecology of the velvet swimming crab, *Necora puber* (Brachyura, Portunidae), at Plymouth. J. Mar. Biol. Assoc. U.K., 73: 379-389.

Olmstead, J.M.P., and J.P. Baumberger. 1923. Form and growth of grapsoid crabs. A comparison of the form of three species of grapsoid crabs and their growth at moulting. J. Morphol., 38: 279-294.

Otsu, T. 1963. Bihormonal control of sexual cycle in the freshwater crab, *Potamon dehaani*. Embryologica, 8: 1-20.

Overstreet, R.M., I. Dyková and W.E. Hawkins. 1992. Branchiura. *In:* Microscopic Anatomy of Invertebrates. Vol. IX, Crustacea. F.W. Harrison (ed.). J. Wiley and Sons, New York.

Panouse, J.B. 1943. Influence de l'ablation de pedoncule oculaire sur la croissance de l'ovarie chez la crevette Leander serratus. C. V. Hebd. Seances Acad. Sci., 217: 553-555.

Panouse, J.B. 1947. The sinus gland and maturation of genital products in shrimp. Bull. Biol. Fr. Bel., Suppl. 33: 160-163.

Passano, L.M. 1960. Moulting and its control. *In:* The Physiology of Crustacea: Metabolism and Growth. Vol. I. T.H. Waterman (ed.). Academic Press, New York, pp. 473-536.

Payen, G. 1974. Morphogenese sexuelle de quelques Brachyoures (Cyclometopes) au cours du developpement embryonnaire, larvaire et postlarvaire. Bull. Mus. Hist. Nat., 209, Zool., 139: 201-262.

Pearson, J. 1908. *Cancer*. *In:* Memoirs on Typical British Marine Plants and Animals, Vol. XVI. W.A. Herdman (ed.). Liverpool Marine Biology Committee. Williams & Norgate, London, pp. 1-209.

Pennak, R.W., and D.J. Zinn. 1943. Mystacocarida, a new order of Crustacea from intertidal beaches in Massachusetts and Connecticut. Smithsonian Misc. Coll., (103): 1-11.

Perez Farfante, I., and B. Kensley. 1997. Penaeoid and Sergestoid Shrimps and Prawns of the World: Keys and Diagnoses for the Families and Genera. Memoires du Museum National d'Histoire Naturelle, Vol. 175, 233 pp.

Persoone, G., P. Sorgeloos, O. Roels and E. Jaspers. 1980. The Brine Shrimp Artemia. Vol. 3, Ecology, Culturing and Use in Aquaculture. Universa Press, Wettern, Belgium.

Pillay, K.K., and N.B. Nair. 1971. The annual reproductive cycles of *Uca annulipes, Portunus pelagicus* and *Metapenaeus affinis* (Decapoda: Crustacea) from the south-west coast of India. Mar. Biol. (Berlin), 11: 152-166.

Pillay, K.K., and Y. Ono. 1978. The breeding cycles of two species of grasped crabs (Crustacea: Decapoda) from the north coast of Kyushu, Japan. Mar. Biol. (Berlin), 45: 237-248.

Pirow, R., F. Wollinger and R.J. Paul. 1999. The sites of respiratory gas exchange in the planktonic crustacean *Daphnia magna*: An in vivo study employing blood haemoglobin as an internal oxygen probe. J. Exp. Zool., 202(22): 3089-3099.

Poore, G.B., and W.F. Humphreys. 1998. First record of Spelaeogriphacea (Crustacea) from Australia – a new genus and species from an aquifer in the arid Pilbara of Western Australia. Crustaceana, 71: 721-742.

Porter, M.L., M. Pérez-Losada and K.A. Crandall. 2005. Model-based multi-locus estimation of decapod phylogeny and divergence times. Mol. Phylogenet. Evol., 37: 355-369.

Rangneker, P.V., and R.D. Deshmukh. 1968. Effect of eyestalk removal on the ovarian growth of the marine crab, *Scylla serrata* (Forskall). J. Anim. Morphol. Physiol., 15: 116-126.

Razouls C., F. de Bovée and N. Desreumaux. 2005. Diversity and Geographical Distribution of Pelagic Copepoda. See http://copepodes.obs-banyuls.fr

Riley, J. 1986. The biology of pentastomids. Adv. Parasitol., (25): 46-128.

Riley, J. 1994. Pentastomids. *In:* Reproductive Biology of Invertebrates, Vol. VI. K.G. Adiyodi and R.G. Adiyodi. Oxford & IBH Publishing, New Delhi.

Rolfe, W.D.I. 1969. Phyllocarida. *In:* Treatise on Invertebrate Paleontology. Part R. Arthropoda 4 Crustacea (except Ostracoda) Myriapoda - Hexapoda. Vol. 1. R.C. Moore (ed.). The University of Kansas and The Geological Society of America, Inc., Laurence, Kansas, pp. R296-R331.

Ross, R.M., and L.B. Quetin. 1986. How productive are Antarctic krill? Bioscience, 36: 264-269.

Ruppert, E.E., and R.D. Barnes. 1994. Invertebrate Zoology, 6th ed. Saunders College Publishing, Fort Worth, Texas.

Ruppert, E.E., S. Richard, R.S. Fox and R.D. Barnes. 2003. Invertebrate Zoology: A Functional Evolutionary Approach, 7th ed. Brooks Cole Pub., New York.

Ryan, E.P. 1967. Structure and function of the reproductive system of the crab *Portunus sanguinolentus* (Herbst) (Brachyure: Portunidae). II. The female system. Proc. Symp. Crust. Mar. Biol. Assoc. India, Ernakulam, 1965 Ser. II, Part 2, pp. 522-524.

Sanders, H.L. 1955. The Cephalocarida, a new subclass of Crustacea from Long Island Sound. Proc. Nat. Acad. Sci., 41: 61-66.

Sastry, A.N. 1983. Ecological aspects of reproduction. *In:* The Biology of Crustacea: Environmental Adaptations, Vol. III. F.J. Vernberg and W.B. Vernberg (eds.). Academic Press, New York, pp. 179-270.

Saxena, A. 2005. Text Book of Crustacea. Discovery Publishing House, New Delhi.

Scholtz, G., and S. Richter. 1995. Phylogenetic systematics of the reptantian Decapoda (Crustacea, Malacostraca). Zool. J. Linn. Soc., 113: 289-328.

Schram, F. 1986. Crustacea. Oxford University Press, Oxford.

Shields, J.D., and J. Gómez-Gutiérrez. 1996. *Oculophryxus bicaulis,* a new genus and species of dajid isopod parasitic on the euphausiid *Stylocheiron affine Hansen*. Intl. J. Parasitol., 26(3): 261-268.

Shimomura, O. 1995. The roles of the two highly unstable components F and P involved in the bioluminescence of euphausiid shrimps. J. Biolumin. Chemilumin., 10(2): 91-101.

Shuster, S.M., and R.L. Caldwell. 1989. Male defense of the breeding cavity and factors affecting the persistence of breeding pairs in the stomatopod, *Gonodactylus bredini* (Crustacea: Hoplocarida). Ethology, 82: 192-207.

Sochasky, J.B. 1973. Failure to accelerate moulting following eyestalk ablation in decapod crustaceans: A review of the literature. Fish. Res. Board Can., 431: 1-127.

Stensmyr, M.C., S. Erland, E. Hallberg, R. Wallén, P. Greenaway and B.S. Hansson. 2005. Insect-like olfactory adaptations in the terrestrial giant robber crab. Curr. Biol., 15: 116-121.

Stevcic, A. 1971. The main features of brachyuran evolution. Syst. Zool., 20: 331-340.

Subramoniam, T. 1993. Spermatophores and sperm transfer in marine crustaceans. *In:* Advances in Marine Biology, Vol. 29. J.H.S. Blaxter and A.J. Southward (eds.). Academic Press, New York, pp. 129-214.

Subramoniam, T. 2000. Crustacean ecdysteroids in reproduction and embryogenesis. Comp. Biochem. Physiol., 125: 135-156.

Tasch, P. 1969. Branchiopoda. *In:* Treatise on Invertebrate Paleontology. Part R. Arthropoda 4. 1: 128-191. R.C. Moore (ed.) Geological Society of America, Inc. and University of Kansas.

Taylor, H.H., and E.W. Taylor. 1992. Gills and lungs: the exchange of gases and ions. Microsc. Anat. Invert., 10: 203-293.

Teytaud, A.R. 1971. The laboratory studies of sex recognition in the blue crab *Callinectes sapidus* Rathbun. Sea Grant Technical Bulletin No, 15, University of Miami Sea Grant Program.

Tricarico, E., and F. Gherardi. 2006. Shell acquisition by hermit crabs: which tactic is more efficient? Behav. Ecol. Sociobiol., 60(4): 492-500.

Tschudy, D. 2003. Clawed lobster diversity through time. J. Crustacean Biol., 20: 178-186.

Vermeij, G.J. 1977. Patterns in crab claw size: the geography of crushing. Syst. Zool., 26: 138-151.

Wagner, H.P. 1994. A monographic review of the Thermosbaenacea (Crustacea: Peracarida). Zoologische Verhandelingen, 291: 1-338.

Walker-Smith, G., and G.C.B. Poore. 2001. A phylogeny of the Leptostraca (Crustacea) with keys to families and genera. Mem. Mus. Vic., 58(2): 383-410.

Wallace, R.L., and W.K. Taylor. 2002. Invertebrate Zoology Lab Manual, 6th ed. Benjamin Cummings, London.

Warner, G.F., and A.R. Jones. 1976. Leverage and muscle type in crab chelae (Crustacea, Brachyura). J. Zool. Lond., 180: 57-68.

Weldon, W.F.R. 1889. Note on the function of the spines of the Crustacean zoea. J. Mar. Biol. Assoc., U.K., 1(2): 169-172.

Whittington, H.B., and W.D.I. Rolfe. 1963. Phylogeny and evolution of Crustacea. Spec. Publ. Museum of Comparative Zoology, Cambridge, Massachusetts.

Williams, A.B. 1974. The swimming crabs of the genus *Callinectes* (Decapoda: Portunidae). Fish. Bull., 72: 685-798.

Wilson, G.D.F. 1989. A systematic revision of the deep-sea subfamily Lipomerinae of the isopod crustacean family Munnopsidae. Bull. Scripps Inst. Oceanogr., 27: 1-138.

Zinn, D.S., B.W. Found and M.G. Kraus. 1982. A bibliography of the Mystacocarida. Crustaceana, 43: 270-274.

Glossary

Acron: Unsegmented head of the ancestral arthropod.

Alga: Aquatic, eukaryotic, photosynthetic organisms, ranging in size from single-celled forms to the giant kelp. Once considered to be plants but now classified separately because they lack true roots, stems, leaves, and embryos.

Antennule: A small antenna or similar organ, especially one of the first pair of small antennae on the head of a crustacean.

Apolysis: Separation of the cuticula from the epidermis in arthropods and related groups.

Apophysis: Natural swelling, projection, or outgrowth of an organ or part, such as the process of a vertebra.

Astatic: Not static or stable.

Bathypelagic: Relating to or living in the depths of the ocean, especially between 1,000 and 4,000 m.

Benthic: Pertaining to or living on the bottom or at the greatest depths of a large body of water. Also called benthonic.

Benthonic: See Benthic.

Benthos: Organisms and habitats of the sea floor; the bottoms of lakes, rivers, and creeks, categorized according to size as macrobenthos (> 1 mm), meiobenthos (between 1 mm and 32 μm) and microbenthos (< 32 μm).

Bilateral symmetry: Symmetrical arrangement, as of an organism or a body part, along a central axis, so that the body is divided into equivalent right and left halves by only one plane. Also called plane symmetry.

Biofouling: The impairment or degradation of something, such as a ship's hull or mechanical equipment, as a result of the growth or activity of living organisms.

Bioluminescence: Emission of visible light by living organisms such as the firefly and various fish, fungi, and bacteria.

Biramous: Arthropod appendage comprising two series of segments.

Blue-green algae: See Cyanobacteria.

Branchial: Relating to or resembling the gills of a fish, their homologous embryonic structures, or the derivatives of their homologous parts in higher animals.

Brood: Offspring of an animal that are the result of one breeding season.

Cambrian: Geologic period that spanned about 60 million years and extends from about 542 to 488.3 million years before the present. It is the first period of the Paleozoic Era.

Carboniferous: Geologic period of the Paleozoic Era following the Devonian and preceding the Permian that extends from about 359.2 to 299.0 million years before the present.

Carcinization: Hypothesized process whereby a crustacean evolves into a crab-like form from a non-crab-like form.

Caridoid escape reaction: Type of escape reaction in some crustaceans (also known as "lobstering") involving very fast backward swimming, reaching speeds of over 60 cm per second.

Carnivore: Flesh-eating animal.

Carrion: Dead and decaying flesh.

Cephalothorax: Anterior section of arachnids and many crustaceans, consisting of the fused head and thorax.

Chelicerae: Mouth parts of the arthropod subphylum Chelicerata, used to grasp food, divided into jackknife, scissor and three-segmented chelate chelicerae. Foelix, Rainer F. (1996). Biology of Spiders (2nd ed.). Oxford University Press. Oxford.

Chelipeds: Pereiopods armed with a claw (chela).

Chironomidae: Family of Nematoceran Diptera with a global distribution (chironomids).

Ciliates: Protozoans of the class Ciliata, characterized by numerous cilia.

Clade: Group of organisms whose members share homologous features derived from a common ancestor.

Cnidaria: Invertebrate animals of the phylum Cnidaria, characterized by a radially symmetrical body with a sac-like internal cavity, and including the jellyfishes, hydras, sea anemones, and corals. Also called coelenterate.

Coccolithophores: Single-celled algae distinguished by special calcium carbonate plates called coccoliths, exclusively marine, found in large numbers throughout the euphotic zone.

Coelom: Cavity within the body (also called body cavity) of all animals higher than the coelenterates and certain primitive worms, formed by the splitting of the embryonic mesoderm into two layers.

Coelenterate: See Cnidaria.

Commensal: Relating to or characterized by a symbiotic relationship in which one species is benefited while the other is unaffected.

Compound eye: Eye composed of many light-sensitive elements, each having its own refractive system and each forming a portion of an image.

Coxa: First segment of the leg of an insect or other arthropod, joining the leg to the body.

Crepuscular: Animals that become active during twilight, before sunrise or after sunset.

Cryptobiosis: State in which the metabolic rate of the organism is reduced to an imperceptible level.

Cuticula: Multi-layered structure outside the epidermis of many invertebrates forming an exoskeleton.

Cyanobacteria: Prokaryotic photosynthetic bacterium, generally blue-green and in some species capable of nitrogen fixation.

Cyst: Small, capsule-like sac that encloses certain organisms in their dormant or larval stage.

Detritivore: Organism that consumes dead organic matter.

Diapause: Period during which growth or development is suspended and physiological activity is diminished, in response to adverse environmental conditions.

Diatoms: Microscopic one-celled or colonial algae of the class Bacillariophyceae, having cell walls of silica consisting of two interlocking symmetrical valves.

Dinoflagellates: Chiefly marine protozoans of the order Dinoflagellata, with two flagella and a cellulose covering comprising one of the chief constituents of plankton, including bioluminescent forms and forms that produce red tide.

Diurnal: Having a 24-hour period or cycle.

Drought: An extended period of abnormally dry weather.

Ecdysis: The moulting of the cuticula in arthropods and related groups.

Echinoderms: Radially symmetrical marine invertebrates of the phylum Echinodermata, which includes the starfishes, sea urchins, and sea cucumbers, having an internal calcareous skeleton and often covered with spines.

Ectoparasite: Parasite that lives on the surface or exterior of the host organism.

Embryo: Organism in its early stages of development, before birth, or hatching not having reached a distinctively recognizable form.

Endemic: Native to or confined to a certain region.

Endite: One of the appendages on the inner aspect of an arthropod limb.

Endoparasite: Parasite that lives inside the body of the host.

Endopod: Inner branch of a biramous appendage.

Endopodite: See Endopod.

Ephemeral pool: Pool lasting or existing only for a short time.

Epibenthic: Living at the surface of the substratum or on the sea bed.

Epidermis: Outermost layer of the skin.

Epipod: Respiratory endite arising from the coxal segment of a pereiopod.

Euphotic zone: See Photic zone.

Exopod: See Exopodite.

Exopodite: Outer branch of a biramous crustacean appendage.

Exoskeleton: Hard external anatomical feature, such as the shell of a crustacean, that provides protection or support for an organism.

Exuvia: Remains of an exoskeleton that are left after an arthropod has moulted.

Eyestalk: Movable stalk-like structure in certain crustaceans, such as crabs and shrimp, that bears an eye at the tip.

Filiform: Having the form of or resembling a thread or filament.

Filter-feeder: Aquatic animal, such as a clam, barnacle, or sponge, that feeds by filtering particulate organic material from water.

Flagellum: Long, thread-like appendage.

Foraminifera: Group of amoeboid protists with reticulating pseudopods with a chambered calcareous shell formed by several united zooids, many with perforated walls.

Gastropods: Largest and most successful class of molluscs, with 60,000-75,000 extant species known, comprising the snails and slugs.

Ghost crab: Any of several light-coloured burrowing crabs of the genus *Ocypoda* frequenting the tide line along sandy shores from the northeast United States to Brazil (also called white crab).

Gnathopod: Gnathopodite or maxilliped. See Maxilliped.

Gonad: Organ in animals that produces gametes, especially a testis or ovary.

Gonopods: First two pairs of pleopods in some taxa, specialized in the males for fertilization.

Gonopore: Reproductive aperture or pore.

Hemocoel: Cavity or series of spaces between the organs of most arthropods and molluscs through which the blood circulates.

Hemolymph: Circulatory fluid composed of water, inorganic salts and organic compounds with primary oxygen transporter molecule hemocyanin. Free-floating cells, the hemocytes, within the hemolymph play a role in the arthropod immune system.

Herbivore: Animal that feeds chiefly on plants.

Hetero-: *prefix.* Different.

Heterodontic: Teeth of the same type that differ in size or form.

Hyperspectral: Sensor system observing a target in very different spectral bands.

Hypogean: Located or operating beneath the earth's surface.

Instar: A discrete stage in a growth series, delimited by successive moulting.

Interstitial: Relating to, occurring in, or affecting interstices.

Intertidal zone: The part of the littoral zone above low-tide mark.

Keystone species: Species that has a disproportionate effect on its environment relative to its abundance.

Lamella: Thin scale, plate, or layer of bone or tissue, as in the gills of a bivalve mollusc or around the minute vascular canals in bone.

Larva: Juvenile form of animal with indirect development, undergoing metamorphosis and differing markedly in form and appearance from the adult.

Lecithotrophic larva: Planktonic-dispersing larva that lives off yolk supplied via the egg, as in most bony fish.

Licinia mobilis: Small, generally toothed, process articulated with incisor process of mandible.

Littoral zone: The biogeographic zone between the high- and low-water marks.

Lobstering: See Caridoid escape reaction.

Manca: Young of some Peracarida lacking last thoracopod at time of release from marsupium.

Mandible: Any of various mouth organs of invertebrates used for seizing and biting food, especially either of a pair of such organs in insects and other arthropods.

Marsupium: Temporary egg pouch in various fishes and crustaceans.

Maxilla: Either of two laterally moving appendages situated behind the mandibles in insects and most other arthropods.

Maxilliped: One of the three pairs of crustacean head appendages located just posterior to the maxillae and used in feeding.

Maxillule: Second pair of mouthparts (or fourth pair of appendages).

Megalopa: Larval stage of a crab.

Meiofauna: Small benthic animals ranging in size between macrofauna and microfauna; can pass through a 1 mm mesh but will be retained by a 45 μm mesh.

Mesopelagic: Organisms relating to or living at ocean depths between 180 and 900 m.

Metamere: See Somite.

Metamerism: Condition of having the body divided into metameres, exhibited in most animals only in the early embryonic stages of development.

Microalgae: Most primitive form of plants with simple cellular structure and mechanism of photosynthesis similar to that of higher plants.

Monogamous: Condition of having only one mate during a breeding season or during the breeding life of a pair.

Monophyletic: Relating to, descended from, or derived from one stock or source.

Monotype: Taxonomic group with a single member (a single species or genus).

Moulting: In arthropods, the periodic shedding of the exoskeleton, which is then replaced by a new growth.

Mutualism: Association between organisms of two different species in which each member benefits.

Nauplius: Free-swimming first stage of the larva of certain crustaceans, having an unsegmented body with three pairs of appendages and a single median eye.

Nematode: Worm of the phylum Nematoda, having unsegmented, cylindrical bodies, often narrowing at each end. Also called roundworm.

Neritic: Relating to or inhabiting a region of shallow water adjoining the coast between the low tide mark and a depth of about 200 m.

Nocturnal: Most active at night.

Nymph: Larval form, usually resembling the adult form but smaller and lacking full development.

Obligate parasite: Organism that cannot live independently of its host, cannot lead a non-parasitic existence.

Omnivorous: Eating both animal and vegetable foods.

Oostegite: Plate-like expansion of the basal segment of a thoracic appendage that aids in forming an egg receptacle.

Ootheca: Egg case of certain insects and molluscs.

Operculum: Lid or flap covering an aperture.

Opisthosoma: Posterior portion of the body behind the prosoma.

Ovary: Usually paired female or hermaphroditic reproductive organ that produces ova.

Oviduct: Tube through which the ova pass from the ovary to the uterus or to the outside.

Ovigerous: Bearing or carrying eggs; oviferous.

Ovisac: Egg-containing capsule, such as an ootheca.

Ovum: Female reproductive cell or gamete of animals; egg. Plural ova.

Palp: Elongated, often segmented appendage usually found near the mouth in invertebrates, the functions of which include sensation, locomotion, and feeding. Also called palpus.

Parasitism: Relationship between two phylogenetically unrelated organisms co-existing over a prolonged period of time, in which one benefits at the expense of the other.

Parthenogenesis: Form of reproduction in which an unfertilized egg develops into a new individual.

Penis: Any of various copulatory organs in males of lower animals.

Pereiopods: Primarily walking legs of a decapod crustacean (subphylum Crustacea, order Decapoda) also used for gathering food.

Permian: Geologic period that extends from about 299.0 to 248.0 million years before the present. It is the last period of the Palaeozoic Era.

Petasma: Modified endopodite of the first abdominal appendage in a male decapod crustacean.

Pheromone: Chemical secreted by an animal that influences the behaviour or development of others of the same species, often functioning as an attractant of the opposite sex.

Photic zone: Uppermost layer of a body of water that receives sufficient light for photosynthesis to occur (also called euphotic zone).

Photophore: Light-producing organ found especially in marine fishes that emits light from specialized structures or derives light from symbiotic luminescent bacteria.

Phyllosoma: A flat, transparent, long-legged larval stage of various spiny lobsters.

Phytoplankton: Minute, free-floating aquatic plants.

Pigmentation: Coloration of tissues by pigment or deposition of pigment by cells.

Plankton: (from Greek for wanderer or drifter) Drifting organisms that inhabit the water column of oceans, seas, and bodies of freshwater.

Plankton bloom: Outburst of planktonic organism, triggered by favourable environmental conditions such as nutrients and light; an event in which conditions are conducive to rapid reproduction of planktonic organisms in the ocean or other body of water.

Pleopods: Primarily swimming legs of decapod crustaceans (also called swimmerets).

Predaceous: Living by seizing or taking prey; predatory.

Preparatory female: An adult female that has developing oostegites and is in the instar just before the brooding condition.

Prosoma: The anterior or cephalic portion of the body of certain invertebrates, such as arachnids, in which segmentation is not evident.

Protozoan: Any of a large group of single-celled, usually microscopic, eukaryotic organisms, such as amoebas, ciliates, flagellates, and sporozoans.

Protist: Any of a group of eukaryotic organisms belonging to the kingdom Protista.

Puerulus: Larval stage of crawfish; a broad, thin schizopod larva.

Raptorial: Adapted for the seizing of prey.

Ramus: Bony process extending like a branch from a larger bone. Plural rami.

Rostrum: Beak-like structure projecting anteriorly from the carapace.

Rotifer: Microscopic, multicellular aquatic organisms of the phylum Rotifera, having at the anterior end a wheel-like ring of cilia.

Salp: Free-swimming tunicate of the genus *Salpa* having a transparent body with an opening at each end.

Scavenger: An animal that feeds on dead or decaying matter.

Sclerite: (Greek for hard) Chitinous or calcareous plate, spicule, or similar part of an invertebrate, especially one of the hard outer plates forming part of the exoskeleton of an arthropod.

Schizopod: Any of various shrimp-like crustaceans of the orders Euphausiacea and Mysidacea (formerly included in the single order Schizopoda), consisting of the krill and the mysids.

Seta: A stiff hair, bristle, or bristle-like process or part on an organism.

Shearwater: Oceanic birds of the genus *Puffinus*, having a short, hooked bill with tube-shaped nostrils and long slender wings that appear to shear the water as the bird flies along the surface.

Sinkhole: Natural depression in a land surface communicating with a subterranean passage, generally occurring in limestone regions and formed by solution or by collapse of a cavern roof.

Somite: One of a series of similar body segments of certain animals, such as earthworms and lobsters, into which they are divided longitudinally.

Southern Ocean: Body of water between 60° S latitude and Antarctica.

Spermatheca: Receptacle in the reproductive tracts of certain female invertebrates, in which spermatozoa are received and stored until needed to fertilize the ova.

Spermatophore: Capsule or compact mass of spermatozoa extruded by the males of certain invertebrates and primitive vertebrates and directly transferred to the reproductive parts of the female.

Spawning: Production or deposition of eggs in large numbers by aquatic animals.

Stalk: Slender or elongated support or structure, as one that holds up an organ or another body part.

Statocyst: Sensory organ of awareness of rotation and position located at base of first antenna.

Stream bed: Channel through which a natural stream of water runs or used to run.

Stylet: Small, stiff, needle-like organ or appendage.

Swimmerets: See Pleopods.

Symbiosis: (Greek for living together) Close, prolonged association between two or more different organisms of different species that may, but does not necessarily, benefit each member.

Tagma: (Greek for something that has been ordered or arranged) Distinct section of an arthropod, consisting of two or more adjoining segments.

Taxon: Taxonomic category or group, such as phylum, order, family, genus, or species.

Telson: Hindmost segment of the body of certain arthropods.

Tegumental glands: Compound glands situated in pereonal cavity that secrete a silk-like substance used in construction of the animal's tube.

Tetrapod: Vertebrate animal with four feet, legs, or leg-like appendages.

Thelycum: External reproductive organ of the female.

Thorax: Second or middle region of the body of an arthropod, between the head and the abdomen.

Thoracopod: Leg of a crustacean used for walking, swimming, feeding or filtration.

Transoceanic: Spanning or crossing the ocean.

Triassic: Geologic period that extends from about 245 to 202 million years before the present. The first period of the Mesozoic Era.

Umbilical cord: Flexible structure connecting a fetus at the abdomen with the placenta, transporting nourishment and removing its wastes.

Uniramus: Arthropod appendage comprising a single series of segments.

Uropods: One of the last pair of posterior abdominal appendages of certain crustaceans, such as the lobster or shrimp.

Uterus: Hollow, muscular womb, an enlarged portion of the oviduct in the adult female.

Vernal pool: Shallow depression in level ground with no permanent above-ground outlet.

White crab: See Ghost crab.

Yolk: Corresponding portion of the egg consisting of protein and fat that serve as the primary source of nourishment for the early embryo.

Zoea: Larval form of crabs and other decapod crustaceans, characterized by one or more spines on the carapace and rudimentary limbs on the abdomen and thorax.

Zooplankton: Plankton that consists of animals, including the corals, rotifers, sea anemones, and jellyfish.

3

Aspects on Population and Aquaculture Genetics of Crustaceans

A. Exadactylos
School of Agricultural Sciences, Department of Ichthyology and Aquatic Environment,
University of Thessaly, Volos 38446, Greece.
E-mail: exadact@apae.uth.gr

INTRODUCTION

The crustaceans (Crustacea, Brünnich 1772) are a large subphylum of arthropods (~55,000 species). They include various familiar animals, such as lobsters, crabs, shrimps, prawns, crayfish and barnacles. The majority are aquatic, living in either freshwater or marine environments, but a few groups have adapted to terrestrial life, such as terrestrial crabs and terrestrial hermit crabs. The majority are motile, moving about independently, though adult barnacles live a sessile life attached head-first to the substrate.

Crustaceans are particularly useful in aquatic environmental studies for several reasons. They are diverse and abundant in many habitats, play important roles in ecosystem processes, are often good indicators of stressed/polluted conditions, are relatively amenable to life history studies and frequently have commercial and cultural significance. Crustaceans constitute a dominant and diverse group of marine animals. Many of them, predominantly the decapods, are of major commercial importance, and studies on their genetic diversity are necessary for the sustainable exploitation and conservation of these bioresources (e.g., Chu et al., 2001; Hellberg et al., 2002). This work is focused on the main commercially important edible marine decapod crustacean species.

Molecular Population Genetics

Following the development of polymerase chain reaction (PCR) methods, molecular techniques have become widely used for detecting genetic variation in natural populations. Most nucleotide changes can be detected by these techniques. Many of these changes probably reflect silent substitutions that are likely to be selectively neutral, making them particularly suitable to population genetic studies.

In this work we focus on molecular genetics of natural populations, with respect to the genome assayed (nuclear or mitochondrial), and the molecular techniques used.

Aquaculture Genetics

Key to the emergence of genomics were advances in DNA marker technology. These advances have resulted in a wealth of genetic markers including allozymes, mtDNA, restriction fragment length polymorphisms (RFLPs), ramdomly amplified polymorphic DNA (RAPDs), amplified fragment length polymorphisms (AFLPs), microsatellites single nucleotide polymorphisms (SNPs), and expressed sequence tags (ESTs), with potentially widespread utility in a variety of contexts for the benefit of aquaculture genomics and aquaculture genetics (Rafalski and Tingey, 1993; Buitkamp and Epplen, 1996; Dodgson et al., 1997; Beuzen et al., 2000; Rafalski, 2002; Vignal et al., 2002).

In this work, the principles, potential power, advantages and disadvantages, and requirements for use are discussed for the various marker types, along with their application in assessments of genetic variability and inbreeding, inheritance, parentage assignment, species and strain identification, hybridization, and marker-assisted identification of quantitative trait loci (QTL) through the construction of genetic linkage maps.

MOLECULAR POPULATION GENETICS

Genes in Populations

Molecular techniques provide a method for identifying differences in DNA composition between individuals. We can determine whether two individuals differ from each other at particular points on their chromosomes (although we do not necessarily know what parts we are looking at). In the same way as for the morphological traits that Mendel worked with, individuals possess two copies of each gene (or fragment of DNA). These can be the same (homozygous) or different (heterozygous). On an electrophoretic gel, the different bands denote differences in DNA composition, such that each individual has two different alleles at a locus (two bands) or two copies of the same allele (one band). Population genetics is a much more recent discipline than Mendelian genetics, especially its application to ecology. Although Sewell-Wright developed the theory of population genetics in the 1920s, it was not until the advent of allozyme electrophoresis in 1956 that geneticists were able to apply the theory to natural populations. Basically, population genetics involves the study of the proportion of different genes or alleles at a single gene locus. For any locus, a diploid individual possesses two alleles (these may or may not be the same). In a population, the relative proportion of each allele is determined by working out the genotype of each individual and then calculating frequencies across the whole population. Gene frequencies are the relative frequencies of each of a number of possible genes or alleles at a single locus.

Processes Governing Genetic Evolution

Genetics seeks to understand the heritable basis of variation and evolutionary change at all levels (Fig. 1) from the fact that the population level is common to all the strata of biodiversity. Genetic differences

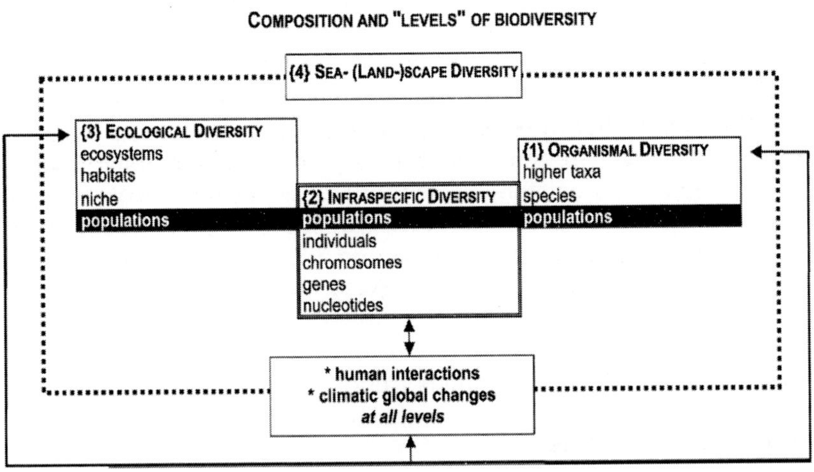

Fig. 1 Composition and levels of biodiversity. Biodiversity is conveniently described at three levels: infra-specific diversity, organismal diversity, and community or ecosystem diversity. A fourth level, the sea- (land-)scape diversity, takes into account global climatic changes and human activity impacts. It also integrates the type, condition, pattern, and connectivity of natural communities or ecosystems (Féral, 2002).

among individuals within a species provide the foundation for diversity among species and ultimately the foundation for diversity among ecosystems. Genetic diversity (i.e., intra-specific diversity) determines the ecological and evolutionary potential of species. It originates by mutation, but different processes determine the extent to which it is maintained within populations. Mutations range from a change in a single nucleotide to a modification of the karyotype. Mutation is ultimately the source of all genetic variation, but it is unpredictable and, so far, an uncontrollable source of diversity. Mutations create new alleles, whose evolutionary fates are governed by three forces: selection, gene flow (i.e., migration), and random drift (see Hartl and Clark, 1989; Weir, 1990; Avise, 1994; Slatkin et al., 1995; Thorpe and Smartt, 1995). Natural (or artificial) selection operates via differential survival and reproductive success of individual organisms. Those individuals with well-adapted phenotypes will pass a greater proportion of their genes on to the next generation. This means that in a population, only a fraction of individuals will produce progeny. Such a sampling of a finite number of genes at each generation means that some may increase in frequency, others may decrease, and some may be lost. Moreover, genetic drift randomly alters gene frequencies. The importance of genetic drift as a source of genetic differentiation is inversely related to population size. It has its maximum effect in small populations. It is thus important to note that effective population sizes (N_e, a theoretical population where every individual has the same probability of contributing genes to the next generation) in marine organisms are the lowest that have been estimated to date (Frankham, 1995a, b). At a steady state, equilibrium is reached between the creation of variants by mutation and the loss of others by genetic drift. The third force, migration, produces gene flow (passive or direct) because of the movement of individuals between established populations or because of the establishment of entire populations. Dispersal tends to homogenize gene frequencies and eliminate local differences. It occurs only during the larval phase in sessile or benthic invertebrates with limited mobility. Gene frequencies may be altered through migration if migrants entering the population have different genotypes and interbreed with residents. It has to be noted that even when organisms reproduce sexually, genetic loci are arranged on one or several chromosomes and sets of alleles tend to be inherited together. Recombination does not itself create genetic diversity at each locus, but it can create new combinations of alleles at different loci by mixing alleles of different individuals. It can therefore also be a source of diversity.

By definition, except in the case of (inter-fertile) hybrids, different species do not exchange genes; therefore, they have accumulated and still accumulate differences. This implies that genetic variation can be analysed at various levels, from the individual (existence of polymorphism = inter-individual variation in a gene product), through populations (gene frequency differences), to species or higher taxa (accumulation of differences = building up divergence).

In most animal species at least some degree of genetic differentiation among geographic regions can be observed (e.g., Ehrlich and Raven, 1969; Nevo, 1978; Avise, 1994, 2004; Slatkin, 1994; Roderick, 1996). Genetic subdivision can be caused directly by the organisms through their mating strategies, reduced dispersal or strict habitat preferences, or the subdivision can be caused by exogenous factors such as changing environments or barriers to dispersal. All these factors will lead to isolation of groups of populations, reduced gene flow and differentiation through genetic drift or differing selection regimes. The genetic diversity of an organism is the basis for its evolutionary potential to adapt to a changing environment, and the loss of this ability may jeopardize its survival. The maintenance of genetic diversity as a part of global efforts to conserve the earth's biodiversity depends critically on knowledge of the genetic architecture of the organism to be protected (Wilson, 1988; Avise, 1994; Avise and Hamrick, 1996). The use of molecular genetic markers to determine the population subdivision in various species has significantly contributed to this task (Avise and Hamrick, 1996), and quite often unexpected structures, not recognized by traditional morphological analysis, have appeared (Avise, 1992, 1994). Molecular methods can also provide indirect estimates of larval dispersal and gene flow in marine organisms with pelagic larvae in which direct estimates are impractical. Furthermore, statistical tests designed for assessing whether nucleotide polymorphisms deviate from expectations under neutral theory can also reveal details on population demographics.

Dispersal and Life Histories

Population genetic techniques can be used to infer the mechanisms and patterns of dispersal of animal species. To do this, we rely on two very simple principles. First, if populations are isolated from one another, over time they will tend to diverge and become genetically differentiated. This will occur even if the populations are not affected by different selection pressures, as a result of chance. This process is called genetic drift. If dispersal occurs between populations, this will tend to homogenize gene frequencies and make them more similar genetically. This process is called gene flow. By using genetic markers, it is possible to calculate gene frequencies at a number of gene loci and thus to infer the levels of dispersal and evaluate potential mechanisms.

A concrete understanding of population genetic structure is of primary importance for the management and conservation of genetic resources in exploited marine organisms. Studies on population genetic structure of marine biota have frequently indicated that organisms with high dispersal capacity have little genetic distinction over large geographic scales. These studies suggest that there are high levels of gene flow between marine populations. However, there is growing evidence that widespread marine organisms are more genetically structured than expected, given their high dispersal potential and apparent lack of barriers to dispersal in the ocean. Thus, there may be limits to the actual dispersal of marine organisms with high dispersal potential. These limits vary widely with species, habitats, local ocean conditions, or historical geomorphological events, and they may produce sufficient opportunities for genetic variation (i.e., Exadactylos et al., 1998, 2003; Tzeng et al., 2004b). Fisheries management and conservation of commercially important marine species, directed to ensure their sustainable exploitation, must rely on a profound knowledge of the biology and ecology of the target species (Exadactylos et al., 2007). In this context, genetic approaches have been long recognized as effective methods for stock identification in marine organisms. Population genetic studies have demonstrated that despite the difficulty of distinguishing physical barriers in the marine environment, and considering that populations are often connected through larval dispersal and free migration of adults, genetic differentiation among populations at different geographical scales occurs in many species (Barcia et al., 2005). The cessation of gene flow between populations, and their subsequent differentiation into biological species, is a fundamental step in the origin of biological diversity.

Population genetics theory predicts that small, isolated populations with low levels of gene flow characteristically show low genetic diversity within populations and high genetic differentiation among populations. These phenomena are often responsible for a genetic fragmentation of species into discrete stocks that could reflect local adaptation to their environment. Information on the degree and distribution of genetic variability in natural endangered species stocks is important for conservation strategies. Conservationists working to reintroduce species to native waters from which they have disappeared need to identify their genetic stocks in order to avoid the mixing of genetically highly differentiated populations within a hydrographic basin. Moreover, in case of population reinforcement, introduction of genetically different individuals could adversely alter the gene pool of locally adapted populations. Thus the application of genetics to conservation issues is a practical endeavour and has produced concrete recommendations for management of animal species (see Ashley et al., 1990; Welsh and McClelland, 1990; Williams et al., 1990; Gray, 1997a, b; Hellberg et al., 2002).

Species and Cryptic Biodiversity

Nevertheless, the relative importance of natural selection versus spatial isolation in driving speciation remains obscure. Comparison of closely related ("sibling") species offers one of the clearest available windows on this process, as recently diverged species are more likely than older species to retain geographic distributions and characters useful in reconstructing the history of their divergence. The more recently a speciation event has occurred, however, the more the resultant sister species are likely to be similar in morphology and in other characters that might be used to identify them. Thus, the taxa most promising for macroevolutionary studies of speciation are frequently the most difficult to recognize (Duffy, 1996; Exadactylos and Thorpe, 2001).

Members of the same species mate with one another, exchange genes and therefore share many morphological and behavioural characteristics in common. Individuals belonging to different species do not interbreed, do not exchange genes and as a consequence are likely to have diverged in morphological and behavioural characteristics. Most taxonomists have assumed that morphology is sufficient to differentiate among species. However, this is not always the case, because the characteristics that individuals of a species use to recognize potential mates differ vastly among broad taxonomic groups. An understanding of genetic population structure and the identification of cryptic species are critical for the conservation and management of exploited fish and crustacean species (Shaklee, 1983; Ward and Grewe, 1994; Ferguson and Danzmann, 1998). It is now widely recognized that molecular genetic data provide fisheries managers with essential information regarding both the existence and distribution of discrete genetic populations and the occurrence of cryptic taxonomic forms (e.g., races, varieties and subspecies) or evolutionarily significant units (Moritz, 1994; Ward and Grewe, 1994).

Fisheries biologists worldwide are constantly being asked about translocation and re-stocking of depleted populations. Without knowledge of the diversity present in natural populations, it is not possible even to predict the effects of these management options. The results of such studies also have implications for our understanding of speciation. Many authors propose that if hybrids between two genetic forms are unviable or less viable than the parental types, natural selection will act to reduce the probability of mating between the forms (e.g.,

Dobzhansky, 1970). The alternative view is that one form is likely to become extinct in the area of overlap (Paterson, 1993). Future monitoring of the areas of co-occurrence will provide an effective field test of the two hypotheses in a natural situation. The use of genetic techniques has enhanced or changed our understanding of ecological and evolutionary processes. Molecular markers have contributed significantly to our understanding of mating systems in animals, patterns of dispersal and connectivity among populations, the importance of microevolutionary processes such as natural selection and genetic drift, the role of historical processes in shaping species distributions, and the recognition of cryptic species.

Genetic Drift and Gene Flow

Population genetic studies have demonstrated restricted gene flow among conspecific marine populations without any obvious barriers to dispersal (i.e., Karl and Avise, 1992; Burton and Lee, 1994; Pogson et al., 1995; Knutsen et al., 2003; Taylor and Hellberg, 2003). These and other observations have led to a re-evaluation of the proposition that planktonic dispersal implies open populations (Warner and Cowen, 2002). Genetic differentiation among populations reflects a dynamic balance between the forces of genetic drift, gene flow, selection, and mutation. Interpretation of the distributions of genetic markers often requires that some of these forces be assumed negligible. For example, it has often been assumed that selection and mutation can be ignored when F_{ST} (genetic diversity coefficient) is used to estimate gene flow (Slatkin and Barton, 1989). Inferences about the strength of gene flow based on F_{ST} also require assumptions about the strength of genetic drift, because F_{ST} is used to estimate the magnitude of gene flow (m) relative to genetic drift ($1/N_e$). If the forces involved are relatively weak, equilibria will be approached very slowly and historical effects can also be important (Crow and Aoki, 1984; Neigel, 1997). This confounding of gene flow and genetic drift is especially problematic for populations of marine species that conceivably could have either very large effective sizes because they have large numbers of individuals or very small effective sizes because reproductive success is limited to a small number of individuals (Hedgecock et al., 1992; Hedgecock, 1994; Neigel, 1994, 2002).

Different types of genetic markers are expected to vary in the way their distributions respond to forces other than gene flow, and in the statistical power they provide for the detection of genetic differentiation among populations (Neigel, 1997). Allozymes have strongly influenced our views of marine larva dispersal (e.g., Burton, 1983). If sufficient numbers of polymorphic allozyme loci are used, they can provide considerable statistical power for detection of genetic differentiation (Slatkin, 1985; Slatkin and Barton, 1989), although it is likely that some early claims of significant allozyme differentiation among marine populations can be attributed to biases in contemporaneous methods for estimation of F_{ST} (Cockerham and Weir, 1993). However, any interpretation of allozyme data should acknowledge the possibility that allozymes are at least occasionally subject to strong selection (e.g., Koehn et al., 1980; Hilbish and Koehn, 1985a, b; Karl and Avise, 1992).

Mitochondrial DNA (mtDNA) variation has been especially useful for revealing the phylogeographic structure of species (Avise, 2000). Because we do not expect to see deep phylogeographic divisions among populations interconnected by gene flow their discovery in marine species has been interpreted as evidence for strong barriers to gene flow (Barber et al., 2000, 2002; Taylor and Hellberg, 2003). However, in such cases we are observing isolation that developed *in situ* under contemporary oceanographic conditions or secondary boundaries between cryptic species that had achieved reproductive isolation under different conditions. This distinction is more than semantic because very different types of mechanisms could be at work, such as dispersal-limiting versus post-dispersal barriers to gene flow.

Microsatellite loci are often highly polymorphic because of high rates of length mutation (Hughes and Queller, 1993). Remarkably precise estimates of migration rates and effective population sizes can be obtained with microsatellite data and maximum likelihood estimators based on coalescent models (Rannala and Mountain, 1997; Waser and Strobeck, 1998; Cornuet et al., 1999; Beerli and Felsenstein, 1999, 2001; Pritchard et al., 2000; Knowles and Maddison, 2002). Microsatellite loci are expected to be less useful for resolving phylogenetic relationships among isolated populations because high mutation rates can quickly saturate measures of population differentiation (Neigel, 1997; Hedrick, 1999) and only a weak correlation is expected between number of mutation steps and allelic differences in length (Slatkin, 1995).

Effective Population Size (N_e)

For most marine species, census and spawning stock estimates have often suggested that stocks retain large population size despite intensive exploitation. This has been one of the primary arguments behind the continuation of fishing on heavily depleted stocks. Of the total census population size (N), however, only a proportion (the effective population size N_e) will pass on their genetic characteristics to the next generation in any reproductive stock. Importantly, there is a connection between the magnitude of random change in populations between generations (genetic drift) and the number of parents that are successful in leaving offspring. A large number of successful parents would limit the amount of genetic drift, and vice versa. Thus, population genetic theory can be used to "count" the number of individuals actively reproducing at any one time; the amount of drift that is occurring can provide an indirect measure of the number of spawning individuals in a population that successfully leave offspring (recruits) in the next generation.

Until recently, it was generally believed that the majority of marine species populations was sufficiently large to be unaffected by random change between generations (drift) during population declines, and that population connectivity for most species was generally high (Ward and Grewe, 1994; Carvalho, 1998; Ward, 2000). Although earlier work had suggested that some marine species might have considerably lower effective population sizes (N_e) than census sizes (Hedgecock et al., 1989, 1992, 1994, 2001; Hedrick and Hedgecock, 1994) because of marked variance in reproductive success, it is only within the past few years that new genetic methods have been employed to provide successful estimates of N_e in marine species populations (Hauser et al., 2002; Turner et al., 2002; Hutchinson et al., 2003). Such discrepancies have profound implications for estimating quantitative change in population size relative to recruitment and harvesting as well as qualitative change, in terms of the nature and speed of genetic change in marine populations. A low ratio of effective population size to census size (N_e/N) suggests greater vulnerability to changes in genetic diversity, patterns of genetic differentiation and responses to environmental change (selection pressures). The relationship between successful spawners and subsequent recruits is critically important to stock assessment models. Genetic estimates of effective population size may assist in the formulation of more accurate, independent measures of these parameters.

Molecular markers, primarily microsatellites and mtDNA, are being applied increasingly for estimating effective population size using approaches based on coalescence theory, and for inferring long-term demographical patterns using mismatch distribution analysis.

There are two fundamentally different methods for investigating N_e:

- Methods for directly estimating effective population size (N_e)
- Methods for detecting changes in N_e without actually providing an estimate of N_e (Garza and Williamson, 2001).

Most estimates of N_e to date are based only on shifts in allele frequencies over the sampling period; temporal method estimates are insensitive to bottlenecks that occurred prior or subsequent to sampling. Application of alternative methods (e.g., Turner et al., 2002), such as estimation by a coalescence approach, can detect historical bottleneck events. Such approaches take account of the differences in mutation rates and loss of alleles, as well as the equilibrium between genetic drift and mutation (Avise, 2000). Such data indicate not only that even very large marine populations may be vulnerable to overexploitation and environmentally induced recruitment failure, but also importantly that fishery models need to take account of genetic estimates of population size if sustainable forecasts are to be achieved.

Molecular Tools

A major concern in the field of population genetics is to understand the causes of differentiation between populations across ranges of geographic distribution. For example, one can study evolutionary forces in action (e.g., natural selection, genetic drift) and note phylogenetic and biogeographic patterns among living organisms through geographical space and time. The knowledge produced by the use of theoretical and experimental tools has also proved useful for more applied purposes, such as aquaculture, fisheries and conservation (reviews by Avise, 1994, 1996; Cipriano and Palumbi, 1999). There are a number of good reviews of molecular techniques applicable to aspects of population biology (e.g., Karl and Avise, 1993; Lessa and Applebaum, 1993; Queller et al., 1993; Avise, 1994; Skibinski, 1994; Hillis et al., 1996).

Allozymes

Since Lewontin and Hubby (1966), allozyme electrophoresis has become widely used among almost all groups of animals and has been applied to almost every sort of problem in population genetics and its branches (Murphy et al., 1990). Such intense use of the method has revealed its limitations for some taxa or for specific problems. These limitations, however, do not reflect the exhaustion of the method but have simply served to narrow the range of possibilities (Murphy, et al., 1996). Allozymes have been widely used as markers to study the genetic variability of several crustacean species populations. However, most allozyme studies have revealed very low levels of genetic variation in crustaceans (Hedgecock et al., 1982) and in crayfish species in particular (Nemeth and Tracey, 1979; Brown, 1980; Albrecht and Von Hagen, 1981; Attard and Vianet, 1985; Austin, 1986; Austin and Knott, 1996; Busackc, 1988, 1989; Fevolden and Hessen, 1989; Agerberg, 1990; Santucci et al., 1997; Lörtscher et al., 1997). These results could be related to the increased fragmentation of habitat that has allowed genetic variation to develop between subpopulations and to decrease within subpopulations as a result of genetic drift. In regard to these results, it seemed interesting to obtain additional data from a more variable genetic marker such as mitochondrial DNA.

DNA Markers

Currently DNA techniques are rapidly replacing allozymes in population genetics. One reason is the development of PCR (Saiki et al., 1988a, b), so that DNA techniques can be applied to tissues preserved with different methods (e.g., alcohol-fixed, air-dried and even fossils) (Lewin, 1994). However, DNA techniques are still expensive and labour-intensive (Skibinski, 1994). These limitations are restrictive, because the number of individuals sampled is inversely proportional to the genetic difference between operational taxonomic units. The genetic difference to be detected is often only slightly higher than that between individuals in a single population; thus, for an accurate assessment, the number of individuals sampled should be high (> 30 individuals). In all cases, both approaches have their advantages and disadvantages, their limits and their idiosyncrasies. For instance, it has been suggested that allozyme electrophoresis is the best technique for alpha systematics (Hillis et al., 1996), but DNA sequencing has proved very useful for a wide range of phylogenetic studies (e.g., Holmes et al., 1999; Maggioni et al., 2001) and microsatellites and Single strand conformational polymorphism (SSCP) may be the methods of choice for some population genetic studies (Goldstein et al., 1999).

Restriction enzymes have been widely used to obtain information on DNA sequence length polymorphism, and a great effort has been put into the sequencing of DNA (both genomic and organellar). Randomly amplified polymorphic DNA (RAPD) and multi-locus satellite screening (fingerprinting) techniques have been used mainly for addressing problems related to reproduction and kinship. The use of techniques involving single-locus minisatellites and microsatellites has recently increased. However, the cost associated with the construction of DNA libraries is still a limiting factor. DNA denaturing techniques (e.g. SSCP, denaturing gradient gel electrophoresis (DGGE), heteroduplexing) are new and their application to population genetic studies could provide an interesting new array of molecular markers.

Other developed methodologies, such as amplified fragment length polymorphisms (AFLP) and sequence-specific oligonucleotides (SSO), are widely used in population biology (Zabeau and Vos, 1993; Vos et al., 1995; Wang et al., 2004). AFLP combines the techniques of restriction endonuclease digestion and PCR amplification of restriction fragments and thus possesses the advantages of both techniques. It has been widely applied to study genotyping, population differentiation and genetic diversity in a wide variety of organisms. In general, scientific efforts are concentrated on groups of economic importance (e.g., fish, molluscs, crustaceans). The management and conservation of these groups has had emphasis, and they have generally shown low levels of variation at the allozyme level (Avise, 1994; Slade et al., 1998).

Microsatellites

Eukaryotic genomes contain a large number of interspersed, tandemly repeated sequences called microsatellites or simple sequence repeats (Tautz, 1989; Weber and May, 1989; Avise, 1994; Toth et al., 2000). These refer to tandem repeats with a motif of less than 6 base pairs (bp) in length. The repeat number of microsatellites has been demonstrated to be highly variable in animals, providing an attractive source of genetic polymorphisms (Stalling et al., 1991; Roder et al., 1995). Based on the unique sequences flanking microsatellites, primers are designed to detect the microsatellite loci through PCR with genomic DNA. Alleles at microsatellite loci can be size fractionated by denaturing polyacrylamide gel electrophoresis. Allelic variants of a microsatellite locus are codominant and show Mendelian inheritance.

Typically, microsatellite loci are highly polymorphic and evolve rapidly, making them ideal genetic markers (Schlotterer, 2000; Balloux and Lugon-Moulin, 2002). Microsatellites are highly replicable and have a great ability to resolve genetic differences compared to other classes of genetic markers (Mueller and Wolfenbarger, 1999; Sunnucks, 2000). For these reasons, they are routinely used for pedigree analysis (McCouch et al., 1997; Itokawa et al., 2003), and population structure determination (Goldstein and Schlotterer, 1999; Schlotterer, 2000). Microsatellite markers have great potential use in aquaculture and fisheries biology. A number of studies have applied microsatellites to genetic tagging and population studies (Brooker et al., 1994; Garcia de Leon et al., 1995; Colbourne and Hebert, 1996; Herbinger et al., 1997).

Simple sequence repeats (SSRs) have a wide range of applications including genetic mapping, QTL association, kinship analysis, population genetics, and evolutionary studies. Traditionally, SSR isolation has relied on the screening of genomic libraries using repetitive probes and sequencing of positive clones in order to develop locus-specific primers. Expressed sequence tags (ESTs) are generated by single-pass sequencing of complementary DNA clones obtained by reverse transcription of

subunit (16S) ribosomal DNA and cytochrome *c* oxidase subunit I (COI) are popular markers used in molecular systematic studies of crustaceans at the species and population levels (e.g., Baldwin et al., 1998; Tam and Kornfield, 1998; Sarver et al., 1998; Chu et al., 1999; Tong et al., 2000; Schubart et al., 2000; Perez et al., 2004). Yet a major criticism of using mitochondrial DNA markers is that the genes are linked and inherited as a single unit. Moreover, 16S rDNA and, to lesser degree, COI are quite conserved to serve as good intraspecific markers in some crustacean species (Chu et al., 2001).

In recent years, a diversity of nuclear mitochondrial-like sequences (*numts*) has been reported in an increasing number of organisms. Although reported from a diverse array of species, *numts* of only two mitochondrial genes have been documented in three crustaceans, comprising the large-subunit ribosomal RNA (16S rRNA) in stone crabs and the COI in planktonic copepods and snapping shrimp (Nguyen et al., 2002). Much more mitochondrial genomic information from various species is required in order to advance these researches, in particular its uses in crustacean biology. It appears that this results from the early availability of "universal" primers for those genes. There is a critical paucity of information on crustacean mitochondrial genomes (Yamauchi et al., 2003, Imai et al., 2004).

Phylogenetics

The determination of species boundaries is perhaps the most important application of phylogeny to conservation biology. Many species concepts have phylogenetic relatedness as the central focus in the determination of species, e.g., phylogenetic species concept (Cracraft, 1983), cohesion species concept (Templeton, 1989) and evolutionary species concept (Wiley, 1978). Furthermore, phylogeny can be used to assess the degree of genetic isolation between two populations (Slatkin and Maddison, 1989), and partition historical and repeated events shaping population structure (Templeton, 1998). It is of primary importance to establish whether or not distinct populations are distinct species within a hypothesis-testing framework (Sites and Crandall, 1997). Furthermore, phylogenetic approaches are used extensively to identify highly differentiated subpopulations of a species in need of conservation action (Crandall et al., 2000). The delineation of species and distinct populations greatly influences conservation policy in terms of introductions and listings as rare and endangered.

Until the implementation of phylogenetic methods for measuring biodiversity, conservation priorities were based solely on methods such as species richness, all species being considered of equal value. Methods for evaluation of biodiversity based on phylogeny have played a large role in recent advances in conservation biology, being particularly useful for a preliminary assessment of conservation priorities and endangered species, and to identify populations or species for which more detailed studies are needed (i.e., Templeton, 1991; Eldredge, 1989, 1992; Cracraft, 1994; Krajewski, 1994; Greene, 1994; Crandall, 1998; Crandall et al., 2000; Whiting et al., 2000; Porter et al., 2005). These phylogenetic methods have typically been separated into two categories: topology dependent and distance dependent (Krajewski, 1994). Topology-dependent methods rely on a rooted phylogeny and reflect the branching order and therefore rank those organisms that evolved earliest with the highest priority regardless of divergence between species (Vane-Wright et al., 1991; Nixon and Wheeler, 1992). Methods dependent on distance or branch length sum the branch lengths to derive a phylogenetic diversity for an organism and strive to represent the genetic diversity or divergence of each organism (Crozier, 1997; Krajewski, 1994; Faith, 1992). It has been generally agreed that it is more appropriate for conservation purposes to use unrooted trees, and that branch length measures more accurately assess genetic diversity (Crozier, 1997).

At a greater observation scale, one should consider as a priority the reconstitution of the historical aspects of the development of extant biodiversity in the world ocean. The geography of genes and molecular phylogenies have shown their remarkable performance in this matter on the continental domain. These biogeographical reconstructions allow a better understanding of the impact on genetic diversity of the palaeoclimatical changes frequently recorded during various glacial episodes. They also address other questions such as the impact of larval dispersal or the origin of vicariance in a dispersive environment.

AQUACULTURE GENETICS

With global population expansion, the demand for high-quality protein, especially from aquatic sources, is rising dramatically. Increased aquaculture production is clearly needed to meet this demand in the third millennium, because capture fisheries are at capacity or showing precipitous declines due to overfishing, habitat destruction and pollution. Further increases in capture fisheries are not anticipated under current global conditions.

Increased demand for aquaculture production means increasing pressure for development of more efficient production systems. Major improvements have already been achieved through enhanced management, nutrition, disease diagnostics and therapeutics, water quality maintenance and genetic improvement of production traits. A common theme through all these is genetics, which, actively and passively, has been used to meet many production challenges such as disease resistance, tolerance of handling, enhanced feed conversion and spawning manipulation, namely all those areas to which wild animals must adapt for productive "domestication". Closely related species are in the wild reproductively isolated and have species status because of their genetic differences. Thus, comparison for culture suitability is a genetic comparison.

Genetic improvement programmes are being applied to more and more aquatic species. The number of farmed taxa reported to FAO has increased for all major groups since 1994: 34% for fishes, 29% for crustaceans, and 31% for molluscs (Garibaldi, 1996; FAO, 1999). As domestication extends to more species, so will the application of genetic improvement technologies. Only a small percentage of aquaculture production currently comes from genetically improved species (Gjedrem, 1997). This refers to directed genetic improvement, not simply to the domestication process. Therefore, there is scope to increase productivity by genetic improvement, involving selective breeding, chromosome manipulation, hybridization, production of mono-sex groups, and gene transfer (see Exadactylos and Arvanitoyannis, 2006).

Modern molecular techniques now allow analysis of an organism's DNA, such that family relationships and pedigrees can often be established and the genetic stock structure of natural populations be determined. The use of DNA probes, specifically microsatellite DNA probes, to identify families and establish pedigrees of farmed animals may greatly facilitate selective breeding programmes where physical tagging is impractical. DNA probes are also being used to identify minute quantities of pathogens such as shrimp viruses, thus helping to ensure that an organism is free of a specific pathogen (Wongteerasupaya et al., 1996).

Long-term management of decapod crustacean populations spatially and temporally must rely on a full understanding of their basic biology, especially the genetic structure of their breeding populations. Few prior genetic studies have been able to discern differences between populations. This lack of differences may be due to the lower sensitivity of the genetic markers used in earlier studies, namely allozymes. Future experiments could attempt to correlate microsatellite size to moulting phenotypes (i.e., fast or slow moulting rate). Loci that have been screened to date are inherited in classical Mendelian fashion (i.e., the parental alleles are the only alleles present in the offspring and are represented equally in the next generation). Further efforts are needed to establish an appropriate management strategy for maintenance of such stocks. This requires a better understanding of (1) the level of genetic diversity in populations and in species as a whole, (2) the existence of discrete and genetically isolated populations along their geographical platform, and (3) the level of gene flow, if any, among these populations.

The use of DNA markers can contribute significantly to the development and implementation of genetic improvement programmes. Properly designed breeding programmes can make appropriate use of additive and non-additive genetic variation for traits of economic importance and minimize the negative effects of inbreeding. DNA markers can be used to verify pedigrees, screen natural populations to maximize diversity in founder animals, and monitor inbreeding levels in breeding populations. Additionally, markers can be used to characterize QTL. Andersson et al. (1994) thereby enabled marker-assisted selection to be applied as an additional component to a selective breeding programme. There are clearly compelling reasons for the development of DNA markers for use in marine crustacean breeding (Moore et al., 1999).

The potential evolutionary response of a character to environmental changes is indicated by the concept of heritability, which is a measure of the genetic diversity affecting a quantitative character, morphological, physiological or behavioural (= characters affected by numerous loci) (Hartl and Clark, 1989; Falconer and MacKay, 1996). Breeding studies allow inferences about the genetic basis of quantitative characters. It is still difficult to establish relationships between variability at the genetic and phenotypic levels. However, such studies may provide means of predicting the distributions of phenotypes among offspring from the parental phenotypes. The chance for success in identifying QTL is much improved by selecting Mendelian markers that may have a functional relation to the trait (candidate genes). Such approaches will underline the regions of the genome undergoing strong selective constraints, either for themselves or in interaction with others. This applies particularly in situations where the populations are subjected to genetic tensions such as adaptive crises, bottlenecks or hybrid zones.

Several marker types are popular in aquaculture genetics. Allozyme and mtDNA markers have long been popular. More recent marker types include restriction fragment length polymorphism (RFLP), randomly amplified polymorphic DNA (RAPD), amplified fragment length polymorphism (AFLP), microsatellites, single nucleotide polymorphism (SNP), and expressed sequence tag (EST) markers. Table 1 summarizes the basic properties of each marker type.

Table 1 Types of DNA markers, their characteristics, and potential applications (Liu and Cordes, 2004).

Marker type	*Acronym*	*Requires prior molecular information?*	*Mode of inheritance*	*Locus under investigation*	*Likely allele numbers*	*Polymorphism or power*	*Major applications*
Allozyme		Yes	Mendelian, codominant	Single	2-6	Low	Linkage mapping, population studies
Mitochondrial DNA	mtDNA	No	Maternal inheritance	Multiple	haplotypes	Maternal lineage	
Restriction fragment length polymorphism	RFLP	Yes	Mendelian, codominant	Single	2	Low	Linkage mapping
Random amplified polymorphic DNA	RAPD, AP-PCR	No	Mendelian, dominant	Multiple	2	Intermediate	Fingerprinting for population studies, hybrid identification
Amplified fragment length polymorphism	AFLP	No	Mendelian dominant	Multiple	2	High	Linkage mapping, population studies
Microsatellites	SSR	Yes	Mendelian, codominant	Single	Multiple	High	Linkage mapping, population studies, paternity analysis
Expressed sequence tags	EST	Yes	Mendelian, codominant	Single	2	Low	Linkage mapping, physical mapping, comparative mapping
Single nucleotide polymorphism	SNP	Yes	Mendelian, codominant	Single	2, but up to 4	High	Linkage mapping, population studies?
Insertions/deletions	Indels	Yes	Mendelian, codominant	Single	2	Low	Linkage mapping

Evolutionary Forces

There is an emerging rapprochement between the population genetic accounts and the developmental genetic accounts of evolution. The population genetic approach has focused on variation within populations, while the developmental genetic approach has focused on variation between populations (Gilbert, 2000; Amundson, 2001, 2005). Similarly, population geneticists have been looking primarily at genes in adults competing for reproductive success, while developmental geneticists have been looking at genes involved in forming embryonic and larval organs. These differences are becoming blurred as both population geneticists and developmental geneticists begin examining the regulatory genes that control development. The two approaches complement each other; while the population genetic approach focuses on the survival of the fittest, the developmental genetic approach to evolution is more concerned with the arrival of the fittest. To explain evolution, both the population genetic and the developmental genetic accounts are required (Fig. 3).

Farmers and hatchery managers should determine the N_e of a hatchery population every generation. A population's N_e is one of the most important items of information about that population because it describes its genetic size. An understanding of N_e and what it means is of major importance because N_e helps determine the genetic stability of the population in that it determines the average inbreeding in the population and influences genetic drift. Once N_e has been quantified, it can be used to determine the average inbreeding value in the population. The average inbreeding value in the population is inversely related to N_e, which means that inbreeding can become a problem in small, closed populations, which are typical of those found at most farms and culture stations. Once inbreeding occurs, the problem can quickly worsen as inbreeding and N_e are linked in a positive feedback loop. Consequently, small N_e will produce levels of inbreeding that will cause growth rate, fecundity, and other traits to decline over succeeding generations. As was the case with average inbreeding, genetic drift is inversely related to N_e. Consequently, small hatchery populations can cause random changes in gene frequency. The ultimate effect of a small N_e is the loss of alleles via genetic drift. Rare alleles will be lost more easily, but common alleles can also be lost. The loss of genetic variance can produce irreversible damage to a population's gene pool.

Chromosomal manipulation can be used to quickly produce highly inbred lines of cultured animals. One generation of mitotic gynogenesis or mitotic androgenesis will produce animals that are 100% homozygous and 100% inbred. A second generation of chromosome set manipulation

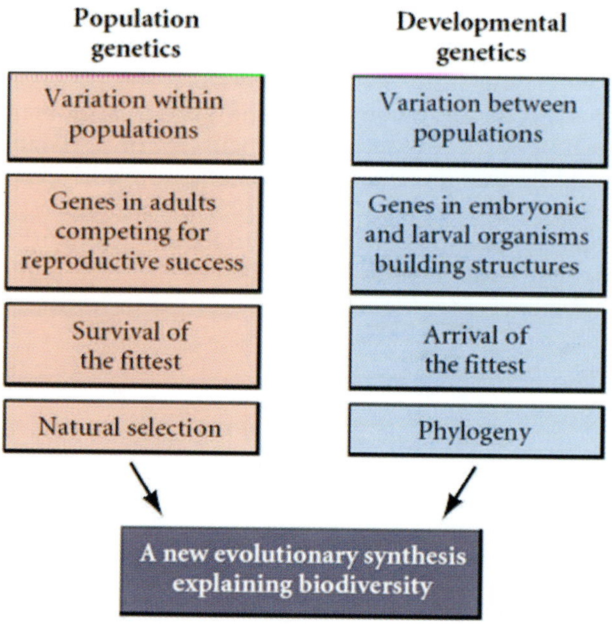

Fig. 3 A newly emerging evolutionary synthesis. The classic approach to evolution has been that of population genetics. This approach emphasized variations within a species that enabled certain adult individuals to reproduce more frequently; thus, it could explain natural selection. The developmental approach looks at variation between populations and it emphasizes the regulatory genes responsible for organ formation. It is better able to explain evolutionary novelty and constraint. Together, these two approaches constitute a more complete genetic approach to evolution (Gilbert, 2006).

is needed to produce 100% inbred specimens that are capable of reproducing. The amount of inbreeding produced by meiotic gynogenesis is variable and depends on crossing over frequencies.

Domestication and Strain Evaluation

When wild species are moved to aquaculture settings, a new set of selective pressures comes into play that often changes gene frequencies. Consequently an organism better suited for the aquaculture environment begins to evolve. This process, termed domestication, occurs even without directed selection by the aquaculturist. Domestication effects can be observed within as few as one to two generations after removal from the natural environment. Although most domesticated strains usually perform better in the aquaculture environment than wild strains, there are exceptions (Dunham, 1996, 1999). The explanation for this appears to be related to a lack of maintenance of genetic quality, and genetic degradation in domesticated strains. Poor performance of some domestic aquatic species is related to poor founding (parental) lines, random genetic drift, inbreeding and introgression with slower-growing species, and slower-growing strains.

Domestication of farmed shrimps (penaeids) has been relatively slow compared to that of fish. This can be attributed to use of wild broodstock and post-larvae, a lack of understanding of shrimp reproductive biology for domestication of the species and perceptions of low potential for genetic improvement. Current reliance on wild broodstock is risky and negates the opportunity to enhance disease resistance (as well as other production traits) through selective breeding. Efforts to domesticate broodstock are now hampered by endemic disease challenges. Economic analysis has demonstrated that domesticated broodstock are more cost-effective than wild broodstock (Preston et al., 1999) and that reproductive performance of a domesticated one can match that of a wild broodstock of a similar size (Crocos and Preston, 1999).

Use of established, high-performance domestic strains is the first step in applying genetic principles to improved aquaculture management. Strain variation is also important, since there is a strain effect on other genetic enhancement approaches, such as intraspecific crossbreeding, interspecific hybridization, sex control and genetic engineering.

Inbreeding and Maintenance of Genetic Quality

Possible deleterious consequences of uncontrolled inbreeding, and engendered attempts to maintain larger effective population sizes in breeding programmes, are well documented (Malecha and Hedgecock, 1989). Sufficient levels of genetic variation in cultured stocks can be maintained over the short term by maximizing the variation in the base populations, and in the longer term by including progeny from a large number of broodstock to prevent inbreeding. It is as important to prevent production losses due to inbreeding as it is to increase production from genetic enhancement. This applies especially to species with high fecundity, where few broodstock are necessary to meet demands for fry and broodstock replacement. The detrimental effects of inbreeding are well documented and can result in decreases of 30% or more in growth production, survival and reproduction (Kincaid, 1976a, b, 1983; Dunham, 1996).

Intraspecific Crossbreeding

Intraspecific crossbreeding (crossing of different strains) may increase growth rate, but heterosis (differences between offspring and parents) may not be obtained in every case. Intraspecific crossbreeding often increases growth rate of aquacultured species, although specific heterotic combinations need to be experimentally identified. Heterosis was also found in survival, disease resistance and reproductive traits.

Interspecific Hybridization

Interspecific hybridization has been used to increase growth rate, manipulate sex ratios, produce sterile animals, improve flesh quality, increase disease resistance, improve tolerance of environmental extremes, and improve a variety of other traits that make aquatic animal production more profitable (Bartley et al., 1992). Although interspecific hybridization rarely results in an F_1 suitable for aquaculture application, there are a few significant exceptions. Hybridization between species can also result in offspring that are sterile or have diminished reproductive capacity. As with monosex production, the production of sterile hybrids can reduce unwanted reproduction or improve growth rate by energy diversion from gametogenesis. Karyotype analysis can be used as a general predictor of potential hybrid fertility (Reddy, 2000). Such hybrids may be of commercial value where reproduction needs to be restricted for ecological reasons. Sometimes an interspecific hybrid does not exhibit heterosis for any trait but is still quite important for aquaculture application as it expresses a good combination of beneficial traits from both parent species.

Genetic Manipulations in Aquaculture

DNA markers replaced allozymes (Utter, 1991) as the major analytical tool in genetic research in the mid-1980s, following Jeffreys et al.'s (1985) finding of the usefulness of the hypervariability of the genome at the DNA level. Aquaculture and fisheries geneticists quickly adopted them as well (e.g., Wright, 1990; Ward and Grewe, 1995; Wright and Bentzen, 1994). Novel genetic technologies involving the use of DNA-based tools are under development for a range of aquaculture species and applications, including species/strain identification and pedigree tracing (e.g., Heath et al., 1995a, b; Thongpan et al., 1997; Perez-Enriquez et al., 1999; Tan et al., 1999; Norris et al., 2000), confirmation of isogenicity in products of chromosome-set manipulations and clones (e.g., Young et al., 1996; Jenneckens et al., 1999, 2001), linkage mapping (e.g., Alcivar-Warren et al., 1997; Kocher et al., 1998; Young et al., 1998; Liu et al., 1999a, b; Agresti et al., 2000; Li et al., 2000, 2003; McConnell et al., 2000; Sakamoto et al., 2000; Nichols et al., 2003), identification of QTLs (e.g. Jackson et al., 1998; Danzmann et al., 1999; Moore et al., 1999; Palti et al.,

1999; Sakamoto et al., 1999; Alcivar-Warren et al., 2002), and other uses in breeding programmes (Davis and Hetzel, 2000). Knowledge of linkages between marker and QTL alleles may be used for marker-assisted selection (MAS), increasing the rate of genetic progress above that for selective breeding alone (Poompuang and Hallerman, 1997). While holding great promise for advancing breeding programmes, these genomic methods have, so far, only marginal commercial application.

DNA-based mapping technologies have now been developed to the point where a genetic linkage map can be developed for any species within a relatively short period of time, provided that some degree of genetic variation exists in the families being used for the mapping and that sufficient family material is available. These DNA-based markers can subsequently be used as markers for the genetic mapping of genes controlling economically important traits, including ones that show quantitative variation identified as QTLs. DNA-based mapping is highly applicable to aquaculture species that are just undergoing domestication and for which genetic selection programmes are at most at the initial stages. In the last few years, large numbers of molecular markers have been developed and evaluated for application in the culture of decapoda, as well as other commercially important species (Kocher et al., 1998; Liu and Dunham, 1998; Waldbieser et al., 1998, 2001; Liu et al., 1999a, b; Lundin et al., 1999; Palti et al., 1999; Li et al., 2000; Agresti et al., 2000; Norris et al., 2000; Espinosa et al., 2001).

Microsatellite Markers: The Tool

A genetic map is much like an atlas that shows the locations of genes or DNA fragments that have been identified, some of which express themselves as desirable traits for the farmer (Hearne et al., 1992). It is possible, however, to find microsatellites throughout the Crustacea genome without knowledge of the genome or genetic functions and mechanisms that make the genes operative. When these markers are identified and characterized, each can be used as a signpost to help locate genes that express the desired trait. Markers are related to traits as a function of the distance from the gene responsible for that trait.

In the future a particular class of nuclear markers, short tandem repeats (STR or microsatellites), may maximize cost-benefit ratio for most aquatic selective breeding programmes. Besides a large abundance throughout the genome of a target species, microsatellites exhibit high levels of allelic variation. Thus, microsatellites confer more information per unit assay than other marker systems. Microsatellite techniques are

THE CRUSTACEANS

Edible Crustaceans

Genetic work on the main commercially important edible marine decapod crustacean species (in terms of catch rates or aquaculture production) is listed below.

Crabs

Crabs are decapod crustaceans of the infraorder Brachyura (Martin and Davis, 2001). They are a very diverse group found in all oceans. Crabs are omnivores, feeding on algae or any other food, depending on availability. For many crabs, a mixed diet of plant and animal matter results in the fastest growth and greatest fitness (Buck et al., 2003). Most crabs show clear sexual dimorphism and so can be easily sexed (Zrzavý and Štys, 1997). Crabs make up 20% of all marine crustaceans caught and farmed worldwide, with over 1.5 million t being consumed annually (FAO, 2006).

Portunus trituberculatus, the Japanese blue crab, is the most widely fished species of crab in the world, with over 300,000 t being caught annually, 98% of it off the coast of China (FAO, 2006). Because of its commercial importance, the genetic structure of *P. trituberculatus* has been investigated by RFLP analysis with respect to stock identification, management, and conservation. These analyses, however, have failed to reveal unambiguous geographic structure of the local populations, partly because of paucity of the information contained in the RFLPs (Imai et al., 1999). Its entire genome was amplified using a long PCR technique, and the products were subsequently used as templates for direct sequencing using a primer walking strategy. The gene order of *P. trituberculatus* was largely identical to those so far obtained for other arthropods such as *Drosophila yakuba* (Yamauchi et al., 2003). The new sequence data from this crab may be of importance not only for macroevolutionary studies, such as crustacean phylogenies, but also for microevolutionary studies, such as population structure and stock identifications.

Portunus pelagicus, also known as sand crab, is a large crab found in intertidal estuaries of the Indian and Pacific oceans and the middle-eastern coast of the Mediterranean Sea. The species is important commercially and highly prized as the meat is almost as sweet as Japanese blue crab, but it is physically much larger. Allozyme analysis was used to examine the species-level systematics and stock structure of *P. pelagicus*, revealing the presence of two species: one the common *P. pelagicus*, the other most likely either an undescribed "cryptic" species, or *P. trituberculatus*. Discrete subpopulations could be discerned. The results support the current recognition of the three south Australian regions as separate stocks (Bryars and Adams, 1999). Furthermore, seven dinucleotide and one tetranucleotide polymorphic microsatellite loci were isolated and characterized, using routine protocols, from *P. pelagicus*. These loci will be useful for investigating the population structure of this broadly distributed and economically important portunid (Yap et al., 2002).

Callinectes sapidus (blue crab) occurs in the Atlantic Ocean and Gulf of Mexico. Research is currently being conducted by several institutions in order to attempt to replenish the dwindling populations of the blue crab. These projects hope to increase the breeding population of blue crabs by raising hatchery-born larvae to juvenile crabs and releasing them. Given the commercial and ecological importance of the blue crab fishery, there is a surprising scarcity of information concerning the molecular ecology of this species. In recent years record low catches of blue crabs have prompted renewed interest in its biological and ecological characteristics. Early studies of protein polymorphisms (allozymes) suggested similar gene frequencies in some populations but differences between others. Heterozygote deficiencies may reflect the mixing of genetically differentiated subpopulations indicating a single, panmictic population along the Atlantic coast (Burton and Feldman, 1982; Kordos and Burton, 1993; McMillen-Jackson et al., 1994). Nowadays, sequences available online (by electronic database accession numbers) are developed for *C. sapidus* (U75267, AJ130813) (Schubart et al., 2001). McMillen-Jackson and Bert (2004) reported a notably high genetic diversity by investigating the population genetic structure of blue crabs using RFLP analysis of the mtDNA molecule. A mismatch analysis of pairwise haplotype differences indicated that *C. sapidus* experienced a historic period of rapid population expansion. The combination of a distinct cline and the geographically widespread distribution of common and closely related haplotypes suggested that short-term gene flow in blue crabs is regional, whereas long-term gene flow occurs over a long distance. Management of this species should focus on maintaining local blue crab populations and habitats to ensure the long-term sustainability of the species and its fishery.

Two separate screenings of the blue crab nuclear genome were performed, using both dinucleotide and tetranucleotide repeat oligonucleotide probes. Genotyping of a captive-mated pair of blue crabs and their offspring showed that some loci are inherited in true Mendelian fashion, some being monomorphic. Additionally, collected samples of blue crab indicated that a majority of the loci isolated will ultimately be useful markers for population genetic studies (Steven et al., 2005). Place et al. (2005) reported the entire nucleotide sequence for the blue crab mtDNA, which is 16,263 bp in length, circular. Gene order and arrangement is similar to that of many other arthropods (e.g., *Artemia*) but dramatically different from that of the hermit crab, which has a unique gene order among arthropods. Genetic variation is matrilineally inherited on the basis of parent/offspring screening for nucleotide variation in the putative control region.

Charybdis cruciata is found in the Indian and Pacific oceans. It is an edible crab and, because of its large size, high quality of meat, and relatively soft exoskeleton, it has a high commercial value. Attempts are being made to farm it (Parado-Estepa et al., 2003). This species is also known as *C. feriata* and *C. feriatus* (Abelló and Hispano, 2006) and has also been found in the Mediterranean Sea. A PCR-based genomic DNA walking technique was used to clone the gene for the moult-inhibiting hormone of the crab, as *C. feriatus* (Chan et al., 1998).

Analysis of the mitochondrial cytochrome oxidase subunit I sequences of four species of *Charybdis* revealed genetic differences between them and phylogenetic affinities among them. In *C. affinis* and *C. japonica*, no sequence divergence was observed between samples from different estuaries. The two species are closely related but genetically distinct and may represent sibling species (Chu et al., 1999). Ryu et al. (1999) determined the full sequence of the ribosomal DNA intergenic spacer of *C. japonica* by long PCR for the first time in crustacean decapods. Mitochondrial DNA sequences were also used to identify an invasive swimming crab found in New Zealand. This was the first record of *C. japonica* establishing populations outside its native range (Smith et al., 2003). Full-length apoCr cDNA sequences, annotated as Vtg in the GenBank™/EBI Data Bank, are currently available for 11 decapod species including *Penaeus monodon, Litopenaeus vannamei, Charybdis feriatus* and *Portunus trituberculatus* (Avarre et al., 2007).

Cancer pagurus, the edible crab, is found in the North Sea, north Atlantic and the Mediterranean Sea. Its natural lifespan is about 20 years, but few individuals reach this age because of heavy commercial exploitation. Edible crabs are heavily exploited commercially throughout their range. In order to study the structural organization of the edible crab's MIH and MO-IH genes, a genomic DNA library was constructed from DNA of an individual female crab and screened with both MO-IH and MIH probes. Southern blot analysis indicated that in *C. pagurus* there are approximately two copies of the gene for MO-IH and three copies of the MO-IH genes. Phylogenetic analysis and gene organization show that MO-IH and MIH genes are closely related. Their relationship suggests that they represent an example of evolutionary divergence of crustacean hormones (Tang et al., 1999; Lu et al., 2000).

Cancer magister (Dungeness crab) is a species that inhabits eelgrass and sea beds from the Aleutian Islands in Alaska to Santa Cruz, California. They are a popular delicacy and are the most commercially important crab in the Pacific Northwest (FAO, 2006). *Cancer magister* is the focus of one of the most intensely harvested fisheries in North America. Given its economic importance, there is considerable interest in assessing the degree and spatial pattern of its genetic structure. To that end, Toonen et al. (2004) developed a series of hypervariable microsatellite loci. Also reported was the inheritance of such loci in two families obtained from artificial crosses in the laboratory (Jensen and Bentzen, 2004). The ability to multiplex multiple loci greatly increases the ease and speed of genotyping for this species. Furthermore, the mitochondrial genetic variability among Dungeness crabs was assessed in order to more closely examine the dispersal capabilities of the species, which will ultimately aid in its effective conservation. A ~315 base pair region of the mitochondrial COI gene was isolated and sequenced. Genetic diversity indicated a high dispersal capability among Dungeness crabs, allowing the species to retain high amounts of gene flow between populations (Lardy and Parr, 2006).

Menippe mercenaria, the Florida stone crab, is found in the western-north Atlantic and the Gulf of Mexico and is widely caught for food. The species was heavily exploited in the past; its popularity has grown again in recent years and it is considered to be one of the best varieties to eat. Electrophoretically detectable variation was used to determine the evolutionary relationships of crabs of the genus *Menippe* (Xanthidae). Allele frequencies (= genotype) showed that *M. mercenaria* is probably a taxonomic supergroup composed of two taxa (semispecies). The taxa appear to have hybridized in discrete regions (Bert, 1986). Amplification from whole genomic DNA by PCR with oligonucleotide primers based on conserved portions of large-subunit mitochondrial rRNA genes amplified two products of similar length in stone crabs. This was the first report of a mitochondrial gene sequence translocated into the nuclear genome of a crustacean (Schneider-Broussard and Neigel, 1997). Evolutionary relationships among stone crabs were investigated by comparisons of restriction sites within anonymous nuclear DNA sequences and nucleotide sequences of both mitochondrial and a duplicated nuclear form of the mitochondrial large subunit ribosomal RNA (LSrDNA) gene. The survey failed to reveal differences between *M. adina* and *M. mercenaria*. However, because identical sequences are shared by the two species, these data are also compatible with a more recent common ancestry, suggesting an alternative scenario based on relatively recent events and ongoing, rather than historical, gene flow (Schneider-Broussard et al., 1998).

Scylla serrata (mud crab) is an economically important crab species found in the estuaries and mangroves of Africa, Australia and Asia. These crabs are highly cannibalistic in nature. They are among the tastiest crab species and have a huge demand in the South Asian market. Gopurenko et al. (1999) investigated the phylogeographic distribution of *S. serrata* mitochondrial DNA haplotypes sampled throughout the species range. The fact that many locations contain a single unique haplotype suggested limited contemporary gene flow between trans-oceanic sites, and that recent historical episodes of population founding and retraction have both determined and affected the current distribution of *S. serrata* populations. However, its population genetic structure was also examined and analysed for mutational differences at a mitochondrial coding gene (COI) (Gopurenko and Hughes, 2002). Analysis of molecular variance indicated mitochondrial haplotypes to be structured regionally. These results could indicate that gene flow might be reduced, even between geographically close sites, despite the high potential of the species for dispersal (Fratini and Vannini, 2002). Gopurenko et al. (2002) identified and characterized polymorphic microsatellite loci from the genome of *S. serrata*. The loci were detected by randomly screening for di- and tri-nucleotide repeat units within a partial genomic library developed for the species. These markers may provide genetic information that will be useful for both aquaculture and studies of natural populations of the genus. Levels of diversity at two independent genetic markers (mtDNA, microsatellites) were also compared to obtain relative estimates of effective population sizes of the species between colonist and suspected source population(s). Differences in diversity among nuclear and mitochondrial loci may reflect different responses to the colonization process (Gopurenko et al., 2003). Furthermore, RAPD-PCR detected high levels of genetic *S. serrata* population variation with a low geographic bias (Davis et al., 2003).

Karyology and RFLP of 12s and 16s rRNA mitochondrial PCR products failed to differentiate two species of mud crabs: *S. serrata* and *S. tranquebarica* (Gopikrishna and Shashi-Shekhar, 2003). Nevertheless, RAPD-PCR detected large genetic differences between three mud crab species, whereas those between populations within each species were much lower. No genotypes were shared among these three species (Klinbunga et al., 2000). The first internal transcribed spacer (ITS-1) of nuclear ribosomal DNA and mt DNA 16S rRNA were amplified by PCR using genomic DNA extracted from adult tissue of four species of *Scylla* spp. These genetic markers can be used for hybridization breeding studies and in field studies of larval and juvenile mud crabs of the genus *Scylla* (Imai et al., 2004).

Chionoecetes is a genus of crabs that live in the cold waters of the northern Pacific and Atlantic oceans. Marketing strategies describe anything in this genus as snow crabs. Urbani et al. (1998) developed primers for detecting allelic variation at several microsatellite loci using PCR in the snow crab *C. opilio* and evaluated their utility in other crab species. Bunch et al. (1998) compared 413 nucleotides of the first subunit of the mitochondrial cytochrome oxidase c gene of *C. bairdi* crabs from coastal waters of southern Alaska. They revealed low haplotype diversity; crab populations were apparently a mixture of both local and upstream haplotypes. Declining stocks of crabs in western management areas are the result of overharvesting in eastern, upstream populations. Novel polymorphic microsatellite loci were isolated from *C. opilio* by screening an enriched genomic library using non-radioactive PCR techniques, and the polymorphisms were examined to estimate genetic variability. The high variabilities revealed suggest that these microsatellite loci should provide useful markers for genetic variation monitoring of *C. opilio* (Puebla et al., 2003; An et al., 2007).

Maja squinado is a migratory crab found in the north-east Atlantic and the Mediterranean Sea. It is the subject of commercial fishery, with over 5,000 t caught annually (FAO, 2006). Neumann (1998) established that Mediterranean spider crab populations are separable from those of the Atlantic by morphological and biometrical characters justifying the recognition of two different species: the Mediterranean *M. squinado* and the Atlantic *M. brachydactyla*. The spiny spider crab *M. brachydactyla* in an important fishery resource throughout its distribution range. Sotelo et al. (2007) recently described the isolation of nine polymorphic microsatellite loci for this species. Some of these loci seem to be better fitted by a multi-step substitution process.

Carcinus maenas is a common littoral crab and an important invasive species. It is native to European and north African coasts as far as the Baltic Sea in the east and Iceland and central Norway in the north and is one of the commonest crabs throughout much of its range. In the Mediterranean Sea, it is replaced by the closely related species *C. aestuarii*. In view of its potentially harmful effects on ecosystems, efforts have been made to control introduced populations of *C. maenas* around the world. *Carcinus maenas* is fished on a small scale in the north-east Atlantic Ocean, with approximately 1,200 t being caught annually, mostly in France and the United Kingdom (FAO, 2006).

Allozyme genetic differentiation among populations of *C. maenas* and *C. aestuarii* was studied in various marine and brackish habitats (Bulnheim and Bahns, 1996; Nissen et al., 2005). Low genetic population variability was detected, but in general, a slight clinal variation in allele frequencies was reported for *C. maenas* coinciding with decreased latitude. Patterns of morphological variability in this species were largely determined by the local environmental conditions; local factors could have a within-generation selective influence on mean trait values or the species may exhibit phenotypic plasticity (Brian et al., 2006). Examination of 16S mtDNA sequences and the presence/absence of a diagnostic Alu I restriction site were used to distinguish between Atlantic and Mediterranean haplotypes of *C. maenas* (Geller and Bagley, 1997).

Geller et al. (1997) reported the cryptic multiple invasion of European crabs of the genus *Carcinus* in five regions globally. Partial 16S ribosomal RNA gene sequences confirmed sibling species status of morphologically similar Atlantic *C. maenas* and Mediterranean *C. aestuarii*. Grosholz et al. (2000) have used molecular phylogeographic methods to understand the pathways of invasion as well as changes in population genetic structure that accompany these invasions. Apparently, each invasion was accompanied by a dramatic loss of genetic diversity in comparison to the native range (Roman, 2006; Petersen, 2006). In order to examine the population structure of this global invader in its native range, Roman and Palumbi (2004) analysed a 502-base-pair fragment of the mitochondrial COI gene. A clear genetic break occurred between the Mediterranean and Atlantic populations, supporting the species-level status of these two forms.

Tepolt et al. (2006) characterized a series of polymorphic microsatellite loci in *C. maenas* and its sister species *C. aestuarii*. Using microsatellite DNA to analyse molecular genetic variation of those invading populations can potentially enable the identification of source populations, an estimation of the number of individuals founding the invasion, and an assessment of changes to population dynamics after the initial invasion (Bagley and Geller, 2000).

Lobsters

Lobsters are economically important seafood considered a delicacy around the world. The world catch of lobsters recorded (FAO, 2006) exceeded 205,000 t, of which about 127,000 t corresponded to true lobsters (Family Nephropidae), about 78,000 t to spiny lobsters (Family Palinuridae) and about 2,100 t to slipper lobsters (Family Scyllaridae). Species of genera such as *Homarus* (about 64,000 t), *Jasus* (about 14,000 t) and *Panulirus* (about 56,000 t) form the subject of specialized fisheries and are the basis for important industries. Other species (e.g., *Nephrops*, *Metanephrops* and *Palinurus*) often form an important part of mixed catches (e.g., with shrimps), and are sold separately on markets. Several of the deep-sea species need specially equipped ships for their capture, and at present most are not commercially exploitable because of the high operating costs, but better knowledge of their biology and ecology might make them of commercial interest in the future.

Homarus americanus, the American lobster, is found on the Atlantic coast of North America. American lobsters are a popular food, very low in fat, but not suitable for low sodium diets.

RAPD-DNA profiles for genetic relationships among *H. americanus* from ecologically disparate and geographically separate regions were tested (Harding et al., 1997). Phenotypic analyses of the RAPD bands showed no significant difference between samples at these geographic locations; this is not surprising given the enormous potential for larval dispersal, the wide ranging movements of adult lobsters within regions, and the level of anthropogenic interference through displacement of both larvae and adults over the past century in the name of conservation. The average genetic variability in European and American lobster populations appear to be equivalent. It is clear that a small but significant genetic divergence separates the European and American lobster. This is consistent with hypotheses relating genetic variability to adaptive strategy. The apparent weakness of reproductive isolating barriers suggests that these populations have evolved allopatrically (Hedgecock et al., 1997). A quantification of genetic differences in the species, together with recent successes in interspecific laboratory matings, implicates species hybridization as a potentially important breeding practice in lobster aquaculture (Hedgecock et al., 1997).

Despite its commercial importance, DNA genetic markers for American lobster (*H. americanus*) are limited. Gosselin et al. (2005) studied the frequency of multiple paternities using four microsatellite loci for the American lobster at three sites differing in exploitation rate and mean adult size. Multiple paternities were observed at the most exploited sites, whereas single paternity was observed at the least exploited one. Jones et al. (2003) developed 12 tetra- and 1 trinucleotide microsatellite loci for American lobster that exhibit little stuttering after PCR amplification. They also tested inheritance and pairwise linkage using embryos from each of two females demonstrating the occurrence of multiple paternities. An apparent occurrence of dispermic androgenesis was observed. Furthermore, eight haplotypes were detected when quantifying mtDNA sequence variation through the use of PCR amplification of two DNA fragments, followed by digestion with restriction enzymes. However, when microsatellite allele frequencies were used to determine genetic differences between *H. americanus* groups in a given area (western Long Island Sound), greater differences were detected that could not be attributed to geographical separation or the earlier lobster fishery collapse (Crivello et al., 2005a, b). Hodgins-Davis et al. (2007) developed eight polymorphic short sequence repeats (SSR) for the American lobster, which were derived from expressed sequence tags.

Homarus gammarus, the European lobster, is a large clawed lobster. Its natural range is the eastern Atlantic Ocean from northwestern Norway to the Azores and Morocco. It can also been found in the Mediterranean Sea west of Crete and in the northwestern parts of the Black Sea, but not in the Baltic Sea. The European lobster is fished throughout its natural range, but on a smaller scale than the American lobster, although it is considered by connoisseurs to have a finer texture and flavour. Most of the fishery is carried out with lobster pots. Although attempts have been made to run lobster farms, they proved to be not feasible because of the lobsters' aggressively territorial habits. FAO (2006) reported an annual catch of over 2,000 t.

Allozymes were used to screen genetic variation in different lobster populations (Jørstad and Farestveit, 1999; Jørstad et al., 2005a). Minor but statistically significant variation in allele frequencies was detected. Despite generally low levels of genetic differentiation, the tests for population differentiation revealed highly significant values for all loci investigated. By means of RAPD analysis, the degree of its genetic variation was also assessed by Ulrich et al. (2001). All populations investigated showed high values of genetic distance, indicating that genetic exchange among European populations of *H. gammarus* may be restricted. The European lobster populations appeared to be isolated by distance since the pattern of genetic relationships reflects their geographic location. A distinct genetic population structure, possibly maintained by genetic drift within relatively isolated local populations and restricted gene flow between them, was suggested. This contrasts with the homogeneous population structure of the American lobster, *H. americanus*. A more detailed investigation including supplementary extensive sampling from Norway and additional allozyme, microsatellite and mtDNA analyses was reported by Jørstad et al. (2004). The results supported the genetic distinctness of geographically close lobster populations, which may be due to the local hydrological conditions, preventing larval dispersal between the fjord systems. Triantafyllidis et al. (2005) showed that the Aegean population of *H. gammarus* differ significantly from the Atlantic ones, as well as those from the Adriatic and western Mediterranean, based on haplotype frequencies and F_{ST} of a 3-kb mtDNA segment. F_{ST} analyses also showed significant heterogeneity among the European lobster population samples. Based on the low degree of differentiation revealed in the European lobster and its limited capacity for dispersal, the most probable hypothesis is that all populations were established from a common refuge after the end of the last Ice Age, that is, within the past 15,000 years.

Lobster stock enhancement is based on selection of wild broodstock and artificial production of juveniles tagged with coded wire that are released into the natural environment (Burton, 2001; Castro et al., 2001; Beal et al., 2002). Several experiments were made on the communal-rearing approach in which the performance of mixed larval groups (families) was evaluated under identical conditions (Jørstad et al., 2005b). Berried females of wild and cultured origin and their respective fertilized eggs were screened by using microsatellite DNA profiling, thereby allowing determination of both parental genotypes. Significant differences in early survival between offspring of wild and cultured origin were detected, suggesting a genetic component for these traits and a potential for selective breeding. Future activities aiming for local stock enhancement should evaluate the risk of unwanted genetic impacts. Commercial ranching operations of lobster, including selective breeding, should be carried out only in areas with low levels of genetic differentiation (Jørstad and Farestveit, 1999).

Nephrops norvegicus, the Norway lobster, is found in the north-eastern Atlantic Ocean and North Sea as far north as Iceland and northern Norway, and south to Portugal and the Mediterranean Sea. The Norway lobster is an important fishery species, mostly by trawling. Around 60,000 t are caught annually, half in UK waters (FAO, 2006).

Allozyme assays of *N. norvegicus* (Passamonti et al., 1997; Maltagliati et al., 1998; Stamatis et al., 2006) revealed three main characteristics of their genetic structure: a highly homogeneous allele pattern, a strong heterozygote deficiency at some loci, and a significant genotypic variance among populations. Population subdivision, intra-unit inbreeding, bottlenecks and selective constraints were taken into account to allow for these genetic features, although none of them proved decisive. Moreover, no clear clines in allelic frequencies were detected; thus, genetic variability seemed to be randomly distributed among populations and *N. norvegicus* seems to follow the island model of genetic structure. To assess the level and spatial pattern of *N. norvegicus*, hypervariable molecular markers were obtained by screening of a partial genomic library enriched for microsatellite dinucleotide motifs by Streiff et al. (2001). Consequently, Streiff et al. (2004) investigated its genetic diversity and mating system in a natural population off the Portuguese coast. High genetic diversity was observed. Paternity within broods was analysed by comparing multilocus genotypes of each egg with the corresponding mother, and the male parent contribution was then deduced. Multiple paternities were observed. When multiple paternities were involved, the comparative reproductive success of the male parents was quite even.

RFLP analysis of two mtDNA segments was performed on 12 populations of Norway lobster (Stamatis et al., 2004). Low levels of differentiation were found among them and there were no signs of an Atlantic-Mediterranean divide or of an isolation-by-distance scheme of differentiation. Possible reasons for these low levels of differentiation can be found in the recent expansion of *N. norvegicus* populations. With a combination of reverse transcription-polymerase chain reaction (RT-PCR) and rapid amplification of cDNA ends approaches, Edomi et al. (2002) determined the cDNA sequence of *N. norvegicus*.

Panulirus japonicus is an important fisheries resource in Japan, with an annual catch reaching 1,600 t (FAO, 2006). Because of its commercial importance, many studies have been conducted regarding larval culture, larval distribution and migration process, reproductive biology, and resource management. Consequently, there has been growing interest in genetic structure of *P. japonicus* with respect to stock identification, management, and conservation (Inoue et al., 2007). Furthermore, DNA information is useful for larval identification, because larvae within *P. japonicus* species-group are morphologically very similar. The complete nucleotide sequence of the mitochondrial genome of *P. japonicus* was determined. The entire genome was amplified using long PCR, and the products were subsequently used as templates for direct sequencing using a primer-walking strategy (Yamauchi et al., 2002; Chow et al., 2006).

Palinurus elephas and ***Palinurus mauritanicus*** are two spiny lobsters commonly present in the Mediterranean Sea. All species of *Palinurus* are of present or potential future commercial interest. In the central and western Mediterranean the species are regularly found at fishmarkets, and in the eastern Atlantic, outside the Mediterranean, they are fished on a minor scale in England, and more intensively in France and Portugal. No catch statisics are known but probably the catches of *Palinurus* spp. reported (Anonymous, 1999) correspond partly to *P. elephas* and partly to *P. mauritanicus* (~5,000 to 8,000 t).

In the last decades, there has been increasing fishing pressure on *P. elephas* and *P. mauritanicus* reaching overexploitation with marked declines in catches that show the need for a new approach to management strategies based on a clear identification of genetic stocks. Sequences of COI gene confirm that they are reliable phylogenetic indicators for spiny lobsters. All methods of phylogenetic reconstruction used consistently recovered the same branching pattern: four major evolutionary lineages were found. The topology of the trees obtained agrees with the accepted classification, showing a clear separation of the three genera *Palinurus, Panulirus* and *Jasus* (Cannas et al., 2006).

Prawns

Prawns are edible, shrimp-like crustaceans, belonging to the suborder Dendrobranchiata. They are distinguished from the superficially similar shrimp by the gill structure, which is branching in prawns but lamellar in shrimp. The sister taxon to Dendrobranchiata is Pleocyemata, which contains all the true shrimp, crabs, and lobsters. Sustainability of penaeid commercial fisheries and shrimp aquaculture industry has been threatened by overfishing, habitat destruction, viral diseases, and chemical pollutants (Rothlisberg, 1998; Naylor et al., 1998, 2000).

Penaeus monodon (giant tiger prawn) is a marine crustacean widely reared for food. The natural distribution is the Indo-West-Pacific, ranging from the eastern coast of Africa and the Arabian Peninsula to as far as South-East Asia and the Sea of Japan. They can also be found in eastern Australia, and a small number have colonized the Mediterranean Sea via the Suez Canal. *Penaeus monodon* is the most widely cultured prawn in the world, although it is gradually losing ground to the whiteleg shrimp, *Litopenaeus vannamei*. Over 900,000 t are consumed annually, two-thirds of it coming from farming, chiefly in South-East Asia.

Allozyme surveys of genetic variation in *P. monodon* populations over wide geographical ranges demonstrated highly significant differences in gene frequencies between them (Lester, 1979; Dall et al., 1990; Benzie et al., 1992; Forbes et al., 1999; Sugama et al., 2002). Klinbunga et al. (1998) observed inter- and intra-individual polymorphisms of rDNA in *P. monodon*. However, DNA-based markers have revealed far greater levels of variation compared with the allozyme data. The isolation of highly polymorphic microsatellite markers in *P. monodon* became useful in population studies and parental determination of the species (Tassanakajon et al., 1998b; Xu et al., 1999; Pongsomboon et al., 2000; Wuthisuthimethavee et al., 2003; Pan et al., 2004). Supungul et al. (2000) and Brooker et al. (2000) indicated high genetic diversity in the species when using such microsatellite loci. RFLP was also used to determine genetic variation and population structure in *P. monodon* and further substantiated the previous findings (Benzie et al., 1993, 2002; Bouchon et al., 1994; Klinbunga et al., 1999). Geographic heterogeneity analysis indicated population differentiation between them. Shekhar and Chandra (2000) amplified mtDNA ribosomal genes by PCR, also suggesting the existence of genetic variation between wild prawn populations. Furthermore, genetic diversity of the giant tiger prawn was examined by RAPD and mtDNA (16S ribosomal DNA and COI-COII) polymorphism. High haplotype and nucleotide diversity were observed (Klinbunga et al., 2001). Wang et al. (2004) explored the use of AFLP technology in species identification and phylogenetic analysis of penaeid shrimp (Lavery et al., 2004; Quan et al., 2004). Successful differentiation of five penaeid shrimps based on PCR-RFLP and SSCP of 16S rDNA has been reported (Vázquez-Bader et al., 2004; Khamnamtong et al., 2005). The segregation of two distinct clades of *Penaeus* as previously documented by mtDNA analysis was evident (Bierne et al., 2000). Finally, microsatellite loci are found in the ribosomal DNA internal transcribed spacer 1 (ITS1) and seem to be associated with sequence divergence and size variation in *Penaeus* species (Wanna et al., 2006), providing a great potential use in population studies.

Half-sib groups of the giant tiger prawn were obtained using artificial insemination of two females by each male. Results suggested large non-additive genetic effects and/or common environmental effects that could not be differentiated on the available data (Benzie, 1997; Benzie et al., 1997). RAPD was used to amplify the genome of tiger prawns by Tassanakajon et al. (1997, 1998a). Analysis of wild populations and broodstocks revealed different levels of genetic variability among them, suggesting a high genetic variability to be used in selective breeding programmes (Benzie, 2000). Cultured populations showed less genetic diversity and were significantly different from the wild populations based on pairwise comparisons of allelic and genotypic frequencies (Xu et al., 2001). *Penaeus monodon* were reared in captivity in tanks over three generations with full pedigree information. The genetic correlations between weight and survival revealed no significant trend, indicating significant concurrent improvements in both growth and survival through selective breeding (Kenway et al., 2006). Wilson et al. (2002) reported the construction of an initial genetic linkage map for the tiger shrimp. Mapping was carded out using polymorphic markers derived from AFLP primer pairs. These were analysed on three reference families of known pedigree. Finally, unique microsatellite-containing expressed sequence tags (ESTs) were designed for *P. monodon* for further analysis of genetic diversity in shrimp breeding programmes (Wolfus et al., 1997; Tong et al., 2002; Klinbunga et al., 2006; Maneeruttanarungroj et al., 2006).

Litopenaeus vannamei (whiteleg shrimp), also known as Pacific white shrimp, is from the eastern Pacific Ocean and is commonly caught or farmed for food. It is the major species of farmed shrimp. Knowledge of its genetic diversity and population structure will be of interest for aquaculture and fisheries management to utilize and preserve aquatic biodiversity.

A fragment of the mitochondrial 12S rRNA-tRNAVal-16S rRNA genes from *P. californiensis* was sequenced and compared with the corresponding regions from *L. vannamei* and *P. stylirostris*. Maximum-likelihood analysis grouped *L. vannamei* and *P. stylirostris* separately from *P. californiensis* (Gutierrez-Millan et al., 2002). Furthermore, the organization and evolution of corresponding α-amylase genes were determined after PCR amplification (van Wormhoudt and Sellos, 2003). Regarding α-amylase gene structure in the shrimp, many recombinants are present from a set of individuals and constitute an important mechanism of evolution of α-amylase function. DNA-based molecular markers for a successful differentiation of five penaeid shrimps (*P. monodon, P. semisulcatus, Feneropenaeus merguiensis, L. vannamei* and *Marsupenaeus japonicus*) were developed on the basis of PCR-RFLP and single-stranded conformation polymorphism (SSCP) of 16S ribosomal (r)DNA (Khamnamtong et al., 2005).

The genetic variation in two farmed strains of *L. vannamei* was examined on the basis of DNA multilocus analyses using a set of VNTR core sequence primers (de Freitas and Galetti, 2002). The results revealed different levels of genetic variation within the strains, although a bottleneck effect could not be discarded. Polymorphic microsatellite loci were characterized for *L. vannamei*, which will be useful for mating analysis in breeding programmes or might prove useful for population genetic studies (Garcia et al., 1996; Bagshaw and Buckholt, 1997; Cruz et al., 2002; Meehan et al., 2003; Zhi-Ying et al., 2006; de Freitas et al., 2007). Genetic variation and population structure of wild white shrimp from different geographic locations were investigated using such microsatellite DNA loci. Genetic differences between localities were detected by pairwise comparison based on allelic and genotypic frequencies. Populations were structured into subpopulations (Valles-Jimenez et al., 2004). However, population differentiation does not follow an isolation-by-distance model. Enzymatic systems were analysed to determine allozyme genetic differentiation among hatchery strains and natural populations of *L. vannamei* (Soto-Hernandez and Grijalva-Chon, 2004). In all samples observed heterozygosity was smaller than expected. High inbreeding values and heterozygote deficiencies were detected in all samples. Additionally, a set of 26 EST-SSRs were evaluated for Mendelian segregation, which showed their utility for population genetic analysis of *L. vannamei* (Perez et al., 2005). *Litopenaeus vannamei* population genetic structure from the eastern Pacific was determined by RFLP analysis of the mtDNA control region (Valles-Jimenez et al., 2006). White shrimp showed high average within-locality haplotype and nucleotide diversities and also high average nucleotide divergence between all pairs of localities. White shrimp does not fit the sudden population expansion model. Population differentiation may be maintained by a combination of physical, oceanographic, and biological factors acting as barriers to gene flow among localities.

To develop genetic and physical maps for prawns, accurate information on the number of chromosomes and a large number of genetic markers is needed. Previous reports have shown two different chromosome numbers for the whiteleg shrimp, one of the most important penaeid shrimp species cultured. Alcivar-Warren et al. (2006) identified a large number of (TAACC/GGTTA)-containing SSRs in *L. vannamei* ovary's genomic library. Arena et al. (2003a) determined whether artificial selection based on body weight and body size affected its adaptation ability to use dietary carbohydrates as a source of energy. No association analyses for either the AMY2 or the CTSL genes with body weight were detected (Glenn et al., 2005). The decrease in genetic variance without an increase in phenotypic variances in an acute response to hypoxia might be related to the known suppression of metabolic pathways that either use or produce ATP, which could result in a decreased expression of additive genes (Pérez-Rostro et al., 2004).

A captive population of *L. vannamei* replicated in two environments was also evaluated for genetic variability and covariability of size traits (Pérez-Rostro and Ibarra, 2003). There was no family by environment interaction for any of the traits. However, both fixed effects, sex and environment, were significant. Heritability estimates indicated that such traits can be subjected to selection to improve reproductive output from females (Arcos et al., 2004; Gitterle et al., 2005). Furthermore, genetic diversity in a shrimp-breeding programme was monitored for two generations by microsatellite DNA markers to establish levels of variation and proceed with a selection programme. Most common alleles and high heterozygosities were maintained through the generations, indicating that there had not been a significant loss of genetic variability in the breeding programme. However, it is recommended that additional genetic variability should be introduced to a breeding stock by crossing it with a different line (Rivera-García and Grijalva-Chon, 2006; Cruz et al., 2004). Genetic variation among and within five generations of an inbred commercial captive line of *L. vannamei* and genetic distance among them were evaluated by RAPD using descriptive and genetic similarity analyses for dominant markers at single- and multi-populational level (de Freitas and Galetti, 2005). Identity and genetic distance analyses indicated high genetic homogeneity within and between broodstock lineages, which suggested the similar genetic structure of Brazilian foundation broodstocks (Gonçalves et al., 2005; de Francisco et al., 2005).

Litopenaeus setiferus (Atlantic white shrimp), also known as white shrimp or green tails, is a prawn species found on the eastern seaboard of North America. The annual average US catch for white shrimp is around 47,000 t. It is sold in a variety of fresh and frozen products, either whole body or tails.

The genetic variation of white shrimp was assessed by using highly polymorphic microsatellite loci (Ball et al., 1998; Ball and Chapman, 2003). Pairwise tests of the similarity of allele frequency distributions and distance measure analyses showed broad-scale genetic homogeneity superimposed over occasional indications of random geographical and temporal differentiation. This large-scale genetic homogeneity was attributed to a consequence of genetic mixing resulting from pelagic larvae and adult migrations. Genetic differentiation and variability data of two shrimp species populations (*L. setiferus* and *L. schmitti*) are also available by electrophoretic analyses and by analysis of 16S mtDNA (Arena et al., 2003b). The greatest differentiation was found within, rather than between, populations. A high similarity was detected between these two species with a bimodal distribution of the loci with respect to genetic identity.

McMillen-Jackson and Bert (2003) compared the population genetic structures and phylogeographies of brown shrimp (*Farfantepenaeus aztecus*) and white shrimp (*L. setiferus*), two sympatric penaeids of the eastern United States, using sequence analysis of the mtDNA control region. Brown shrimp showed no significant phylogenetic structure or population subdivision, and closely related haplotypes were geographically dispersed. In contrast, white shrimp had a complex haplotype phylogeny consisting of two distinct lineages and two less well-defined sublineages, and the haplotypes and lineages were geographically structured. These disparate patterns may have developed as a result of species-specific differences in physiological tolerances and habitat preferences that caused greater fluctuations in white shrimp population sizes and reductions in long-term effective population size relative to that of the brown shrimp, and thereby increased the susceptibility of the white shrimp populations to stochastic genetic change.

Penaeus japonicus, the kumura prawn, is an important Indo-West Pacific species ranging widely from South Africa into the Red Sea, Japan and northern Australia. It has migrated through the Suez Canal into the Mediterranean and is now caught off southern Turkey. In Japan its fishery is of major importance; it not only is trawled but also plays a vital role in pond culture. It is highly regarded in Japan, where it is used head-on fresh as well as in more usual processed forms.

The genetic structure and variability of natural populations of kuruma prawn in Japan was examined by estimating relatedness among individuals by microsatellite DNA markers and PCR-RFLP analysis (Sugaya et al., 2002a). No spatial differentiations occurred because of the geographic or historical effects between the localities and there was a possibility of a mixture of hatchery populations. Sequence analyses on the complete mtDNA control region (992 bp) were conducted to elucidate its population structure in eastern Asia. F_{ST} values showed significant differences. AMOVA showed clear genetic differences between populations (Tzeng et al., 2004a). AFLP analysis confirmed that two morphologically similar varieties of *P. japonicus* are genetically distinct, suggesting that these two varieties have different geographical distribution (Tsoi et al., 2005). Moreover, a phylogenetic study revealed that the two varieties are more closely related to each other than to the other phylogenetically related *Penaeus* species (cryptic species) (Tsoi et al., 2007). Further studies on the genetic structure of the species complex are needed not only to understand its evolutionary history, but also to improve the knowledge-based fishery management and aquaculture development programmes of this important biological resource.

Microsatellite and AFLP DNA markers have been characterized in establishing pedigrees, linkage mapping and identifying QTLs influencing commercially important traits in *P. japonicus* (Moore et al., 1999). The allelic inheritance mode of microsatellite DNA markers was examined using mated wild females and their offspring (Sugaya et al., 2002b). At almost all family/locus combinations, one sire was determined and distributions of genotypes in offspring were consistent with the Mendelian segregation ratio. Li et al. (2003) used AFLP markers in a two-way pseudo-testcross strategy to generate genetic maps of a *P. japonicus* family. A two-stage selective genetic mapping strategy was applied. The sex marker initially mapped on the maternal parent map was also confirmed in the second-stage mapping with more progeny information. The linkage data developed in these maps proved useful in detecting loci controlling commercially important traits in a full-sib family (Li et al., 2006). The magnitude of phenotypic variation explained by the joint action of the QTL markers indicated that genetic factors of large effect were involved in the control of these traits. Moreover, Keys et al. (2004) quantified the effects of inbreeding on growth and survival (and associated yield) of domesticated *P. japonicus* in selective breeding schemes. Although a level of inbreeding can be tolerated by penaeids it can cause a degree of inbreeding depression and reduced production (Li and Wu, 2003). Microsatellites, coupled with DNA parentage analyses, were also used to determine the relative performance of 22 families reared in commercial production ponds (Jerry et al., 2006). Parentage analyses based on individual genotypic data demonstrated that some families were over-represented, while others were under-represented owing to slower growth rates. The results also revealed some weak, but significant, male genotype × environment (G × E) interactions in the expression of shrimp growth for some families.

Shrimps

True shrimps are small decapod crustaceans classified in the infraorder Caridea, found widely around the world in both fresh and salt water. They are basically benthic and swim only occasionally. A number of more or less unrelated crustaceans also have the word "shrimp" in their common name.

Crangon crangon (brown shrimp, common shrimp and sand shrimp) is a commercially important species fished mainly in the southern North Sea, although it is also found in the Irish, Baltic, Mediterranean and Black seas, as well as off much of Scandinavia and parts of Morocco's Atlantic coast. The FAO (2006) reported catches over 37,000 t of *C. crangon* in the North Sea. Population structure of this important commercial brown shrimp was investigated using a multidisciplinary approach based on morphometrics, allozymes and AFLPs (Bulnheim and Schwenzer, 1993; Beaumont and Croucher, 2006). Multivariate analysis and linear discriminant analysis of the morphometric data revealed variation in its tail shape. Allozyme allele frequencies showed evidence of population subdivision. Data from AFLP loci confirmed population differentiation. Random genetic drift and selection in populations with restricted gene flow are likely to be the main causes of the differentiation observed.

Pandalus borealis is found in cold parts of the Atlantic and Pacific oceans. It is an important food resource and has been widely fished since the early 1900s in Norway. Shrimp alkaline phosphatase, an enzyme used in molecular biology, is obtained from this species. Allozyme analyses revealed low mean heterozygosity and proportion of polymorphic loci, as is common for decapods in general. However, significant genetic differences were reported among shrimps inhabiting different areas, especially between inshore and offshore ones (Kartavtsev et al., 1992a, b, c; Jónsdóttir et al., 1998; Sévigny et al., 2000; Drengstig et al., 2000). Martinez et al. (2006) postulated that the large genetic variability found at an individual level, when using RAPD analysis as a molecular tool, may provide the total *P. borealis* population with a diverse genetic pool from which traits can be selected to respond to variations in local environmental conditions, and that this local selection may be the cause of the subpopulation structure observed.

Aristeus antennatus, the blue shrimp, is encountered in deep-water beds. It is a highly esteemed commercial species, extensively exploited in the western and central Mediterranean Sea. The state of the stocks in the Mediterranean is not known. Populations of *A. antennatus* from different areas of the Mediterranean Sea and adjacent Atlantic waters were subjected to morphometric and electrophoretic analyses but could not be differentiated (Sarda' et al., 1998). Historical but local records in some areas indicate that these resources show large fluctuations in stock abundance. The commercial importance of this species, as well as other by-catch deep-water species, makes such deep-water resources of increasing interest.

Crustacean Farming

Crustacean farming is a rapidly expanding aquaculture industry. In 2000 the farmed production was 1,300,000 t, representing about 40% of the total production. Crustacean farming established a rapidly growing activity and its production reached 54% of the total world crustacean production (Gusmao et al., 2000; Keys et al., 2004; Wanna et al., 2004). A major weakness of this industry is that domestication has not been fully achieved in many of the species cultured. Devastating disease epidemics in shrimp farms worldwide have demonstrated that genetic improvement and other biotechnology applications are crucial to the future development of this industry.

As the crustacean aquaculture industry aims to meet growing world demand, there is a need to dramatically increase production. With the increasing prevalence of disease and limited supplies of wild broodstock, domestication and selective breeding approaches are being increasingly used to improve production efficiency. The natural populations of crustaceans are under pressure from intensive fishery and the introduction of cultured exotic species (Maggioni et al., 2003; Borrell et al., 2003, 2004). In recent years, a decrease has been noticed in the capture of wild post-larvae for stocking because of a gradual increase in the market of laboratory-produced post-larvae. The continuous use of wild post-larvae and adults for broodstock has transferred pathogens, which has caused difficulties in farming domesticated crustaceans. The high fecundity of some crustaceans, together with their short generation time and reasonable response to selection, provides an opportunity for enhancing production in most commercial species. The domestication process selects crustaceans better adapted to the artificial environment in which they are grown and allows the establishment of genetic improvement programmes coupled with selective breeding programmes under biosecure conditions, in order to provide an opportunity for enhancing production in most commercial species. These genetic improvement and selective breeding programmes aim at increasing growth, survival, and resistance to diseases (Soto-Hernández and Grijalva-Chon, 2004). Through selection of performance traits such as growth, survival and disease resistance, selective breeding programmes can produce reliable supplies of healthy seedstock with improved production performance (Keys et al., 2004).

Assessment of genetic resource has been considered a desirable starting point for the development of breeding and genetic improvement programmes of reared species, as well as assessment of new species with potential for culture (Velez et al., 1999). The growing interest in establishing crustacean culture farms along the littoral zone of tropical and subtropical countries has motivated studies on the natural populations. The crustacean aquaculture industry still relies largely on wild broods to seed farmed populations. Consequently, gravid female brooders have been transported between regions of high aquaculture intensity, in particular southeastern Asian countries and Australia. The exchange of female brooders creates a situation that is prone to the admixture of local and introduced populations once cultured specimens escape (Pan et al., 2004). The power of artificial selection to significantly improve animal performance in culture could be counterbalanced by a rapid growth in the inbreeding coefficient, which can affect fitness-related traits such as survival, reproduction, growth, presence of deformities, and a decrease of genetic variability, resulting from careless genetic management of the populations. The negative effects of these traits could also have a significant negative impact on production in the form of inbreeding depression (Donato et al., 2005).

Compared to fish production, the annual yield of cultured crustaceans is relatively small, currently measuring about 1.4 million t. This is a sector that continues to grow, consistently and substantially, although some fluctuations are observed during the process. A concurrent growth of crab culture, mainly a fattening process, has occurred, particularly in Asia, which has assisted the overall growth of crustacean aquaculture. Marine shrimp culture dominates crustacean culture, representing 96% in brackish water and 73% of all crustacean aquaculture (Rosenberry, 1998). Tiger prawn, *P. monodon*, contributes over 50%, followed by the whiteleg shrimp, *L. vannamei* (18%), and the oriental or fleshy prawn, *P. chinensis* (10%). Shrimp culture is essentially confined to Asia and South America and, interestingly, the production share of the latter has continued to increase steadily throughout the decade, rising from around 15% to nearly 20% of global production. It is envisaged that Africa may become an important player in this sector in the future.

The application of biotechnology to crustacean aquaculture is increasing in importance. Early doubts that such approaches were unnecessary, given the crude state of development of this aquaculture industry, have been replaced by the recognition that sophisticated molecular approaches will be required to tackle key problems of disease recognition and control, and to achieve higher production through the development of domesticated strains. Progress in penaeid genetic and biotechnological research has been slow because of a lack of knowledge on fundamental aspects of their biology. However, data are beginning to emerge from research projects started in the last decade and are likely to increase rapidly in the future. Highly variable markers have been developed that allow the genetic variation in crustacean stocks to be assessed, even in highly inbred cultured lines. The extent to which growth rate is under genetic control has now been assessed in some species using rigorous experimental designs to estimate heritabilities. Highly sensitive techniques that allow the isolation and characterization of very small quantities of peptides are being used to investigate the endocrine control of reproduction, and work has begun on the isolation and characterization of genes that play an important role in growth and reproduction. Greater attention is being paid to developing cell lines for crustaceans and to the development of molecular probes for the identification of pathogens. These tools will provide a basis for a more detailed description of the penaeid genome, for understanding disease resistance, and for developing effective control of reproduction and the development of disease-free or disease-resistant domesticated lines.

These newly emerged molecular techniques are beginning to transform the research and practice of animal agriculture and will be instrumental in developing profitable, environment-friendly agriculture and in meeting the demands of growing world populations. The complete genomic sequences and DNA microarrays featuring the whole genome of major farm animal species are expected to be available in the near future. Novel approaches to enhancing animal productivity and health will be developed, and new biotherapeutics and new animal breeds with desired characteristics will be produced. In addition, transgenic farm animals hold great promise to provide much needed pharmaceuticals and immunologically suitable organs for transplantation. However, it should be kept in mind that use of transgenic animals for the production of foods and pharmaceuticals and for xenotransplantation will require extensive characterization before release to the market. It is our scientists' responsibility to ensure the safety of genetically modified animals and animal products, and also to educate the general public for its increased acceptance of animal biotechnology (Exadactylos and Arvanitoyannis, 2006).

References

Abelly, P., and C. Hispano. 2006. The capture of the Indo-Pacific crab *Charybdis feriata* (Linnaeus, 1758) (Brachyura: Portunidae) in the Mediterranean Sea. Aquatic Invasions, 1: 13-16.

Agerberg, A. 1990. Genetic variation in three species of freshwater crayfish, *Astacus astacus* L., *Astacus leptodactylus* Aesch. and *Pacifastacus leniusculus* (Dana), revealed by isozyme electrophoresis. Hereditas, 113: 101-108.

Agresti, J.J., S. Seki, A. Cnaani, S. Poompuang, E.M. Hallerman, N. Umiel, G. Hulata, G.A.E. Gall and B. May. 2000. Breeding new strains of tilapia: development of an artificial center of origin and linkage map based on AFLPs and microsatellite loci. Aquaculture, 185: 43-56.

Albrecht, H., and H.O. Von Hagen. 1981. Differential weighting of electrophoretic data in crayfish and fiddler crabs (Decapoda; Astacidae and Ocypodidae). Comp. Biochem. Physiol., 70B: 393-399.

Alcivar-Warren, A., R. Overstreet, A.K. Dhar, K. Astrofsky, W. Carr, J. Sweeney and J. Lotz. 1997. Genetic susceptibility of cultured shrimp (*Penaeus vannamei*) to Infectious Hypodermal and Hematopoeitic Necrosis Virus and *Baculovirus penaei*: Possible relationship with growth status and metabolic gene expression. J. Invert. Pathol., 70: 190-197.

Alcivar-Warren, A., Z. Xu, D. Meehan, Y. Fan and L. Song. 2002. Shrimp genomics: development of a genetic map to identify QTLs responsible for economically important traits in *Litopenaeus vannamei*. *In:* Aquatic Genomics: Steps Toward a Great Future. N. Shimizu, T. Aoki, I. Hirono and F. Takashima (eds.). Springer-Verlag, Tokyo, pp. 61-72.

Alcivar-Warren, A., D. Meehan-Meola, Y. Wang, X. Guo, L. Zhou, J. Xiang, S. Moss, S. Arce, W. Warren, Z. Xu and K. Bell. 2006. Isolation and mapping of telomeric pentanucleotide (TAACC)n repeats of the pacific whiteleg shrimp, *Penaeus vannamei*, using fluorescence in situ hybridization. Mar. Biotechnol., 8(5): 467-480.

Amundson, R. 2001. Adaptation and development: on the lack of common ground. *In:* Adaptationism and Optimality. S.H. Orzack and E. Sober (eds.). Cambridge University Press, New York, pp. 303-334.

Amundson, R. 2005. The Changing Rule of the Embryo in Evolutionary Biology: Structure and Synthesis. Cambridge University Press, New York.

An, H.S., J.H. Jeong and J.Y. Park. 2007. New microsatellite markers for the snow crab *Chionoecetes opilio* (Brachyura: Majidae). Mol. Ecol. Notes, 7: 86-88.

Andersson, L., C.S. Haley, H. Ellegren, S.A. Knott, M. Johansson, K. Andersson, L. Andersson-Eklund, I. Edfors-Lilja, M. Fredholm, I. Hansson, J. Håkansson and K. Lundström. 1994. Genetic mapping of quantitative trait loci for growth and fatness in pigs. Science, 263: 1771-1774.

Anonymous. 1999. National Research Council Sustaining Marine Fisheries (National Academy Press, Washington DC).

Arcos, F.G., I.S. Racotta and A.M. Ibarra. 2004. Genetic parameter estimates for reproductive traits and egg composition in Pacific white shrimp *Penaeus* (*Litopenaeus*) *vannamei*. Aquaculture, 236(1-4): 151-165.

Arena, L., G. Cuzon, C. Pascual, G. Gaxiola, C. Soyez, A. van Wormhoudt and C. Rosas. 2003a. Physiological and genetic variations in domesticated and wild populations of *Litopenaeus vannamei* fed with different carbohydrate levels. J. Shellfish Res., 22(1): 269-279.

Arena, L., M. Montalvan, G. Espinosa, G. Gaxiola, A. Sánchez, A van Wormhoudt, D. Hernández, R. Díaz and C. Rosas. 2003b. Genetic relationship between *Litopenaeus setiferus* (L.) and *L. schmitti* (Burkenroad) determined by using 16S mitochondrial sequences and enzymatic analysis. Aquacult. Res., 34(12): 981-990.

Ashley, M.V., D.J. Melnick and D. Western. 1990. Conservation genetics of the Black Rhinoceros (*Diceros bicornis*). I: Evidence from the mitochondrial DNA of three populations. Conserv. Biol., 4: 71-77.

Attard, J., and R. Vianet. 1985. Variabilité génétique et morphologique de cinq populations de l'écrevisse européenne *Austropotamobius pallipes* (Lereboullet, 1858) (Crustacea, Decapoda). Can. J. Zool., 63: 2933-2939.

Austin, C. 1986. Genetic considerations in freshwater crayfish farming. *In:* Aquaculture in Australia Fish Farming for Profit. P. Owen and J. Bowden (eds.). Rural Press, Brisbane, pp. 73-78.

Austin, C.M., and B. Knott. 1996. Systematics of the freshwater crayfish genus *Cherax* Erichson (Decapoda: Parastacidae) in South Western Australia: electrophoretic, morphological and habitat variation. Austral. J. Zool., 44: 223-258.

Avarre, J-C., E. Lubzens and P.J. Babin. 2007. Apolipocrustacein, formerly vitellogenin, is the major egg yolk precursor protein in decapod crustaceans and is homologous to insect apolipophorin II/I and vertebrate apolipoprotein B. BMC Evol. Biol., 7: 3.

Avise, J.C. 1992. Molecular population structure and the biogeographic history of a regional fauna: a case history with lessons for conservation biology. Oikos, 63: 62-76.

Avise, J.C. 1994. Molecular markers, natural history and evolution. Chapman & Hall, New York.

Avise, J.C. 1996. The scope of conservation genetics. *In:* Conservation Genetics. Case Histories from Nature. J.C. Avise and J.L. Hamrich (eds.). Chapman & Hall, New York, pp. 1-9.

Avise, J.C. 2000. Phylogeography: The History and Formation of Species. Harvard University Press, Cambridge, Mass.

Avise, J.C. 2004. Molecular markers, natural history and evolution. Sinauer Associates, Inc Publishers, Sunderland, Mass.

Avise, J.C., and J.L. Hamrick. 1996. Conservation Genetics: Case Histories from Nature. Chapman & Hall, New York.

Bagley, M.J., and J.B. Geller. 2000. Microsatellite DNA analysis of native and invading populations of European green crabs. *In:* Marine Bioinvasions: Proceedings of the First National Conference, J. Pederson (ed.), pp. 241-243.

Bagshaw, J.C., and M.A. Buckholt. 1997. A novel satellite/microsatellite combination in the genome of the marine shrimp, *Penaeus vannamei*. Gene, 184: 211-214.

Baldwin, J.D., A.L. Bass, B.W. Bowen and W.H. Clark. 1998. Molecular phylogeny and biogeography of the marine shrimp *Penaeus*. Mol. Phylogenet. Evol., 10: 399-407.

Ball, A.O., and R.W. Chapman. 2003. Population genetic analysis of white shrimp, *Litopenaeus setiferus*, using microsatellite genetic markers. Mol. Ecol., 12(9): 2319-2330.

Ball, A.O., S. Leonard and R.W. Chapman. 1998. Characterization of (GT)n from native white shrimp (*Penaeus setiferus*). Mol. Ecol., 7: 1251-1253.

Balloux, F., and N. Lugon-Moulin. 2002. The estimation of population differentiation with microsatellite markers. Mol. Ecol., 11: 155-165.

Barber, P.H., S.R. Palumbi, M.V. Erdmann and M.K. Moosa. 2000. A marine wallace's line? Nature, 406: 692-693.

Barber, P.H., S.R. Palumbi, M.V. Erdmann and M.K. Moosa. 2002. Sharp genetic breaks among populations of *Haptosquilla pulchella* (Stomatopoda) indicate limits to larval transport: patterns, causes, and consequences. Mol. Ecol., 11(4): 659-674.

Barcia, A.R., G.E. López, D. Hernández and E. García-Machado. 2005. Temporal variation of the population structure and genetic diversity of *Farfantepenaeus notialis* assessed by allozyme loci. Mol. Ecol., 14(10): 2933-2942.

Bartley, D., M. Bagley, G. Gall and B. Bentley. 1992. Use of linkage disequilibrium data to estimate effective size of hatchery and natural fish populations. Conserv. Biol., 6: 365-375.

Beal, B.F., J.P. Mercer and A.Ó Conghaile. 2002. Survival and growth of hatchery-reared individuals of the European lobster, *Homarus gammarus* (L.), in field-based nursery cages on the Irish west coat. Aquaculture, 210(1-4): 137-157.

Beaumont, A.R., and T. Croucher. 2006. Limited stock structure in UK populations of the brown shrimp, *Crangon crangon*, identified by morphology and genetics. J. Mar. Biol. Assoc. UK, 86(5): 1107-1112.

Beerli, P., and J. Felsenstein. 1999. Maximum-likelihood estimation of migration rates and effective population numbers in two populations using a coalescent approach. Genetics, 152: 763-773.

Beerli, P., and Felsenstein J. 2001. Maximum likelihood estimation of migration rates and effective population numbers in two populations using a coalescent approach. Proc. Natl Acad. Sci., 98: 4563-4568.

Benzie, J.A.H. 1997. A review of the effect of genetics and environment on the maturation and larval quality of the giant tiger prawn *Penaeus monodon*. Aquaculture, 155(1-4): 69-85.

Benzie, J.A.H. 2000. Population genetic structure in penaeid prawns. Aquac. Res., 31: 95-119.

Benzie, J.A.H., S. Frusher and E. Ballment. 1992. Geographical variation in allozyme frequencies of *Penaeus monodon* (Crustacea: Decapoda) populations in Australia. Austral. J. Mar. Freshwater Res., 43: 715-725.

Benzie, J.A.P., E. Ballment and S. Frusher. 1993. Genetic structure of *Penaeus monodon* in Australia: concordant results from mtDNA and allozymes. Aquaculture, 111: 89-93.

Benzie, J.A.H., M. Kenway and L. Trott. 1997. Estimates for the heritability of size in juvenile *Penaeus monodon* prawns from half-sib matings. Aquaculture, 152(1-4): 49-53.

Benzie, J.A.H., E. Ballment, A.T. Forbes, N.T. Demetriades, K. Sugama, Haryantic and S. Moria. 2002. Mitochondrial DNA variation in Indo-Pacific populations of the giant tiger prawn, *Penaeus monodon*. Mol. Ecol., 11(12): 2553-2569.

Bermingham, E., S. McCafferty and A. Martin. 1997. Fish biogeography and molecular clocks: perspectives from the Panamanian Isthmus. *In:* Molecular Systematics of Fishes. T. Kocher and C. Stepien (eds.). Academic Press, New York, pp. 113-126.

Bert, T.M. 1986. Speciation in western Atlantic stone crabs (genus *Menippe*): the role of geological processes and climatic events in the formation and distribution of species. Mar. Biol., 93: 157-170.

Beuzen, N.D., M.J. Stear and K.C. Chang. 2000. Molecular markers and their use in animal breeding. Vet J., 160: 42-52.

Bierne, N., A. Lehnert, E. Bédier, F. Bonhomme and S.S. Moore. 2000. Screening for intron length polymorphisms in penaeid shrimps using exon primed intron crossing (EPIC) PCR. Mol. Ecol., 9: 233-235.

Boore, J.L. 1999. Animal mitochondrial genomes. Nucl. Acids Res., 27: 1767-1780.

Boore, J.L., T.M. Collins, D. Stanton, L.L. Daehler and W.M. Brown. 1995. Deducing the pattern of arthropod phylogeny from mitochondrial DNA rearrangements. Nature, 376: 163-165.

Boore, J.L., D. Lavrov and W.M. Brown. 1998. Gene translocation links insects and crustaceans. Nature, 392: 667-668.

Borrell, Y., G. Blanco, E. Vásquez, J. Alvarez, H. Pineda, G. Espinosa, C. Fernandez-Pato, C. Martinez and J.A. Sánchez. 2003. The variability of microsatellite and their application into aquaculture. CIVA, 2003: 104-1056.

Borrell, Y., G. Espinoza, J. Romo, G. Blanco, E. Vazquez and J.A. Sanchez. 2004. DNA microsatellite variability and genetic differentiation among natural populations of the Cuban white shrimp *Litopenaeus schmitti*. Mar. Biol., 144: 327-333.

Bouchon, D., C. Souty-Grosset and R. Raimond. 1994. Mitochondrial DNA variation and markers of species identity in two penaeid shrimp species: *Penaeus monodon* Fabricius and *Penaeus japonicus* Bate. Aquaculture, 127: 131-144.

Brasher, D.J., J.R. Ovenden and R.W.G. White. 1992a. Mitochondrial DNA variation and phylogenetic relationships of *Jasus* spp. J. Zool. Lond., 227: 1-16.

Brasher, D.J., J.R. Ovenden, J.D. Booth and R.W.G. White. 1992b. Genetic subdivision of Australian and New-zealand populations of *Jasus verreauxi* (Decapoda: Palinuridae) – preliminary evidence from the mitochondrial genome. New Zealand J. Mar. Freshwater Res., 26: 53-58.

Brian, J.V., T. Fernandes, R.J. Ladle and P.A. Todd. 2006. Patterns of morphological and genetic variability in UK populations of the shore crab, *Carcinus maenas* Linnaeus 1758 (Crustacea: Decapoda: Brachyura). J. Exp. Mar. Biol. Ecol., 329(1): 47-54.

Brooker, A.L., D. Cook, P. Bentzen, J.M. Wright and R.W. Doyle. 1994. Organization of microsatellites differs between mammals and cold-water teleost fishes. Can. J. Fish Aquat. Sci., 51: 1959-1966.

Brooker, A.L., J.A.H. Benzie, D. Blair and J.J. Versini. 2000. Population structure of the giant tiger prawn *Penaeus monodon* in Australian waters, determined using microsatellite markers. Mar. Biol., 136: 149-157.

Brown, K. 1980. Low genetic variability and high similarities in the crayfish genera *Cambarus* and *Procambarus*. Am. Midl. Nat., 105: 225-232.

Bryars, S.R., and M. Adams. 1999. An allozyme study of the blue swimmer crab, Portunus pelagicus (Crustacea: Portunidae), in Australia: stock delineation in southern Australia and evidence for a cryptic species in northern waters. Mar. Freshwater Res., 50(1): 15-26.

Buck, T.L., G.A. Breed, S.C. Pennings, M.E. Chase, M. Zimmer and T.H. Carefoot. 2003. Diet choice in an omnivorous salt-marsh crab: different food types, body size, and habitat complexity. J. Exp. Mar. Biol. Ecol., 292(1): 103-116.

Buitkamp, J., and J. Epplen. 1996. Modern genome research and DNA diagnostics in domestic animals in the light of classical breeding techniques. Electrophoresis, 17: 1-11.

Bulnheim, H.P., and S. Bahns. 1996. Genetic variation and divergence in the genus *Carcinus* (Crustacea, Decapoda). Intl. Rev. gesamten Hydrobiologie, 81(4): 611-619.

Bulnheim, H.P., and D.E. Schwenzer. 1993. Population genetic studies on *Crangon crangon* and *C. allmanni* (Crustacea, Decapoda) from European coastal areas. Zool. Jahrb. Abt. Allg. Zool. Physiol. Tiere, 97(4): 327-347.

Bunch, T., R.C. Highsmith and G.F. Shields. 1998. Genetic evidence for dispersal of larvae of Tanner crabs (*Chionoecetes bairdi*) by the Alaskan Coastal Current. Mol. Mar. Biol. Biotechnol., 7(2): 153-159.

Burton, C.A. 2001. The role of lobster (*Homarus* spp.) hatcheries in ranching, restoration and remediation programmes. Hydrobiologia, 465(1): 45-48.

Burton, R.S. 1983. Protein polymorphisms and genetic differentiation of marine invertebrate populations. Mar. Biol. Lett., 4: 193-206.

Burton, R.S., and M.W. Feldman. 1982. Population genetics of coastal and estuarine invertebrates: does larval behavior influence population structure. *In:* Estuarine Comparisons. V.S. Kennedy (ed.). Academic Press, New York, pp. 537-551.

Burton, R.S., and B.N. Lee. 1994. Nuclear and mitochondrial gene genealogies and allozyme polymorphism across a major phylogeographic break in the copepod *Tigriopus californicus*. Proc. Natl. Acad. Sci. USA, 91: 5197-5201.

Busackc, A. 1988. Electrophoretic variation in the red swamp (*Procambarus clarkii*) and White river crayfish (*P. acutus*) (Decapoda: Cambaridae). Aquaculture, 69: 211-226.

Busackc, A. 1989. Biochemical systematics of crayfishes of the genus *Procambarus*, subgenus *Scapulicambarus* (Decapoda: Cambaridae). J. N. Am. Benthol. Soc., 8: 180-186.

Cannas, R., A. Cau, A.M. Deiana, S. Salvadori and J. Tagliavini. 2006. Discrimination between the Mediterranean spiny lobsters *Palinurus elephas* and *P. mauritanicus* (Crustacea: Decapoda) by mitochondrial sequence analysis. Hydrobiologia, 557: 1-4.

Carvalho, G.R. 1998. Advances in Molecular Ecology. NATO ASI Series, IOS Press, Amsterdam.

Castro, K.M., J.S., Cobb, R.A. Wahle and J. Catena. 2001. Habitat addition and stock enhancement for American lobsters, *Homarus americanus*. Mar. Freshwater Res., 52(8): 1253-1261.

Chan, S.M., X.G. Chen and P.L. Gu. 1998. PCR cloning and expression of the molt-inhibiting hormone gene for the crab (*Charybdis feriatus*). Gene, 224(1-2): 23-33.

Chow, S., N. Suzuki, H. Imai and T. Yoshimura. 2006. Molecular species identification of spiny lobster phyllosoma larvae of the genus *Panulirus* from the northwestern Pacific. Mar. Biotechnol., 8(3): 260-267.

Chu, K.H., J.G. Tong and T.Y. Chan. 1999. Mitochondrial cytochrome oxidase I sequence divergence in some Chinese species of *Charybdis* (Crustacea: Decapoda: Portunidae). Biochem. Syst Ecol., 27: 461-468.

Chu, K.H., C.P. Li and H.Y. Ho. 2001. The first internal transcribed spacer (ITS-1) of ribosomal DNA as a molecular marker for phylogenetic and population analyses in Crustacea. Mar. Biotechnol., 3: 355-361.

Cipriano, F., and S.R. Palumbi. 1999. Genetic tracking of a protected whale. Nature, 397(6717): 307-308.

Cockerham, C.C., and B.S. Weir. 1993. Estimation of gene flow from F-statistics. Evolution, 47: 855-863.

Colbourne, J.K., and P.D.N. Hebert. 1996. The systematics of North American *Daphnia* (Crustacea: Anomopoda): a molecular phylogenetic approach. Phil. Trans. Roy. Soc., Lond., B 351: 349-360.

Cornuet, J., S. Piry, G. Luikart, A. Estoup and M. Solignac. 1999. New methods employing multilocus genotypes to select or exclude populations as origins of individuals. Genetics, 153: 1989-2000.

Cracraft, C. 1983. Species concepts and speciation analysis. Curr. Ornithol., 1: 159-187.

Cracraft, J. 1994. Species diversity, biogeography, and the evolution of biotas. Amer. Zool., 34: 33-47.

Crandall, K.A. 1998. Conservation phylogenetics of Ozark crayfish: assigning priorities for aquatic habitat protection. Biol. Conserv., 84: 107-117.

Crandall, K.A., O. Bininda-Edmonds, G. Mace and R.K. Wayne. 2000. Considering evolutionary processes in conservation biology. Trends Ecol. Evol., 15: 290-295.

Crivello, J.F., D.F. Landers Jr. and M. Keser. 2005a. The contribution of egg-bearing female American lobster (*Homarus americanus*) populations to lobster larvae collected in Long Island Sound by comparison of microsatellite allele frequencies. J. Shellfish Res., 24(3): 831-839.

Crivello, J.F., D.F. Landers Jr. and M. Keser. 2005b. The genetic stock structure of the American lobster (*Homarus americanus*) in Long Island Sound and the Hudson Canyon. J. Shellfish Res., 24(3): 841-848.

Crocos, P.J., and N.P. Preston. 1999. Genetic improvement of farmed prawns in Australia. Global Aquacult. Advocate, 2(6): 62-63.

Crow, J.F., and K. Aoki. 1984. Group selection for a polygenic behavioural trait: estimating the degree of population subdivision. Proc. Natl. Acad. Sci. USA, 81: 6073-6077.

Crozier, R.H. 1997. Preserving the information content of species: genetic diversity, phylogeny, and conservation worth. Ann. Rev. Ecol. Syst., 28: 243-268.

Cruz, P., C.H. Mejia-Ruiz, R. Perez-Enriquez and A.M. Ibarra. 2002. Isolation and characterization of microsatellites in Pacific white shrimp *Penaeus* (*Litopenaeus*) *vannamei*. Mol. Ecol. Notes, 2(3): 239-241.

Cruz, P., A.M. Ibarra, H. Mejia-Ruiz, P.M. Gaffney and R. Pérez-Enríquez. 2004. Genetic variability assessed by microsatellites in a breeding program of pacific white shrimp (*Litopenaeus vannamei*). Mar. Biotechnol., 6(2): 157-164.

Curole, J.P., and T.D. Kocher. 1999. Mitogenomics: digging deeper with complete mitochondrial genomes. Trends Ecol. Evol., 14: 394-398.

Dall, W., B.J. Hill, P.C. Rothlisberg and D.J. Staples. 1990. The biology of Penaeidea. Adv. Mar. Biol., 27: 1-489.

Danzmann, R.G., T.R. Jackson and M.M. Ferguson. 1999. Epistasis in allelic expression at upper temperature tolerance QTL in rainbow trout. Aquaculture, 173: 45-58.

Davis, G.P., and D.J.S. Hetzel. 2000. Integrating molecular genetic technology with traditional approaches for genetic improvement in aquaculture species. Aquacult. Res., 31: 3-10.

Davis, J.A., L.L. van Blerk, R. Kirby and T. Hecht. 2003. Genetic variation in the mud crab *Scylla serrata* (Forskål, 1775) (Crustacea: Portunidae) in South African estuaries. Afr. Zool., 38(2): 343-350.

de Francisco, A.K., and P.M. Galetti Jr. 2005. Genetic distance between broodstocks of the marine shrimp *Litopenaeus vannamei* (Decapoda, Penaeidae) by mtDNA analyses. Genet. Mol. Biol., 28(2): 258-261.

de Freitas, P.D., and M.P. Galetti Jr. 2002. PCR-based VNTR core sequence analysis for inferring genetic diversity in the shrimp *Litopenaeus vannamei*. Genet. Mol. Biol., 25(4): 431-434.

de Freitas, P.D., and M.P. Galetti Jr. 2005. Assessment of the genetic diversity in five generations of a commercial broodstock line of *Litopenaeus vannamei* shrimp. Afr. J. Biotechnol., 4(12): 1362-1367.

de Freitas, P.D., C.M. Jesus and P.M. Galetti Jr. 2007. Isolation and characterization of new microsatellite loci in the Pacific white shrimp *Litopenaeus vannamei* and cross-species amplification in other penaeid species: Primer note. Mol. Ecol. Notes, 7(2): 324-326.

Dobzhansky, T. 1970. Genetics of the Evolutionary Process. Columbia University Press, New York.

Dodgson, J.B., H.H. Cheng and R. Okimoto. 1997. DNA marker technology: A revolution in animal genetics. Poult. Sci., 76(8): 1108-1114.

Donato, M., R. Manrique, R. Ramirez, L. Mayer and C. Howell. 2005. Mass selection and inbreeding effects on a cultivated strain of *Penaeus* (*Litopenaeus*) *vannamei* in Venezuela. Aquaculture, 247: 159-167.

Drengstig, A., S.-E. Fevolden, P.E. Galand and M.M. Aschan. 2000. Population structure of the deep-sea shrimp (*Pandalus borealis*) in the north-east Atlantic based on allozyme variation. Aquat. Living Resour., 13(2): 121-128.

Duffy, J.E. 1996. Eusociality in coral reef shrimp. Nature, 381: 512-514.

Dunham, R.A. 1996. Contribution of genetically improved aquatic organisms to global food security. Paper presented at the International Conference on Sustainable Contribution of Fisheries to Food Security. Government of Japan and FAO, Rome, Italy. 4-9 Dec. 1995, Kyoto, Japan.

Dunham, R.A. 1999. Utilization of transgenic fish in developing countries: potential benefits and risks. J. World Aquacult. Soc., 30: 1-11.

Edomi, P., E. Azzoni, R. Mettulio, N. Pandolfelli, E.A. Ferrero and P.G. Giulianini. 2002. Gonad-inhibiting hormone of the Norway lobster (*Nephrops norvegicus*): cDNA cloning, expression, recombinant protein production, and immunolocalization. Gene, 284(1-2): 93-102.

Ehrlich, P.R., and P.H. Raven. 1969. Differentiation of populations. Science 165: 1228-1232.

Eldredge, N. 1989. Macroevolutionary Dynamics: Species, Niches and Adaptive Peaks, McGraw-Hill, New York.

Eldredge, N. 1992. Systematics, Ecology and the Biodiversity Crisis. Columbia University Press, New York.

Espinosa, G., M. Jager, E. Machado, Y. Borrell, N. Corona, A. Robainas and J. Deutsh. 2001. Microsatellites from the white shrimp *Litopenaeus schmitti* (Crustacea, Decapoda). Biotechnol. Appl., 18: 19-22.

Exadactylos, A., and I. Arvanitoyannis. 2006. Aquaculture Biotechnology for enhanced fish production for food consumption. *In:* Microbial Biotechnology in Agriculture and Aquaculture, Vol. II. R.C. Ray (ed.). Science Publishers, Enfield, New Hampshire, USA, pp. 453-510.

Exadactylos, A., and J.P. Thorpe. 2001. Allozyme variation and genetic inter-relationships between seven flatfish species (Pleuronectiformes). Zool. J. Linn. Soc., 132: 487-499.

Exadactylos, A., A.J. Geffen and J.P. Thorpe. 1998. Population structure of the Dover sole, *Solea solea* L. in a background of high gene flow. J. Sea Res., 40: 117-129.

Exadactylos, A., A.J. Geffen, P. Panagiotaki and J.P. Thorpe. 2003. Population structure of Dover sole, *Solea solea*: RAPD and allozyme data indicate divergence in European stocks. Marine Ecology-Progress Series, 246: 253-264.

Exadactylos, A., M.J. Rigby, A.J. Geffen and J.P. Thorpe. 2007. Conservation aspects of natural populations and captive bred stocks of turbot (*Scophthalmus maximus*) and Dover sole (*Solea solea*) using estimates of genetic diversity. ICES Journal of Marine Science, 64: 1173-1181.

Faith, D.P. 1992. Conservation evaluation and phylogenetic diversity. Biol. Conserv., 61: 1-10.

Falconer, D.S., and T.F.C. Mackay. 1996. Introduction to quantitative genetics. 4th ed. Longman, New York.

FAO. 1999. The State of World Fisheries and Aquaculture 1998 (Food and Agricultural Organization, Rome).

FAO. 2006. FIGIS (Fisheries global information system). http://www.fao.org/

Féral, J-P. 2002. How useful are the genetic markers in attempts to understand and manage marine biodiversity? J. Exp. Mar. Biol. Ecol., 268: 121-145.

Ferguson, M.M., and R.G. Danzmann. 1998. Role of genetic markers in fisheries and aquaculture: useful tools or stamp collecting? Can. J. Fish. Aquat. Sci., 55: 1553-1563.

Fevolden, S.E., and D.O. Hessen. 1989. Morphological and genetic differences among recently founded populations of noble crayfish (*Astacus astacus*). Hereditas, 110(2): 149-158.

Forbes, A.T., N.T. Demetriades, J.A.H. Benzie and E. Ballment. 1999. Allozyme frequencies indicate little geographic variation among stocks of giant tiger prawn *Penaeus monodon* in the south-west Indian Ocean. S. Afr. J. Mar. Sci., 21: 271-277.

Frankham, R. 1995a. Conservation genetics. Annu. Rev. Genet., 29: 305-327.

Frankham, R. 1995b. Effective population size/adult population size ratios in wildlife: a review. Genet. Res. Camb., 66: 95-106.

Fratini, S., and M. Vannini. 2002. Genetic differentiation in the swimming crab *Scylla serrata* (Decapoda: Portunidae) within the Indian Ocean. J. Exp. Mar. Biol. Ecol., 272: 103-116.

Garcia, D.K., A.K. Dhar and A. Alcivar-Warren. 1996. Molecular analysis of a RAPD marker (B20) reveals two microsatellites and differential mRNA expression in *Penaeus vannamei*. Mol. Mar. Biol. Biotechnol., 5: 71-83.

Garcia de Leon, F.J., J.F. Dallas, B. Chatain, M. Canonne, J.J. Versini and F. Bonhomme. 1995. Development and use of microsatellite markers in sea bass

Dicentrarchus labrax (Linnaeus, 1758) (Perciformes: Serranidae). Mol. Mar. Biol. Biotechnol., 4(1): 62-68.

Garibaldi, L. 1996. List of animal species used in aquaculture. FAO Fish. Circ., N° 914, FAO, Rome, 38 p.

Garza, J.C., and E.G. Williamson. 2001. Detection of reduction in population size using data from microsatellite loci. Mol. Ecol., 10: 305-318.

Geller, J., and M. Bagley. 1997. Genetic and systematic studies of *Carcinus maenas*. *In:* R. Thresher (ed.). Australia: CSIRO, Centre for Research on Introduced Marine Pests Technical Report 11: 19-20.

Geller, J.B., E.D. Walton, E.D. Grosholz and G.M. Ruiz. 1997. Cryptic invasions of the crab *Carcinus* detected by molecular phylogeography. Mol. Ecol., 6(10): 901-906.

Gilbert, S.F. 2000. Diachronic biology meets evo-devo: C. H. Waddington's approach to evolutionary developmental biology. Am. Zool., 40: 729-737.

Gilbert, S.F. 2006. Developmental Biology, 8th ed. Sinauer Associates, Inc., Publishers, Sunderland, Mass.

Gitterle, T., M. Rye, R. Salte, J. Cock, H. Johansen, C. Lozano, J. Arturo Suárez and B. Gjerde. 2005. Genetic (co)variation in harvest body weight and survival in *Penaeus* (*Litopenaeus*) *vannamei* under standard commercial conditions. Aquaculture, 243(1-4): 83-92.

Gjedrem, T. 1997. Selective breeding to improve aquaculture production. World Aquacult., 28: 33-45.

Glenn, K.L., L. Grapes, T. Suwanasopee, D.L. Harris, Y. Li, K. Wilson and M.F. Rothschild. 2005. SNP analysis of AMY2 and CTSL genes in *Litopenaeus vannamei* and *Penaeus monodon* shrimp. Anim. Genet., 36(3): 235-236.

Goldstein, D.B., and C. Schlötterer. 1999. Microsatellites: Evolution and Applications. Oxford University Press, Oxford.

Goldstein, D.B., G.W. Roemer, D.A. Smith, D.E. Reich, A. Bergman and R.K. Wayne. 1999. The use of microsatellite variation to infer population structure and demographic history in a natural model system. Genetics, 151: 797-801.

Gonçalves, M.M., M.V.F. Lemos, P.M. Galetti Jr., P.D. de Freitas and M.A.A.F. Neto. 2005. Fluorescent amplified fragment length polymorphism (fAFLP) analyses and genetic diversity in *Litopenaeus vannamei* (Penaeidae). Genet. Mol. Biol., 28(2): 267-270.

Gopikrishna, G., and M. Shashi-Shekhar. 2003. Karyological and PCR-RFLP studies of the mud crabs *Scylla serrata* and *Scylla tranquebarica*. J. Fish. Soc. Taiwan, 30(4): 315-320.

Gopurenko, D., and J.M. Hughes. 2002. Regional patterns of genetic structure among Australian populations of the mud crab, *Scylla serrata* (Crustacea: Decapoda): evidence from mitochondrial DNA. Mar. Freshwater Res., 53(5): 849-857.

Gopurenko, D., J.M. Hughes and C.P. Keenan. 1999. Mitochondrial DNA evidence for rapid colonization on the Indo-West Pacific by the mudcrab *Scylla serrata*. Mar. Biol., 134: 227-233.

Gopurenko, D., J.M. Hughes and J. Ma. 2002. Identification of polymorphic microsatellite loci in the mud crab *Scylla serrata* (Brachyura: Portunidae). Mol. Ecol. Notes, 2: 481-483.

Gopurenko, D., J.M. Hughes and L. Bellchambers. 2003. Colonization of the south-west Australian coastline by mud crabs: Evidence for a recent range expansion or human-induced translocation? Mar. Freshwater Res., 54(7): 833-840.

Gosselin, T., B. Sainte-Marie and L. Bernatchez. 2005. Geographic variation of multiple paternity in the American lobster, *Homarus americanus*. Mol. Ecol., 14: 1517-1525.

Grandjean, F., C. Souty-Grosset and D.M. Holdich. 1997. Geographical variation of mitochondrial DNA between European populations of the white-clawed crayfish *Austropotamobius pallipes*. Freshwater Biol., 37: 493-501.

Grandjean, F., N. Gouin, M. Frelon and C. Souty-Grosset. 1998. Genetic and morphological systematic studies on the crayfish Austropotamobius pallipes. J. Crustacean Biol., 18: 549-555.

Gray, J.S. 1997a. Marine biodiversity: patterns, threats and conservation needs. Biodiversity Conserv., 6: 153-175.

Gray, J.S. 1997b. Gradients in marine biodiversity. *In:* Marine Biodiversity: patterns and processes. R.F.G. Ormond, Gage J.D. and Angel M.V. (eds). Cambridge University Press, Cambridge, pp. 18-34.

Greene, H.W. 1994. Systematics and natural history: foundations for understanding and conserving biodiversity. Am. Zool., 34: 48-56.

Grosholz, E., G. Ruiz, J. Geller, M. Bagley and R. Thresher. 2000. Increased body size and reduced genetic diversity in invading populations of the European green crab. Ecological Consequences of Adaptive Evolution Among Invasive Species in Terrestrial and Marine Systems. The Ecological Society of America, 85th Annual Meeting, Snowbird, Utah.

Gusmão, J., C. Lazoski and A.M. Sole-Cava. 2000. A new species of *Penaeus* (Crustacea: Penaeidae) revealed by allozyme and cytochrome oxidase I analyses. Mar. Biol., 137: 435-446.

Gutiérrez-Millán, L.E., A.B. Peregrino-Uriarte, R. Sotelo-Mundo, F. Vargas-Albores and G. Yepiz-Plascencia. 2002. Sequence and conservation of a rRNA and tRNAVal mitochondrial gene fragment from *Penaeus californiensis* and comparison with *P. vannamei* and *P. stylirostris*. Mar. Biotechnol., 4(4): 392-398.

Harding, G.C., E.L. Kenchington, C.J. Bird, D.S. Pezzack and D.C. Landry. 1997. Genetic relationships among subpopulations of the American lobster (*Homarus americanus*) as revealed by random amplified polymorphic DNA. Can. J. Fish. Aquat. Sci., 54(8): 1762-1771.

Hartl, D.L., and A.G. Clark. 1989. Principles of Population Genetics, 2nd ed. Sinauer Associates, Sunderland, Mass.

Hauser, L., G. Adcock, P.J. Smith, J.H. Bernal Ramirirez and G.R. Carvalho. 2002. Loss of microsatellite diversity and low effective population size in an overexploited population of New Zealand snapper (*Pagrus auratus*). PNAS, 99: 11742-11747.

Hearne, C.M., S. Ghosh and J.A. Todd. 1992. Microsatellites for linkage analysis of genetic traits. Trends Genet., 8: 288-294.

Heath, D.D., N.J. Bernier and T.A. Mousseau. 1995a. A single-locus minisatellite discriminates chinook salmon (*Oncorhynchus tshawytscha*) populations. Mol. Ecol., 4: 389-393.

Heath, D.D., P.D. Rawson and T.J. Hilbish. 1995b. PCR based nuclear markers identify alien blue mussel (*Mytilus*) genotypes on the west coast of Canada. Can. J. Fish. Aquat. Sci., 52: 2621-2627.

Hedgecock, D., M.L. Tracey and K. Nelson. 1982. Genetics. *In:* The Biology of Crustacea. L.G. Abele (ed.). Academic Press, New York, pp. 283-403.

Hedgecock, D., E.S. Hutchinson, G. Li, F.L. Sly and K. Nelson. 1989. Genetic and morphometric variation in the Pacific Sardine, *Sardinops sagax caerulea*: comparisons and contrasts with historical data and with variability in the Northern Anchovy, *Engraulis mordax*. U.S. Fish. Bull., 87: 653-671.

Hedgecock, D., V. Chow and R.S. Waples. 1992. Effective population numbers of shellfish broodstocks estimated from temporal variance in allelic frequencies. Aquaculture, 108(3-4): 215-232.

Hedgecock, D., P. Siri and D.R. Strong. 1994. Conservation biology of endangered Pacific salmonids: Introductory remarks. Conserv. Biol., 8: 863-864.

Hedgecock, D., K. Neslon, J. Simons and R. Shleser. 1997. Genic similarity of American and European species of the lobster *Homarus*. Biol. Bull., 152(1): 41-50.

Hedgecock, D., M.A. Banks, V.K. Rashbrook, C.A. Dean and S.M. Blankenship. 2001. Applications of population genetics to conservation of chinook salmon diversity in the Central Valley. Fish Bulletin 179, Contributions to the Biology of Central Valley Salmonids: 45-70.

Hedrick, P.W. 1999. Perspective: highly variable loci and their interpretation in evolution and conservation. Evolution, 53: 313-318.

Hedrick, P.W., and D. Hedgecock. 1994. Effective population size in winter-run chinook salmon. Conserv. Biol., 8: 890-892.

Hellberg, M.E., R.S. Burton, J.E. Neigel and S.R. Palumbi. 2002. Genetic assessment of connectivity among marine populations. Bull. Mar. Sci., 70(1): 273-290.

Herbinger, C.M., R.W. Doyle, C.T. Taggart, S.E. Lochmann, A.L. Brooker, J.M. Wright and D. Cook. 1997. Family relationships and effective population size in a natural cohort of Atlantic cod (*Gadus morhua*) larvae. Can. J. Fish. Aquat. Sci., 54(Suppl. 1): 11-18.

Heywood, V.H., and R.T. Watson. 1995. Global Biodiversity Assessment. Cambridge University Press, New York.

Hilbish, T.J., and R.K. Koehn. 1985a. Dominance in physiological phenotypes and fitness at an enzyme locus. Science, 229: 52-54.

Hilbish, T.J., and R.K. Koehn. 1985b. The physiological basis of natural selection as the *Lap* locus. Evolution, 39: 1302-1317.

Hillis, D.M., C. Moritz and B.K. Mable. 1996. Molecular Systematics. Sinauer Associates, Inc., USA.

Hodgins-Davis, A., S. Roberts, D.F. Cowan, J. Atema, M. Bennie, C. Avolio, J. Defaveri and G. Gerlach. 2007. Characterization of SSRs from the American lobster, *Homarus americanus*: Primer note. Mol. Ecol. Notes, 7(2): 330-332.

Holmes, E.C., M. Worobey and A. Rambaut. 1999. Phylogenetic evidence for recombination in dengue virus. Mol. Biol. Evol., 16: 405-409.

Hughes, C.R., and D.C. Queller. 1993. Detection of highly polymorphic microsatellite loci in a species with little allozyme polymorphism. Mol. Ecol., 2: 131-137.

Hutchinson, W.F., C. van Oosterhout, S.I. Rogers and G.R. Carvalho. 2003. Temporal analysis of archived samples indicates marked genetic changes in declining North Sea cod (*Gadus morhua*). Proc. Roy. Soc. London, Ser. B: Biol. Sci., 270: 2125-2132.

Imai, H., Y. Fujii and J. Karakawa. 1999. Analysis of the population structure of the swimming crab, *Portunus trituberculatus* in the coastal waters of Okayama Prefecture, by RFLPs in the whole region of mitochondrial DNA. Fish. Sci., 65: 655-656.

Imai, H., J.H. Cheng, K. Hamasaki and K.I. Numachi. 2004. Identification of four mud crab species (genus *Scylla*) using ITS-1 and 16S rDNA markers. Aquat. Living Resour., 17: 31-34.

Inoue, N., H. Watanabe, S. Kojima and H. Sekiguchi. 2007. Population structure of Japanese spiny lobster *Panulirus japonicus* inferred by nucleotide sequence analysis of mitochondrial COI gene. Fish. Sci., 73(3): 550-556.

Itokawa, M., K. Yamada, Y. Iwayama-Shigeno, Y. Ishitsuka, S. Detera-Wadleigh and T. Yoshikawa. 2003. Genetic analysis of a functional GRIN2A promoter (GT) N repeat in bipolar disorder pedigrees in humans. Neurosci. Lett., 345: 53-56.

Jackson, T.R., M.M. Ferguson, R.G. Danzmann, A.G. Fishback, P.E. Ihssen., M. O'Connell and T.J. Crease. 1998. Identification of two QTL influencing upper temperature tolerance in three rainbow trout (*Onchorhynchus mykiss*) half-sib families. Heredity, 80: 143-151.

Jeffreys, A.J., V. Wilson, and S.L. Thein. 1985. Hypervariable "minisatellite" regions in human DNA. Nature, 314: 67-73.

Jenneckens, I., A. Muller-Belecke, G. Horstgen-Schwark and J. Meyer. 1999. Proof of the successful development of Nile tilapia (*Oreochromis niloticus*) clones by DNA fingerprinting. Aquaculture, 173: 377-388.

Jenneckens, I., J.N. Meyer., G. Hörstgen-Schwark, B. May, L. Debus, H. Wedekind and A. Ludwig. 2001. A fixed allele at microsatellite LS-39 is characteristic for the black caviar producer *Acipenser stellatus*. J. Appl. Ichthyol., 17: 39-42.

Jensen, P.C., and P. Bentzen. 2004. Isolation and inheritance of microsatellite loci in the Dungeness crab (Brachyura: Cancridae: *Cancer magister*). Genome, 47: 325-331.

Jerry, D.R., N.P. Preston, P.J. Crocos, S. Keys, J.R.S. Meadows and Y. Li. 2006. Application of DNA parentage analyses for determining relative growth rates of *Penaeus japonicus* families reared in commercial ponds. Aquaculture, 254(1-4): 171-181.

Jónsdóttir, O.D.B., A.K. Imsland and G. Naevdal. 1998. Population genetic studies of northern shrimp, *Pandalus borealis*, in Icelandic waters and the Denmark Strait. Can. J. Fish. Aquat. Sci., 55(3): 770-780.

Jørstad, K.E., and E. Farestveit. 1999. Population genetic structure of lobster (*Homarus gammarus*) in Norway, and implications for enhancement and sea-ranching operation. Aquaculture, 173(1-4): 447-457.

Jørstad, K.E., P.A. Prodöhl, A.L. Agnalt, M. Hughes, A.P. Apostolidis, A. Triantafyllidis, E. Farestveit, T.S. Kristiansen, J. Mercer and T. Svåsand. 2004. Sub-arctic populations of European lobster, *Homarus gammarus*, in northern Norway. Environ. Biol. Fish., 69(1-4): 223-231.

Jørstad, K.E., E. Farestveit, E. Kelly and C. Triantaphyllidis. 2005a. Allozyme variation in European lobster (*Homarus gammarus*) throughout its distribution range. New Zealand J. Mar. Freshwater Res., 39(3): 515-526.

Jørstad, K.E., P.A. Prodöhl, T.S. Kristiansen, M. Hughes, E. Farestveit, J.B. Taggart, A.L. Agnalt and A. Ferguson. 2005b. Communal larval rearing of European lobster (*Homarus gammarus*): Family identification by microsatellite DNA profiling and offspring fitness comparisons. Aquaculture, 247(1-4): 275-285.

Jones, M.W., P.T. O'Reilly, A.A. McPherson, T.L. McParland, D.E. Armstrong, A.J. Cox, K.R. Spence, E.L. Kenchington, C.T. Taggart and P. Bentzen. 2003. Development, characterization, inheritance, and cross-species utility of American lobster (*Homarus americanus*) microsatellite and mtDNA PCR-RFLP markers. Genome, 46: 59-69.

Karl, S.A., and J.C. Avise. 1992. Balancing selection at allozyme loci in oysters: implications from nuclear RFLPs. Science, 256: 100-102.

Karl, S.A., and J.C. Avise. 1993. PCR-based assays of Mendelian polymorphisms from anonymous single-copy nuclear DNA: techniques and applications for population genetics. Mol. Biol. Evol., 10: 342-361.

Kartavtsev, Yu.F., B.I. Berenboim and K.I. Zgurovsky. 1992a. Population genetic differentiation of the pink shrimp *Pandalus borealis* kroyer from the Barents and Bering seas. Genetika, 28(5): 114-123.

Kartavtsev, Yu.F., A.V. Sitnikov, P.I. Komissarov and D.M. Eggers. 1992b. Allozyme variability and differentiation of the pink shrimp *Pandalus borealis* Kroyer from the Gulf of Alaska and the Bering Sea. Genetika, 28(12): 58-72.

Kartavtsev, Yu.F., K.A. Zgurovsky and Zh.M. Fedina. 1992c. Allozyme variability and differentiation of pink shrimp *Pandalus borealis* from three Far East seas. Genetika, 28(2): 110-122.

Kenway, M., M. Macbeth, M. Salmon, C. McPhee, J. Benzie, K. Wilson and W. Knibb. 2006. Heritability and genetic correlations of growth and survival in black tiger prawn *Penaeus monodon* reared in tanks. Aquaculture, 259(1-4): 138-145.

Keys, S.J., P.J. Crocos, C.Y. Burrdge, G.J. Coman, G.P. Davis and N.P. Preston. 2004. Comparative growth and survival of inbred and outbred *Penaeus* (*Marsupenaeus*) *japonicus*, reared under controlled environment conditions: indications of inbreeding depression. Aquaculture, 241: 151-168.

Khamnamtong, B., S. Klinbunga and P. Menasveta. 2005. Species identification of five penaeid shrimps using PCR-RFLP and SSCP analyses of 16S ribosomal DNA. J. Biochem. Mol. Biol., 38(4): 491-499.

Kincaid, H.L. 1976a. Effects of inbreeding on rainbow trout populations. Trans. Am. Fish. Soc., 105: 273-280.

Kincaid, H.L. 1976b. Effects of inbreeding in rainbow trout. J. Fish. Res. Board Can., 33: 2420-2426.

Kincaid, H.L. 1983. Inbreeding in fish populations used in aquaculture. Aquaculture, 3: 215-227.

Klinbunga, S., D.J. Penman and B.J. McAndrew. 1998. A preliminary study of ribosomal DNA polymorphism in the tiger shrimp, *Penaeus monodon*. J. Mar. Biotechnol., 6(3): 186-188.

Klinbunga, S., D.J. Penman, B.J. McAndrew and A. Tassanakajon. 1999. Mitochondrial DNA diversity in three populations of the giant tiger shrimp *Penaeus monodon*. Mar. Biotechnol., 1(2): 113-121.

Klinbunga, S., A. Boonyapakdee and B. Pratoomchat. 2000. Genetic diversity and species-diagnostic markers of mud crabs (genus *Scylla*) in eastern Thailand determined by RAPD analysis. Mar. Biotechnol., 2(2):180-187.

Klinbunga, S., D. Siludjai, W. Wudthijinda, A. Tassanakajon, P. Jarayabhand and P. Menasveta. 2001. Genetic heterogeneity of the giant tiger shrimp (*Penaeus monodon*) in Thailand revealed by RAPD and mitochondrial DNA RFLP analyses. Mar. Biotechnol., 3(5): 428-438.

Klinbunga, S., R. Preechaphol, S. Thumrungtanakit, R. Leelatanawit, T. Aoki, P. Jarayabhand and P. Menasveta. 2006. Genetic diversity of the giant tiger shrimp (*Penaeus monodon*) in Thailand revealed by PCR-SSCP of polymorphic EST-derived markers. Biochem. Genet., 44(5-6): 222-236.

Knowles, L.L., and W.P. Maddison. 2002. Statistical phylogeography. Mol. Ecol., 11: 2623-2635.

Knutsen, H., P.E. Jorde, C. André and N.C. Stenseth. 2003. Fine-scaled geographic population structuring in a highly mobile marine species: the Atlantic cod. Mol. Ecol., 12: 385-394.

Kocher, T.D., W. Lee, H. Sobolewska D. Penmanand and B. McAndrew. 1998. A genetic linkage map of a cichlid fish, the tilapia (*Oreochromis niloticus*). Genetics, 148: 1225-1232.

Koehn, R.K., R.I. Newell and F. Immerman. 1980. Maintenance of aminopeptidase allelic frequency cline by natural selection. Proc. Natl. Acad. Sci. USA, 77(9): 5385-5389.

Komm, B., A. Michaels, J. Tsokos and J. Linton. 1982. Isolation and characterization of the mitochondrial DNA from the Florida spiny lobster, *Panulirus argus*. Comp. Biochem. Physiol., 73b: 923-929.

Kordos, L.M., and R.S. Burton. 1993. Genetic differentiation of Texas Gulf coast populations of the blue crab, *Callinectes sapidus*. Mar. Biol., 117: 227-233.

Krajewski, C. 1994. Phylogenetic measures of biodiversity: a comparison and critique. Biol. Conserv., 69: 33-39.

Lardy, C., and L. Parr. 2006. Clinal and biogeographical genetic variation of the Dungeness crab (*Cancer magister*): implications for marine reserve design. Conservation without Borders, Society for Conservation Biology, 20th Annual Meeting, 24-28 June 2006, San Jose, California, USA.

Lavery, S., T.Y. Chan, Y.K. Tam and K.H. Chu. 2004. Phylogenetic relationships and evolutionary history of the shrimp genus *Penaeus s.l.* derived from mitochondrial DNA. Mol. Phylogenet. Evol., 31: 39-49.

Lessa, E.P., and G. Applebaum. 1993. Screening techniques for detecting allelic variation in DNA sequences. Mol. Ecol., 2: 119-129.

Lester, L.J. 1979. Population genetics of penaeid shrimp from the Gulf of Mexico. J. Hered., 70(3): 175-180.

Lewin, B. 1994. Genes IV, 4th ed. John Willey & Sons, New York.

Lewontin, R.C., and J.L. Hubby. 1966. A molecular approach to the study of genic heterozygosity in natural populations. II. Amount of variation and degree of heterozygosity in natural populations of *Drosophila pseudoobscura*. Genetics, 54: 595-609.

Li, Y., K.J. Wilson, K. Byrne, V. Whan, D. Iglesis, S.A. Lehnert, J. Swan, B. Ballment, Z. Fayazi, M. Kenway, J. Benzie, S. Pongsomboon, A. Tassanakajon and S.S. Moore. 2000. International collaboration on genetic mapping of the black tiger shrimp, *Penaeus monodon*: progress update. Plant and Animal Genome VIII, p. 8. San Diego, 9-12 January 2000.

Li, Y., K. Byrne, E. Miggiano, V. Whan, S. Moore, S. Keys, P. Crocos, N. Preston and S. Lehnert. 2003. Genetic mapping of the kumura prawn *Penaeus japonicus* using AFLP markers. Aquaculture, 219: 143-156.

Li, Y., L. Dierens, K. Byrne, E. Miggiano, S. Lehnert, N. Preston and R. Lyons. 2006. QTL detection of production traits for the Kuruma prawn *Penaeus japonicus* (Bate) using AFLP markers. Aquaculture, 258(1-4): 198-210.

Li, Z., and Z. Wu. 2003. Study on genetic diversity of cultivated populations in 4 species of shrimps. Acta Oceanologica Sinica, 22(1): 97-101.

Liu, Z.J., and J.F. Cordes. 2004. DNA marker technologies and their applications in aquaculture genetics. Aquaculture, 238: 1-37.

Liu, Z.J., and R.A. Dunham. 1998. Genetic linkage and QTL mapping of ictalurid catfish. Alabama Agric. Exp. Stn. Circ. Bull., 321: 1-19.

Liu, Z.J., A. Karsi and R.A. Dunham. 1999a. Development of polymorphic EST markers suitable for genetic linkage mapping of catfish. Mar. Biotechnol., 1: 437-447.

Liu, Z.J., P. Li, B.J. Argue and R.A. Dunham. 1999b. Random amplified polymorphic DNA markers: usefulness for gene mapping and analysis of genetic variation of catfish. Aquaculture, 174: 59-68.

Lörstcher, M., T.P. Stucki, M. Clalüna and A. Scholl. 1997. Phylogeographic structure of *Austropotamobius pallipes* populations in Switzerland. Bull. Fr. Pêche Pisc., 347: 649-661.

Lu, W., G. Wainwright, S.G. Webster, H.H. Rees and P.C. Turner. 2000. Clustering of mandibular organ-inhibiting hormone and moult-inhibiting hormone genes in the crab, *Cancer pagurus*, and implications for regulation of expression. Gene, 253(2): 197-207.

Lundin, M., B. Mikkelsen, P. Moran, J.L. Martinez and M. Syed. 1999. Cosmid clones from Atlantic salmon: physical genome mapping. Aquaculture, 173: 59-64.

Machida, R.J., M.U. Miya, M. Nishida and S. Nishida. 2002. Complete mitochondrial DNA sequence of Tigriopus japonicus (Crustacea: Copepoda). Mar. Biotechnol., 4: 406-417.

Machida, R.J., M.U. Masaki, N. Mutsumi and N. Shuhei. 2004. Large-scale gene rearrangements in the mitochondrial genomes of two calanoid copepods *Eucalanus bungii* and *Neocalanus cristatus* (Crustacea), with notes on new versatile primers for the srRNA and COI genes. Gene, 332: 71-78.

Maggioni, R., A.D. Rogers, N. Maclean and D' Incao. 2001. Molecular phylogeny of Western Atlantic *Farfantepenaeus* and *Litopenaeus* shrimp based on mitochondrial 16S partial sequences. Mol. Phylogenet. Evol., 18: 66-73.

Maggioni, R., A.D. Rogers and N. Maclean. 2003. Population structure of *Litopenaeus schmitti* (Decapoda: Penaeidae) from the Brazilian coast identified using six polymorphic microsatellite loci. Mol. Ecol., 12: 3213-3217.

Malecha, S.R., and D. Hedgecock. 1989. Prospects for domestication and breeding of marine shrimp. Hawaii Sea Grant College Report, University of Hawaii, Manoa. UNIHI-SEAGRANT-TR-89-01. 40 pp.

Maltagliati, F., L. Camilli, F. Biagi and M. Abbiati. 1998. Genetic structure of Norway lobster, *Nephrops norvegicus* (L.) (Crustacea: Nephropidae), from the Mediterranean Sea. Scientia Marina, 62(Suppl. 1): 91-99.

Maneeruttanarungroj, C., S. Pongsomboon, S. Wuthisuthimethavee, S. Klinbunga, K.J. Wilson, J. Swan, Y. Li, V. Whan, K.H. Chu, C.P. Li, J. Tong, K. Glenn, M. Rothschild, D. Jerry and A. Tassanakajon. 2006. Development of polymorphic expressed sequence tag-derived microsatellites for the extension of the genetic linkage map of the black tiger shrimp (*Penaeus monodon*). Anim. Genet., 37(4): 363-368.

Martin, J.W., and G.E. Davis. 2001. An Updated Classification of the Recent Crustacea. Natural History Museum of Los Angeles County, 132 p.

Martinez, I., M. Aschan, T. Skjerdal and S.M. Aljanabi. 2006. The genetic structure of *Pandalus borealis* in the Northeast Atlantic determined by RAPD analysis. ICES J. Mar. Sci., 63(5): 840-850.

McConnell, S.K.J., C. Beynon, J. Leamon and D.O.F. Skibinski. 2000. Microsatellite marker based genetic linkage maps of *Oreochromis aureus* and *O. niloticus* (Cichlidae): extensive linkage group segment homologies revealed. Anim. Genet., 31: 214-218.

McCouch, S.R., Y.G. Cho, M. Yano, E. Paul and M. Blinstrub. 1997. Report on QTL nomenclature. Rice Genet. Newsletter, 14: 11-13.

McLean, M., C.K. Okubo and M.L. Tracey. 1983. MtDNA heterogeneity in *Panulirus argus*. Experientia, 39: 536-538.

McMillen-Jackson, A.L., and T.M. Bert. 2003. Disparate patterns of population genetic structure and population history in two sympatric penaeid shrimp species (*Farfantepenaeus aztecus* and *Litopenaeus setiferus*) in the eastern United States. Mol. Ecol., 12(11): 2895-2905.

McMillen-Jackson, A.L., and T.M. Bert. 2004. Mitochondrial DNA variation and population genetic structure of the blue crab *Callinectes sapidus* in the eastern United States. Mar. Biol., 145(4): 769-777.

McMillen-Jackson, A.L., T.M. Bert and P. Steele. 1994. Population genetics of the blue crab, *Callinectes sapidus*: modest population structuring in a background of high gene flow. Mar. Biol., 118: 53-65.

Meehan, D., Z. Xu, G. Zuniga and A. Alcivar-Warren. 2003. High frequency and large number of polymorphic microsatellites in cultured shrimp, *Penaeus* (*Litopenaeus*) *vannamei* [crustacea: decapoda]. Mar. Biotechnol., 5(4): 311-330.

Moore, S.S., V. Whan, G.P. Davis, K. Byrne, D.J.S. Hetzel and N. Preston. 1999. The development and application of genetic markers for the kuruma prawn *Penaeus japonicus*. Aquaculture, 173: 19-32.

Moore, W.S. 1995. Inferring phylogenies from mtDNA variation: Mitochondrial-gene trees versus nuclear gene trees. Evolution, 49: 718-726.

Moritz, C. 1994. Applications of mitochondrial DNA analysis in conservation: a critical review. Mol. Ecol., 3: 401-411.

Mueller, U.G., and L.L. Wolfenbarger. 1999. AFLP genotyping and fingerprinting. Trends Ecol. Evol., 14: 389-394.

Murphy, R.W., J.W. Sites, D.G. Buth and C.H. Haufler. 1990. Proteins I: isozyme electrophoresis. *In:* Molecular Systematics. D.M. Hillis and C. Moritz (ed.). Sinauer, Sunderland, Mass., pp. 45-126.

Murphy, R.W., J.W., Sites, D.G. Buth and C.H. Haufler. 1996. Proteins: isozyme electrophoresis. In: Molecular Systematics Vol. 2. D.M. Hillis, C. Moritz and B.K. Mable (eds). Sinaeur Associates Inc., Sunder land, Massachusetts, USA, pp. 51-120.

Naylor, R.L., R.J. Goldburg, H. Mooney, M. Beveridge, J. Clay, C. Folke, N. Kautsky, J. Lubchenco, J. Primavera and M. Williams. 1998. Nature's subsidies to shrimp and salmon farming. Science, 282: 883-884.

Naylor, R.L, R.J. Goldburg, J.H. Primavera, N. Kautsky, M.C.M. Beveridge, J. Clay, C. Folke, J. Lubchenco, H. Mooney and M. Troell. 2000. Effect of aquaculture on world fish supplies. Nature, 405: 1017-1024.

Neigel, J.E. 1994. Analysis of rapidly evolving molecules and DNA sequence variants: Alternative approaches for detecting genetic structure in marine populations. CalCOFI Rep., 35: 82-89.

Neigel, J.E. 1997. A comparison of alternative strategies for estimating gene flow from genetic markers. Annu. Rev. Ecol. Syst., 28, 105-128.

Neigel, J.E. 2002. Is F_{ST} obsolete? Conserv. Genet., 3: 167-173.

Nemeth, S.T., and M.L. Tracey. 1979. Allozyme variability and relatedness in six crayfish species. J. Hered., 70: 37-43.

Neumann, V. 1998. A review of the *Maja squinado* (Crustacea: Decapoda: Brachyura) species complex with a key to the eastern Atlantic and Mediterranean species of the genus. J. Nat. Hist., 32: 1667-1684.

Nevo, E. 1978. Genetic variation in natural populations: patterns and theory. Theor. Pop. Biol., 13: 121-177.

Nguyen, T.T.T., N.P. Murphy and C.M. Austin. 2002. Amplification of multiple copies of mitochondrial Cytochrome b gene fragments in the Australian freshwater crayfish, Cherax destructor Clark (Parastacidae: Decapoda). Anim. Genet., 33: 304-308.

Nichols, K.M., W.P. Young, R.G. Danzmann, B.D. Robison, C. Rexroad, M. Noakes, R.B. Phillips, P. Bentzen, I. Spies, K. Knudsen, F.W. Allendorf, B.M. Cunningham, J. Brunelli, H. Zhang, S. Ristow, R. Drew, K.H. Brown, P.A. Wheeler and G.H. Thorgaard. 2003. A consolidated linkage map for rainbow trout (*Oncorhynchus mykiss*). Anim. Genet., 34(2): 102-115.

Nissen, L.R., P. Bjerregaard and V. Simonsen. 2005. Interindividual variability in metal status in the shore crab *Carcinus maenas*: the role of physiological condition and genetic variation. Mar. Biol., 146(3): 571-580.

Nixon, K.C., and Q.D. Wheeler. 1992. Measures of phylogenetic diversity. *In:* Extinction and Phylogeny. M.J. Novacek and Q.D. Wheeler (eds.). Columbia University Press, New York, pp. 217-234.

Norris, A.T., D.G. Bradley and E.P. Cunningham. 2000. Parentage and relatedness determination in farmed Atlantic salmon (*Salmo salar*) using microsatellite markers. Aquaculture, 182: 73-83.

Ovenden, J.R., D.J. Brasher and R.W.G. White. 1992. Mitochondrial DNA analyses of the red rock lobster *Jasus edwardsii* supports an apparent absence of population subdivision throughout Australia. Mar. Biol., 112: 319-326.

Page, R.D.M., and E.C. Holmes. 1998. Molecular Evolution: A Phylogenetic Approach. Blackwell Science, Oxford.

Palti, Y., J.E. Parsons and G.H. Thorgaard. 1999. Identification of candidate DNA markers associated with IHN virus resistance in backcrosses of rainbow (*Oncorhynchus mykiss*) and cutthroat trout (*O. clarki*). Aquaculture, 173: 81-94.

Pan, Y.W., H.H. Chou, E.M. You and H.T. Yu. 2004. Isolation and characterization of 23 polymorphic microsatellite markers for diversity and stock analysis in tiger shrimp (*Penaeus monodon*). Mol. Ecol., Notes 4: 345-347.

Parado-Estepa, F.D., E.T. Quinitio and E.M. Rodriguez. 2003. Seed Production of the Crucifix Crab *Charybdis feriatus*. Aqua KE Government Documents, VII(3): 37.

Passamonti, M., Mantovani, B., Scali, V., Froglia, C. 1997. Allozymic characterization of Scottish and Aegean populations of *Nephrops norregicus*. Journal of Marine Biological Association of the United Kingdom 77: 727-735

Paterson, H.E.H. 1993. The Recognition Concept of Species. The Johns Hopkins University Press, London.

Perez, F., C. Erazo, M. Zhinaula, F. Volckaert and J. Calderon. 2004. A sex specific linkage map of the white shrimp *Penaeus* (*Litopenaeus*) *vannamei* based on AFLP markers. Aquaculture, 242: 105-118.

Perez, F., J. Ortiz, M. Zhinaula, C. Gonzabay, J. Calderon and F.A.M.J. Volckaert. 2005. Development of EST-SSR markers by data mining in three species of shrimp: *Litopenaeus vannamei*, *Litopenaeus stylirostris*, and *Trachypenaeus birdy*. Mar. Biotechnol., 7: 554-569.

Perez-Enriquez, R., M. Takagi and N. Taniguchi. 1999. Genetic variability and pedigree tracing of a hatchery-reared stock of red sea bream (*Pagrus major*) used for stock enhancement, based on microsatellite DNA markers. Aquaculture, 173: 413-423.

Pérez-Rostro, C.I., and A.M. Ibarra. 2003. Heritabilities and genetic correlations of size traits at harvest size in sexually dimorphic Pacific white shrimp (*Litopenaeus vannamei*) grown in two environments. Aquacult. Res., 34(12): 1079-1085.

Pérez-Rostro, C.I., I.S. Racotta and A.M. Ibarra. 2004. Decreased genetic variation in metabolic variables of *Litopenaeus vannamei* shrimp after exposure to acute hypoxia. J. Exp. Mar. Biol. Ecol., 302(2): 189-200.

Petersen, C. 2006. Range Expansion in the Northeast Pacific by an Estuary Mud Crab - A Molecular Study. Biol. Invasions, 8(4): 565-576.

Place, A.R., X. Feng, C.R. Steven, H.M. Fourcade and J.L. Boore. 2005. Genetic markers in blue crabs (*Callinectes sapidus*) II. Complete mitochondrial genome sequence and characterization of genetic variation. J. Exp. Mar. Biol. Ecol., 319: 15-27.

Pogson, G.H., K.A. Mesa and R.G. Boutilier. 1995. Genetic population structure and gene flow in the Atlantic cod *Gadus morhua*: a comparison of allozyme and nuclear RFLP loci. Genetics, 139: 375-385.

Pongsomboon, S., V. Whan, S.S. Moore and A. Tassanakajon. 2000. Characterization of tri- and tetranucleotide microsatellites in the black tiger prawn *Penaeus monodon.* Sci. Asia, 26: 1-8.

Poompuang, S., and E.M. Hallerman. 1997. Toward detection of quantitative trait loci and marker-assisted selection in fish. Rev. Fish. Sci., 5: 253-277.

Porter, M.L., M. Pérez-Losada and K.A. Crandall. 2005. Model-based multi-locus estimation of decapod phylogeny and divergence times. Mol. Phylogenet. Evol., 37: 355-369.

Preston, N.P., D.C. Brennan and P.J. Crocus. 1999. Comparative costs of postlarvae production from wild or domesticated Kuruma shrimp, *Penaeus japonicus* (Bate), broodstock. Aquacult. Res., 30: 191-197.

Pritchard, J.K., M. Stephens and P. Donnelly. 2000. Inference of population structure using multilocus genotype data. Genetics, 55: 945-959.

Puebla, O., I. Partent and J.M. Sévigny. 2003. New microsatellite markers for the snow crab *Chionoecetes opilio* (Brachyura: Majidae). Mol. Ecol. Notes, 3: 644-646.

Quan, J., Z. Zhuang, J. Deng, J. Dai and Y.P. Zhang. 2004. Phylogenetic relationships of 12 Penaeoidea shrimp species deduced from mitochondrial DNA sequences. Biochem. Genet., 42: 331-345.

Queller, D.C., J.E. Strassmann and C.R. Hughes. 1993. Microsatellites and kinship. Trends Ecol. Evol., 8: 285-288.

Rafalski, A. 2002. Applications of single nucleotide polymorphisms in crop genetics. Curr. Opin. Plant Biol., 5: 94-100.

Rafalski, J.A., and S.V. Tingey. 1993. Genetic diagnostics in plant breeding: RAPDs, microsatellites and machines. Trends Genet., 9: 275-280.

Rannala, B., and J.L. Mountain. 1997. Detecting immigration by using multilocus genotypes. Proc. Natl. Acad. Sci. USA, 94: 9197-9201.

Reddy, P.V.G.K. 2000. Genetic resources of Indian major carps. FAO Fish. Tech. Pap. No. 387, 76 pp.

Rivera-García, M., and J.M. Grijalva-Chon. 2006. Genetic variability and differentiation in cultured white shrimp *Penaeus* (*Litopenaeus*) *vannamei* with low and high growth. Ciencias Marinas 32(1): 1-11.

Roder, M.S., J. Plaschke, S.U. Konig, A. Borner, M.E. Sorrells, S.D. Tanksley and M.W. Ganal. 1995. Abundance, variability and chromosomal location of microsatellites in wheat. Mol. Gen. Genet., 246: 327-333.

Roderick, G.K. 1996. Geographic structure of insect populations: gene flow, phylogeography, and their uses. Annu. Rev. Entomol., 41: 325-352.

Roman, J. 2006. Diluting the founder effect: cryptic invasions expand a marine invader's range. Proc. R. Soc. B. available from http://joeroman.com/images/roman.foundereffect.06.pdf

Roman, J., and S.R. Palumbi. 2004. A global invader at home: population structure of the green crab, *Carcinus maenas*, in Europe. Mol. Ecol., 13: 2891-2898.

Rosenberry, B. (ed.) 1998. World Shrimp Farming (Shrimp News International, San Diego, California).

Rothlisberg, P.C. 1998. Aspects of penaeid biology and ecology of relevance to aquaculture: A review. Aquaculture, 164(1-4): 49-65.

Ryu, S.H., Y.K. Do, U.W. Hwang, C.P. Choe and W. Kim. 1999. Ribosomal DNA intergenic spacer of the swimming crab, *Charybdis japonica*. J. Mol. Evol., 49: 806-809.

Saiki, R., D.H. Gelfand, S. Stoffel, S.J. Scharf, R. Higuchi, G.T. Horn, K.B. Mullis and H.A. Erlich. 1988a. Primer-directed enzymatic amplification of DNA with thermostable DNA polymerase. Science, 239: 487-491.

Saiki, R.K., V.B. Gyllensten and H.A. Erlich. 1988b. The polymerase chain reaction. *In:* Genome Analysis, A Practical Approach. K.E. Davis (ed.). IRL, Oxford, pp. 141-152.

Sakamoto, T., R.G. Danzmann, N. Okamoto, M.M. Ferguson and P.E. Ihssen. 1999. Linkage analysis of quantitative trait loci associated with spawning time in rainbow trout (*Oncorhynchus mykiss*). Aquaculture, 173: 33-43.

Sakamoto, T., R.G. Danzmann, K. Gharbi, P. Howard, A. Ozaki, S.K. Khoo, R.A. Woram, N. Okamoto, M.M. Ferguson, L.E. Holm, R. Guyomard and B. Hoyheim. 2000. A microsatellite linkage map of rainbow trout (*Oncorhynchus mykiss*) characterized by large sex-specific differences in recombination rates. Genetics, 155: 1331-1345.

Santucci, F., M. Iaconnelli, P. Andreani, R. Cianchi, G. Nascetti and L. Bullini. 1997. Allozyme diversity of European freshwater crayfish of the genus *Austropotamobius*. Bull. Fr. Pêche Pisc., 347: 663-676.

Sardá, F., C. Bas, M.I. Roldán, C. Pla and J. Lleonart. 1998. Enzymatic and morphometric analyses in mediterranean populations of the rose shrimp, *Aristeus antennatus* (Risso, 1816). J. Exp. Mar. Biol. Ecol., 221(1): 131-144.

Sarver, S.K., J.D. Silberman and P.J. Walsh. 1998. Mitochondrial DNA sequence evidence supporting the recognition of two subspecies or species of the Florida spiny lobster *Panulirus argus*. J. Crustacean. Biol., 18: 177-186.

Schlötterer, C. 2000. Evolutionary dynamics of microsatellite DNA. Chromosoma, 109: 365-371.

Schneider-Broussard, R., and J.E. Neigel. 1997. A large-subunit mitochondrial ribosomal DNA sequence translocated to the nuclear genome of two stone crabs (*Menippe*). Mol. Biol. Evol., 14(2): 156-165.

Schneider-Broussard, R., D.L. Felder, C.A. Chlan and J.E. Neigel. 1998. Tests of phylogeographic models with nuclear and mitochondrial DNA sequence variation in the stone crabs, *Menippe adina* und *Medina mercenaria*. Evolution, 52(6): 1671-1678.

Schubart, C.D., J.A. Cuesta, R. Diesel and D.L. Felder. 2000. Molecular phylogeny, taxonomy, and evolution of non-marine lineages within the American grapsoid crabs (Crustacea: Brachyura). Mol. Phylogenet. Evol., 15: 179-190.

Schubart, C.D., J.A. Cuesta and A. Rodríguez. 2001. Molecular phylogeny of the crab genus *Brachynotus* (Brachyura: Varunidae) based on the 16S rRNA gene. Hydrobiologia, 449: 41-46.

Segawa, R.D., and T. Aotsuka. 2005. The mitochondrial genome of the Japanese freshwater crab, *Geothelphusa dehaani* (Crustacea: Brachyura): evidence for its evolution via gene duplication. Gene, 355: 28-39.

Sévigny, J.-M., L. Savard and D.G. Parsons. 2000. Genetic characterization of the northern shrimp, *Pandalus borealis*, in the Northwest Atlantic using electrophoresis of enzymes. J. Northwest Atlantic Fish. Sci., 27: 161-175.

Shaklee, J.B. 1983. The utilization of isozymes as gene markers in fisheries management and conservation. *In:* Isozymes. Current Topics in Biological and Medical Research, Vol. 110. M. Rattazzi, J.G. Scandalios and G.S. Whitt (eds.). Alan R. Liss, Inc., New York, pp. 213-247.

Shekhar, M.S., and P.K. Chandra. 2000. Restriction profile of amplified mitochondrial DNA of wild shrimp (*Penaeus monodon* Fabricius) populations of India. Indian J. Mar. Sci., 29(1): 65-68.

Sites, J.W., Jr., and K.A. Crandall. 1997. Testing species boundaries in biodiversity studies. Conserv. Biol., 11: 1289-1297.

Skibinski, D.O.F. 1994. The potential of DNA techniques in the population and evolutionary genetics of aquatic invertebrates. *In:* Genetics and Evolution of Aquatic Organisms. A.R. Beaumont (ed.). Chapman & Hall, London, pp. 177-199.

Slade, R.W., C. Moritz, A.R. Hoelzel and H.B. Burton. 1998. Molecular population genetics of the southern elephant seal *Mirounga leonine*. Genetics, 149: 1945-1957.

Slatkin, M. 1985. Gene flow in natural populations. Annu. Rev. Ecol. Syst., 16: 393-430.

Slatkin, M. 1994. Linkage disequilibrium in growing and stable populations. Genetics, 137: 331-336.

Slatkin, M. 1995. A measure of population subdivision based on microsatellite allele frequencies. Genetics, 139: 457-462.

Slatkin, M., and N.H. Barton. 1989. A comparison of three indirect methods for estimating the average level of gene flow. Evolution, 43: 1349-1368.

Slatkin, M., and W.P. Maddison. 1989. Cladistic measure of gene flow inferred from the phylogenies of alleles. Genetics, 123: 603-613.

Slatkin, M., K. Hindar and Y. Michalakis. 1995. Processes of genetic diversification. *In:* V.H. Heywood and R.T. Watson (eds.). Global Biodiversity Assessment. UNEP-Cambridge Univ. Press, Cambridge, pp. 213-225.

Smith, P.J., W.R. Webber, S.M. McVeagh, G.J. Inglis and N. Gust. 2003. DNA and morphological identification of an invasive swimming crab, *Charybdis japonica*, in New Zealand waters. New Zealand J. Mar. Freshwater Res., 37: 753-762.

Sotelo, G., P. Morán and D. Posada. 2007. Identification and characterization of microsatellite loci in the spiny spider crab *Maja brachydactyla*. Conserv. Genet., 8: 245-247.

Soto-Hernández, J., and J.M. Grijalva-Chon. 2004. Genetic differentiation in hatchery strains and wild white shrimp *Penaeus* (*Litopenaeus*) *vannamei* (Boone, 1931) from northwest Mexico. Aquacult. Intl. 12(6): 593-601.

Stalling, L., F. Ford, D. Torney, C. Hildebrand and R. Moyzis. 1991. Evolution and distribution of (GT)n repetitive sequences in mammalian genomes. Genomics, 10: 807-815.

Stamatis, C., A. Triantafyllidis, K.A. Moutou and Z. Mamuris. 2004. Mitochondrial DNA variation in Northeast Atlantic and Mediterranean populations of Norway lobster, *Nephrops norvegicus*. Mol. Ecol., 13(6): 1377-1390.

Stamatis, C., A. Triantafyllidis, K.A. Moutou and Z. Mamuris. 2006. Allozymic variation in Northeast Atlantic and Mediterranean populations of Norway lobster, *Nephrops norvegicus*. ICES J. Mar. Sci., 63(5): 875-882.

Steven, C.R., J. Hill, B. Masters and A.R. Place. 2005. Genetic markers in blue crabs (*Callinectes sapidus*) I: Isolation and characterization of microsatellite markers. J. Exp. Mar. Biol. Ecol., 319: 3-14.

Streiff, R., T. Guillemaud, F. Alberto, J. Magalhaes, M. Castro and M.L. Cancela. 2001. Isolation and characterization of microsatellite loci in the Norway lobster (*Nephrops norvegicus*). Mol. Ecol. Notes 1: 71-72.

Streiff, R., S. Mira, M. Castro and M.L. Cancela. 2004. Multiple paternity in Norway lobster (*Nephrops norvegicus* L.) assessed with microsatellite markers. Mar. Biotechnol., 6: 60-66.

Sugama, K., Haryanti, J.A.H. Benzie and E. Ballment. 2002. Genetic variation and population structure of the giant tiger prawn, *Penaeus monodon*, in Indonesia. Aquaculture, 205(1-2): 37-48.

Sugaya, T., M. Ikeda and N. Taniguchi. 2002a. Relatedness structure estimated by microsatellites DNA and mitochondrial DNA polymerase chain reaction - Restriction fragment length polymorphisms analyses in the wild population of kuruma prawn *Penaeus japonicus*. Fish. Sci., 68(4): 793-802.

Sugaya, T., M. Ikeda, H. Mori and N. Taniguchi. 2002b. Inheritance mode of microsatellite DNA markers and their use for kinship estimation in kuruma prawn *Penaeus japonicus*. Fish. Sci., 68(2): 299-305.

Sunnucks, P. 2000. Efficient genetic markers for population biology. Trends Ecol. Evol., 15: 199-203.

Supungul, P., P. Sootanan, S. Klinbunga, W. Kamonrat, P. Jarayabhand and A. Tassanakajon. 2000. Microsatellite polymorphism and the population structure of the black tiger shrimp (*Penaeus monodon*) in Thailand. Mar. Biotechnol., 2(4): 339-347.

Tam, Y.K., and I. Kornfield. 1998. Phylogenetic relationships of clawed lobster genera (Decapoda: Nephropidae) based on mitochondrial 16S rRNA gene sequences. J. Crustacean Biol., 18: 138-146.

Tan, G., A. Karsi, P. Li, S. Kim, X. Zheng, H. Kucuktas, B.J. Argue, R.A. Dunham and Z.J. Liu. 1999. Polymorphic microsatellite markers in *Ictalurus punctatus* and related catfish species. Mol. Ecol., 8: 1758-1760.

Tang, C., W. Lu, G. Wainwright, S.G. Webster, H.H. Rees and P.C. Turner. 1999. Molecular characterization and expression of mandibular organ-inhibiting hormone, a recently discovered neuropeptide involved in the regulation of growth and reproduction in the crab *Cancer pagurus*. Biochem. J., 343: 355-360.

Tassanakajon, A., S. Pongsomboon, V. Rimphanitchayakit, P. Jarayabhand and V. Boonsaeng. 1997. Random amplified polymorphic DNA (RAPD) markers for determination of genetic variation in of the black tiger prawn (*Penaeus monodon*) in Thailand. Mol. Mar. Biol. Biotechnol., 6(2): 110-115.

Tassanakajon, A., S. Pongsomboon, P. Jarayabhand, S. Klinbunga and V. Boonsaeng. 1998a. Genetic structure in wild population of black tiger shrimp (*Penaeus monodon*) using randomly amplified polymorphic DNA analysis. J. Mar. Biotechnol., 6: 249-254.

Tassanakajon, A., A. Tiptawongnukul, P. Supungul, V. Rimpanitchayakit, D. Cook, P. Jarayabhand, S. Klinbunga and V. Boonsaeng. 1998b. Isolation and characterization of microsatellite markers in the black tiger prawn *Penaeus monodon*. Mol. Mar. Biol. Biotechnol., 7: 55-61.

Tautz, D. 1989. Hypervariability of simple sequences as a general source for polymorphic DNA markers. Nucl. Acids Res., 17: 6463-6471.

Taylor, M.S., and M.E. Hellberg. 2003. Genetic evidence for local retention of pelagic larvae in a Caribbean reef fish. Science, 299: 107-109.

Templeton, A.R. 1981. Mechanisms of speciation - a population genetic approach. Annu. Rev. Ecol. Syst., 12: 23-48.

Templeton, A.R. 1989. The meaning of species and speciation: a genetic prospective. In: Speciation and its consequences, D. Ofte and J.A. Endler (eds). Sunderland, Massachusetts, Sinaner Associates Inc., USA, pp. 3-27.

Templeton, A.R. 1991. Off-site breeding of animals and implications for plant conservation strategies. *In:* Genetics and Conservation of Rare Plants. D.A. Falk and K.E. Holsinger (eds.). Oxford University Press, New York.

Templeton, A.R. 1998. Nested clade analysis of phylogeographic data: Testing hypotheses about gene flow and population history. Mol. Ecol., 7: 381-397.

Tepolt, C.K., M.J. Bagley, J.B. Geller and M.J. Blum. 2006. Characterization of microsatellite loci in the European green crab (*Carcinus maenas*). Mol. Ecol. Notes, 6: 343-345.

Thongpan, A., M. Mingmuang, S. Thinchant, R. Cooper, T. Tiersch and K. Mongkonpunya. 1997. Genomic identification of catfish species by polymerase chain reaction and restriction enzyme analysis of the gene encoding the immunoglobulin M heavy chain constant region. Aquaculture, 156: 129-137.

Thorpe, J.P., and J. Smartt. 1995. Genetic diversity as a component of biodiversity. *In:* Global Biodiversity Assessment. V.H. Heywood and R.T. Watson (eds.). UNEP-Cambridge Univ. Press, Cambridge, pp. 57-88.

Tong, J.G., T.Y. Chan and K.H. Chu. 2000. A preliminary phylogenetic analysis of *Metapenaeopsis* (Decapoda: Penaeidae) based on mitochondrial DNA sequences of selected species from the Indo-West Pacific. J. Crustacean Biol., 20: 541-549.

Tong, J., S. Lehnert, K. Byrne, H.S. Kwan and K.H. Chu. 2002. Development of polymorphic EST markers in *Penaeus monodon*: applications in penaeid genetics. Aquaculture, 208: 69-79.

Toonen, R.J., M. Locke and R. Grosberg. 2004. Isolation and characterization of polymorphic microsatellite loci from the Dungeness crab, *Cancer magister*. Mol. Ecol. Notes, 4: 30-32.

Toth, G., Z. Gaspari and J. Jurka. 2000. Microsatellites in different eukaryotic genomes; survey and analysis. Genome Res., 10: 967-981.

Triantafyllidis, A., A.P. Apostolidis, V. Katsares, E. Kelly, J. Mercer, M. Hughes, K.E. Jorstad, A. Tsolou, R. Hynes and C. Triantaphyllidis. 2005. Mitochondrial DNA variation in the European lobster (*Homarus gammrus*) throughout the range. Mar. Biol., 146: 223-235.

Tsoi, K.H., Z.Y. Wang and K.H. Chu. 2005. Genetic divergence between two morphologically similar varieties of the kuruma shrimp *Penaeus japonicus*. Mar. Biol., 147(2): 367-379.

Tsoi, K.H., T.Y. Chan and K.H. Chu. 2007. Molecular population structure of the kuruma shrimp *Penaeus japonicus* species complex in western Pacific. Mar. Biol., 150(6): 1345-1364.

Turner, G.F., R.L. Robinson, B.P. Ngatunga, P.W. Shaw and G.R. Carvalho. 2002. Pelagic cichlid fishes of Lake Malawi/Nyasa: biology, management and conservation. *In:* Management and Ecology of Lake and Reservoir Fisheries. I.G. Cowx (ed.). Fishing News Books, Blackwell Science Ltd., Oxford, pp. 353-367.

Tzeng, T., S. Yeh and C. Hui. 2004a. Population genetic structure of the kuruma prawn (*Penaeus japonicus*) in East Asia inferred from mitochondrial DNA sequences. ICES J. Mar. Sci., 61: 913-920.

Tzeng, Y.-H., R. Pan and W.H. Li. 2004b. Comparison of three methods for estimating rates of synonymous and non-synonymous nucleotide substitutions. Mol. Biol. Evol., 21(12): 2290-2298.

Ulrich, I., J. Müller, C. Schütt and F. Buchholz. 2001. A study of population genetics in the European lobster, *Homarus gammarus* (Decapoda, Nephropidae). Crustaceana, 74(9): 825-837.

Urbani, N., J.M. Sevigny, B. Sainte-Marie, D. Zadworny and U. Kuhnlein. 1998. Identification of microsatellite markers in the snow crab *Chionoecetes opilio.* Mol. Ecol., 7: 357-363.

Utter, F.M. 1991. Biochemical genetics and fishery management: an historical perspective. J. Fish Biol., 39(Suppl. A): 1-20.

Valles-Jimenez, R., P. Cruz and R. Perez-Enriquez. 2004. Population genetic structure of Pacific white shrimp (*Litopenaeus vannamei*) from Mexico to Panama: Microsatellite DNA variation. Mar. Biotechnol., 6(5): 475-484.

Valles-Jimenez, R., P.M. Gaffney and R. Perez-Enriquez. 2006. RFLP analysis of the mtDNA control region in white shrimp (*Litopenaeus vannamei*) populations from the eastern Pacific. Mar. Biol., 148(4): 867-873.

van Wormhoudt, A., and D. Sellos. 2003. Highly variable polymorphism of the α-amylase gene family in *Litopenaeus vannamei* (Crustacea Decapoda). J. Mol. Evol., 57(6): 659-671.

Vane-Wright, R.I., C.J. Humphries and P.H. Williams. 1991. What to protect? Systematics and the agony of choice. Biol. Conserv. 55: 235-254.

Vázquez-Bader, A.R., J.C. Carrero, M. Gárcia-Varela, A. Gracia and J.P. Laclette. 2004. Molecular phylogeny of superfamily Penaeoidea Rafinesque-Schmaltz, 1815, based on mitochondrial 16s partial sequence analysis. J. Shellfish Res., 23: 911-917.

Velez, J.R., R. Escobar, F. Correa and E. Felix. 1999. High allozyme variation and genetic similarity of two populations of commercial penaeids, *Penaeus brevirostris* (Kingsley) and *P. vannamei* (Boone), from the Gulf of California. Aquacult. Res., 30: 459-463.

Vignal, A., D. Milan, M. San Cristobal and A. Eggen. 2002. A review on SNP and other types of molecular markers and their use in animal genetics. Gen. Sel. Evol., 34: 275-305.

Vos, P., R. Hogers, M. Bleeker, M. Reijans, T. Van De Leet, M. Hornes, A. Fijters, J. Pot, J. Peleman, M. Kuiper and M. Zabeau. 1995. AFLP: A new technique for DNA fingerprinting. Nucl. Acids Res., 23: 4407-4414.

Waldbieser, G., Z.L. Liu, W. Wolters and R.A. Dunham. 1998. Genetic linkage and QTL mapping of catfish. Proceedings of the National Aquaculture Species Gene Mapping Workshop, USA.

Waldbieser, G.C., B.G. Bosworth, D.J. Nonneman and W.R. Wolters. 2001. A microsatellite-based genetic linkage map for channel catfish, Ictalurus punctatus. Genetics, 158: 727-734.

Wang, Z.Y., K.H. Tsoi and K.H. Chu. 2004. Applications of AFLP technology in genetic and phylogenetic analysis of penaeid shrimp. Biochem. Syst. Ecol., 32(4): 399-407.

Wanna, W., J. Rolland, F. Bonhomme and A. Phongdara. 2004. Population genetic structure of *Penaeus merguiensis* in Thailand based on nuclear DNA variation. J. Exp. Mar. Biol. Ecol., 311: 63-78.

Wanna, W., W. Chotigeat and A. Phongdara. 2006. Sequence variations of the first ribosomal internal transcribed spacer of *Penaeus* species in Thailand. J. Exp. Mar. Biol. Ecol., 331(1): 64-73.

Ward, R.D. 2000. Genetics in fisheries management. Hydrobiologia, 420: 191-201.

Ward, R.D., and P.M. Grewe. 1994. Appraisal of molecular genetic techniques in fisheries. Rev. Fish Biol. Fish., 4: 300-325.

Ward, R., and P.M. Grewe. 1995. Appraisal of molecular genetic techniques in fisheries. *In:* Molecular Genetics in Fisheries. G.P. Carvalho and J.J. Pitcher (eds.). Chapman & Hall, London.

Warner, R.R., and R.K. Cowen. 2002. Local retention of production in marine populations: evidence, mechanisms and consequences. Bull. Mar. Sci., 70(1): 245-249.

Waser, P.M., and C. Strobeck. 1998. Genetic signature of interpopulation dispersal. Trends Ecol. Evol., 13: 43-44.

Weber, J., and P. May. 1989. Abundant class of human DNA polymorphisms which can be typed using the polymerase chain reaction. Am. J. Human Genet., 44: 388-396.

Weir, B.S. 1990. Genetic data analysis. Sinauer Sunderland, Mass.

Welsh, J., and M. McClelland. 1990. Fingerprinting genomes using PCR with arbitrary primers. Nucl. Acids Res. 18: 7213-7218.

Whiting, A.S., S.H. Lawler, P. Horwitz and K.A. Crandall. 2000. Biogeographic regionalisation of Australia: Assigning conservation priorities based on endemic freshwater crayfish phylogenetics. Anim. Conserv., 3: 155-163.

Wiley, E.O. 1978. The evolutionary species concept reconsidered. Syst. Zool., 27: 17-26.

Williams, J.G.K., A.R. Kubelik, K.J. Livak, J.A. Rafalski and S.V. Tingey. 1990. DNA polymorphism amplified by arbitrary primers are useful as genetic markers. Nucl. Acids Res., 18: 6531-6535.

Wilson, A.C., R.L. Cann, S.M. Carr, M. George, U.B. Gyllensten, K.M. Helm-Bychowski, R.G. Higuchi, S.R. Palumbi, E.M. Prager, R.D. Sage and M. Stoneking. 1985. Mitochondrial DNA and two perspectives on evolutionary genetics. Biol. J. Linn. Soc., 26: 375-400.

Wilson, E.O. 1988. The current state of biological diversity. *In:* Biodiversity. E.O. Wilson and F.M. Peter (eds.). National Academy Press, Washington, D.C., pp. 3-18.

Wilson, K., Y. Li, V. Whan, S. Lehnert, K. Byrne, S. Moore, S. Pongsomboon, A. Tassanakajon, G. Rosenberg, E. Ballment, Z. Fayazi, J. Swan, M. Kenway and J. Benzie. 2002. Genetic mapping of the black tiger shrimp *Penaeus monodon* with fragment length polymorphism. Aquaculture, 204: 297-309.

Wolfus, G.M., G.K. García and A. Alcivar-Warren. 1997. Application of the microsatellites technique for analyzing genetic diversity in shrimp breeding programs. Aquaculture, 152: 35-47.

Wolstenholme, D.R. 1992. Animal Mitochondrial DNA: Structure and Evolution. Academic Press, New York.

Wongteerasupaya, C., S. Wongwisansri, V. Boonsaeng, S. Panyim, P. Pratanpipat, G. Nash, B. Withyachumnarnkul and T. Flegel. 1996. DNA fragment of *Penaeus monodon* baculovirus PmNOBII gives positive in situ hybridization with white-spot viral infections in six penaeid shrimp species. Aquaculture, 143: 23-32.

Wright, J.M. 1990. DNA fingerprinting in fishes. *In:* Biochemistry and Molecular Biology of Fishes, Vol. 2. P. Hochachka and T Mommsen (eds.). Elsevier, Amsterdam, The Netherlands, pp. 57-91.

Wright, J.M., and P. Bentzen. 1994. Microsatellites: genetic markers for the future. Rev. Fish Biol. Fish., 4: 384-388.

Wuthisuthimethavee, S., P. Lumubol, A. Vanavichit and S. Tragoonrung. 2003. Development of microsatellite markers in black tiger shrimp (*Penaeus monodon* Fabricius). Aquaculture, 224(1-4): 39-50.

Xu, Z., A.K. Dhar, J. Wyrzykowski and A. Alcivar-Warren. 1999. Identification of abundant and informative microsatellites from shrimp (*Penaeus monodon*). Anim. Genet., 30: 150-156.

Xu, Z., J.H. Primavera, L. De la Pena, P. Petit, J. Belak and A. Alcivar-Warren. 2001. Genetic diversity of wild and cultured black tiger shrimp (*Penaeus monodon*) in the Philippines using microsatellites. Aquaculture, 199: 13-40.

Yamauchi, M.M., M.U. Miya and M. Nishida. 2002. Complete mitochondrial DNA sequence of the Japanese spiny lobster, *Panulirus japonicus* (Crustacea: Decapoda). Gene, 295: 89-96.

Yamauchi, M.M., M.U. Miya, and M. Nishida. 2003. Complete mitochondrial DNA sequence of the swimming crab, *Portunus trituberculatus* (Crustacea: Decapoda: Brachyura). Gene, 311: 129-135.

Yamauchi, M.M., M.U. Miya, R.J. Machida and M. Nishida. 2005. PCR-based approach for sequencing mitochondrial genomes of Decapod crustaceans, with a practical example from Kuruma prawn (*Marsupenaeus japonicus*). Mar. Biotechnol., 6: 419-429.

Yap, E.S., E. Sezmis, J.A. Chaplin, I.C. Potter and P.B.S. Spencer. 2002. Isolation and characterization of microsatellite loci in *Portunus pelagicus* (Crustacea: Portunidae). Mol. Ecol. Notes, 2: 30-32.

Young, W.P., P.A. Wheeler, R.D. Fields and G.H. Thorgaard. 1996. DNA fingerprinting confirms isogenicity of androgenetically-derived rainbow trout lines. J. Hered., 87: 77-81.

Young, W.P., P.A. Wheeler, V.H. Coryell, P. Keim and G.H. Thorgaard. 1998. A detailed linkage map of rainbow trout produced using doubled haploids. Genetics, 148: 839-850.

Zabeau, M., and P. Vos. 1993. Selective restriction fragment amplifications: a general method for DNA fingerprinting. European Patent Application 92402629 (Pub 1. No. 0534 858 A1).

Zhi-Ying, J., S. Xiao-Wen, L. Li-Qun, L. Da-Yu and L. Qing-Quan. 2006. Isolation and characterization of microsatellite markers from Pacific white shrimp (*Litopenaeus vannamei*). Mol. Ecol. Notes, 6(4): 1282-1284.

Zrzavý, J., and P. Štys. 1997. The basic body plan of arthropods: insights from evolutionary morphology and developmental biology. J. Evol. Biol., 10: 353-367.

4

Reproduction of Crustaceans in Relation to Fisheries

H.A. Lizárraga-Cubedo[1], G.J. Pierce[1] and M.B. Santos[1,2]

[1]School of Biological Sciences [Zoology], University of Aberdeen, Tillydrone Avenue, Aberdeen AB24 2TZ, UK

[2]Instituto Español de Oceanografía, Centro Oceanográfico de Vigo, P.O. Box 1552, 36200 Vigo, Spain

E-mail(s): halizarraga@yahoo.com; g.j.pierce@abdn.ac.uk; m.b.santos@vi.ieo.es; g.j.pierce@vi.ieo.es

Current postal address: *337E First Street, Calexico, CA, USA. 92231 PMB 21249*

Tel: 0052 983 1028662 (Mexico)

OVERVIEW

For many decades, decapods have represented a valuable source of food for human consumption and income for fishermen around the world. Crustaceans, unlike fish, are characterized by discontinuous growth, making it difficult to apply traditional growth models. Studying reproduction in commercially exploited populations is important for fisheries management purposes.

The main objective of this chapter is to present, in detail, some of the most important aspects of crustacean reproduction in relation to their exploitation by fisheries.

First, we briefly describe historical catch trends (commercial value and catch weight) in some of the most commercially important decapod fisheries. The purpose of this exercise is to highlight the importance of crustacean fisheries as a constant source of food and revenue around the world. We emphasize some of the fishery-related factors that influence the fluctuations in crustacean populations through time. We focus in particular on Homarid lobster species, especially *Homarus americanus* and *H. gammarus*, which are among the best-known and most studied species.

The second topic concerns management and regulatory measures that can be adopted to ensure efficient and sustainable exploitation of crustacean fisheries. These include catch-effort restrictions, minimum legal landing size, a ban on landings of berried females, closed seasons, and protected areas or no-take zones. We focus in particular on berried females, the protection of which is essential to ensure survival of the population.

In addition, we examine the changes in growth that accompany sexual maturation, and how maturation and reproductive success can be altered by environmental factors. Information on reproductive biology and growth in lobsters is included in Chapter 10 of this book; here we simply describe the relevance of these parameters to studies on fisheries of other commercially important crustaceans. Instead, we focus our investigations on sexual maturity.

Information on physical, functional and behavioural maturity is based on the study of gonad development, *vasa deferentia*, ovaries, maturity ogives, and the use of secondary sexual characters to detect the size at onset of sexual maturity (SOM). We discuss the different statistical procedures available to detect the onset of sexual maturity.

Resilience of populations to fishery-induced or environmental factors can be observed through changes in egg production. Therefore, the analysis of fecundity is essential. Crustaceans, like fish, are susceptible to high levels of exploitation and this can have detrimental effects on the exploited populations, expressed as recruitment failure. Egg production is dependent on size-class and strongly related to size at maturity, frequency of spawning, fecundity and numbers of animals in specific size-classes. We review the relationships between size-class and fecundity, egg development and their spatial variability.

Reproductive parameters of some stocks may vary spatially because of migration patterns, which are important for fisheries management purposes and are therefore reviewed here.

Once biological parameters have been obtained for putative populations, these can be useful in stock discrimination. Lastly, we discuss stock assessment methods.

INTRODUCTION

Crustacean Fisheries

The effects of fishery processes on the reproduction of commercially exploited crustaceans are complex. The population responses to exploitation can be manifested in changes in the reproductive biology, with implications for the fisheries in the short and long term. Studying the factors that affect fluctuations in abundance, recruitment, and spawning success (i.e., egg production) is of high priority in fisheries management. Considerable differences exist between crustaceans and finfish, not only in their reproductive and behavioural characteristics, but also in their observed responses to fishing activity. As a consequence, strategies, practices, management objectives, management plans and management tools that are common in finfish fisheries do not necessarily apply to crustacean fisheries (Hilborn, 1986). However, there are some management regulations that can be seen in both finfish and crustacean fisheries. This is the case for catch limits, effort limits, protection of the reproductive components of the population (e.g., a ban on landings of gravid females), and protected areas, among others.

Crustaceans, like many other fishery resources, have been subjected to high levels of exploitation for decades (FAO, 2004). This has led scientists to express concerns about possible stock depletion and recruitment failure (Ennis, 1986; Tully et al., 2001), occurring as a consequence of over-exploitation, inadequate (or inappropriate) management strategies, variation in the patterns of ocean-atmospheric processes (e.g., decadal climate events), habitat alteration, and fluctuations in egg and larval production in the exploited populations (Polovina et al., 1995). Therefore, describing the actual state of exploited populations and the factors affecting them will indicate where more research effort is needed.

Figure 1 shows a general picture of landings from crustacean fisheries around the world, as provided by the Fisheries Global Information System (FIGIS) of FAO for the period 1990-2003, divided into four different components: (1) shrimps and prawns, (2) lobsters (spiny and clawed), (3) spider crabs, and (4) king crabs and squat lobsters. The figure indicates both the relevance of crustacean fisheries around the world and the importance of specific groups in each continent.

Shrimps and prawns present the highest recorded catches worldwide, with increasing trends for the last decade (Fig. 1b). The highest landings, consisting mainly of Penaeid shrimps (*Penaeus* spp.), are recorded for the Asian continent and the second highest in North

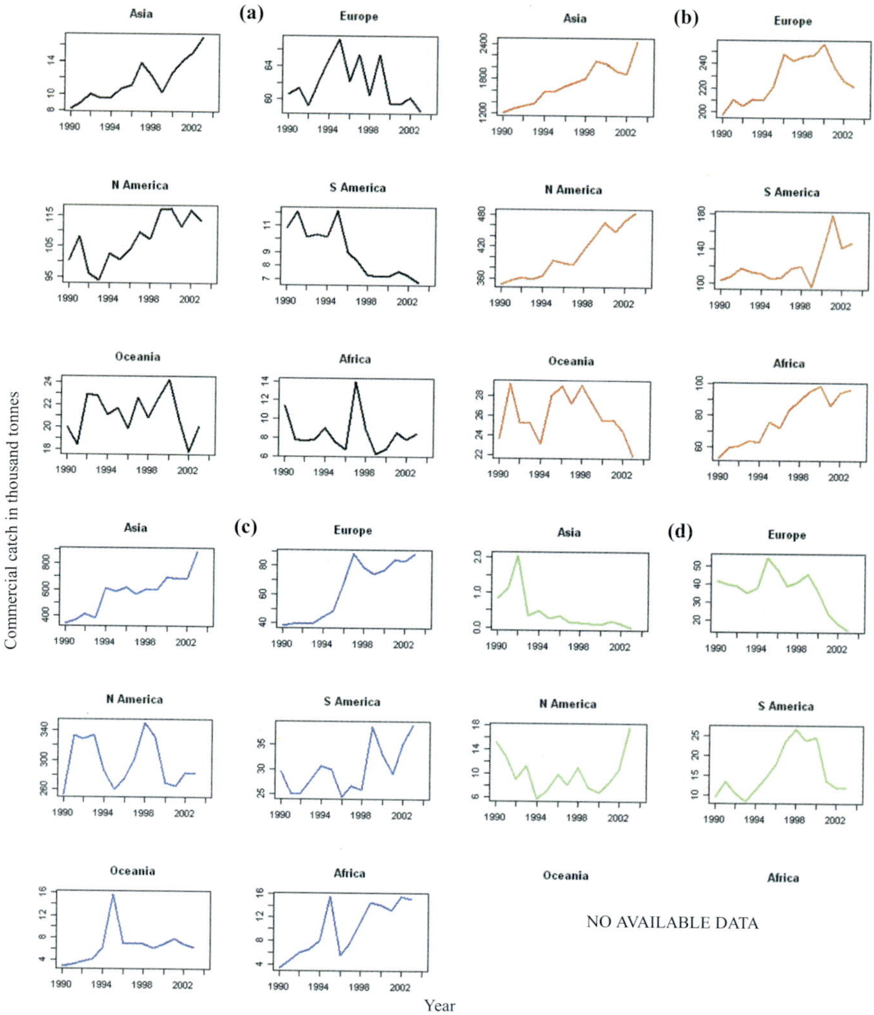

Fig. 1 Catch time series for some of the most commercially important categories of crustaceans in marine waters around the world: (a) lobsters, (b) shrimp and prawns, (c) spider crabs, and (d) king crabs and squat lobsters. Data were taken from the FIGIS (Fisheries Global Information System) FAO database for the period 1990-2003.

America, where exploitation of Pandalid shrimps in cold waters (*Pandalus* spp.) dominates the records.

Fisheries for king crabs and squat lobsters show consistent decreases in catches in recent years (Fig. 1d). There are marked fluctuations in landings in North America, where fisheries for king crabs (Lithodid crabs) are important, and these stocks have faced high levels of exploitation over the last two decades (Briand et al., 2001). In South America, Galatheid (squat lobsters) species have been traditionally exploited (Arancibia et al., 2005). No records of landings of king crabs and squat lobsters were obtained for Oceania and Africa. From this information, it can be deduced that the fisheries for these groups around the world are in a developing state or that their abundance is limited.

Spider crab fisheries are commercially important around the world. Figure 1c shows the recorded steady increases in landings during 1990-2003 for Asia, Europe, South America and Africa, contrasting with marked oscillations for North America. The high landings in Europe may be caused, in part, by an increase in the exploitation of Majid crabs (i.e., *Maja crispata*). On the other hand, lobster catches are highest for North America, with a sustained increase in recent years, and lowest for South America, where there has been a consistent decline (Fig. 1a).

Catch information can be better interpreted if other variables are considered, namely the commercial value of the catch, the fishing gear used, fishing effort, and an appropriate separation of catch into the different component species. Searching for explanations for fluctuations in landings for all the species included in Fig. 1 would, however, be of little use. Instead, we have selected two specific examples that we use as case studies: the lobster fisheries in North America and Europe (Fig. 1a), which exploit two clawed lobster species, the American lobster (*Homarus americanus*) and the European lobster (*H. gammarus*) respectively, for which there is abundant information in the literature.

Management Measures

Homarus americanus and *H. gammarus* show remarkable resemblances both in their reproductive characteristics and the associated fishery practices. However, while lobster catches in North America increased steadily during the period 1990-2003, commercial catches of lobsters in Europe showed negative trends over the same period of time (Fig. 1a). Furthermore, lobster catches are much higher in North America than in Europe, with maximum values of nearly 120,000 t in 1999-2000 and 2002. The value of lobster catches in Canadian waters was US$37.4 million in 1999 according to Lawton et al. (2001). In Europe, lobster landings reached a peak of 67,000 t in 1995. According to Scottish Sea Fisheries Statistical Tables, the reported landings of lobsters in 1999 for all coasts of Scotland were about 509 metric t, with a value of about US$8 million. This indicates evident differences, suggesting that the processes involved in the exploitation of both species have unique and particular characteristics.

The high historical exploitation of the American lobster in the United States and Canada has forced the vigilant application and careful selection of management strategies and regulations. For the European counterpart, despite high catch levels, management regulations are limited. Throughout its range of occurrence, regulation of the European lobster fishery is mainly confined to setting a minimum landing size (87 mm body length or CL in the UK), with no limitations on the number of fishing vessels, total catch, duration of fishing season, fishing effort or gear type (Annala and Sullivan, 1997). In some areas, however, the lobster fishery season is restricted by the prevailing environmental conditions, acting as a limiting factor or a sort of subtle seasonal regulation (Lizárraga-Cubedo, 2004). Some further management measures have been proposed, such as fishing effort limits and protection of berried females (i.e., females carrying eggs) (Annala and Sullivan, 1997).

The American lobster fisheries in Canada and the United States are regulated by limited entry, trap limits, closed seasons, minimum landing size, protection of berried females, and fishing quotas (Annala and Sullivan, 1997). Closed areas, an increase in landing size limit and introduction of a maximum size limit are also being considered (Annala and Sullivan, 1997).

Establishment of particular geographic areas for lobster exploitation in Canadian waters, such as Lobster Fishing Areas (LFAs) (Lawton et al., 2001), allows close monitoring of lobster populations, aimed at revealing the effects of established management strategies and regulations. Throughout the world, the application and selection of fishery management measures vary according to the knowledge available about the species under exploitation. Annala and Sullivan (1997) give a brief description of management regulations in some of the most important spiny and clawed lobster fisheries in the world. From their work, the lack of regulations that apply to lobster fisheries in the United Kingdom is apparent, contrasting with the numerous and diverse management regulations that are applied in other fisheries in the world. Canadian and Australian fisheries, for clawed lobsters and spiny lobsters respectively, are subject to a considerable number of regulations.

Fishing Gear

Capture of crustaceans is a complex process. Different types of fishing gear are employed and their use is strongly dependent on the biological characteristics of the targeted species. Trawling is commonly used for the capture of shrimps (*Penaeus* spp.) and some prawns (Annala and

Sullivan, 1997). Although *Nephrops norvegicus* are normally called prawns, they are clawed lobsters and are also caught by trawling and trapping. In some countries, management regulations for *Nephrops* are based on square mesh panels, as is the case in the United Kingdom (Annala and Sullivan, 1997; Tuck et al., 1997a). In general, lobsters are caught in traps, as are the majority of commercially exploited species of crabs around the world.

The European lobster fishery is a mixed fishery, with both *H. gammarus* and the crab *Cancer pagurus* found in the catches. Similarly, the American lobster is also exploited by a trap-based fishery, in which crabs, *Cancer irroratus*, *C. borealis*, and some species of spider crabs (*Libnia* spp.), can also be simultaneously targeted (Richards and Cobb, 1987). Catch success, catch probability and catch abundance are affected by interspecific and intraspecific behavioural interactions within the traps (Richards et al., 1983). In general, capture processes in crustacean fisheries depend on the agonistic behaviour, sexual maturity and moulting processes of the species under exploitation, as well as food availability, type of bait, soak time, trap design, fishermen's knowledge, time of year, habitat, and other environmental variables, among other factors (Thomas, 1959a, b; Caddy, 1989).

Catch-effort and Size Limits

Successful implementation of management regulations must in general follow the principles of conservation, aiming at achieving the sustainability of the resource. The inappropriate application of management regulations may have detrimental impacts on the biological and reproductive parameters of exploited populations.

In trap-based fisheries, e.g., the *Homarus* spp. lobster fishery, reproductive parameters may be affected by factors such as the degree of exploitation to which a population is subjected over time, intraspecific interactions, gear design, gear selectivity and substrate (Addison et al., 1995; Addison and Lovewell, 1985).

Traps are passive fishing gear that target animals within a certain range of sizes. Intraspecific behavioural interactions and gear design ensure that *Homarus* spp. lobsters as big as 60 mm CL are found in the traps (Addison et al., 1995) with an overall retention rate of about 6% (Jury et al., 2001). The lobsters found in the traps may not necessarily reflect the size composition of the stock under exploitation because of the low retention rate, and the fact that bigger lobsters impede the entry of smaller animals into the traps (Addison and Lovewell, 1985; Lovewell and Addison, 1986). In addition, from trap catches, Homarid lobster distribution appears to be uniform or random among traps, but this may

the response of berried females to different factors such as fishing gear design, intra- and interspecific behavioural interactions in the traps, and seasonality. This may pose a problem if the actual reproductive state of the population is going to be estimated (Bennett, 1995; Smith et al., 2004). Berried females of the edible crab *Cancer pagurus* very rarely enter the traps (Bennett, 1995). Identifying these spawning components is generally possible only through larval sampling (Thompson et al., 1995).

From a management point of view, knowledge of the state of the spawning components of a fishery is necessary at different geographical and spatial scales. If information on the reproductive state of the population is limited, for example, because of the absence of berried females in the catches, then it is likely that stock-recruitment relationships will be difficult to obtain.

On the other hand, closed areas, marine reserves or no-take zones are considered to be one of the most recent and effective crustacean management regulations applied around the world (Childress, 1997). Positive results have been obtained from studies on biological and reproductive parameters of formerly exploited lobster stocks inside no-take zones. Comparisons of highly exploited areas, open to fishing, with closed areas have shown increases in lobster abundance and size, for both males and females, in the closed areas and, in some cases, an increase in the frequency of berried lobsters has been registered over time (Childress, 1997; Rowe, 2002).

The positive impacts that closed areas have on lobster populations are considerable, but more investigations at larger spatial scales for small refuges and good monitoring programmes are necessary (Rowe, 2002). From all this information it is deduced that the success of management regulations depends on knowledge of the biological-reproductive parameters of the population under investigation and that, in most cases, there is a need for a combination of simultaneously implemented regulations.

Growth, Moulting and Reproduction

Moulting, Growth and Size Structure

From a fisheries point of view, acquiring knowledge of the structure or composition of a population under exploitation is highly relevant for successful management. One option is the implementation of analytical models, which consider a population dynamics as a function of recruitment, growth and mortality of individuals composing the population (Gulland, 1987).

For crustacean fisheries, analysing size might be more relevant than analysing age (and in most cases is the only option), and size is often used to describe demographic parameters of a population in relation to fishing pressure (Gulland, 1987; Pezzack and Duggan, 1995). This type of analysis is carried out on different temporal and spatial bases with emphasis on aspects such as sex ratio, presence of gravid females, growth and exploitation level.

Growth parameters can be obtained by catch and release (tagging) experiments or from historical length-frequency data. Crustacean growth is discontinuous and involves a series of progressive changes in the exoskeleton (moulting) as the animal reaches larger sizes. Growth is dependent on moult frequency and moult increment. It can be measured, quantitatively, as increase in body size (fresh and dry weight, carapace length, total body length, and/or telson length). Tagging studies are useful in obtaining information on growth, which can be calculated by comparing moult increments and moult frequencies (Shelton et al., 1978a).

Several investigators (e.g., Thomas, 1958a; Simpson, 1961) have found that growth increments, maximum size attained (L_∞, the asymptotic parameter of the von Bertalanffy growth equation) and the rate at which this size is reached (growth parameter k or intrinsic growth rate of the von Bertalanffy growth equation) differ significantly between sexes in decapods. A possible explanation for these differences in growth between sexes is the fact that females and males use their energy into reproduction at different rates (Thomas, 1958a; Grandjean et al., 1997).

From tagging, behavioural differences between the American and the European lobster have been found. *Homarus americanus* undergoes long migratory movements from offshore to inshore grounds, or vice versa, which can occur gradually depending on season and region (Pezzack, 1987; Pezzack and Duggan, 1995; Robichaud and Lawton, 1995; Tremblay et al., 1998), while *H. gammarus* appears to be non-migratory (Simpson, 1961; Hepper and Gough, 1978; Shelton et al., 1978a; Bannister and Lovewell, 1985; Bannister et al., 1990; Collins, 1998). Moult increments in weight and length of sampled individuals of some populations have shown variations both geographically and by sex and these are generally larger for *H. americanus* than *H. gammarus* (Thomas, 1958a, b; Simpson, 1961; Bennett et al., 1978; Conan, 1978). Growth at moult under natural conditions in *H. americanus* is progressive (i.e., positively allometric) (see section on Maturity indices, below) for males and females, while for *H. gammarus* it is retrogressive (negatively allometric) for the females and progressive for the males (Conan, 1978).

Recruitment processes are directly related to moulting, with pre-recruits being one, two or three moults away from entering the fishery. Post-recruits are those animals attaining a size around the MLS mark or bigger (Addison et al., 1995; Zheng and Kruse, 2000). Once maturity is reached, size increments and moult frequency decrease considerably. Regional differences in moult increments and frequencies may be attributed to different levels of exploitation and/or difference in temperature (Thomas, 1958b; Bennett et al., 1978). For the European lobster, mature males are assumed to moult annually, irrespective of their length, while mature females moult only every second year, being in a berried condition in the second year after moulting.

Knowing when moulting will occur, especially in females, is fundamental to determine the proportion of mature animals in a population, allowing the application and resultant interpretation of stock assessment models. In studies on maturity of female lobsters, use of a combination of criteria to estimate size at maturity for a particular population is recommended (Waddy and Aiken, 2005).

In some lobster populations there are marked regional differences in size structure. For example, in the American lobster, the size range present in heavily exploited populations from inshore fishing grounds is mainly confined to small lobsters, contrasting with populations in offshore fishing grounds that present a greater size range with a dome-shaped size-frequency distribution (Pezzack and Duggan, 1995), the latter presumably being closer to the natural (unexploited) size-frequency distribution.

In some American and Canadian lobster populations, there is no conclusive evidence that fishing effort is the driving influence affecting changes in size structure; instead it is suggested that a combination of factors is involved (Fogarty et al., 1982; Pezzack and Duggan, 1995). For Scottish lobster populations there are also regional differences in size (Thomas, 1958a, 1969; Shelton et al., 1978b; Lizárraga-Cubedo et al., 2003).

Figure 2 shows the typical length frequency distribution for two populations of European lobster, in the Hebrides (West of Scotland) and the Southeast of Scotland. Density plots for the Hebrides show larger sizes for both sexes as well as a higher presence of berried females in the catches than for the Southeast. The size structure indicates a higher proportion of recruits around the MLS mark (87 mm CL), of both sexes, for the Southeast fishery than for the Hebrides (Fig. 2b). Although these two lobster populations have experienced different historical levels of exploitation (Lizárraga-Cubedo, 2004), differences in their size structures and the maximum size attained by individual lobsters may be related to

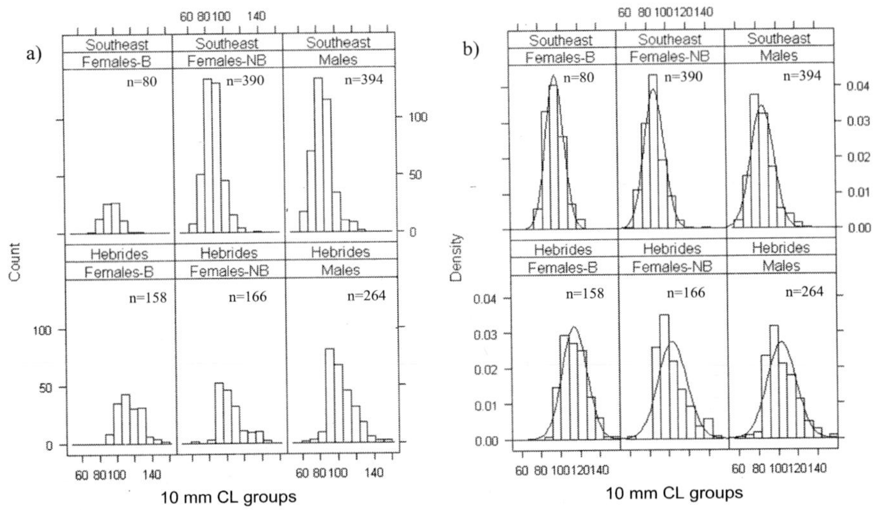

Fig. 2 Length-frequency distribution of lobster *H. gammarus* by 10 mm CL size group for berried females, non-berried females and males from two fisheries in Scotland, the Hebrides (data for 2000-01) and Firth of Forth (Southeast, data for 1999-2001). Minimum legal landing size during the sampled period was 87 mm CL in all Scottish lobster fisheries. Frequencies (a) and density plots (b) are presented. Modified from Lizárraga Cubedo et al. (2003).

substrate and size of the available shelter on the ground (Howard, 1980). This may also be the case for other crustaceans.

The sex ratio of an exploited population is also relevant in stock assessment. In many crustacean populations the ratio of males to females in the catches is unequal. Females invest more energy to produce offspring than do males and this can skew the sex ratios towards males, suggesting implications for mating success on the ground (Debuse et al., 1999).

In lobster fisheries, the sex composition of the catches varies seasonally, with changes associated with behaviour, moulting, shelter availability, fishing strategy (Thomas, 1955a, 1958b, 1965; Collins, 1998), and exploitation rates (Skud, 1979; Debuse et al., 1999). A higher proportion of females than males is found in the catches for large size-classes and this is markedly increased with the onset of maturity in the females (Templeman, 1936; Skud and Perkins, 1969; Thomas, 1955a). This is the case when there is a high number of berried females in the catches. Sex ratio after maturity is a function of the mortality rate and, because of their behavioural characteristics, gravid females are not as heavily fished as non-gravid females and males (Thomas, 1955a, 1958a). Investigations on sex ratio and length distributions have suggested that moulting frequency slows dramatically with increasing size and that, since the males moult twice as often as the females, marked disparities in the sex ratio of individual length groups will result (Thomas, 1955a).

Analysing the sex ratio of a population is relevant for management purposes. Since there are marked differences in growth between the sexes, it is necessary to apply stock assessment techniques separately for each sex, which will allow a better comprehension of the respective growth progression for each sex. Information on modal progressions (i.e., the shifts in size modes over time) may lead to a better understanding of recruitment processes, mortality, and growth parameters. However, modal progression is extremely hard to visualize in long-lived crustacean species.

Traditional methods to obtain growth parameters that are applied and designed for fish are, in many cases, difficult to apply for crustaceans because growth, unlike in fish, is not continuous. This poses problems when applying the von Bertalanffy growth equation. There have been cases of misapplications of length frequencies to study growth parameters in shellfish such as squid, which could be categorized as a rapidly growing organism with multiple (micro-)cohorts (see Pierce et al., 1994 and Jackson et al., 2000 for further discussion of this point). In the past, size modes in squid length frequency data have been interpreted as separate annual cohorts even though the animals may live only 12 months whereas, in fact, these modes seem to represent animals from different pulses of recruitment within the same year, which may also grow at different rates. If modal analysis is carried out, the results should be carefully interpreted and considered as a provisional approach for detecting year-classes in a lobster population. This approach has been widely used and has proved to be most effective in relatively short-lived animals such as Penaeids (Garcia, 1983; Cobb and Caddy, 1989).

It has been possible to obtain demographic parameters for some clawed lobsters from length frequency data (Bailey and Chapman, 1983). Tuck et al. (1997b) studied density and growth for several Scottish *Nephrops norvegicus* populations using a combination of underwater television surveys (for burrow counting) and length frequency data obtained from trawling, analysed in the MULTIFAN program (a more recent version of Electronic Length Frequency Analysis (ELEFAN, Gayanilo et al., 1989) to calculate growth. The authors found geographic variations between populations in growth and density, and it was suggested that growth may be density-dependent.

Many stock assessment methods are based on the study of length, which is often one of the easiest parameters to measure in crustacean fisheries and can thus be used for the analysis of growth parameters, mortality and recruitment in lobsters. Clearly, length-based assessment methods are particularly important for species in which age determination is difficult or impossible. Caddy (1986) gives account of the diverse techniques and software available for the study of crustaceans from length data. Although this is a theme that deserves special attention, it falls beyond the scope of the present chapter.

Alternative approaches exist for the analysis of growth parameters in long-lived crustaceans, for example, through studies on eye pigment deposition in relation to age (O'Donovan and Tully, 1996; Sheehy et al., 1996, 1999; Comeau and Mallet, 2001). In crustaceans, a compound called auto-fluorescent lysosoma, better known as lipofuscin, accumulates in the eyestalks of the animals and has been used to discern age-related parameters in some crustacean species. Sheehy et al. (1999) studied lipofuscin concentration in *H. gammarus* on English coasts (Yorkshire) and derived length-at-age relationships. Lipofuscin accumulation in tagged lobsters of known age, reared in the laboratory and released into the wild, as well as data from literature studies on *H. gammarus* and other short-lived crustaceans were used to estimate the age of wild lobsters from the fishery. The results indicated that lobsters recruit to the fishery at a wide range of sizes (aged from about 3 to 15-20 years). Some sampled specimens exceeded 50 years of age, with females reaching greater ages at smaller sizes. Although lipofuscin data provided no evidence of differences between the sexes in rates of physiological aging, it was observed on the basis of independent estimates of age that lipofuscin concentration was a better aging criterion than size (Sheehy et al., 1999). For lobsters of similar age, no significant correlation between concentration of lipofuscin and size was discerned (O'Donovan and Tully, 1996; Sheehy et al., 1999).

Reproduction, Fecundity and Egg Development

Information of growth is useful to quantify the number of eggs that a female can produce (Paul and Paul, 2000). In fisheries, studying the incidence and presence of berried females in the catches is important and has been used to establish the SOM or the size at which animals, in this case females, are capable of reproducing (see section on Sexual maturity below). Reproductive output in females (egg production) for a given size-class is a function of the number of individuals in that size group, size at maturity, fecundity and frequency of spawning (MacDiarmid, 1989; Tully et al., 2001).

The number of mates available on the ground, rather than sperm produced, is a limiting factor for reproduction in males (MacDiarmid,

1989). Female reproductive output (fecundity) is estimated in various ways, all depending on oocyte development stage. Potential fecundity is considered the number of oocytes in mature ovaries, actual or real fecundity the number of eggs attached to the pleopods at time of capture, and effective or realized fecundity the number of eggs attached to the pleopods of gravid females at time of hatching (Corey, 1991 in Anger and Moreira, 1998; Tuck et al., 2000). The aforementioned definitions are of great relevance for the accurate estimation of fecundity.

It is important to account for any egg loss during the capture process (berried females entering the traps) prior to egg mass collection (Tuck et al., 2000). If egg loss is unknown then the resulting fecundity estimates should be considered minimum estimates. The investigator must also consider the time of year at which sampling is carried out in order to determine when hatching will occur and to avoid miscalculations of fecundity. Realized fecundity can be estimated from the ratio between egg mass size (number of eggs) and female body weight (Anger and Moreira, 1998). This approach has been applied to tropical Caridean shrimps, resulting in lower estimates for marine species than for freshwater species (Anger and Moreira, 1998). In lobsters, fecundity is measured from the relationship between number of eggs and carapace length.

Actual fecundity estimates have been obtained for both the American and European lobster (Perkins, 1972; Estrella and Cadrin, 1995; Tully et al., 2001; Lizárraga-Cubedo et al., 2003). It is also common to calculate egg size by measuring egg length and width, which gives egg diameter. Egg development, as the eye spot index, is also used and is calculated by measuring egg eye length and width. For geographical comparisons of fecundity, eye spot index and egg size are used (see Estrella and Cadrin, 1995).

In crustaceans, brood size is inversely related to egg size (Pollock and Melville-Smith, 1993). Based on this, fecundity in the European lobster appears to be similar throughout its range of occurrence (Hepper and Gough, 1978; Latrouite et al., 1984; Bennett and Howard, 1987; Tully et al., 2001; Lizárraga-Cubedo et al., 2003). In contrast, for the American lobster, geographic variations in historical fecundity data have been stressed although these could have been an artefact of discrepancies in the sample size and egg counting techniques (Estrella and Cadrin, 1995).

Waddy and Aiken (2005) describe the biological processes involved in the reproductive cycle of the female American lobster. The authors suggest that females are in a berried condition every two years depending on moulting, ovary stage, vitellogenesis and previous spawning. They also describe different maturity stages based on cement gland development, moult stage, vitellogenesis, and other factors. For the European lobster, similar information is lacking and more research is needed. However, assumptions of annual or biennial spawning have been made (Latrouite et al., 1981; Tully et al., 2001). Studies suggest that gravid *H. gammarus* females can be found at any time of the year and that the incubation period is between 10 and 11 months. Spawning begins around July-August and is most intensive in September, and hatching takes place in May before the moulting period (June-July) (Thomas, 1958b; Lizárraga-Cubedo et al., 2003).

In many crustaceans, length-fecundity relationships are best described by a power function (MacDiarmid, 1989). Size-fecundity relationships have been compared for *H. americanus* and *H. gammarus* (Tully et al., 2001; Fig. 2c). In most cases, the curves obtained were best represented by a power function with exponents of around 3, which indicate that egg numbers are a function of body mass. According to the intercepts of the fitted models, a female American lobster produces more eggs for a given size than its European counterpart. A few studies on size-fecundity relationships for the European lobster have suggested a linear relationship between fecundity and length (Bennett and Howard, 1987). There is, however, a general consensus that the relationship is normally best represented by a power function (Hepper and Gough, 1978; Tully et al., 2001) and this is consistent with results on fecundity obtained from a small number of samples for the Firth of Forth in Scotland (Lizárraga-Cubedo et al., 2003).

It is argued that clawed lobsters have a relatively low fecundity in comparison to other crustaceans (Cobb et al., 1997) and this relates to differences in reproductive strategies (Pollock, 1997). In spiny lobsters, fecundity is high, egg size is small and several moults are needed before the hatched larvae settle to the bottom. In contrast, clawed lobsters have a relative low fecundity and bigger eggs and larvae settle down to the seabed after about four moults (Pollock, 1997).

Environmental Factors and Reproductive Parameters

Environmental factors play an important role in recruitment and reproductive strategies of decapods. The influence of temperature on the reproductive parameters of the American and European lobster, such as SOM, egg production, time of spawning, co-ordination of the moulting and spawning cycles, time of hatching, and egg attachment and incubation success, is believed to be strong (Hepper and Gough, 1978; Aiken and Waddy, 1986; Waddy and Aiken, 1992; Pollock, 1995a; Tully et al., 2001). Tagging studies on gravid female American lobsters have demonstrated the importance of sea temperature for seasonal migratory patterns.

In Canadian waters, there is a noticeable migration of berried females associated with variations in water temperature that are correlated with water depth. These factors may help maximize successful completion of egg development (Campbell, 1986). Although temperature is an important factor, size at maturity, survival and juvenile growth are mainly regulated by density-dependent processes (Pollock, 1993). Food availability is also an important factor (Pollock, 1995a). Under adverse circumstances, such as sustained environmental fluctuations or overfishing, crustacean populations can respond to the new biological requirements through changes in size at maturity, growth and survival (Pollock, 1993).

Sexual Maturity

Fisheries regulations are mostly based on size limits with the aim to protect immature individuals from exploitation and to ensure an adequate number of reproductively active animals to support the total reproductive output, or egg production, of the population (Grey, 1979; Hobday and Ryan, 1997). Estimates of size at maturity may vary depending on the maturity indicators chosen and the approaches used to measure them. Selection of appropriate criteria for defining size at maturity is important to avoid use of misleading values of reproductive parameters in stock assessment models. Studies on maturity are mostly targeted at female crustaceans.

The term "sexual maturity" may refer to sexual, gonad, reproductive, physiological, expressed, functional, morphometric or behavioural maturity, and misuse (or imprecise use) of terminology is therefore a source of confusion (see Waddy and Aiken, 2005 for a complete review of the terminology concerning female lobster maturity).

The definition of maturity is usually based on physical (morphological) or physiological (reproductive) characteristics, or a combination of the two (Aiken and Waddy, 1980; Muiño et al., 1999; Conan et al., 2001; Waddy and Aiken, 2005). For example, in Brachyuran crabs, males may be considered mature when they are able to carry females during the pre-copulatory embrace and copulate successfully (Hartnoll, 1969; Conan and Comeau, 1986; González-Gurriarán and Freire, 1994), and mature females are those capable of copulating and extruding eggs (Campbell and Eagles, 1983; González-Gurriarán and Freire, 1994).

To avoid the confusion in the terminology concerning crustacean maturity that can be found in the literature (Waddy and Aiken, 2005), throughout the present text we refer to two definitions only, morphometric maturity and gonad maturity. The former relates to development of secondary sexual characters, while the latter concerns development of the reproductive organs.

Maturity Indicators: Secondary Sexual Characters

A wide range of internal maturity indicators (related directly with reproduction) and external (morphometric) maturity indicators can be listed and these vary depending on the characteristics of the species under consideration (Aiken and Waddy, 1980). The purpose of using maturity indicators is to differentiate between the growth patterns of immature and mature individuals and choose a stage at which at least 50% of the population is in a reproductive state or able to reproduce (Montgomery, 1992).

Maturity indicators (Tables 1-3) are more easily identifiable in female crustaceans than in male. In fisheries, indicators of female maturity include size of the smallest berried females in the samples (BF), the smallest size class in which 50% of females are berried (for a given measure of body size) ($BF_{50\%}$), cement gland (Cem) development (to indicate when spawning will occur), gonad (ovary) development stages (GO), presence of sperm plugs (spermatophores) in spermatheca (genital orifices) (Sp), presence of ovigerous setae (OvS), evidence of previous spawning (EvPS), moult stages (Moult), shell condition and changes in morphometric allometry (e.g., changes in abdomen width in relation to body size). A complete review of some of the aforementioned indicators and concepts used in maturity studies for female American lobster is provided by Waddy and Aiken (2005).

Maturity indicators in males include presence of spermatophores (Sp), gonad development (GO, i.e., presence of spermatozoa in vasa deferentia) and morphometric dimensions of the crusher propodite length (PL), width (PW), thickness or height (PH), length of first pair of pereiopods or walking legs (Leg), and length of first pair of pleopods (PLEOP). A list of such indicators as used for some commercially important crustacean fisheries around the world is presented in Tables 1-3.

External indicators of maturity, based on morphometric characteristics, are easy to collect in the field and animals do not have to be sacrificed. Although potentially more accurate, recording of internal indicators is time consuming for large sample sizes and the animals have

Table 1 The most commonly used maturity indicators, and the associated abbreviations as used in Tables 2-4, in maturity studies to determine size at onset of sexual maturity (SOM) for commercially important crustacean species.

Abbreviation	*Maturity indicator*	*Sex*	*Determines SOM*	*Type*
AL	Abdomen length	♀		External
AW	Abdomen width	♀, ♂	♀	External
BF	Size of smallest berried females	♀		External
$BF_{50\%}$	50% berried females per size class	♀		External
Cem	Cement gland development stages	♀,		Internal
CL	Carapace length	♀, ♂		External
CW	Carapace width	♀, ♂		External
EvPS	Evidence of previous spawning	♀		Internal
GO	Gonad development stages	♀, ♂	♀, ♂	Internal
Legs	Length of walking legs (pereiopods)	♂	♂	External
Molt	Moult stages	♀		Internal
OvS	Ovigerous setae	♀	♀	External
PH	Propodus height (crusher and/or pincer)	♀, ♂	♂	External
PL	Propodus length (crusher and/or pincer)	♀, ♂	♂	External
PW	Propodus width (crusher and/or pincer)	♀, ♂	♂	External
PLEOP	Length of 1st pleopods	♀, ♂	♂	External
PoL	Post-orbital length (Grandjean et al., 1997a)	♀, ♂		External
RCL	Rostral carapace length	♀, ♂		External
Shell	Shell condition	♀, ♂		External
Sp	Spermatophores	♂	♂	External
Sps	Sperm plugs in genital orifices or tar spot	♀		External
TaL	Tail length	♀, ♂		External
TaW	Tail width	♀, ♂		External
TL	Total length	♀, ♂		External

to be sacrificed (Conan et al., 2001). Discrepancies exist between estimates of physical maturity (morphometric) and physiological maturity (gonad) and have been reported for crabs (Muiño et al., 1999), clawed lobsters (Free et al., 1992; Free, 1998; Conan et al., 2001), and spiny lobsters (Fielder, 1964; Heydorn, 1965; Hobday and Ryan, 1997). In many cases, estimates of size at maturity obtained from gonad development are much

Table 2 Female maturity indicators from the literature, arranged according to family and species, geographical area of study and author. Asterisks indicate that both crusher and pincer propodus were measured, and variables in bold indicate the main measure of body size or independent variables.

Scientific name	*Area of study*	*Author*	♀*Maturity indicator*
Penaeidae[0] (Tropical shrimps)			
[0] *Penaeus chinensis*	Yellow Sea	Cha et al., 2002	GO (size), **CL**
[0] *Litopenaeus stylirostris*	Gulf of California	López-Martínez et al., 2005	GO (colour, shape), **AL**
[0] *Crangon crangon*	Germany	Temming et al. 1993	$BF_{50\%}$, **CL**
BRACHYURA: Calappidae[1] (Box crabs); Portunidae[2] (Swimming crabs); Majidae[3] (Spider crabs); Raninidae[4] (Spanner crabs); Cancridae[5] (Edible crabs); Eriphiidae[6] (Champagne crabs); Geryonidae[7] (Crystal crabs); Xanthidae[8]; Grapsidae[9]; ANOMURA: Lithodidae[10] (Stone crabs)			
[1] *Calappa convexa.*	Gulf of California	Ayón Parente and Hendrickx, 2001	CW, GO, **CL**
[2] *Liocarcinus depurator*	Spain	Muiño et al., 1999	BF, $BF_{50\%}$, Sps, GO, AW, PL*, PH*, PW*, **CL**, **CW**
[2] *Necora puber*	Spain	González-Gurriarán and Freire, 1994	GO, Sps, $BF_{50\%}$, AW, **CL**, **CW**
[2] *Portunus pelagicus*	Australia	de Lestang et al., 2003	GO, PL, PH, **CW**
[2] *Arenaeus cribbrarius*	Brazil	Pinheiro and Fransozo, 1998	PPL, GO, **CW**
[3] *Chionoecetes bairdi*, [3] *C. opilio*	Alaska	Somerton,1980	AW, Shell, **CW**
[3] *Maja crispata*	Italy	Carmona-Suárez, 2003	AW, GO, **CL**
[3] *Scyra acutifrons*	East-West USA	Hines, 1989	CW, AW, PLEOP, **CL**
[4] *Ranina ranina*	Thailand	Krajangdara and Watanabe, 2005	CW, GO (colour-shape), BF, $BF_{50\%}$, **CL**
[5] *Cancer magister*	Pacific coast, USA	Hankin et al., 1997	Sps, GO, **CW**
[5] *C. pagurus*	Bay of Biscay	Le Foll, 1986	GO **CW**
[8] *Panopeus herbstii*	East-West USA	Hines, 1989	CW, AW, PLEOP, **CL**
[9] *Hemigrapsus nudus*	East-West USA	Hines, 1989	CW, AW, PLEOP, **CL**
[9] *H. oregonensis*	East-West USA	Hines, 1989	CW, AW, PLEOP, **CL**
[9] *Pachygrapsus crassipes*	East-West USA	Hines, 1989	CW, AW, PLEOP, **CL**

Contd.

Contd.

[10] *Paralomis spinosissima*, [10] *P. anamerae*, [10] *P. formosa*	South Atlantic, South Georgia	Purves et al., 2003	$BF_{50\%}$, **CL**, **CW**
Palinuridae[11] (Spiny lobsters), Nephropidae[12] (Clawed lobsters), Scyllaridae[13] (Slipper lobsters); Astacidea[14] (Clawed crayfish)			
[11] *Jasus edwardsii*	Australia	MacDiarmid, 1989	BF, OvS, **CL**
		Hobday and Ryan, 1997	$BF_{50\%}$, OvS, **CL**
		Gardner et al., 2005	$BF_{50\%}$, OvS, **CL**
[11] *J. tristani*	South Africa	Pollock, 1991	OvS, **CL**
[11] *J. verreauxi*	Australia	Montgomery, 1992	$BF_{50\%}$, OvS, **CL**
[11] *Panulirus. cygnus*	Australia	Grey, 1979	$BF_{50\%}$, Leg (1-5), **CL**
[11] *P. guttatus*	Florida Keys	Sharp et al., 1997	$BF_{50\%}$, **CL**
[11] *P. homarus*	Sri Lanka	Jayakody, 1989	RCL, TL, TaL, **CL**
	Oman	Mohan, 1997	BF, Sps, **CL**
[11] *P. marginatus*	Hawaiian Isles	DeMartini et al., 2005	GO, PLEOP, **CL**
[11] *P. versicolor*	Australia	George & Morgan, 1979	Leg (1-5), **CL**
[11] *P. japonicus*	Japan	Minagawa, 1997	GO, **CL**
[12] *Nephrops norvegicus*	Irish Sea	Farmer, 1974	TL, AW, PL, **CL**
	Scotland	Tuck et al., 2000	BF, GO (size-colour), AW, PL, **CL**
[12] *Homarus americanus*	East Canada	Conan et al., 1985	Cem, GO (colour-size), CW, AW, P(L-W-H), **CL**
	East USA	Landers et al., 2001	AW, **CL**
[12] *H. gammarus*	Ireland	Tully et al., 2001	GO (size, colour), AW, Cem, Molt, EvPS, **CL**
	England & Wales	Free et al., 1992	GO (size-colour), AW, **CL**
	Scotland	Lizárraga-Cubedo et al., 2003	AW, **CL**
[13] *Scyllarides squammosus*,	Hawaiian Isles	DeMartini et al0., 2005	GO, PLEOP, **TaW**
[14] *Austropotamobius pallipes*	France (freshwater)	Grandjean et al., 1997a	TL, CL, AW, PL, PW, **PoL**

Table 3 Male maturity indicators from the literature, arranged according to family and species, geographical area of study, and author. Asterisks indicate that both crusher and pincer propodus were measured, and variables in bold indicate the main measure of body size or independent variables.

Scientific name	Area of study	Author	♂Maturity indicator
BRACHYURA: Calappidae[1] (Box crabs); Portunidae[2] (Swimming crabs); Majidae[3] (Spider crabs); Raninidae[4] (Spanner crabs); Cancridae[5] (Edible crabs); Eriphiidae[6] (Champagne crabs); Geryonidae[7] (Crystal crabs); Xanthidae[8]; Grapsidae[9]; ANOMURA: Lithodidae[10] (Stone crabs)			
[2] *Liocarcinus depurator*	Spain	Muiño et al., 1999	PL*, PH*, PW*, PLEOP, **CL**, **CW**
[2] *Necora puber*	Spain	González-Gurriarán and Freire, 1994	Sp, PLEOP, PL, PH, PW, **CL**, **CW**
[2] *Portunus pelagicus*	Australia	de Lestang et al., 2003	PL, PH, **CW**
	Australia	Hall et al., 2006	Sp, GO, PL, **CL**
[2] *Arenaeus cribbrarius*	Brazil	Pinheiro and Fransozo, 1998	PPL, GO, **CW**
[3] *Chionoecetes bairdi*	Alaska	Somerton, 1980	PH (right), **CW**
[3] *C. opilio*	Alaska	Somerton, 1980	PH (right), **CW**
	East Canada	Conan and Comeau, 1985	Sp, CL, PL, PW, PH, **CW**
[3] *Maja crispata*	Italy	Carmona-Suárez, 2003	PL, GO, **CL**
[3] *Scyra acutifrons*	East-West USA	Hines, 1989	CW, AW, PLEOP, **CL**
[4] *Ranina ranina*	Thailand	Krajangdara and Watanabe, 2005	CW, GO, **CW**
[4] *Limulus polyphemus*	Atlantic, USA	Hata and Berkson, 2003	GO, **CW**
[5] *Cancer pagurus*	UK	Edwards, 1979	Sp, GO, **CW**
[6] *Hypothalassia acerba*	Australia	Hall et al., 2006	Sp, GO, PL, **CW**
[7] *Chaceon bicolor*	Australia	Hall et al., 2006	Sp, GO, PL, **CW**
[8] *Panopeus herbstii*,	East-West USA	Hines, 1989	CW, AW, PLEOP, **CL**
[9] *Hemigrapsus nudus*,	East-West USA	Hines, 1989	CW, AW, PLEOP, **CL**
[9] *H. oregonensis*,	East-West USA	Hines, 1989	CW, AW, PLEOP, **CL**
[9] *Pachygrapsus crassipes*,	East-West USA	Hines, 1989	CW, AW, PLEOP, **CL**

Contd.

Contd.

[10] *Hapalogaster dentata*	Japan	Goshima et al., 2000	PH (right), GO, **CL**
[10] *Paralomis spinosissima*	South Atlantic,	Purves et al., 2003	PL, **CL**
[10] *P. anamerae*, [10] *P. formosa*	South Georgia		
Palinuridae[11] (Spiny lobsters), Neprhopidae[12] (Clawed lobsters), Scyllaridae[13] (Slipper lobsters); Astacidea[14] (Clawed crayfish)			
[11] *Panulirus cygnus*	Australia	Grey, 1979	GO, Leg (1-5), **CL**
[11] *P. guttatus*	Florida Keys	Sharp et al., 1997	Leg (2), **CL**
[11] *P. homarus*	Sri Lanka	Jayakody, 1989	RCL, TL, TaL, **CL**
[11] *P. versicolor*	Australia	George and Morgan, 1979	Leg (1-5), **CL**
[12] *Nephrops norvegicus*	Irish Sea	Farmer, 1974	TL, AW, PL, **CL**
	Scotland	Tuck et al., 2000	PL, **CL**
[12] *Homarus americanus*	East Canada	Conan et al., 1985	AW, Sp, PL, PW, PH, **CL**
	East Canada	Conan et al., 2001	GO, PL, PW, PH
	East Canada	Aiken and Waddy, 1989	PL, PW, PH, **CL**
[12] *H. gammarus*	Scotland	Lizárraga-Cubedo et al., 2003	PL, PH, PW, **CL**
[14] *Austropotamobius pallipes*	France (freshwater)	Grandjean et al., 1997a	TL, CL, AW, PL, PW, **PoL**

smaller than those obtained using morphometric characteristics. It is argued that discriminating between functional and physiological maturity is important (Muiño et al., 1999; Conan et al., 2001; Waddy and Aiken, 2005) but a combination of the two is suggested.

A detailed knowledge of the biology of moulting and reproduction of the animal is necessary in order to validate and correctly interpret estimates of maturity (Waddy and Aiken, 2005). Studies on female spiny lobsters (*Panulirus longipes* and *Jasus verreauxi*) in Australian and New Zealand waters have shown a delay between the appearance of pleopod setae (maturity indicator) and spawning (Jayakody, 1989). Gardner et al. (2005), working with female *Jasus edwardsii*, stressed that recording of ovigerous setae can give reliable estimations of size at maturity but the sampling season should be consistent in order to avoid recording different stages of ovigerous setae, which could result in misleading estimates of maturity.

Hankin et al. (1997), investigating maturity in Dungeness crabs, found that remnants of sperm plugs or spermatophores in females are definite indicators of mating and these can be found in 97.5% of recently moulted large females, regardless of moulting status and size. Although moult frequency in very large females (> 155 mm carapace width) is reduced, as indicated by tag-recovery data (Hankin et al., 1989), viable sperm can be retained for at least 2.5 years and this allows females to extrude successive egg masses without annual moulting and mating. Fecundity is reduced in skip-moult crabs (i.e., animals that have not moulted) as compared to crabs that have moulted (Hankin et al., 1997).

Retention of sperm and biennial cycles of spawning have been observed for female American lobster but strongly depend on factors such as moulting, sea temperature, and multiple mating (Waddy and Aiken, 1995, 2005). Waddy and Aiken (2005) give a very good example of the criteria needed to validate maturity estimates. This involves consideration of ovary stage development (1-6 stages), cement gland development (stages 0-4), time of moulting (10 stages for C_4, D_0-D_3), time of spawning and hatching. Based on these characteristics and criteria, four phases can be observed, prepubertal, pubertal, adult I and adult II, over a 4-year period.

Pubertal females are females one year before their first egg extrusion, adult I are females in their first year of egg production, and adult II are females in their second year of adulthood, and so on (Waddy and Aiken, 2005). The authors concluded that the success of maturity studies on the female American lobster depends on "the criteria used to discriminate between mature and immature lobsters, the time of year the study is conducted, and the sizes and numbers of lobsters sampled". We suggest that similar criteria should be considered when designing maturity studies for females of other commercially important crustacean fisheries.

Morphometry: Allometric Growth

Growth in crustaceans, unlike in vertebrates, is discontinuous. A series of growth increments occurs through a process known as ecdysis or moulting (Mariappan et al., 2000). There are three main growth phases: larval, juvenile, and adult. Puberty moult, which separates the last juvenile and adult instars, is the most important, since secondary sexual characters are fully developed (Pinheiro and Franzoso, 1998).

Male and female decapods have allometric growth of some of their somites (i.e., chelipeds, abdominal cavity, or pleopods) in relation to body size. The growth rates of such body parts can, in many cases, be used as growth phase indicators (Pinheiro and Franzoso, 1998).

Two types of allometry are seen in crustaceans: (1) **positive allometry or progressive geometric growth** and (2) **negative allometry or retrogressive geometric growth.** In positive allometry, a dependent variable (i.e., crusher propodite length or pleopod length) grows more rapidly than body size (i.e., CL or CW) and it can be described by a curvilinear relationship between the variables. In negative allometry or retrogressive geometric growth, the dependent variable (i.e., CL or CW) grows more slowly than body size and the changes in growth from the juvenile to adulthood phase can be better detected by plotting two separate lines. In some crabs, allometric growth is positive in juveniles and negative in adults (Muiño et al., 1999).

Chelipeds of the male Homarid lobsters are the best example of structures that show positive allometric growth in crustaceans (Templeman, 1935; Farmer, 1974; Aiken and Waddy, 1980, 1989; Conan et al., 1985, 2001; Free, 1994, 1998; Tuck et al., 2000; Lizárraga-Cubedo et al., 2003) and it is normally more apparent in the major (crusher) than in the minor (cutter) cheliped. A dimorphic pattern of development begins at the time of the puberty moult (Hartnoll, 1974; Pinheiro and Fransozo, 1993, 1998; Hall et al., 2006) and it is associated with functional (morphometric) sexual maturity (González-Gurriarán and Freire, 1994; Mariappan et al., 2000). Factors such as feeding, mate-guarding, and fighting (Vermeij, 1977; Hughes, 1989 in Mariappan et al., 2000) may influence such dimorphism.

Female cheliped size growth at moulting is minimal and this could be attributed to the different roles that the chelae play in males and females (Paul and Paul, 2000). Some Anomuran crabs (Lithodidae) also present positive allometry of their chelae (Jewett et al., 1985; Somerton and Otto,

1986; Paul and Paul, 2000). In Portunid crabs, growth observed in pleopods is positively allometric in juveniles and slightly negatively allometric in adults (Hartnoll, 1974; Davidson and Marsden, 1987; Abelló et al., 1990; González-Gurriarán and Freire, 1994; Muiño et al., 1999).

There are other body parts, secondary sexual characters, with allometric growth, which have been used as maturity indicators. In male Palinurids (spiny lobsters), changes in allometry can be perceived in the growth of the pereiopods (walking legs), which have the function of grasping the female during the copulatory process (George and Morgan, 1979; Grey, 1979; Sharp et al., 1997). Although positive allometric growth of the legs is also seen in females, this is not as pronounced as for males (George and Morgan, 1979; Grey, 1979).

Female crustaceans also have distinctive morphometric characters that show allometric growth. The best-known case is abdominal width, which is frequently used in studies of SOM for Homarid lobsters (Farmer, 1974; Conan et al., 1985; Free et al., 1992; Tuck et al., 2000; Landers et al., 2001; Tully et al., 2001; Lizárraga-Cubedo et al., 2003), Astacidae freshwater crayfish (Grandjean et al., 1997a,b), Majid crabs (Hartnoll, 1978; Somerton, 1980), and Portunid crabs (González-Gurriarán and Freire, 1994; Muiño et al., 1999).

Maturity Indices: Size at Onset of Sexual Maturity and Detection of Inflection Points

Establishing or determining the size at which maturity is reached, normally referred to as the size at onset of sexual maturity (SOM), is a priority goal for fishery managers, allowing them to recommend minimum landing sizes that protect immature animals. This also allows geographic comparisons to be made of the reproductive status between populations.

The allometry of the largest chela of male crabs and lobsters has often been assumed to undergo a change at the pubertal moult, i.e., the moult to maturity, and could therefore be used as a basis for determining the size of males at morphometric maturity (Conan et al., 2001; Hall et al., 2006). Conan et al. (2001) considered three models. Model I refers to a progressive differentiation from the smallest sizes after every moult, with no indication of an "inflection" point. Model II refers to differentiation of sexual characters starting over a small range of individual sizes and progressing over several moults. In Model III, differentiation occurs over a wide range of individual sizes and is achieved after a single final moult. The SOM in Majid crabs and some Homarid lobsters, *Nephrops norvegicus*, has a wide range. It can be detected by two distinctive groups of points, separating immature from mature animals, that can be plotted as two parallel rectilinear curves (Farmer, 1974; Hartnoll, 1974; Conan et al., 2001).

Aiken and Waddy (1989) pointed out that male lobsters present changes in allometric growth that can be measured by plotting claw morphometrics, or a measure of volume of the large cheliped, against a measure of body length (CL). It is normally assumed that immature and mature lobsters have different growth curves. Thus, growth phase transition is marked by an abrupt change in the allometric growth function of these structures (Hartnoll, 1974, 1978, 1982), and this point, where onset of sexual maturity occurs, is often called the inflection point (Aiken and Waddy, 1980, 1989).

Some authors argue that, for a better detection of an inflection point, based on morphometric data, *a priori* criteria are needed to separate immature from mature animals (González-Gurriarán and Freire, 1994). For example, González-Gurriarán and Freire (1994), working on maturity for the velvet swimming crab *Necora puber*, selected *a priori* maturity values based on male gonad development, which allowed them to obtain an approximate estimate of transition sizes between juvenile and adult animals. A complementary study based on morphometric information produced similar estimates of size at maturity. However, the authors concluded that, for more reliable detection of the inflection point that separates immature from mature animals, different methodologies are needed.

Hall et al. (2006), working on male maturity of three species of crabs, *Portunus pelagicus*, *Hypothalassia acerba*, and *Chaceon bicolor*, found that morphometric relationships between chela length and carapace width revealed no conspicuous change in allometry. They explored several approaches (e.g., an information-theory approach and a range of models) to detect allometric changes.

The SOM and associated inflection point can be determined through various statistical methods based on establishing the bivariate relationships between the morphometric variables. Identifying which morphometric variables can be used to detect the changes in allometry related to sexual maturity is a priority goal in maturity studies (González-Gurriarán and Freire, 1994).

There are two general approaches to detect SOM in crustaceans using morphometrics: those requiring **log-transformed** data and those using **untransformed** data. Reduced major axis (RMA) regression is a multivariate regression technique that does not require log-transformation of the data. It allows fitting single lines that summarize the data and delineates a transition point (or point of inflection) more sharply than techniques requiring log-transformed data (e.g., least squares regression) (Lovett and Felder, 1989; Lizárraga-Cubedo et al.,

2003). This technique has previously been used to determine SOM in shrimp *Callianassa lousianensis* (Felder and Lovett, 1989), crabs *Necora puber* (González-Gurriarán and Freire, 1994) and *Liocarcinus depurator* (Muiño et al., 1999), and lobsters *Nephrops norvegicus* (Tuck et al., 2000) and *Homarus gammarus* (Lizárraga-Cubedo et al., 2003). Similar estimates of SOM can be obtained with the Lovett and Felder (1989) technique (Lizárraga-Cubedo et al., 2003) and with classical approaches, i.e., least square regressions.

Log-transformation of morphometric data has been widely used among crustacean researchers (Saila and Flowers, 1969; Conan et al., 1985, 2001). A range of statistical approaches and models can be applied to log-transformed data. Linear models, quadratic models, cubic models, broken-stick models (BSM), broken-stick models with logistic transitions (BSMLT), and two-line-segment models with logistic transitions (TLSMLT) have all been applied to male claw allometry of three different species of crabs (see Hall et al., 2006). A graphical display of four different models, Somerton's (1980), BSM, BSMLT and TLSMLT, applied and compared by Hall et al. (2006), is shown in Fig. 3. Application of these models, along with analysis of gonad material, indicated that in the Portunid crab *Portunus pelagicus*, the chela undergoes a subtle change in allometry, with morphometric and gonad maturity estimates overlapping. For Eriphid and Geryonid crabs, there was not enough evidence of a conspicuous change in chela allometry, and no visible inflection point; hence, the authors suggested that management decisions should be mainly based on estimates of physiological maturity (gonad maturity).

The best-known method that employs log-transformed data is least squares regression analysis (Ricker 1973), which has been applied in various maturity studies on crustaceans, using different morphometric variables (George and Morgan, 1979; Grey, 1979; Jayakody, 1989; Tuck et al., 2000; Carmona-Suárez, 2003; Lizárraga-Cubedo et al., 2003). Simple bivariate regression can be used to detect differences in growth patterns (from pubertal to adulthood), for example, by fitting separate regressions to morphometric data from juvenile or adult phases (González-Gurriarán and Freire, 1994).

Another widely used method is principal components analysis (PCA). It is an ordination technique that analyses two or more linearly related variables. If morphometric variables are going to be analysed, then log-transformation of the data is suggested. The output from PCA includes an ordination diagram or bi-plot, which can help to detect whether the variables analysed are positively or negatively related and whether the correlation between them is significant. The technique can be

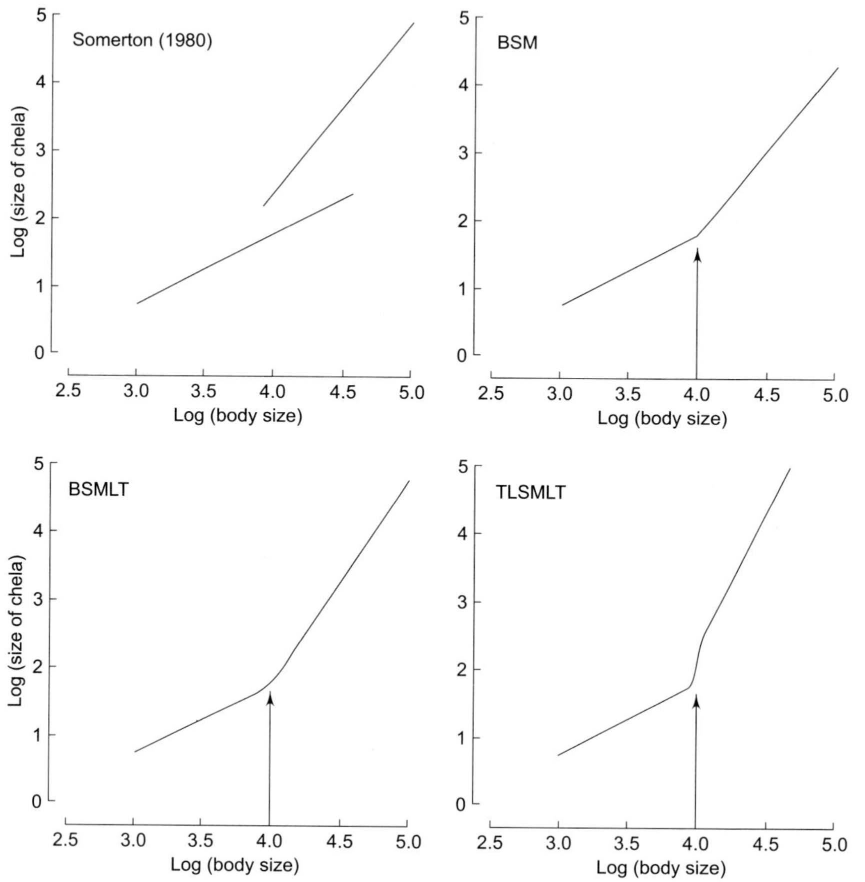

Fig. 3 The forms of the relationships between log-transformed chela length and carapace length of different species of male crabs for four models: Somerton's (1980) model, BSM model, BSMLT model, and TLSMLT model. Arrows denote the sizes at which allometry changes. Modified from Hall et al. (2006).

used to detect discrete groups of individuals sharing common morphometric characteristics (Conan et al., 1985). Information on the use of PCA in crustaceans can be found in the literature (Conan and Comeau, 1985; Conan et al., 1985, 2001; González-Gurriarán and Freire, 1994; Tzeng et al., 2001).

A priority goal in maturity studies is to indicate the size group or age at which at least 50% of the fished population is in a reproductive state. It is important, however, to first obtain an estimate of transition size of maturity. Then the proportion of mature animals is calculated. A logistic model, also called a maturity ogive, is applied to the whole range of sizes and the size-class at which 50% of animals are mature animals is estimated (Somerton, 1980). According to Hobday and Ryan (1997), the logistic model takes the following form (although other formulations are possible):

$$P = 1/[1 + \exp(-S * (CL - CL))] \qquad ...(1)$$

where P is the percentage of mature animals in a size-class, S is the parameter controlling the slope of the curve at the inflection point, CL is the carapace length (mm), and CL_{50} is the length at which 50% of individuals are mature (SOM).

The result is a sigmoid curve and it shows a progressive increase in the proportion of mature animals with increasing size (Fig. 4). Maturity ogives can be applied to both female and male data (Table 4). Female maturity ogives have been obtained with a series of different maturity indicators for shrimps (Temming et al., 1993; Cha et al., 2002), Brachyuran crabs (González-Gurriarán and Freire, 1994; Pinheiro and Fransozo, 1998; Muiño et al., 1999; Ayón-Parente and Hendrickx, 2001; de Lestang et al., 2003), Anomuran crabs (Purves et al., 2003), spiny lobsters (Grey, 1979; Montgomery, 1992; Hobday and Ryan, 1997; Minagawa, 1997; Mohan, 1997; DeMartini et al., 2005), clawed lobsters (Gibson, 1967; Latrouite et al., 1981; Conan et al., 1985; Tuck et al., 2000; Tully et al., 2001), and slipper lobsters (DeMartini et al., 2005).

For males, maturity ogives have been obtained for swimming crabs (González-Gurriarán and Freire, 1994; Pinheiro and Fransozo, 1998; Muiño et al., 1999; de Lestang et al., 2003; Hall et al., 2006), spider crab

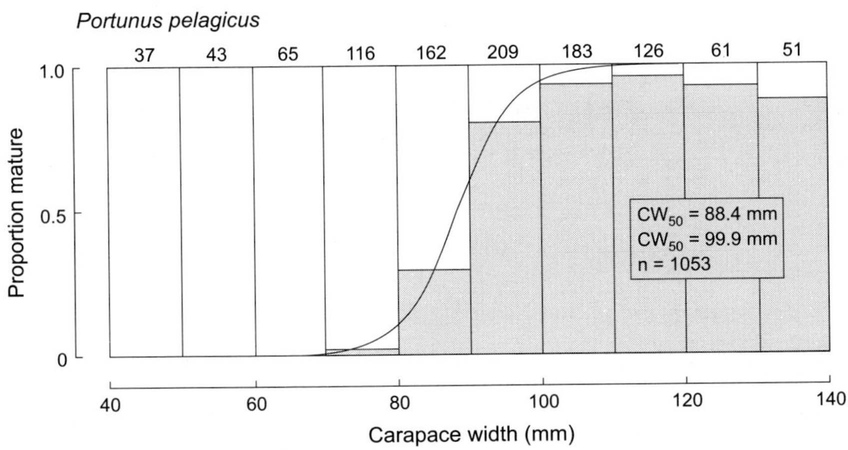

Fig. 4 Fitted maturity ogive and proportion of animals that are physiologically mature (based on gonad development) as a function of body size (recorded as CW) in males of *Portunus pelagicus*. Modified from Hall et al. (2006). The second (upper) x-axis refers to sample sizes.

Table 4 Studies that fit maturity ogives (including applications of logistic or non-linear regression models) crustacean maturity data. Variables in bold indicate the main measure of body size or independent variables.

Scientific name	*Author*	*Maturity ogives per species and sex*
Penaeidae[0] (Tropical shrimps)		
[0] *Penaeus chinensis*	Cha et al., 2002	♀ GO(size) vs **CL**
[0] *Crangon crangon*	Temming et al., 1993	♀ $BF_{50\%}$ vs **CL**
BRACHYURA: Calappidae[1] (Box crabs); Portunidae[2] (Swimming crabs); Majidae[3] (Spider crabs); Raninidae[4] (Spanner crabs); Cancridae[5] (Edible crabs); Eriphiidae[6] (Champagne crabs); Geryonidae[7] (Crystal crabs); Xanthidae[8]; Grapsidae[9]; ANOMURA: Lithodidae[10] (Stone crabs)		
[1] *Calappa convexa.*	Ayón Parente and Hendrickx, 2001	♀ $BF_{50\%}$ vs **CL**
[2] *Liocarcinus depurator*	Muiño et al., 1999	♀ AW vs **CW**; ♂ PL vs **CW**, PLEOP vs **CW**
[2] *Necora puber*	González-Gurriarán and Freire, 1994	♀ AW vs **CW**, $BF_{50\%}$, vs **CW**; ♂ PL vs **CW**, PLEOP vs **CW**
[2] *Portunus pelagicus*	de Lestang et al., 2003	♀ GO vs **CW**; ♂ GO vs **CW**, PL vs **CW**
	Hall et al., 2006	♂ GO vs **CW**
[2] *Arenaeus cribbrarius*	Pinheiro and Fransozo, 1998	♀ AW vs **CW**; ♂ PL vs **CW**
[3] *Chionoecetes bairdi,*	Somerton, 1980	♂ PL vs **CW**
[6] *Hypothalassia acerba*	Hall et al., 2006	♂ GO vs **CW**
[7] *Chaceon bicolor*	Hall et al., 2006	♂ GO vs **CW**
[10] *Hapalogaster dentata*	Goshima et al., 2000	♂ GO vs **CL**
[10] *Paralomis spinosissima,* [10] *P. anamerae,* [10] *P. formosa*	Purves et al., 2003	♀ $BF_{50\%}$, vs **CW**

Contd.

Contd.

Palinuridae[11] (Spiny lobsters), Nephropidae[12] (Clawed lobsters), Scyllaridae[13] (Slipper lobsters); Astacidea[14] (Clawed crayfish)		
[11] *Jasus edwardsii*	Hobday and Ryan, 1997	♀ $BF_{50\%}$ vs **CL** and/or OvS vs **CL**
[11] *J. verreauxi*	Montgomery, 1992	♀ $BF_{50\%}$ vs **CL** and/or OvS vs **CL**
[11] *Panulirus cygnus*	Grey, 1979	♀ Sps vs **CL**; ♂ GO vs **CL**
[11] *P. homarus*	Mohan, 1997	♀ Sps vs **CL** and/or $BF_{50\%}$ vs **CL**
[11] *P. marginatus*	DeMartini et al., 2005	♀ $BF_{50\%}$, vs **TaW**, GO vs **TaW**
[11] *P. japonicus*	Minagawa, 1997	♀ GO vs **CL**
[12] *Nephrops norvegicus*	Tuck et al., 2000	♀ GO(colour) vs **CL**
[12] *Homarus americanus*	Conan et al., 1985	♀ Cem and GO vs **CL**; ♂ of Sp vs **CL**
	Conan et al., 2001	♀GO vs **CL**
[12] *H. gammarus*	Tully et al., 2001	♀ combined criteria: GO, Cem, Molt, EvPS, ratio AW/CL vs **CL**
[13] Scyllarides squammosus	Demattini et al., 2005	♀ $BF_{50\%,}$ vs **TaW**, GO vs **TaW**

(Somerton, 1980), champagne crabs (Hall et al., 2006), crystal crabs (Hall et al., 2006), stone crabs (Goshima et al., 2000), spiny lobsters (Grey, 1979), and clawed lobsters (Conan et al., 1985, 2001). In most of the cases, maturity estimates were obtained from gonad material (Table 4; Fig. 4).

Maturity indices, which are sex-specific, can help in emphasizing manifested changes in growth from immature to mature animals with respect to body size. Maturity indices have both historic and geographic implications. A good example of the application of maturity indices in crustaceans is found for Homarid lobsters and is based on a marked differentiation in growth of different somites between sexes. Males have larger claws and a narrower abdomen than females at a given carapace length (Templeman, 1935). Male maturity indices can be calculated from claw morphometry, namely relative size, weight, or volume, often compared to total length, weight, or carapace length (Templeman, 1939, 1944; Squires, 1970; Aiken and Waddy, 1980, 1989; Ennis, 1980; Waddy et al., 1995). Indices of maturity in females are mostly based on allometric changes of the abdomen.

Two indices of male size at maturity can be obtained, the Anderson Index and the Crusher Propodite Index (CPI). The Anderson Index, described by Aiken and Waddy (1980), is:

$$AI = (PL * PW * PH/CL) * 100 \quad \text{....(2)}$$

and the CPI, described by Aiken and Waddy (1989), is:

$$CPI = ((PL * PW * PH) * 100/CL) \quad \text{...(3)}$$

Further description of both indices can be found in Aiken and Waddy (1980, 1989); abbreviations are given in Table 1. The Anderson cheliped index (Aiken and Waddy, 1980) and the CPI (Aiken and Waddy, 1989) have been used to obtain functional (morphometric) maturity indices in male Homarid lobsters (Lizárraga-Cubedo et al., 2003). In both cases, visual identification of an "inflection point" on a fitted curve was used to identify the size at functional maturity (Aiken and Waddy, 1980, 1989; Free, 1994, 1998; Conan et al., 2001; Lizárraga-Cubedo et al., 2003).

A female maturity index can be estimated from the ratio of abdominal width to CL (Templeman, 1935, 1939; Simpson 1961), although careful interpretation of the results is needed (Simpson, 1961; Ennis, 1980; Free, 1998; Conan et al., 1985, 2001). Figure 5 shows the typical graphical relationships between the Anderson Index, CPI, Maturity Index and body size group for *Homarus gammarus*. From Fig. 5 no clear inflection point can be detected with the Anderson Index and Maturity Index, for either males or females. In contrast, the same figure shows an inflection point when using CPI, i.e., at the intersection of the regression lines for males and females (the latter being viewed as the baseline, with a slope approximately equal to zero).

Aiken and Waddy (1989) showed, for American lobster, that changes in allometric growth can be detected in individual males by applying the CPI. As in Fig. 5, regressions of CPI on CL are plotted for both males and females on the same graph. In their work, Aiken and Waddy (1989) observed that the male CPI regression line intersected with that of the female. For geographical comparison purposes the resulting intersection point is used as the CPI value at which maturity is reached for males and it may be representative of a group of animals from a specific geographic range.

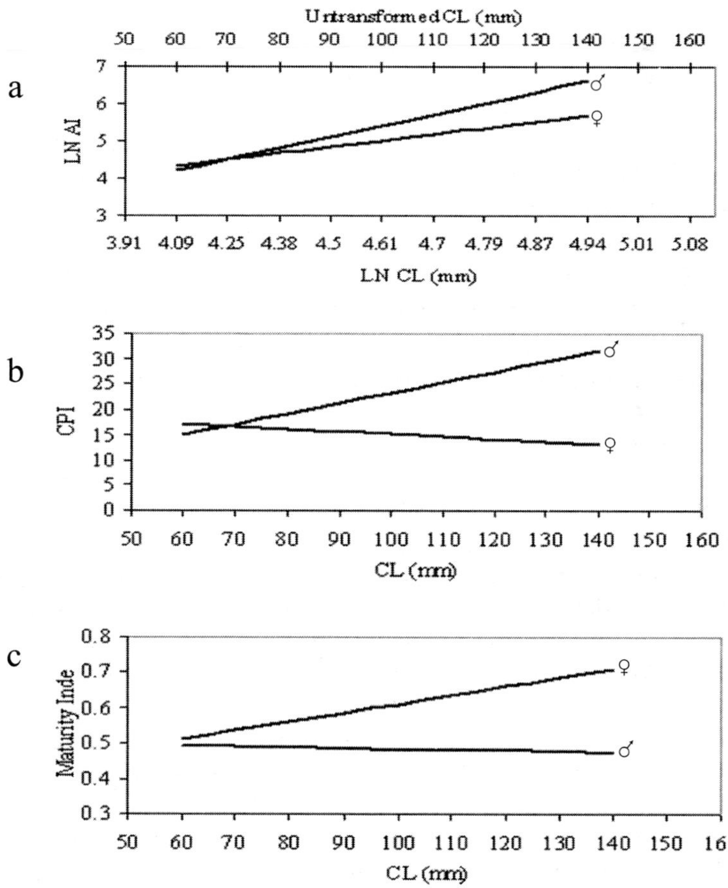

Fig. 5 Examples of the diverse maturity indices that are used for maturity studies in Homarid lobsters: (a) Anderson Index, (b) Crusher Propodite Index and (c) Maturity Index. Values are plotted by 5 mm CL groups for male (a, b) and female (c) *Homarus gammarus* in the Firth of Forth, Scotland. Modified from Lizárraga-Cubedo et al. (2003). Maturity index is the ratio of abdomen width to carapace length (AW/CL).

For American lobsters from Canadian coasts, Aiken and Waddy (1989) obtained values of CPI greater than 22-24, indicative of a positive influence of maturity on the product of claw measurements. The *H. americanus* female CPI baseline was similar for North Rustico and Grand Manan, Canada; therefore, for regional comparisons of male size at maturity, the female CPI baseline was set at approximately 22. However, this estimate represents a baseline only for the SOM of the male American lobster (Aiken and Waddy, 1989).

Similar approaches have been followed for *H. gammarus* lobsters on the English and Welsh coasts (Free, 1994) and two areas in Scotland (Firth of Forth and Hebrides, Lizárraga-Cubedo et al., 2003). Lizárraga-Cubedo et al. (2003) found that the *H. gammarus* female CPI baseline lay between 15 and 17 for the Firth of Forth (Fig. 4b) and Hebrides, which coincided with the estimates of Free (1994) and differed from the values of Aiken and Waddy (1989) for the American lobster. However, Free (1994) did not observe any inflection point. In contrast, for the Firth of Forth and Hebrides, inflection points were detected at approximately 70 mm and 80 mm CL respectively. The results suggested that careful interpretation of CPI, as an index of SOM of male *H. gammarus*, is required, and also that, for *H. gammarus*, the female CPI baseline may be set between 15 and 17 (Fig. 5b).

Maturity studies are the foundation of good management decisions in fisheries, by allowing the setting of appropriate minimum landing sizes that are based on size at maturity for a population over a defined geographic range. Choosing the most appropriate maturity criteria, considering the time of year when the studies should be conducted, accurate discrimination between immature and mature animals, and number of samples are specific points that would permit valid estimates (Waddy and Aiken, 2005).

Geographic variation in SOM has been noted for Homarid lobsters (Aiken and Waddy, 1980; Thomas, 1965; Bailey, 1984; Free, 1998; Tuck et al., 2000; Tully et al., 2001; Lizárraga-Cubedo et al., 2003), as well as in the crab families Cancridae (Hines, 1989), Grapsidae (Hines, 1989), Majidae (Somerton, 1980; Hines, 1989) and Portunidae (Hines, 1989; Muiño et al., 1999). Possible factors causing such variations are population density and the availability of mates (Annala et al., 1980; MacDiarmid, 1989).

Stock Discrimination and Stock Assessment

Crustacean reproductive parameters vary geographically among populations, which may lead to the question of how to set appropriate boundaries to define populations that share similar characteristics, for the purposes of stock assessment. In many heavily exploited crustacean populations, the size structure is skewed towards small sizes, with the recruits and post-recruits being most abundant, presenting a "knife-edge" form. Knowledge of size at maturity is relevant, from a management viewpoint, to increase egg production and eggs per recruit. Estimates of fecundity, egg size, spawning season and hatching period may also vary geographically and between populations. Therefore, for stock assessment purposes, defining discrete stock units and geographic stock boundaries (Cushing, 1968 in Cadrin, 1995) is a priority and biological characteristics can be used for such purposes (Booke, 1981; Ennis, 1986; Caddy, 1989).

In crustacean fisheries, stock discrimination has been possible with the use of morphometric information (Templeman, 1935; Saila and Flowers, 1969; Campbell and Mohn, 1982; Davidson et al., 1982; Harding et al., 1993; Cadrin, 1995; Debuse et al., 2001), including size at morphometric maturity (Somerton, 1981; Aiken and Waddy, 1989; Grandjean et al., 1997b), as well as studies of genetic population structure (Harding et al., 1997; Jørstad and Farestveit, 1999).

One of the disadvantages of using morphometrics in stock discrimination is that significant morphometric differences may be ephemeral or restricted to specific cohorts (Cadrin and Friedland, 1999). Morphometric stock discrimination also depends both on the consistent mapping of morphometric (phenotypic) differences on to underlying genetic differences and on the reliability of the multivariate analytical techniques employed, such as discrimination function analysis, which may maximize accuracy in the separation of groups with similar characteristics.

Studies on selected morphometric characteristics for female and male American lobsters in Canadian waters have made possible the identification of separate lobster stocks (Cadrin, 1995). Debuse et al. (2001) found that exoskeleton damage was a major distinctive morphometric character discriminating males, and exoskeleton damage, claw spines and rostrum teeth could be used to discriminate between female lobsters from nine different sampled sites on English coasts.

The existence of several stock units of the American lobster in American waters, based on adult and larval distribution, has been identified from a combination of approaches including electrophoretic analysis of allozyme frequencies, morphometrics, parasite presence, information from tagging studies, and trends in long-term landings (Pezzack, 1987). In all cases it is recommended that, when possible, several approaches should be applied simultaneously and results compared. If the study is based on morphometrics alone, in general at least two independent character sets should be analysed (Thorpe, 1987).

The aim of stock assessment is to provide the information necessary to make appropriate management decisions to preserve the exploited stocks in a sustainable state. The application of some kind of biological reference point or range of biological reference points, normally levels of catch rates, is necessary to allow the selection of appropriate future levels of exploitation such as total allowable catch (Hilborn, 2003).

Information on catch and length is easy to collect in the field and in most cases provides the foundation of stock assessment for crustaceans (Addison, 1997) and other shellfish species, in which age determination is difficult. Reviews of the application of stock assessment models in crustacean fisheries can be consulted in Caddy (1986), Addison (1997), and Hilborn (1997). Data on catch, fishing effort and length frequencies represent, in most of the cases, the only available tools to describe qualitative trends of the fishery (Addison, 1997).

Estimates of exploitation rates have been obtained from fishery-independent surveys for American lobsters (Comeau and Mallet, 2001; Mallet and Comeau, 2001; Smith and Tremblay, 2003) and spiny lobsters (Frusher et al., 1997; Punt and Kennedy, 1997). If information on growth and size of the animals is lacking then production models are an option because data input is minimal (Hilborn, 1997). However, data must come from a unit of stock, the application of these models needs assumptions of constant recruitment and catchability, biological aspects of the species are not considered, and the interpretation of the resulting information depends on the reliability of CPUE estimates (Hilborn, 1997).

Other models, requiring length data, are more flexible, like dynamic pool models, but knowledge of growth, mortality and recruitment parameters is needed, which may be limited in crustacean fisheries (Morgan, 1979). Stock assessment in lobsters has been carried out to obtain reference points (maximum yield and optimal fishing effort) by implementing stock production models, along with comparisons with dynamic pool models, emphasizing the impact of spawning stock size on recruitment and yield per recruit in Homarid species (Marchessault et al., 1976) and Palinurid species (Morgan, 1979; Saila et al., 1979). Although results may suggest good fits to models, caution in the interpretation is necessary because the assumptions of constant recruitment, fishing and natural mortality may not be justified.

For the European lobster, yield-per-recruit analysis, based on length cohort analysis, has been applied to stocks in English and Scottish waters (Shelton et al., 1978a; Bannister et al., 1983; Bannister and Addison, 1984). In most cases, length cohort analysis has provided the only assessment used for stocks of the European lobster, with known limitations because biological information such as values for the von Bertalanffy growth parameters $L_{-\infty}$ and k is lacking or poor and needs updating (Lizárraga-Cubedo, 2004). In any case, biological information (such as mean size) and fishing effort are affected by environmental influences on the stock. Fishing effort data may not take into account changes in catchability through time. Concerns over the use of size frequencies relate to the fact that changes in growth and recruitment often cannot be detected (Caddy, 1986). Therefore, for a complete analysis of the stocks it is recommended that data on both mean size and fishing effort be used.

In stock assessment it is desirable to establish precautionary levels of exploitation or reference points that may avoid recruitment overfishing. Many fishery management strategies are based on calculating minimum levels of egg production per recruit (E/R), mostly desired to be above 10% of virgin biomass production. Analysis of E/R must be accompanied by short- and long-term predictions about the incorporation of recruits to the adult fishery (Ennis and Fogarty, 1997). Information about the number of eggs produced, obtained from size-fecundity relationships, and number of recruits, estimated from discard surveys or log-book information, is the minimal requirement for the application of these models. From a fisheries management viewpoint, the application of these models is relevant for the adequate protection of the mature female component of a population.

CONCLUSIONS

In this chapter we have dealt with fundamental aspects that stress the relationships of the fisheries process to the reproduction of crustaceans. Crustaceans represent valuable sources of income and employment in many fisheries around the world with historical levels of exploitation that, in many cases, are increasing because of the increasing demand for food for human consumption. Describing the fishery processes that relate to reproduction in crustaceans, for crustacean species that vary in their biological characteristics and fishery practices, is a difficult task. For this reason, two case examples were chosen, the American and European lobster fisheries, in view of the vast information available in the literature. When appropriate, information of other crustacean species was included, especially in the section on Sexual maturity.

In crustaceans, management strategies are based on size, a parameter that may include information about reproductive status. Size-frequencies and catch data are parameters that are easy to measure, hence their use in stock assessment. Viable management strategies include applications of catch-effort restrictions, selection of areas for enhancement of egg production or protection of reproductive females, and closed seasons.

In trap-based fisheries, fishery and environmental factors such as gear selectivity, trap immersion time, selection of fishing grounds, inter- and intraspecific behavioural interactions inside and outside the traps, sexual maturity, food availability, type of bait, ecology, and time of year, all affect fishery success and size structure of the exploited populations. Size structure may reflect the level of exploitation but may also provide information on recruitment. It is necessary, however, to consider that catch and size information from traps may not truly represent the distribution of lobsters on the ground.

Reproductive biology of crustaceans differs considerably from that of fish. Unlike in fish, growth in crustaceans is discontinuous, carried out through a series of changes in the exoskeleton, and this poses some problems when applying stock assessment models that have been traditionally applied and mostly designed for fish. Growth can be determined from moult increment and frequency, in terms of both body weight and size. Relevant information on growth, mortality estimates and migratory movements can be obtained from tagging experiments. Despite the problems that these methods face, their contribution in providing new information is unquestionable.

Information on the presence of gravid females in the catches is essential for the analysis of size-fecundity relationships. In clawed lobsters, fecundity is a function of female size. For geographical comparison purposes, egg or embryo size (eye spot index) should be determined and fecundity estimates per size class calculated. This latter aspect is important for stock assessment purposes. Spatial comparisons, when possible, are desirable. Environmental factors affect crustacean reproduction through modifications of size at maturity, delay in time of moulting, and spawning and hatching periods. Temperature is an important factor affecting catch rates in the short term and this must be considered when using CPUE as an abundance index.

Our study was focused mainly on the analysis of sexual maturity. From the literature review, it becomes relevant to stress that, when designing maturity studies for female crustaceans, selection criteria must be homogeneous among researchers and based on three characteristics: discrimination between mature and immature animals, the time of year the study is conducted, and the sizes and numbers of lobsters sampled (Waddy and Aiken, 2005). Several criteria exist to define maturity including those concerning physiological characteristics (gonad development), here considered as physiological maturity, and those based on the growth of secondary sexual characters (morphometrics) or morphometric maturity. In view of the diversity and sometimes contradictory use of maturity concepts in the literature, we have decided to base our analysis on two main concepts: morphometric and gonad maturity.

The presence of external and internal maturity indicators is more evident in females than in males, and female maturity is more relevant for stock assessment. Therefore, analysis of maturity in females is a more common practice in crustacean maturity studies. We presented a list of maturity indicators, and their respective relationships to body size and maturity ogives, for different species of crustaceans around the world, for both males and females. Although the list may be incomplete, we intended to present how different maturity studies are carried out. Additional information on the diverse maturity indices that can be applied to crustacean fisheries is included. Their interpretation depends on the particular biological, ecological and fishery-related characteristics of the species under assessment. This was observed in the CPI estimates obtained for the American and European lobsters, for which differences in the estimates of female CPI baseline and CPI intercept values were perceived.

The appropriate selection of biological information regarding reproduction is necessary for the identification of different units of stock. From a fisheries management viewpoint, it is important to ensure identification of groups that share similar biological characteristics and are subjected to similar fishery practices and levels of exploitation over a confined geographical area. Once different stock units are identified then stock assessment models can be applied. Length-frequency and catch data are the basis of stock assessment in crustaceans because these variables are easy to measure and in most cases represent the only available information.

Egg per recruit and yield per recruit models are options that consider maturity components of the population under study. From these models, information concerning levels of exploitation at diverse temporal scales can be estimated and incorporated into management decision making. Throughout this chapter, we have emphasized the relevance of knowledge of maturity and, in general, reproductive biology in relation to fishery practices and argue that this information is essential to achieve the sustainable management of the species.

References

Abelló, P., J.P. Pertiera and D.G. Reid. 1990. Sexual size dimorphism, relative growth and handedness in *Liocarcinus depurator* and *Micrpipus tuberculatus* (Brachyura: Portunidae). Sci. Mar., 54: 195-202.

Addison, J.T. 1995. Influence of behavioural interactions on lobster distribution and abundance as inferred from pot-caught samples. ICES Mar. Sci. Symp., 199: 294-300.

Addison, J.T. 1997. Lobster stock assessment: report from a workshop. Mar. Freshwater Res., 48: 941-944.

Addison, J.T., and M.C. Bell. 1997. Simulation modelling of capture processes in trap fisheries for clawed lobsters. Mar. Freshwater Res., 48: 1035-1044.

Addison, J.T., and S.R.J. Lovewell. 1985. Pot selectivity and the size composition of lobster (*Homarus gammarus* L.) and crab (*Cancer pagurus* L.) on the East coast of England. ICES CM 1985/K:35, 1-16.

Addison, J.T., R.C.A. Bannister and S.R.J. Lovewell. 1995. Variation and persistence of pre-recruit catch rates of *Homarus gammarus* in Bridlington Bay, England. Crustaceana, 68: 267-279.

Aiken, D.E., and S.L. Waddy. 1980. Reproductive biology. *In*: The Biology and Management of Lobsters. Vol. 1. Physiology and Behaviour. J.S. Cobb and B.F. Phillips (eds.). Academic Press, London, pp. 215-276.

Aiken, D.E., and S.L. Waddy. 1986. Environmental influence on recruitment of the American lobster, *Homarus americanus*: a perspective. Can. J. Fish. Aquat. Sci., 43: 2258-2270.

Aiken, D.E., and S.L. Waddy. 1989. Allometric growth and onset of maturity in male American lobsters (*Homarus americanus*): The crusher propodite index. J. Shellfish Res., 8: 7-11.

Anger, K., and G.S. Moreira. 1998. Morphometric and reproductive traits of tropical Caridean shrimps. J. Crustacean. Biol., 18: 823-838.

Annala, J.H., and K.J. Sullivan. 1997. Management strategies in lobster fisheries: report from a workshop. Mar. Freshwater Res., 48: 1081-1083.

Annala, J.H., J.D. McKoy, J.D. Booth and R.B. Pike. 1980. Size at the onset of sexual maturity in female *Jasus edwardsii* (Decapoda: Palinuridae) in New Zealand. N.Z. J. Mar. Freshwater Res., 14: 217-227.

Arancibia, H., L.A. Cubillos and E. Acuña. 2005. Annual growth and age composition of the squat lobster Cervimunida johni off northern-central Chile (1996-97). Sci. Mar., 69: 113-122.

Ayon Parente, M., and M.E. Hendrickx. 2001. Biology and fishery of the arched box crab *Calappa convexa* de saussure (Crustacea, Brachyura, Calappidae) in the Southeastern Gulf of California, Mexico. Ciencias Marinas, 27: 521-541.

Bailey, N. 1984. Some aspects of reproduction in Nephrops. ICES CM 1984/K:33.

Bailey, N., and C.J. Chapman. 1983. A comparison of density, length composition and growth of two *Nephrops* populations off the West coast of Scotland. ICES C.M. 1983/K:42, 20 pp.

Bannister, R.C.A. 1986. Assessment and population dynamics of commercially exploited shellfish in England and Wales. *In:* North Pacific Workshop on Stock Assessment and Management of Invertebrates. G.S. Jamieson and N. Bourne (eds.). Can Spec. Publ. Aquat. Sci., 92: 182-194.

Bannister, R.C.A., and J.T. Addison. 1984. The effect on yield and biomass per recruit of changes in fishing mortality and size at first capture in the fisheries for lobsters, *Homarus gammar* L., in England and Wales. ICES CM 1984/K:10, 13 pp.

Bannister, R.C.A., and J.T. Addison. 1995. Investigating space and time variation in catches of lobster (*Homarus gammarus* (L.)) in a local fishery on the east coast of England. ICES Mar. Sci. Symp., 199: 334-348.

Bannister, R.C.A., and S.R Lovewell. 1985. Sampling the size composition of catches of the lobster (*Homarus gammarus* L.) on the East coast of England. ICES CM 1985/K:36.

Bannister, R.C.A., J.T. Addison, S.J. Lovewell and A.E. Howard. 1983. Results of a recent minimum size assessment for the fisheries for lobsters, *Homarus gammarus*, in England and Wales. ICES CM 1983/K:4, 9 pp.

Bannister, R.C.A., B.M. Thompson, J.T. Addison, S.J. Lovewell and A.E. Howard. 1990. The 1989 results from a lobster stock enhancement experiment on the East coast of England. ICES CM 1990/K:13.

Bannister, R.C.A., J.T. Addison and S.R.J. Lovewell. 1994. Growth, movement, recapture rate and survival of hatchery-reared lobsters (*Homarus gammarus* (Linnaeus, 1758)) released into the wild on the English East coast. Crustaceana, 67: 156-172.

Bell, M.C., J.T. Addison and R.C.A. Bannister. 2001. Estimating trapping areas from trap-catch data for lobsters and crabs. Mar. Freshwater Res., 52: 1233-1242.

Bennett, D.B. 1995. Factors in the life history of the edible crab (*Cancer pagurus* L.) that influence modelling and management. ICES Mar. Sci. Symp., 199: 89-98.

Bennett, D.B., and A.E. Howard. 1987. Estimates of lobster (*Homarus gammarus*) fecundity from East and West Britain. ICES, CM 1987/K:47, 14 pp.

Bennett, D.B., C.G. Brown, A.E. Howard and S.R.J. Lovewell. 1978. Comparison of lobster (*Homarus gammarus*) growth rates in Norfolk and Yorkshire, England. ICES CM 1978/K:6.

Booke, H.E. 1981. The conundrum of stock concept - are nature and nurture definable in fishery science? Can. J. Fish. Aquat. Sci., 38: 1479-1480.

Briand, G., S.C. Matulich and R.C. Mittelhammer. 2001. A catch per unit effort - soak time model for the Bristol Bay red king crab fishery, 1991-1997. Can. J. Fish. Aquat. Sci., 58: 334-341.

Briggs, R.P. 1989. Temporal and spatial variation in Western Irish Sea Nephrops. ICES CM 1989/K:36, 11 pp.

Butler, M.J., IV. 1997. Benthic processes in lobster ecology: report from a workshop. Mar. Freshwater Res., 48: 659-662.

Caddy, J.F. 1986. Stock assessment in data-limited situations - The experience in tropical fisheries and its possible relevance to evaluation of invertebrate resources. *In:* North Pacific Workshop on Stock Assessment and Management for Invertebrates. G.S. Jamieson and N. Bourne (ed.). Can Spec. Publ. Fish. Aquat. Sci., 92: 379-392.

Caddy, J.F. 1989. Overview of crustacean fisheries: assessment and population dynamics. *In:* Marine Invertebrate Fisheries: Their Assessment and Management. J.F. Caddy (ed.). John Wiley & Sons, Inc. 752 pp.

Cadrin, S.X. 1995. Discrimination of American lobster (*Homarus americanus*) stocks off southern New England on the basis of secondary sex character allometry. Can. J. Fish. Aquat. Sci., 52: 2712-2723.

Cadrin, S.X., and K.D. Friedland. 1999. The utility of image processing techniques for morphometric analysis and stock identification. Fish. Res., 49: 129-139.

Campbell, A. 1986. Migratory movements of ovigerous lobsters, *Homarus americanus,* tagged off Grand Manan, Eastern Canada. Can. J. Fish. Aquat. Sci., 43: 2197-2205.

Campbell, A., and R.K. Mohn. 1982. The quest for lobster stock boundaries in the Canadian Maritimes. NAFO SCR Doc. 82/IX/107.

Campbell, A., and M.D. Eagles. 1983. Size at maturity and fecundity of rock crabs *Cancer irroratus,* from the bay of Fundy and southwestern Nova Scotia. Fish. Bull., 81: 357-362.

Carmona-Suárez, C.A. 2003. Reproductive biology and relative growth in the spider crab *Maja crispata* (Crustacea: Brachyura: Majidae). Sci. Mar., 67: 75-80.

Cha, H.K., C.W. Oh, S.Y. Hong and K.Y. Park. 2002. Reproduction and population dynamics of *Penaeus chinensis* (Decapoda: Penaeidae) on the western coast of Koea, Yellow Sea. Fish. Res., 56: 25-36.

Childress, M.J. 1997. Marine reserves and their effects on lobster populations: report from a workshop. Mar. Freshwater Res., 48: 1111-1114.

Cobb, J.S., and J.F. Caddy. 1989. The population biology of decapods.. *In:* Marine Invertebrate Fisheries: Their Assessment and Management. J.F. Caddy (ed.). Wiley Interscience Publication, New York, pp. 327-374.

Cobb, J.S., J.D. Booth and M. Clancy. 1997. Recruitment strategies in lobsters and crabs: a comparison. Mar. Freshwater Res., 48: 797-806.

Collins, K.J. 1998. Habitat selection and mobility in adult lobsters. *In:* The European Lobster *Homarus gammarus* (L.). G.I. van der Meeren and O. Soldar (eds.). Proceedings from the Seminar at KVITSØY 1995, Havforskningsinstitutet. Fisken og Havet, 13: 46-55.

Comeau, M., and F. Savoie. 2002. Movement of American lobster (*Homarus americanus*) in the southwestern Gulf of St. Lawrence. Fish. Bull., 100: 181-192.

Comeau, M., and M. Mallet. 2001. Difficulties estimating mortality rates by capture-recapture, catch-effort and change-in-ratio models for a spring American lobster (*Homarus americanus*) fishery in Canada. Abstracts of the conference on Life Histories, Assessment and Management of Crustacean Fisheries, A Coruña, Galicia, Spain, 8-12 October.

Conan, G.Y. 1978. Comparison between growths at moult of the European and American lobsters in different geographic locations, according to their sex and state of maturity. ICES CM 1978/K:18, 36 pp.

Conan, G.Y., and M. Comeau. 1985. Functional maturity of the male snow crab, *Chionecetes opilio*. ICES CM 1985/K:28, 30 pp.

Conan, G.Y., and M. Comeau. 1986. Functional maturity and terminal molt of male snow crab, *Chionoecetes opilio*. Can. J. Fish. Aquat. Sci., 43: 1710-1719.

Conan, G.Y., M. Comeau and M. Moriyasu. 1985. Functional maturity of the American lobster *Homarus americanus*. ICES CM 1985/K:29, 55 pp.

Conan G.Y., M. Comeau and M. Moriyasu. 2001. Are morphometrical approaches appropriate to establish size at maturity for male American lobster, *Homarus americanus*? J. Crustacean. Biol., 21: 937-947.

Davidson, K., R.W. Elner and J. Roff. 1982. Discrimination of the Atlantic snow crab, *Chionecetes opilio*, populations: a problem of management application. NAFO SCR Doc. 82/IX/86.

Davidson, R.J., and I.D. Marsden. 1987. Size relationships and relative growth of the New Zealand swimming crab *Ovalipes catharus* (White 1843). J. Crustacean. Biol., 7: 308-317.

de Lestang, S., N.G. Hall and I.C. Potter. 2003. Reproductive biology of the blue swimmer crab (*Portunus pelagicus*, Decapoda: Portunidae) in five bodies of water on the west coasts of Australia. Fish. Bull., 101: 745-757.

Debuse, V.J., J.T. Addison and J.D. Reynolds. 1999. The effects of sex ratio on sexual competition in the European lobster. Anim. Behav., 58: 973-981.

Debuse, V.J., J.T. Addison and J.D. Reynolds. 2001. Morphometric variability in UK populations of the European lobster. J. Mar. Biol. Assoc. U.K., 81: 469-474.

DeMartini, E.E., D.M. Ellis and V.A. Honda. 1993. Comparisons of spiny lobster *Panulirus marginatus* fecundity, egg size, and spawning frequency before and after exploitation. Fish. Bull., 91: 1-7.

DeMartini, E.E., G.T. DiNardo and H.A. Williams. 2003. Temporal changes in population density, fecundity, and egg size of the Hawaiian spiny lobster (*Panulirus marginatus*) at Necker Bank Northwestern Hawaiian Islands. Fish. Bull., 101: 22-31.

DeMartini, E.E., M.L. McCracken, R.B. Moffitt and J.A. Wetherall. 2005. Relative pleopods length as an indicator of size at sexual maturity in slipper (*Scyllarides squammosus*) and spiny Hawaiian (*Panulirus marginatus*) lobsters. Fish. Bull., 103: 23-33.

Edwards, E. 1979. The edible crab and its fishery in British waters. Fishing New Books Ltd., Farnham, Surrey. 142pp.

Ennis, G.P. 1980. Size-maturity relationships and related observations in Newfoundland populations in the lobster (*Homarus americanus*). Can. J. Fish. Aquat. Sci. 37: 945-956.

Ennis, G.P. 1983. Variation in annual growth in two Newfoundland lobster (*Homarus americanus*) populations in relation to temperature conditions. ICES CM 1983/K:24.

Ennis, G.P. 1986. Stock definition, recruitment variability, and larval recruitment processes in the American lobster, *Homarus americanus*: a review. Can. J. Fish. Aquat. Sci., 43: 2072-2084.

Harding, G.C., E.L. Kenchington, C.J. Bird, D.S. Pezzack and D.C. Landry. 1997. Genetic relationships among subpopulations of the American lobster (*Homarus americanus*) as revealed by random amplified polymorphic DNA. Canada. Can. J. Fish. Aquat. Sci., 54: 1762-1771.

Hartnoll, R.G. 1969. Mating in Brachyura. Crustaceana, 16: 161-181.

Hartnoll, R.G. 1974. Variation in growth pattern between some secondary sexual characters in crabs (Decapoda Brachyura). Crustaceana, 27: 131-136.

Hartnoll, R.G. 1978. The determination of relative growth in Crustacea. Crustaceana, 34: 281-293.

Hartnoll, R.G. 1982. Growth. In: Abele, L,G. (Ed.), The Biology of Crustacea, vol. 1. Academic Press, New York, pp. 111-196.

Hata, D., and J. Berkson. 2003. Abundance of horseshoe crabs (*Limulus polyphemus*) in the Delaware Bay area. Fish Bull. 101: 933-938.

Hepper, B.T., and C.J. Gough. 1978. Fecundity and rate of embryonic development of the lobster, *Homarus americanus* (L), off the coast of North Wales. J. Cons. Int. Explor. Mer., 38: 54-57.

Heydorn, A.E.F. 1965. The rock lobster of the South African west coast, *Jasus lalandii* (H. Milne-Edwards). Notes on the reproductive biology and the determination of minimum size limits for commercial catches. S. Afr. Div. Sea Fish. Invest. Rep., 53: 1-32.

Hilborn, R. 1986. A comparison of alternative harvest tactics for invertebrate fisheries. *In:* North Pacific Workshop on Stock Assessment and Management of Invertebrates. G.S. Jamieson and N. Bourne (ed.). Can Spec. Publ. Fish. Aquat. Sci. 92, pp. 311-317.

Hilborn, R. 1997. Lobster stock assessment: report from a workshop; II. Mar. Freshwater Res., 48: 945-947.

Hilborn, R. 2003. The state of the art in stock assessment: where we are and where we are going. Sci. Mar., 67: 15-20.

Hines, A.H. 1989. Geographic variation in size at maturity in Brachyuran crabs. Bull. Mar. Sci., 45: 356-368.

Hobday, D.K., and T.J. Ryan. 1997. Contrasting sizes at sexual maturity of southern rock lobsters (*Jasus edwardsii*) in the two Victorian fishing zones: implications for total egg production and management. Mar. Freshwater Res., 48: 1009-1014.

Howard, A.E. 1980. Substrate controls on the size composition of lobster (*Homarus gammarus*) populations. J. Cons. Int. Explor. Mer., 39: 130-133.

Jackson, G.D., R.A. Alford and J.H. Choat. 2000. Can length frequency analysis be used to determine squid growth? An assessment of ELEFAN. ICES J. Mar. Sci., 57: 948-954.

Jayakody, D.S. 1989. Size at onset of sexual maturity and onset of spawning in female *Panulirus homarus* (Crustacea: Decapoda: Palinuridae) in Sri Lanka. Mar. Ecol. Prog. Ser., 57: 83-87.

Jensen, A.C., E.K. Free and K.J. Collins. 1993. Lobster (*Homarus gammarus*) movement in the Poole Bay fishery, Dorset, UK. ICES CM 1993/K:49.

Jewett, S.C., N.A. Sloan and D.A. Somerton. 1985. Size at sexual maturity and fecundity of the fjord-dwelling golden king crab *Lithodes aequispina*. J. Crustacean. Biol., 5: 377-385.

Jørstad, K.E., and E. Farestveit. 1999. Population genetic structure of lobster (*Homarus gammarus*) in Norway, and implications for enhancement and sea-ranching operation. Aquaculture, 173: 447-457.

Jury, S.H., H. Howell, D.F. O'Grady and W.H. Watson III. 2001. Lobster trap video: *in situ* video surveillance of the behaviour of *Homarus americanus* in and around traps. Mar. Freshwater Res., 52: 1125-1132.

Koeller, P. 1999. Influence of temperature and effort on lobster catches at different temporal and spatial scales and the implications for stock assessments. Fish. Bull., 97: 62-70.

Krajangdara, T., and S. Watanabe. 2005. Growth and reproduction of the red frog crab, *Ranina ranina* (Linnaeus, 1758), in the Andaman Sea off Thailand. Fish. Sci., 71: 20-28.

Krouse, J.S. 1989. Performance and selectivity of trap fisheries for crustaceans. *In:* Marine Invertebrate Fisheries: Their Assessment and Management. J.F. Caddy (ed.). John Wiley & Sons, Inc. 752 pp.

Landers D.F., M. Keser and S.B. Saila. 2001. Changes in female lobster (*Homarus americanus*) size at maturity and implications for the lobster resource in Long Island Sound, Connecticut. Mar. Freshwater Res., 52: 1283-1290.

Latrouite, D., M. Leglise and G. Raguenes. 1981. Données sur la reproduction et la taille de première maturité du homard *Homarus gammarus* de la Mer d'Iroise et du Golfe de Gascogne. ICES CM K:8, 8 pp.

Latrouite, D., Y. Morizur and G. Raguenes. 1984. Fecundites individuelles et par recrue du homard europeen *Homarus gammarus* (L.) des cotes francaises. ICES CM K:38, 18 pp.

Lawton, P., D.A. Rochibaud, M.B. Strong, D.S. Pezzack and C.F. Frail. 2001. Spatial and temporal trends in the American lobster, *Homarus americanus*, fishery in the Bay of Fundy (lobster fishing areas 36, 36 and 38). DFO Canadian Stock Assessment Secretariat Research Document-2001/094.

Le Foll, A. 1986. Contribution à l'étude de la biologie du crabe-tourteau *Cancer pagurus* sur les côtes de bretagne sud. Reu. Trau. Inst. Pêches marit., 48 (1 et 2): 5-22, 1984 (1986).

Lizárraga-Cubedo, H.A. 2004. Factors affecting the fluctuations of the European lobster populations in Scottish coasts. PhD Thesis. University of Aberdeen, Aberdeen, UK, 415 pp.

Lizárraga-Cubedo, H.A., I. Tuck, N. Bailey, G.J. Pierce and J.A.M. Kinnear. 2003. Comparisons of size at maturity and fecundity of two Scottish populations of the European lobster, *Homarus gammarus*. Fish. Res., 65: 137-152.

López-Martínez, J., C. Rábago Quiroz, M.N. Nevárez Martínez, A.R. García Juárez, G. Rivera Parra and J. Chávez Villalba. 2005. Growth, reproduction, and size at first maturity of blue shrimp *Litopenaeus stylirostris* (simpson 1874) along the east coast of the Gulf of California, Mexico. Fish. Res., 71: 93-102.

Rowe, S. 2002. Population parameters of American lobster inside and outside no-take reserves in Bonavista Bay, Newfoundland. Fish. Res., 56: 167-175.

Saila, S.B., and J.M. Flowers. 1969. Geographic morphometric variation in the American lobster. Syst. Zool., 18: 330-338.

Saila, S.B., J.H. Annala, J.L. McKoy and J.D. Booth. 1979. Application of yield models to the New Zealand rock lobster fishery. New Zealand J. Mar. Freshwater Res., 13: 1-11.

Sharp, W.C., J.H. Hunt and W.G. Lyons. 1997. Life history of the spiny lobster, *Panulirus guttatus*, an obligate reef-dweller. Mar. Freshwater Res., 48: 687-698.

Sheehy, M.R.J. 1989. Crustacean brain lipofuscin: an examination of the morphological pigment in the freshwater crayfish *Cherax cuspidatus* (Parastacidae). J. Crustacean Biol., 9: 387-391.

Sheehy, M.R.J., P.M.J. Shelton, J.F. Wickins, M. Belchier and E. Gaten. 1996. Ageing the European lobster *Homarus gammarus* by the lipofuscin in its eyestalk ganglia. Mar Ecol Prog. Ser 143: 99-111.

Sheehy, M.R.J. 2001. Implications of protracted recruitment for perception of the spawner-recruit relationship. Can. J. Fish. Aquat. Sci., 58: 641-644.

Sheehy, M.R.J., and R.C.A. Bannister. 2002. Year-class detection reveals climatic modulation of settlement strength in the European lobster, *Homarus gammarus*. Can. J. Fish. Aquat. Sci., 59: 1132-1143.

Sheehy, M.R.J., R.C.A. Bannister, J.F. Wickins and P.M.J. Shelton. 1999. New perspectives on the growth and longevity of the European lobster, *Homarus gammarus*. Can. J. Fish. Aquat. Sci., 56: 1904-1915.

Shelton, R.G.J., R. Jones, J. Mason, J.A.M. Kinnear and K. Livingstone. 1978a. The lobster fishery at Eyemouth (S.E. Scotland) – A brief review of its port-war history and the prospects for increasing its long term yield. ICES CM 1978/K:24.

Shelton, R.G.J., J.A. Sinclair, J.A.M. Kinnear and K. Livingstone. 1978b. The state of the Scottish lobster stocks. Scottish Fish. Bull., 44: 29-33.

Simpson, A.C. 1958. Some seasonal variations in the catch and stock composition of the lobster (*Homarus gammarus* Edw.) on the coast of Wales. Fisheries Investigations, 2(22)3: 1-33.

Simpson, A.C. 1961. A contribution to the bionomics of the lobster (*Homarus vulgaris* Edw.) on the coast of North Wales. Fisheries Investigations, 2: 1-28.

Skud, B.E. 1979. Soak time and the catch per pot in an offshore fishery for lobsters (*Homarus americanus*). Rapp. P.-v. Reun. Const. Int. Explor. Mer., 175: 190-196.

Skud, B.E., and H.C. Perkins. 1969. Size composition, sex ratio, and size at maturity of offshore northern lobsters. Spec. Scient. Rep. Fish. US Fish Wildl. Serv. 598, 10 pp.

Smith, I.P., K.J. Collins and A.C. Jensen. 1999. Seasonal changes in the level and diel patterns of activity in the European lobster *Homarus gammarus*. Mar. Ecol. Prog. Ser., 186: 255-264.

Smith, K.D., N.G. Hal, S. de Lestang and I.C. Potter. 2004. Potential bias in estimates of the size of maturity of crabs derived from trap samples. ICES J. Mar. Sci., 61: 906-912.

Smith, S.J., and M.J. Tremblay. 2003. Fishery-independent trap surveys of lobsters (*Homarus americanus*): design considerations. Fish. Res., 1483: 1-11.

Somerton, D.A. 1980. Fitting straight lines to Hiatt growth diagrams: a re-evaluation. J. Cons. Int. Explor. Mer., 39: 15-19.

Somerton, D.A. 1981. Regional variation in the size at maturity of two species of tanner crab (*Chionecetes bairdi* and *C. opilio*) in the eastern Bering Sea, and its use in defining Management sub areas. Can. J. Fish. Aquat. Sci., 38: 163-174.

Somerton, D.A., and R.A. Otto. 1986. Distribution and reproductive biology of the golden king crab, *Lithodes aequispina,* in the eastern Bering Sea. Fish. Bull. US., 84: 571-584.

Squires, H.J. 1970. Lobster (*Homarus americanus*) fishery and ecology in Port au Port Bay, Newfoundland, 1960-65. Proc. Natl. Shellfish. Assoc., 60: 22-39.

Temming, A., U. Damm and T. Neudecker. 1993. Trends in the size composition of commercial catches of brown shrimp (*Crangon crangon* L.) along the German coast. ICES CM 1993/K:53, 14 pp.

Templeman, W. 1935. Local differences in the body proportions of the lobster, *Homarus americanus*. J. Biol. Board. Can., 1: 213-226.

Templeman, W. 1936. Local differences in the life history of the American lobster (*Homarus americanus*) on the coast of the Maritime Provinces of Canada. J. Biol. Board. Can., 2: 41-88.

Templeman, W. 1939. Investigations into the life history of the lobster (*Homarus americanus*) on the west coast of Newfoundland 1938. Res. Bull. Div. Fish. Res. Newfoundland, 7, 52 pp.

Templeman, W. 1944. Abdominal width and sexual maturity of the female lobsters on Canadian Atlantic Coast. J. Fish. Res. Board. Can., 6: 281-290.

Thomas, H.J. 1955a. Observations on the sex ratio and mortality rates in the lobster (*Homarus vulgaris* EDW.). J. Cons. Perm. Int. Explor. Mer., 22: 295-305.

Thomas, H.J. 1955b. Lobsters and their fishery in Scotland. Scottish Fish. Bull., 3: 1-9.

Thomas, H.J. 1958a. Observations on the increase in size at moulting in the lobster (*Homarus vulgaris* M. Edw.). J. Mar. Biol. Assoc. UK., 38: 603-606.

Thomas, H.J. 1958b. Some seasonal variations in the catch and stock composition of the lobster. J. Cons. Int. Explor. Mer., 24: 147-154.

Thomas, H.J. 1959a. Creel selectivity in the capture of lobsters and crabs. J. Cons., Cons. Int. Explor. Mer., 24: 342-348.

Thomas, H.J. 1959b. A comparison of some methods used in lobster and crab fishing. Scottish Fish Bull., 12: 3-8.

Thomas, H.J. 1965. The lobster fishery of the South East Scottish coast. Rapp. P. V. Reun. Cons. Perm. Int. Explor. Mer., 156: 13-20.

Thomas, H.J. 1969. Observations on the seasonal variations in the catch composition of the lobster around the Orkneys. ICES C.M. 1969/K:35, 9 pp.

Thompson, B.M., A.R. Lawler and D.B. Bennett. 1995. Estimation of the spatial distribution of the spawning crabs (*Cancer pagurus* L.) using larval surveys in the English channel. ICES Mar. Sci. Symp., 199: 139-150.

Thorpe, R.S. 1987. Geographic variation: a synthesis of cause, data, pattern and congruence in relation to subspecies, multivariate analysis and phylogenesis. Bull. Zool., 54: 3-11.

Tremblay, M.J., M.D. Eagles and G.A.P. Black. 1998. Movements of the lobster, *Homarus americanus*, off northeastern Cape Breton Island, with notes on lobster catchability. Can Tech. Rep. Fish. Aquat. Sci., 2220: iv + 32 p.

Tuck, I.D., C.J. Chapman, R.J.A. Atkinson, N. Bailey and R.S.M. Smith. 1997a. A comparison of methods for stock assessment of the Norway lobster, *Nephrops norvegicus*, in the Firth of Clyde. Fish. Res., 32: 89-100.

Tuck, I.D., C.J. Chapman and R.J.A. Atkinson. 1997b. Population biology of the Norway lobster, *Nephrops norvegicus* (L.) in the Firth of Clyde, Scotland I: Growth and density. ICES J. Mar. Sci., 54: 125-135.

Tuck, I.D., R.J.A. Atkinson and C.J. Chapman. 2000. Population biology of the Norway lobster, *Nephrops norvegicus* (L.) in the Firth of Clyde, Scotland II: fecundity and size at onset of sexual maturity. ICES J. Mar. Sci., 57: 1227-1239.

Tully, O., V. Roantree and M. Robinson. 2001. Maturity, fecundity and reproductive potential of the European lobster (*Homarus gammarus*) in Ireland. J. Mar. Biol. Assoc. U.K., 81: 61-68.

Tzeng, T.D., C.S. Chiu and S.Y. Yeh. 2001. Morphometric variation in red-spot prawn (*Metapenaeopsis barbata*) in different geographic waters off Taiwan. Fish. Res., 53: 211-217.

Uzmann, J.R., R.A. Cooper and K.J. Pecci. 1977. Migration and dispersion of tagged American lobsters, *Homarus americanus*, on the southern New England continental shelf. NOAA Tech. Rep. NMFS SSRF 705.

Vermeij, G.J. 1977. The Mesozoic marine revolution: Evidence from snails, predators and grazers. Paleobiology 3: 245-258.

Waddy, S.L., and D.E. Aiken. 1992. Egg production in the American lobster, *Homarus americanus*. *In:* Crustacean Egg Production, Vol. 7, Crustacean Issues. A.M. Wenner and A.M. Kuris (eds.). A.A. Balkema, The Netherlands, pp. 267-290.

Waddy, S.L., and D.E. Aiken. 1995. Temperature regulation of reproduction in female American lobsters (*Homarus americanus*). ICES Mar. Sci. Symp., 199: 54-60.

Waddy, S.L., D.E. Aiken and D.P.V. De Klein. 1995. Control of growth and reproduction. In J.R. Factor (ed.), Biology of the lobster *Homarus americanus*, pp. 217-266. Academic Press, San Diego.

Waddy, S.L., and D.E. Aiken. 2005. Impact of invalid biological assumptions and misapplication of maturity criteria on size at maturity estimates for American lobster. Trans. Am. Fish. Soc., 134: 1075-1090.

Zheng, J., and G.H. Kruse. 2000. Recruitment patterns of Alaskan crabs in relation to decadal shifts in climate and physical oceanography. ICES J. Mar. Sci., 57: 438-451.

5

Mating Behaviour

A. Barki

Aquaculture Research Unit, Institute of Animal Science, Agricultural Research Organization, Volcani Center, P.O. Box 6, Bet Dagan 50250, Israel. E-mail: barkia@agri.gov.il

INTRODUCTION

The reproductive biology of crustaceans encompasses a myriad complex physiological and regulatory mechanisms within the animal, some of which are reviewed in this book, leading to the production of fertilizable gametes. However, reproduction would not be accomplished if the male and female gametes did not meet for fertilization to occur. This important component of the reproductive biology of animals involves their mating behaviour.

From the evolutionary perspective, the particular mating behaviour of a species is a product of selection for increased reproductive success acting on individuals of each sex. Darwin (1871) recognized that for many traits, including behavioural traits, selection differentially acts on males and females; it is usually the males who compete for mates, possess exaggerated traits and exhibit larger morphological and behavioural variation. Darwin coined the term "sexual selection" to account for the evolution of traits that may be unfavourable for the survival of individuals and thus cannot be accounted for by natural selection. The first part of this chapter attempts to provide a general overview of the main mechanisms through which sexual selection operates, as reflected in the mating behaviour of crustaceans. The focus is mainly on the male sex. Several aspects of male mating competition (also termed intrasexual selection) are reviewed and mate choice (intersexual selection) in crustaceans is briefly referred to. Both of these topics are vast and have been extensively studied in crustaceans. Several exhaustive reviews are available for particular crustacean groups; no attempt is made to cover them all but rather examples from various groups are used in order to demonstrate the diversity of mating behaviour among crustaceans.

An observable result of the differences in the action of sexual selection between the two sexes is sexual dimorphism. It is expected to be pronounced in species in which the variance in reproductive success among individuals is much higher in one sex (usually males) than in the opposite sex, i.e., where there is a sex difference in the opportunity for selection (ΔI) (Shuster and Wade, 2003). Crustaceans may exhibit pronounced sexual dimorphism, which involves morphological traits (e.g., body size, weaponry and male attachment structures) as well as behavioural traits (e.g., exaggerated courtship displays and male aggression). A well-known example is the large major claw possessed by male fiddler crabs and the claw waving display used for attracting females (Murai and Backwell, 2005; Backwell et al., 2006). A brief review of the expression of sexual dimorphism in crustaceans is carried out in the subsequent part of this chapter.

Sex differences in mating behaviour are also evident in crustacean species that do not exhibit considerable morphological dimorphism, for example in snapping shrimp, which live in a monogamous mating system (Knowlton, 1980), and in stomatopods (Caldwell, 1991). Furthermore, sexual selection mechanisms may even operate in hermaphroditic species (Leonard, 2006), for example in protandric hermaphrodite caridean shrimp (Bauer, 2006). Proximate mechanisms underlying sex differences in secondary characteristics such as mating behaviour are usually mediated by sex hormones. It is well known that androgens influence sexual and aggressive behaviour in vertebrates (Kelley, 1988; Hull et al., 2002; Arnold, 2004; Ball and Balthazart, 2006). In crustaceans, a male-specific endocrine gland, the androgenic gland, is responsible for the development and maintenance of sexually dimorphic traits in males (see Chapter 7 for a review), including behaviour traits. The hormonal control of mating behaviour, a topic on which there is very limited knowledge in crustaceans, is reviewed in the final part of this chapter with special reference to the role of the androgenic gland.

MATING COMPETITION

Competition occurs when two or more individuals are dependent on the same resource and this resource is limited. While this term is broadly used in ecology, behavioural ecologists have been specifically interested in the underlying behaviour involved in competitive interactions and the consequences for the individual animal. There are generally two modes of competition (Nicholson, 1954): contest competition, in which one individual gains access to the resource while denying access from other competitors, and scramble competition, in which each individual tries to maximize its share in the limited resource without direct interference from competitors. In fact, "pure" contest (i.e., all losers get nothing) and "pure" scramble (i.e., all competitors obtain a share of resource) are uncommon. Rather, they are two extremes in a continuum of competitive interactions found in nature (Parker, 2000).

Owing to anisogamy and sex differences in investment in gametes and in parental care, the potential reproductive rate of males is higher than that of females (Williams, 1966; Trivers, 1972; Parker and Simmons, 1996). Consequently, females are typically the limiting sex and males compete for access to receptive females. Emlen and Oring (1977) provided a very useful behavioural-ecological framework that considers one sex as a critical resource for which individuals of the other sex compete. Ecological constraints (e.g., the distribution of essential resources for reproduction) and the temporal pattern of availability of receptive females (e.g., the synchronization and duration of female receptivity) strongly influence the spatial and temporal distribution of receptive females, which in turn influences the number of potential mates for males and the ability to monopolize them (Emlen and Oring, 1977). For instance, in species in which females are spatially clumped and female receptivity is asynchronous, the possibility for a small portion of the males to monopolize a large number of females and to obtain multiple mating opportunities is high. Under such conditions, a polygynous mating system is expected with intense male competition and high variation in male mating success. Emlen and Oring (1977) further classified mating systems according to the behavioural means through which individuals of the limited sex obtain mates.

Based on the Emlen and Oring (1977) mating system classification, Christy (1987) distinguished eight kinds of mating associations in brachyuran crabs that fall into the following three general categories: "males may (1) search for or attract individual receptive females that they defend directly from other males, (2) defend resources that females require for breeding or survival and mate with the females associated with these resources, or, (3) compete in ways that maximize the rate they encounter females but neither defend females nor resources" (Christy, 1987: 177-178). In fact, these categories hold true for most other crustacean groups, each of which includes species with diverse competitive modes. Nevertheless, in some groups, e.g., in brachyuran crabs, the first two categories, which involve an element of contest, are prevalent (Christy, 1987), whereas in other groups, e.g., in caridean shrimps in which "pure" mate searching is predominant (Bauer and Abdalla, 2001; Correa and Thiel, 2003; Zhang and Lin, 2005), the typical tendency is more towards scramble competition.

Contest Competition

In general, contest for mating opportunities in crustaceans may be resource centred or female centred (Christy, 1987), and it may be manifested in resource defence, direct competition for females, mate guarding, sperm competition and alternative male mating strategies.

Female-centred Competition

Female-centred competition is widespread among crustaceans and it may take various forms, ranging in intensity from intense aggressive fights over females to brief male-male aggressive interaction, and from prolonged mate guarding to mating without attending the female longer than necessary for copulation.

In light of the fact that male competition is related to the spatial and temporal distribution of receptive females and to the female reproductive life history (Shuster and Wade, 2003), studies in crustaceans have attempted to elucidate how details in the ecology, morphology, and female reproductive traits of a species may explain its particular mode of mating competition. Comparative accounts of mating behaviour in crustaceans are particularly useful in this regard. Regarding the temporal aspect of female availability, in many crustacean species copulation occurs only when the female is soft shortly after moulting. The duration of female receptivity in species in which females copulate hard-shelled during the intermoult stage may also be restricted, usually not extending more than a few days. For instance, in a number of brachyuran crabs copulation occurs after local decalcification of the hinge of immobile opercula sealing the gonopores, which makes them temporarily flexible and mobile (Hartnoll, 1969). Thus, the temporal distribution of receptive females is mainly dependent on the degree of synchrony of short receptivity episodes in the population. The effect of the temporal distribution of females on male competition can be demonstrated by a comparative study on the breeding ecology of cancrid crabs (Orensanz and Gallucci, 1988; Orensanz et al., 1995). *Cancer magister* and *C. gracilis* are sympatric cancrid species that resemble each other in many aspects of their reproductive ecology and biology. Both species inhabit open habitats in which receptive females aggregate at breeding sites. In both species multiple copulations are possible during a single receptive period and sperm competition occurs. The combination of soft-shell mating and sperm competition in these crabs leads to direct defence of mobile females with pre- and post-copulatory mate guarding. However, in *C. gracilis* females can produce multiple broods during a rather protracted breeding season and female receptivity is asynchronous, whereas *C. magister* females produce one annual brood and they are more synchronous in their receptivity period. Consequently, *C. gracilis* males fight intensely for the scarce receptive females and attempt takeovers of guarded females, whereas *C. magister* males invest less in aggressive contests and tend towards scramble competition, since the benefits (in terms of mating opportunities) gained from searching rather than staying and fighting for receptive females increase when another receptive female is likely to be found.

The picture becomes more complex when we consider the combination of factors that might influence mating competition. Recently, Brockerhoff and McLay (2005a, b, c) provided comparative data including many details regarding the mating biology, ecology and behaviour of grapsid crabs from New Zealand, *Cyclograpsus lavauxi, Helice crassa* and *Hemigrapsus sexdentatus*. These data (compiled in Table 1) illustrate how different female-centred competitive modes evolved in related species that are seemingly similar in their mating traits, and conversely, how similar competitive modes evolved in related species that differ in mating characteristics, owing to different combinations of influencing factors. The females in these grapsid species mate during the intermoult period but mating is morphologically restricted to certain times when the gonopore opercula of the female become mobile. The temporal characteristics of female availability in *C. lavauxi* and *H. sexdentatus* are very similar (Table 1); despite their short and synchronous mating season the operational sex ratio (OSR, the number of sexually active males to receptive females at a given time) is male-biased in both species, owing to the very short receptive period of females, which decreases the likelihood of many receptivity episodes overlapping in time. Males of both species were frequently observed to fight over receptive females. However, *C. lavauxi* males do not guard females (or guard only for a brief period prior to and during copulation (2 h), as observed in the lab), whereas *H. sexdentatus* males exhibit post-copulatory mate guarding that lasts for several days (Table 1), which is expected from a species with sperm competition involving last male sperm precedence. This difference was accounted for by the different habitats occupied by these crabs. Although both species occur in rocky

Table 1 Behavioural, ecological and reproductive characteristics of the mating system of three grapsid species in New Zealand. Data compiled from Brockerhoff and McLay (2005a, b, c).

Characteristic		*Cyclograpsus lavauxi*	*Helice crassa*	*Hemigrapsus sexdentatus*
OSR (male/female)		highly male-biased > 18	highly male-biased 3-16 (non-ovigerous + ovigerous females) 13-67 (non-ovigerous females only)	highly male-biased >14
Spatial distribution	density	—	—	1.3-2.0 males m^{-2} 1.9-3.3 females m^{-2}
	Synchronization	**synchronous**	**asynchronous**	**synchronous**
Temporal distribution	breeding season	peaks for 4 weeks	several months (~6)	3 weeks
	receptivity duration	< 1 day in the field	10-15 days in the lab	< 1 day in the field
No. of copulations		1-3	24 (5-51)	8
No. of broods		1	2	1
Sperm competition		possible; last male sperm precedence	possible; last male sperm precedence	possible; last male sperm precedence
Habitat		**rocky, high-intertidal zone**	**sandy or muddy, construct temporary burrows above mid-tide level**	**rocky, mid-intertidal zone**
	pre-copulatory guarding	no (~1h in the lab)	no	no
	elaborate courtship	no	no	no
Male mating behaviour	copulation duration	2 h (in the lab)	10-15 min	10-15 min
	post-copulatory guarding	**no**	**no** (~2-3 h in the lab)	**yes**
	competition	frequent fighting over females; large male advantage: more takeovers, gain more copulations, more likely to be the last to mate before oviposition	frequent fighting over females; large male advantage: more takeovers, gain more copulations, more likely to be the last to mate before oviposition	frequent fighting over females; large male advantage: more takeovers, gain more copulations, more likely to be the last to mate before oviposition

intertidal habitats, *C. lavauxi* crabs occupy the high-intertidal zone where they are exposed to wave action during high spring tides and to long periods of aerial exposure during low neap tides. Such a harsh environment does not favour female guarding by males. *Hemigrapsus sexdentatus* crabs occupy the mid-intertidal zone where they are fairly protected from desiccation, wave action and predators and this allows them to search for and defend receptive females for a prolonged period. The above habitat difference is likely to influence the spatial distribution of receptive females and thereby to shape the male competitive modes of these species; however, there was no available data on the pattern of female distribution within these habitats. In addition, the ability of *H. sexdentatus* females to extend their receptivity period (which was demonstrated in the laboratory in response to the absence of males) may enable them to be more selective, thus increasing male competition and enhancing mate guarding. In contrast, a comparison of *C. lavauxi* and *H. crassa* crabs reveals two species that differ in the temporal characteristics of female receptivity and in the type of habitat they occupy, but in both species the males directly compete for females and do not exhibit mate guarding (Table 1). *Helice crassa* crabs occur in open mud flats, where they dig short-lived burrows. The lack of mate guarding in males of this species was attributed to a high predation risk in open habitat coupled with surface mating (arising from the temporary nature of the burrow), which restricts prolonged guarding on the surface.

In species in which mating is coupled with moulting, post-copulatory mate guarding functions to protect the recently moulted female from predation in addition to preventing sperm competition. For example, in the portunid crab *Callinectes sapidus*, males copulate with newly moulted females and guard them during their vulnerable period, and indeed the duration of post-copulatory guarding increased under high predation risk (Jivoff, 1997). The above effects of predation pressure on male mating behaviour in species in which the females mate when they are hard-shelled (*H. crassa*) as opposed to when they are soft-shelled (*C. sapidus*) illustrate the importance of the female reproductive traits in shaping the mode of male competition in crustaceans.

Variation in mating competition also occurs between and even within populations of the same species. Variations due to population characteristics such as density, local sex ratio and OSR have often been shown in various species mainly with regard to the duration of mate guarding (e.g., Iribarne et al., 1995; Dick and Elwood, 1996; Wada et al., 1999; Rondeau and Sainte-Marie, 2001) (see section on Mate guarding, below). In some species, these factors may yield alterations in the mode of competition. For example, in the fiddler crab *Uca beebei*, a species in

Resource defence also occurs in species of a solitary nature such as clawed lobsters and stomatopods. American lobsters, *Homarus americanus*, live in solitary shelters where they take refuge from predators and strong water currents. The males compete for territory and defend shelters. Most females mate immediately after moulting, although mating may occur at any moult stage (Waddy and Aiken, 1991). Females approaching their moult actively search for mates and prefer to mate with dominant males. Males chemically advertise their presence in the shelter and probably their dominance status using urine signals in combination with visual and behavioural cues (Atema, 1986; Cowan and Atema, 1990; Bushmann and Atema, 1997, 2000). Shelter cohabitation occurs only in the context of mating and may last for up to a few weeks. It begins as intermittent cohabitation, then continuous cohabitation just prior to moulting and mating (which occurs inside the male's shelter) (Bushmann and Atema, 1997; Gosselin et al., 2003) and throughout the period of the female's exoskeleton hardening until the hard-shelled female leaves the shelter. During this time the male provides protection to the female through shelter defence.

A similar resource defence mating system exists in stomatopod species that occupy cavities in rock or coral, usually in those possessing the "smasher" type of raptorial appendages (Caldwell, 1991). In contrast to lobsters, mating in *Gonodactylus bredini* may take place in the cavity of either sex. Nevertheless, it is the male who assumes the role of cavity defence (Shuster and Caldwell, 1989) and who leaves the cavity after the female spawns, regardless of previous ownership.

The persistence of breeding pairs has been extended beyond the period of mating in monogamous species associated with a refuge or symbiotically associated with an invertebrate host (Correa and Thiel, 2003, and references therein). In these species (e.g., in snapping shrimp), both partners typically participate in the defence of the specific refuge against same-sex intruders, thus intrasexual competition is not restricted to males. Selection for resource defence cooperation (Mathews, 2002a) as well as selection on males for extended mate guarding, influenced by ecological conditions, predation risk (Knowlton, 1980), sex ratio (Mathews, 2002b) and the cryptic nature of the female's moult cycle coupled with her short post-moult receptivity period (Rahman et al., 2003), have contributed to the evolution of social monogamy in pair-living snapping shrimp.

Mate Guarding

Mate guarding can be defined as a temporary male-female association that exceeds the time required for insemination. Mate guarding in crustaceans typically takes the form of the male carrying the female, sometimes for a period of days ("precopula" in Isopoda, e.g., Ridley and Thompson, 1979; "amplexus" in Amphipoda, e.g., Borowsky, 1991; Brachyura, e.g., Hartnoll, 1969), or the male performing non-contact guarding by closely attending the female and in some cases caging her between the chelipeds (Ra'anan and Sagi, 1985; Conlan, 1991; Correa et al., 2003).

Pre-copulatory mate guarding is considered a male's competitive strategy to monopolize a female before copulation, in order to ensure his priority of access to the female at the moment she is ready to copulate, and thus to increase his mating success. This strategy is expected to evolve in species in which female receptivity is time-limited and predictable (Parker, 1974), which may explain why it is common among crustaceans. Several theoretical models have been developed with regard to mate guarding in crustaceans (for a review of these models, see Jormalainen, 1998) and empirical studies in various crustaceans have tested their predictions. The central questions addressed by many of these studies concern the timing and duration of guarding and the factors that influence it. From the male's point of view, guarding duration reflects a balance between a selective force favouring prolonged guarding to ensure his monopolization of the female and a contradictory selective force favouring minimization of guarding duration (by starting to guard close to the female's sexual moult or receptivity period) to increase the number of matings and to decrease guarding costs (e.g., energetic cost of pair formation and carrying, feeding limitation and predation costs; Jormalainen, 1998). The decision to start guarding is based on the ability of the male to assess whether the female is in an appropriate reproductive state (Jormalainen, 1998). Males may be attracted to receptive females by distant waterborne pheromones (e.g., for recent studies, Hardege et al., 2002; Mathews, 2003; Ekerholm and Hallberg, 2005), by contact chemical stimuli (Caskey and Bauer, 2005; Herborg et al., 2006), by both waterborne and contact stimuli (Zhang and Lin, 2006) or by visual cues (Díaz and Thiel, 2004). The male also benefits from assessing the size of the female and thereby her potential fecundity especially when fecundity contributes much more to his reproductive success than the number of mating opportunities (Shuster, 1981; Dick and Elwood, 1990, 1996). Guarding duration has been shown to be influenced by factors related to the availability of receptive females and to the intensity of competition among males. For example, in snow crabs (Rondeau and Sainte-Marie, 2001) and in isopods (Jormalainen and Shuster, 1999), males guarded females significantly longer as the OSR became more male-biased. Similarly, male hermit crabs guarded females earlier, and consequently longer, when the sex ratio was more male-biased, when male-female encounter rates were low, and when size differences between competitors were small (Wada et al., 1999).

Recent studies recognized that mate-guarding by males may impose a fitness cost on females, and when the guarding costs for males and females differ, intersexual conflict over guarding duration may arise (Jormalainen, 1998). For example, Cothran (2004) demonstrated that costs of pre-copulatory pairing due to predation by fish are higher for females than for males in an amphipod species. The female resistance to the guarding attempts of males observed in this species, as in various other amphipods and isopods (Jormalainen, 1998), is viewed as a female behaviour that serves to shorten the duration of male guarding and thus to reduce guarding costs for the female. When the female's ability to resist guarding attempts was experimentally reduced, guarding duration substantially increased, whereas reduction of the male guarding ability had no effect on guarding duration (Jormalainen and Merilaita, 1995; Jormalainen and Shuster, 1999). Thus, in these species with sexual conflict, guarding duration appears to be strongly influenced by the female interest (see also Jivoff and Hines, 1998). However, female resistance by itself has been shown to impose energy and fecundity costs on the female (Jormalainen et al., 2001); thus, resistance is expected to have an upper limit at the point that the costs exceed its benefits. In theory, under conflicting interests a compromised guarding duration would be set somewhere between the optimal durations for the male and the female, and the main factors that determine it would be the relative power of the sexes and sex ratio (Yamamura and Jormalainen, 1996).

Post-copulatoy mate guarding is considered a behavioural adaptation in males for the avoidance of sperm competition. In fact, sperm competition is an extension of male-male competition, in which sperm of two or more males compete for the fertilization of ova. The occurrence of internal fertilization, multiple mating by females and their ability to store sperm makes sperm competition a common phenomenon among brachyuran crustaceans (reviewed by Diesel, 1991). Recent studies using molecular techniques have confirmed that multiple paternity, and hence sperm competition, is common also in crustacean species with external fertilization such as lobsters (Streiff et al., 2004; Gosselin et al., 2005), crayfish (Walker et al., 2002), anomuran crabs (Toonen, 2004) and ghost shrimp (Bilodeau et al., 2005). Sexual selection in males via sperm competition is expected to favour adaptations that prevent the sperm of future males from reaching and/or fertilizing an inseminated female as well as adaptations for displacing or removing sperm from previous mating. These include morphological and physiological adaptations (modified pleopods (gonopods) used for sperm delivery into the site of fertilization in the female's receptacle, hardening seminal plasma that forms sperm plugs, increased ejaculate size, see Diesel, 1991) and behavioural adaptations, i.e., post-copulatory mate guarding. Such guarding is expected in species in which (1) female receptivity continues for some time after copulation, (2) the interval between copulation and fertilization (or end of receptivity) is not long, (3) mate searching efficiency is high, and (4) there is last male sperm precedence (Simmons, 2001). An additional function attributed to post-copulatory guarding by males is the protection of mates from predation, which is of high adaptive value particularly for crustaceans in which the females moult prior to copulation. This could in part explain why post-copulatory guarding is particularly common among brachyuran crabs in which copulation is linked to the female moult (e.g., portunids and cancrids, Orensanz et al., 1995; Jivoff, 1997) and why it may be absent in some species in which the females copulate in the intermoult stage, despite the occurrence of sperm competition (e.g., in crayfish, Berrill and Arsenault, 1984; Snedden, 1990; but see, e.g., in a grapsid crab, Brockerhoff and McLay, 2005b). Wilber (1989) experimentally demonstrated in stone crabs (*Xanthidae*) that males guarded post-moult females for a longer duration under conditions of male-male competition than under conditions of predation risk, and concluded that in these crabs post-copulatory guarding is primarily driven by sperm competition. However, the finding that the females survived predation only in trials in which the guarding durations were the longest indicated that post-copulatory guarding effectively prevents predation in stone crab. Thus, the above lack of response in guarding males to predation risk might have been specific to the type of predator used (male blue crabs) rather than a general lack of response to any predator (Wilber, 1989). In a similar study, Jivoff (1997) demonstrated that post-copulatory guarding by male blue crabs prevented additional inseminations as well as predation on females. It appears that the effective male strategy arising from sperm competition in crustaceans, whether guarding, leaving a sperm plug without guarding or both, is species-specific and depends on the relative costs of these strategies, the female's mating traits, encounter rates and searching risks.

Alternative Mating Strategies and Tactics

In various crustaceans, males exhibit discontinuous variation in mating behaviour and morphology, known as alternative mating strategies. Such discontinuous variations are mainly known in species in which sexual selection through male contest competition is strong (Andersson, 1994). Males that are excluded from mating benefit from adopting an alternative strategy instead of engaging in contest for acquiring mates. There is still a debate over how alternative strategies evolve and persist in natural populations and specifically whether the alternative strategies are mainly genetically determined (i.e., of equal fitness values) or environmentally determined conditional strategies that do not necessarily yield equal fitness (see Shuster and Wade, 2003).

Alternative mating strategies in crustaceans can be manifested in inflexible phenotypes or flexible phenotypes. Shuster (1987, 1992) described three genetically discrete male morphs in the marine isopod *Paracerceis sculpa*, which breeds within spongocoels. These morphs differ in mating behaviour and morphology (Fig. 2A). The largest α-male possesses elongated uropods and defends harems within spongocoels. The smaller β-male resembles females in morphology and size and invades the spongocoels by mimicking the behaviour of females. The γ-male is tiny and secretive and invades the harems by stealth. When isolated with females the three male morphs did not differ in their ability to sire young. However, relative fertilization success among male morphs varied with the density of females as well as with the frequency of other male morphs within the spongocoels; α-males could effectively defend females and obtained nearly all fertilizations when there was only one female and one more β- or γ-male in the spongocoel, whereas in the presence of two or three females the fertilization success of β- and γ-males increased and even exceeded the success of α-males (Shuster, 1989). The tendency of females to aggregate in spongocoels already containing gravid females (see section on Mate choice below) thus sets the stage for the co-existence of the discrete male morphs in this isopod species by creating a "mating niche" for the β- and γ-males (Shuster and Wade, 1991a). On the basis of these results, calculations from monthly samples of sponges in the field over 2 years, and the life history and the relative contribution of each morph to the population, Shuster and Wade (1991b) showed that the average mating success was equivalent among the three male morphs over time and only a small fraction of the total opportunity for sexual selection (the total variance in male mating success divided by the square grand mean) was attributed to variance among morphs. Thus, there is no selection favouring one male mating strategy and the conditions exist for maintaining genetic polymorphism in male mating behaviour and morphology in this isopod. Indeed, it was demonstrated that the three morphs are distinct at a single genetic locus (Shuster and Sassaman, 1997).

In two caridean species, the freshwater prawn *Macrobrachium rosenbergii* and the marine rock shrimp *Rhynchocinetes typus*, three distinct sexually mature male morphotypes coexist, representing successive stages in the developmental pathway of adult males (Kuris et al., 1987; Correa et al., 2000). In *M. rosenbergii,* the first adult stage, termed Small Male (SM) morphotype, is considerably smaller than the other morphotypes and possesses short delicate claws. Only part of the SM in the population grow through to the next stage, termed Orange Clawed (OC) morphotype, characterized by orange claws and rapid growth, and ultimately transform into the Blue Clawed (BC) morphotype, which possess extremely long blue claws (Fig. 2B). The three morphotypes coexist within prawn populations in a density-dependent proportion (Karplus et al., 1986). Growth of SM individuals is suppressed in the presence of the BC morphotype, but when the BC is removed some of them proceed along the developmental pathway as described above (Ráanan and Cohen, 1985; Karplus et al., 1992). The heritability of size variation in males was found to be negligible (Malecha et al., 1984). Although the developmental expression of phenotypic plasticity in *M. rosenbergii* reflects conditional strategies, it may arise from genetic architectures (i.e., inherited threshold traits) that are sensitive to social-environmental cues and allow the males to adjust their mating phenotypes in response to the changing social environment (Shuster and Wade, 2003). In a similar manner, *R. typus* males first become mature as the female-like "typus" morphotype, after which they moult through various intermediate stages ("intermedius" morphotype) to the final "robustus" morphotype (Fig. 2C) (Correa et al., 2000). In both species, the three morphotypes readily fertilize the female when isolated with her. When grouped together, a linear morphotype-related dominance hierarchy emerged, dominated by the largest morphotype (Barki et al., 1991, 1992; Correa et al., 2003). Two alternative mating tactics are practised by males of each species; the dominant morphotype courts and defends receptive females and the subordinate morphotype exhibits a sneak mating and rapidly attaches his spermatophore on the female's sternum while the dominant male is trying to defend the female from takeovers by other males (Telecky, 1984; Ráanan and Sagi, 1985; Correa et al., 2003).

Alternative male strategies involving small sneaker males (minors) and large fighter males (majors) were also described in the marine amphipod *Jassa marmorata* (Clark, 1997; Kurdziel and Knowles, 2002). Juvenile amphipods that grow on high quality food will mature later, at larger size, and become majors possessing claws (gnathopods) with large thumbs, whereas those grown on low quality food will be minors (Fig. 2D). Heritability analyses did not reveal genetic differences between dimorphic males (Kurdziel and Knowles, 2002). Thus, alternative male tactics in *J. marmorata* are conditional tactics whose expression depends on the interaction between individual genotypes and their environment.

Although females are considered the limiting sex, there is growing evidence that the reproductive success of females can be constrained by sperm limitation. Thus, by preferring dominant males females may sometimes reduce their fertilization rate since such males are more likely to be sperm depleted than others because of their competitive advantage in obtaining multiple mating. For example, Sato and Goshima (2006) recently demonstrated the occurrence of sperm limitation in populations of the stone crabs, *Haplogaster dentata*, owing to multiple mating and a low sperm recovery rate in males. These authors also demonstrated that, on the basis of waterborne chemical cues, female stone crabs choose larger males, but they also prefer to mate with males that are not sperm depleted (Sato and Goshima, 2007). Males, in turn, were demonstrated to judiciously allocate sperm among females and economize sperm to increase mating success (MacDiarmid and Butler, 1999; Rondeau and Sainte-Marie, 2001; Gosselin et al., 2003; Rubolini et al., 2006; Sato et al., 2006).

In species in which the reproductive success of the female depends on a resource, the characteristics of the resource that a male defends are likely to be used by the female to choose a mate. Active mate choice by females has been demonstrated in fiddler crab species in which females search for males. Females typically sample several male burrows before they select a male and enter to mate in his burrow. While they search the males wave their enlarged major claws to attract females into their burrows. Thus, females may base their mate selection upon male physical traits, male displays and burrow characteristics. *Uca pugilator* females base their choice on burrow stability and not on male size, although by selecting a stable burrow they are likely to select a large male because such males are more abundant in locations where burrows are stable (Christy, 1983). In *U. annulipes*, females use at least two criteria for mate choice, since there is no relationship between burrow quality and male size (Backwell and Passmore, 1996). Females entered the burrows of up to 24 males and traveled up to 28 m before selecting a mate. They first sampled the larger males in the population but the final mate acceptance appeared to involve a relatively invariant threshold criterion based on burrow structure. Backwell and Passmore (1996) showed that females decreased their acceptance criteria, in terms of whom they chose to sample and mate with (and not burrow quality), as the time to the end of the mating cycle decreased. The difference between *U. pugilator* and *U. annulipes* was accounted for by the lower burrow density coupled with higher predation risk in *U. pugilator*, which imposes higher searching cost and restricts prolonged sampling using multiple criteria. In populations of California fiddler crabs (*U. crenulata*) living in muddy-sand substrate at high density under low predation risk, male burrow characteristics vary considerably. Female crabs conducted extensive searches (sampling an average of 22.9 and up to 106 male burrows) and, like *U. annulipes* crabs, used multiple criteria to select males; they selected mates on the basis of burrow characteristics that are important for successful incubation and release of larvae (long burrows with entrances that matched the size of the female), and on correlated male characteristics, namely burrows defended by males near to their size and with small claws given their body size (DeRivera, 2005). It appears that in all of these *Uca* species in which the burrow is crucial for reproductive success, female mate choice depends primarily on the physical features of the burrow.

In some species with male mate guarding, female resistance to guarding attempts by males was suggested to act as a means for mate choice, in addition to its function in reducing female costs of being guarded (see section on Mate guarding, above). It was suggested that by resisting mating attempts the females directly obtain information regarding the male's vigour and condition. Female resistance in isopods and amphipods selects for large male size (Ward, 1984; Jormalainen and Merilaita, 1995) and for males with high energy reserves (Sparkes et al., 2002).

Instead of directly probing the male's vigour, resource holding power or condition, the female can choose a mate on the basis of the variability in the expression of secondary sex characteristics among potential mates. According to theory, elaborate male signals reliably reflect the genetic quality of the male because they are condition-dependent signals reflecting the level of the male's resistance to parasites (Hamilton and Zuk, 1982) or because they are costly to produce and maintain so that low-quality males cannot express as exaggerated signals as high-quality males (Zahavi, 1975, 1977, 1987). The conspicuous courtship displays of male fiddler crabs is an example of this kind of signal on which females may base their choice. When a female approaches a male or a cluster of males, the males wave their enlarged claw up and down to attract the female into their burrows for mating. The major claw at its maximum size may reach almost half the total weight of the crab, so waving is energetically costly. Variation in wave rate is partly due to variation in male condition combined with the energy costs of waving (Jennions and Backwell, 1998). Recent studies revealed that, in addition to claw size and position (Oliveira and Custodio, 1998), female choice through the male's waving display in fiddler crabs is based on features such as display leadership in synchronously waving crabs (Backwell et al., 1998, 1999) and fine-scaled spatio-temporal features of characteristics. Referring to Crustacea, Darwin stated in *The Descent of Man, and Selection in Relation to Sex*: "In this great class we first meet with undoubted secondary sexual characters, often developed in a remarkable manner. Unfortunately the habits of crustaceans are very imperfectly known, and we cannot explain the uses of many structures peculiar to one sex" (Darwin, 1871, chapter IX).

There are many categories of sexual dimorphisms arising from the different mechanisms of sexual selection (reviewed by Andersson, 1994), e.g., in morphological and behavioural characteristics related to competition and attractiveness. Size dimorphism is common among crustaceans, and its degree and direction may vary among species with different mating systems. Male contest competition favours large male size and structures that enhance the male's ability to fight and guard females, as well as increased male aggressiveness. Numerous studies have demonstrated the advantage of large body or cheliped size in contests. Body and cheliped size is usually positively correlated such that a larger individual also has larger chelipeds. The relative importance of body size and cheliped size in determining the fighting ability of males could be evaluated in species in which the male population comprises distinct male morphs of comparable body size that differ in cheliped size relative to body size (Barki et al., 1997; Guiaşu and Dunham, 1998). A remarkable example is provided by the orange-clawed and the blue-clawed male morphotypes of the freshwater prawn *M. rosenbergii* (see section on Alternative mating strategies and tactics, above). Owing to their larger chelipeds, small blue-clawed males dominated large orange-clawed males in mixed male groups (Barki et al., 1992). Blue-clawed males having only a 10% advantage in cheliped length almost invariably won in pair-wise contests despite being 44% smaller in body mass than their orange-clawed opponents, and those having equal cheliped size were as likely as their opponents to win contests despite being almost half their body mass (Barki et al., 1997). The overriding importance of cheliped size has been shown in various decapod species in laboratory experiments, for example, in shore crab *Carcinus maenas* (Sneddon et al., 1997), hermit crab *Pagurus bernhardus* (Neil, 1985), crayfish *Orconectes rusticus* (Rutherford et al., 1995), and American lobster *Homarus americanus* (Vye et al., 1997), as well as in naturally occurring contests in the natural habitat (Jaroensutasinee and Tantichodok, 2003; Pratt et al., 2003; Morrell et al., 2005). Sexual dimorphism in these weapons may arise from female choice as well, as was demonstrated in the previous section.

In species in which scramble competition occurs, large body size may be less important for males. Rather, pure searching is likely to favour mobile males with small body size and sensory capability to rapidly detect and acquire receptive females (Andersson, 1994). Indeed, size dimorphism with smaller males than females is common among pure-searching caridean shrimp, owing as well to a fecundity advantage for large females (reviewed in Correa and Thiel, 2003). Lack of a significant sexual dimorphism has been demonstrated in species in which sexual selection is weak, for example, in monogamous alpheid shrimp with persistent pairs (Correa and Thiel, 2003). Slight sexual dimorphism might also be evident in some species in which both contest (female guarding) and scramble are important components of male mating competition (e.g., *Cancer magister*, Orensanz et al., 1995), perhaps because of contradictory sexual selection for large and small male size. However, in some cases scramble competition may favour large males, as was suggested in *Asellus aquaticus*, an isopod with pre-copulatory mate guarding (Bertin and Cezily, 2003). In this species, sexual dimorphism is evident in both body and antenna length, and both characteristics are closely correlated. Males with longer antennae were better able to detect and pair with receptive females and apparently have an advantage in scramble competition. Thus, large body size in male isopods may also be indirectly favoured by scramble competition, in addition to its obvious advantage in female guarding.

Sexual dimorphism with more developed structures in the male may also involve structures specifically used for holding a female; for example, the posterior gnathopods of the amphipod *Gammarus pulex* were necessary for the male to accomplish a successful copulation (Hume et al., 2005). Sexual dimorphism in sensory organs has been found in olfactory sensillae on the antennules, which are sensitive to waterborne sex pheromones (Hallberg et al., 1997), and in sensillae on the second antennae, which are sensitive to sex pheromones via contact (Bauer and Caskey, 2006). Interestingly, male-specific olfactory sensillae are found in peracarid groups such as mysids, cumaceans and some amphipods, but not in decapods. In Peracarida there are putative pheromone sensors in the males and general olfactory sensors in the two sexes, whereas in Decapoda there is evidently one type of olfactory sensillum. Sexual dimorphism in the latter group is observed mainly in the number of sensillae (Hallberg et al., 1997), which may be positively correlated with antenna length.

Closely related species and even populations of the same species may differ in the degree or direction of sexual dimorphism, as a result of differing ecological and population factors. For example, Wellborn and Bartholf (2005) explored two closely related amphipod species within the *Hyeaella azteca* species complex. The larger of these species inhabit a fishless habitat with a low-predation risk, and mortality decreased with mating success of male crayfish through their effect on mating competition.

CONCLUSIONS

This chapter has reviewed the different aspects of sexual selection and demonstrated the diversity of mating behaviour and mating systems in Crustacea. Numerous studies have investigated various factors influencing the mating behaviour, strategies and systems in crustaceans. Nevertheless, studies investigating the interplay of multiple ecological, reproductive and life history factors underlying the observed diversity in mating behaviour and mating systems are still needed. Crustaceans offer an excellent model for investigating the complex network of relationships caused by the interplay of multiple factors, as well as for studying current issues related to sexual selection such as sexual conflict, mate choice (male and female), and male sperm limitation and sperm allocation.

While the fascinating mating behaviour of crustaceans has attracted a wealth of behavioural studies on the ultimate processes shaping mating behaviour, surprisingly little is known about the proximate hormonal mechanisms underlying these behaviours. Current advances in the identification of the AGH in decapods (e.g., Manor et al., 2007; see Chapter 7) open a new avenue for exploring questions regarding the neuroendocrine mechanisms mediating the organization and activation of sexual behaviour in crustaceans.

Acknowledgments

I am grateful to Prof. Ilan Karplus for introducing me to the world of crustacean behaviour and for the many years of joint research thereafter, and to Prof. Amir Sagi for fruitful collaboration which allowed us to link the study of the androgenic gland and male behaviour in crustaceans. I thank Steve Shuster, Colin McLay and Ilan Karplus for their helpful comments and suggestions.

References

Andersson, M. 1994. Sexual Selection. Princeton University Press, Princeton.

Arnold, A.P. 2004. Sex chromosome and brain gender. Nat. Rev. Neurosci. 5: 701-708.

Atema, J. 1986. Review of sexual selection and chemical communication in the lobster, *Homarus americanus*. Can. J. Fish. Aquat. Sci., 43: 2283-2390.

Backwell, P.R.Y., and M.D. Jennions. 2004. Coalition among male fiddler crabs. Nature, 430: 317.

Backwell, P.R.Y., and N.I. Passmore. 1996. Time constraints and multiple choice criteria in the sampling behaviour and mate choice of the fiddler crab, *Uca annulipes*. Behav. Ecol. Sociobiol., 38: 407-416.

Backwell, P.R.Y., M.D. Jennions, N.I. Passmore and J.H. Christy. 1998. Synchronized courtship in fiddler crabs. Nature, 391: 31-32.

Backwell, P.R.Y., M.D. Jennions, J.H. Christy and N.I. Passmore. 1999. Female choice in the synchronously waving fiddler crab *Uca annulipes*. Ethology, 105: 415-421.

Backwell, P.R.Y., M.D. Jennions, K. Wada, M. Murai and J.H. Christy. 2006. Synchronous waving in two species of fiddler crabs. Acta. Ethol., 9: 22-25.

Ball, G.F., and J. Balthazart. 2006. Androgen metabolism and the activation of male sexual behavior: it's more complicated than you think! Horm. Behav., 49: 1-3.

Barki, A., and I. Karplus. 1999. Mating behavior and a behavioral assay for female receptivity in the red claw crayfish, *Cherax quadricarinatus* (von Martens). J. Crust. Biol., 19: 493-497.

Barki, A., I. Karplus and M. Goren. 1991. Morphotype related dominance hierarchies in males of *Macrobrachium rosenbergii* (Crustacea, Palaemonidae). Behaviour, 117: 145-160.

Barki, A., I. Karplus and M. Goren. 1992. Effects of size and morphotype on dominance hierarchies and resource competition in the freshwater prawn *Macrobrachium rosenbergii*. Anim. Behav., 44: 547-555.

Barki, A., S. Harpaz and I. Karplus. 1997. Contradictory asymmetries in weapon and body size, and assessment in fighting male prawns, *Macrobrachium rosenbergii*. Aggress. Behav., 23: 81-91.

Barki, A., I. Karplus, I. Khalaila, R. Manor and A. Sagi. 2003. Male-like behavioral patterns and physiological alterations induced by androgenic gland implantation in female crayfish. J. Exp. Biol., 206: 1791-1797.

Barki, A., I. Karplus, R. Manor and A. Sagi. 2006. Intersexuality and behavior in crayfish: the de-masculinization effects of androgenic gland ablation. Horm. Behav., 50: 322-331.

Bauer, R.T. 2006. Same sexual system but variable sociobiology: evolution of protandric simultaneous hermaphroditism in *Lysmata* shrimps. Integr. Comp. Biol., 46: 430-438.

Bauer, R.T., and J.H. Abdalla. 2001. Male mating tactics in the shrimp *Palaemonetes pugio* (Decapoda, Caridea): precopulatory mate guarding vs. pure searching. Ethology, 107: 185-199.

Bauer, R.T., and J.L. Caskey. 2006. Flagellar setae of the second antennae in decapod shrimps: sexual dimorphism and possible role in detection of contact sex pheromones. Invertebr. Reprod. Dev., 49: 51-60.

Belk, D. 1991. Anostracan mating behavior: a case of scramble-competition polygyny. *In:* Crustacean Sexual Biology. R.T. Bauer and J.W. Martin (eds.). Columbia University Press, New York, pp. 111-125.

Berrill, M., and M. Arsenault. 1982. Spring breeding of a northern temperate crayfish, *Orconectes rusticus*. Can. J. Zool., 60: 2641-2645.

Berrill, M., and M. Arsenault. 1984. The breeding behaviour of a northern temperate orconectid crayfish, *Orconectes rusticus*. Anim. Behav., 32: 333-339.

Bertin, A., and F. Cezilly. 2003. Sexual selection, antennae length and the mating advantage of large males in *Asellus aquaticus*. J. Evol. Biol., 16: 698-707.

Bertin, A., and F. Cezilly. 2005. Density-dependent influence of male characters on male-locating efficiency and pairing success in the waterlouse *Asellus aquaticus*: an experimental study. J. Zool. Lond., 265: 333-338.

Bilodeau, A.L., D.I. Felder and J.E. Neigel. 2005. Multiple paternity in the ghost shrimp, *Callichirus islagrande* (Crustacea: Decapoda: Callianassidae). Mar. Biol., 146: 381-385.

Bolingbroke, M., and G. Kass-Simon. 2001. 20-hydroxyecdysone causes increased aggressiveness in female American lobsters, *Homarus americanus*. Horm. Behav., 39: 144-156.

Borowsky, B. 1991. Pattern of reproduction of some amphipod crustaceans and insights into the nature of their stimuli. *In:* Crustacean Sexual Biology. R.T. Bauer and J.W. Martin (eds.). Columbia University Press, New York, pp. 33-49.

Brockerhoff, A.M., and C.L. McLay. 2005a. Comparative analysis of the mating strategies in grapsid crabs with special references to the intertidal crabs *Cyclograpsus lavauxi* and *Helice crassa* (Decapoda: Grapsidae) from New Zealand. J. Crust. Biol., 25: 507-520.

Brockerhoff, A.M., and C.L. McLay. 2005b. Mating behaviour, female receptivity and male-male competition in the intertidal crab *Hemigrapsus sexdentatus* (Brachyura: Grapsidae). Mar. Ecol. Prog. Ser., 290, 179-191.

Brockerhoff, A.M., and C.L. McLay. 2005c. Factors influencing the onset and duration of receptivity of female purple rock crabs, *Hemigrapsus sexdentatus* (H. Milne Edwards, 1837) (Brachyura: Grapsidae). J. Exp. Mar. Biol. Ecol., 314: 123-135.

Bushmann, P.J. 1999. Concurrent signals and behavioral plasticity in blue crab (*Callinectes sapidus* Rathbun) courtship. Biol. Bull., 197: 63-71.

Bushmann, P.J., and J. Atema. 1997. Shelter sharing and chemical courtship signals in the lobster *Homarus americanus*. Can. J. Fish. Aquat. Sci., 54: 647-654.

Bushmann, P.J., and J. Atema. 2000. Chemically mediated mate location and evaluation in the lobster, *Homarus americanus*. J. Chem. Ecol., 26: 883-899.

Caiger, K.M., and A.J. Alexander. 1973. The control of dominance in the brachyuran crustacean, *Cyclograpsus punctatus* M. Edw. Zool. African., 8: 138-140.

Caldwell, R.L. 1991. Variation in reproductive behavior in stomatopod crustacea. *In:* Crustacean Sexual Biology. R.T. Bauer and J.W. Martin (eds.). Columbia University Press, New York, pp. 67-90.

Carpenter, M.B., and R. deRoos. 1970. Seasonal morphology and histology of the androgenic gland of the crayfish, *Orconectes nais*. Gen. Comp. Endocrinol., 15: 143-157.

Caskey, J.L., and R.T. Bauer. 2005. Behavioral tests for a possible contact sex pheromone in the caridean shrimp *Palaemonetes pugio*. J. Crust. Biol., 25: 571-576.

Christy, J.H. 1983. Female choice in the resource-defense mating system of the sand fiddler crab, *Uca pugilator*. Behav. Ecol. Sociobiol., 12: 169-180.

Christy, J.H. 1987. Competitive mating, mate choice and mating association of brachyuran crabs. Bull. Mar. Sci., 41: 177-191.

Clark, R.A. 1997. Dimorphic males display alternative reproductive strategies in the marine amphipod *Jassa marmorata* Holmes (Corophioidea: Ischyroceridae). Ethology, 103: 531-553.

Conlan, K.E. 1991. Precopulatory mating behavior and sexual dimorphism in the amphipod Crustacea. Hydrobiologia, 223: 255-282.

Correa, C., and M. Thiel. 2003. Mating systems in caridean shrimp (Decapoda: Caridea) and their evolutionary consequences for sexual dimorphism and reproductive biology. Rev. Chil. Hist. Nat., 76: 187-203.

Correa, C., J.A. Baeza, E. Dupré, I.A. Hinojosa and M. Thiel. 2000. Mating behavior and fertilization success of three ontogenetic stages of male rock shrimp *Rhynchocinetes typus* (Decapoda: Caridea). J. Crust. Biol., 20: 628-640.

Correa, C., J.A. Baeza, I.A. Hinojosa and M. Thiel. 2003. Dominance hierarchy and mating tactics in the rock shrimp *Rhynchocinetes typus* (Decapoda: Caridea). J. Crust. Biol., 23: 33-45.

Cothran, R.D. 2004. Precopulatory mate guarding affects predation risk in two freshwater amphipod species. Anim. Behav., 68: 1133-1138.

Cowan, D.F., and J. Atema. 1990. Moult staggering and serial monogamy in American lobsters, *Homarus americanus*. Anim. Behav., 39: 1199-1206.

Cromarty, S.I., and G. Kass-Simon. 1998. Differential effects of a molting hormone, 20-hydroxyecdysone, on the neuromuscular junctions of the claw opener and abdominal flexor muscles of the American lobster. Comp. Biochem. Physiol., A 120: 289-300.

Darwin, C. 1871. The Descent of Man, and Selection in Relation to Sex. Murray, London.

DeRivera, C.E. 2005. Long searches for male-defended breeding burrows allow female fiddler crabs, *Uca crenulata*, to release larvae on time. Anim. Behav., 70: 289-297.

DeRivera, C.E., P.R.Y. Backwell, J.H. Christy and S.L. Vehrencamp. 2003. Density affects female and male searching in the fiddler crab, *Uca beebei*. Behav. Ecol. Sociobiol., 53: 72-83.

Díaz, E.R., and M. Thiel. 2003. Female rock shrimp prefer dominant males. J. Mar. Biol. Assoc. UK, 83: 941-942.

Díaz, E.R., and M. Thiel. 2004. Chemical and visual communication during mate searching in rock shrimp. Biol. Bull., 206: 134-143.

Dick, J.T.A., and R.W. Elwood. 1990. Symmetrical assessment of female quality by male *Gammarus pulex* (Amphipoda) during struggles over precopula females. Anim. Behav., 40: 877-883.

Dick, J.T.A., and R.W. Elwood. 1996. Effects of natural variation in sex ratio and habitat structure on mate-guarding decisions in amphipods (Crustacea). Behaviour, 133: 985-996.

Diesel, R. 1991. Sperm competition and the evolution of behavior in Brachyura, with special reference to spider crabs (Decapoda, Majidae). *In:* Crustacean Sexual Biology. R.T. Bauer and J.W. Martin (eds.). Columbia University Press, New York, pp. 145-163.

Ekerholm, M., and E. Hallberg. 2005. Primer and short-range releaser pheromone properties of premolt female urine from the shore crab *Carcinus maenas*. J. Chem. Ecol., 31: 1845-1864.

Emlen, S.T., and L.W. Oring. 1977. Ecology, sexual selection and the evolution of mating systems. Science, 197: 215-223.

Gherardi, F., B. Renai, P. Galeotti and D. Rubolini. 2006. Nonrandom mating, mate choice, and male-male competition in the crayfish *Austropotamobius italicus*, a threatened species. Arch. Hydrobiol., 557-576.

Gleeson, R.A. 1991. Intrinsic factors mediating pheromone communication in the blue crab, *Callinectes sapidus*. *In:* Crustacean Sexual Biology. R.T. Bauer and J.W. Martin (eds.). Columbia University Press, New York, pp. 17-32.

Gleeson, R.A., M.A. Adams and A.B. Smith III. 1987. Hormonal modulation of pheromone-mediated behaviour in a crustacean. Biol. Bull., 172: 1-9.

Goshima, S., T. Kawashima and S. Wada. 1998. Mate choice by males of the hermit crab *Pagurus filholi*: Do males assess ripeness and/or fecundity of females? Ecol. Res., 13: 151-161.

Gosselin, T., B. Sainte-Marie and L. Bernatchez. 2003. Patterns of sexual cohabitation and female ejaculate storage in the American lobster (*Homarus americanus*). Behav. Ecol. Sociobiol., 55: 151-160.

Gosselin, T., B. Sainte-Marie and L. Bernatchez. 2005. Geographic variation of multiple paternity in the American lobster, *Homarus americanus*. Mol. Ecol., 14: 1517-1525.

Guiaşu, R.C., and D.W. Dunham. 1998. Inter-form agonistic contests in male crayfishes, *Cambarus robustus* (Decapoda, Cambaridae). Invert. Biol., 117: 144-154.

Hallberg, E., K.U.I. Johansson and R. Wallén. 1997. Olfactory sensilla in crustaceans: morphology, sexual dimorphism, and distribution patterns. Int. J. Insect Morphol. Embryol., 26: 173-180.

Hamilton, W.D., and M. Zuk. 1982. Heritable true fitness and bright birds: a role for parasites? Science, 218: 384-387.

Hardege, J.D., A. Jennings, C.T. Hayden, C.T. Müller, D. Pascoe, M.G. Bentley and A.S. Clare. 2002. Novel behavioural assay and partial purification of a female-derived sex pheromone in *Carcinus maenas*. Mar. Ecol. Prog. Ser., 244, 179-189.

Hartnoll, R.G. 1969. Mating in the brachyuran. Crustaceana, 16: 161-181.

Hatziolos, M.E., and R.L. Caldwell. 1983. Role reversal in courtship in the stomatopod *Pseudosquilla ciliata* (Crustacea). Anim. Behav., 31: 1077-1087.

for the synthesis of steroids could, however, be detected in the hepatopancreas and it was proposed that the hepatopancreas, rather than being a tissue of biosynthesis for steroids, acted as a metabolizing tissue by utilizing steroid precursors produced in the ovary.

Other Neuropeptides

The involvement of opioid-like neuropeptides (enkephalins) in the stimulation of reproductive systems has been demonstrated in several decapod species (Jaros et al., 1985; Lüschen et al., 1991; Rothe et al., 1991; Sarojini et al., 1995a; Reddy and Basha, 2001; Kishori and Reddy, 2000; Reddy and Kishori, 2001; Kishori et al., 2001; Lorenzon et al., 2004). While leucine-enkephalin produced dose-dependent ovarian maturation in the prawn *Penaeus indicus* (Reddy, 2000) and stimulated vitellogenesis in the crab *Oziotelphusa senex senex* (Kishori and Reddy, 2003), the injection with methionine-enkephalin had an antagonistic effect and slowed maturation in a dose-dependent manner in *P. indicus* and *Uca pugilator* (Sarojini et al., 1995a; Reddy, 2000).

Pyrokinins are a family of neuropeptides that regulate a variety of physiological and behavioural processes related to reproduction in insects (Imai et al., 1991; Schoofs et al., 1991; Raina, 1993). In insects these peptides are found in and around the corpora allata, the site of juvenile hormone production, which in turn influences reproduction. The crustacean equivalent of this was identified as methyl farnesoate synthesized by the mandibular organ. Torfs et al. (2001) suspected pyrokinin-like peptides to be responsible for the regulation of urine signals crustaceans use for chemical communication involved in courtship behaviour. The authors were able to isolate two pyrokinin peptides from the central nervous system of the shrimp *Pennaeus vannamei*. Both peptides were found to stimulate spontaneous contractions in the hindgut of *Astacus leptodactylus* and *Leucophaea maderae*, indicating a possible function as pheromones, influencing sexual behaviour in crustaceans.

METHYL FARNESOATE

The mandibular organ (MO) synthesizes and secretes methyl farnesoate (MF), a sesquiterpenoid compound carried through the hemolymph by a specific binding protein to its tissue of destination (Li and Borst, 1991). It is considered to be the functional crustacean equivalent of the insect juvenile hormone and in fact injections with juvenile hormone (JH III) or its mimics enhance reproduction in adult Crustacea (Hinsch, 1981; Linder and Tsukimura, 1999). Studies on the function of the mandibular organ et al. (2006) determined a differential distribution between the neurons within the organs and suggested that the release of these hormones may be subject to tissue-specific cues and therefore may fulfil different physiological roles, depending on the tissue of origin. The functions of CHH produced from extra-eyestalk locations are not yet identified.

Regarding their influence on crustacean reproduction, CHH together with GIH seem to synchronize molting and reproductive activities (De Kleijn et al., 1998). Two groups of peptides comprising more than one isoform have been identified in Penaeid shrimp; i.e., CHH-A and CHH-B, and at least two forms in Astacidaea and Palinura, i.e., CHH-I and CHH-II (for review see Chan et al., 2003). In *H. americanus*, Meusy and Soyez (1991) isolated the two separate CHH molecules, CHH-A and CHH-B. De Kleijn et al. (1998) determined that CHH accumulation in the hemolymph alternated with GIH levels in that total CHH (A+B) was high during maturation, whereas GIH levels were high during the immature and previtellogenic stages. Crustacean hyperglycemic hormone A and B both were involved in triggering the onset of vitellogenesis, while CHH-B only stimulated oocyte growth. Gonad-inhibiting hormone in turn prevented the onset of vitellogenesis. Molting is regulated through the inhibitory influence of both CHH (CHH-A) and MIH on the Y-organ and thus ecdysteroid synthesis (Webster and Keller, 1986; Chang et al., 1990; review in Webster, 1998). Chang et al. (2001) found that CHH distribution in the sinus gland of *H. americanus* changed throughout the molt stages, while concentrations in the hemolymph remained at a steady level. Increased levels of CHH in the sinus gland during mid-premolt therefore indicate either an increase in synthesis and storage in the sinus gland or a decrease in the release of CHH into the hemolymph. To date this regulation has not been clarified.

Crustacean hyperglycemic hormone release in turn seems to be regulated by bioamines acting as neurotransmitters. Evidence exists that serotonin, dopamine, octopamine and enkephalins influence the secretory activity of CHH-producing cells in the sinus gland complex (Sarojini et al., 1995a; Ollivaux et al., 2002; Basu and Kravitz, 2003; Zou et al., 2003; Lorenzon et al., 2004). However, the regulatory function of these neurotransmitters seems to be species and transmitter dependent.

The injection of dopamine increased CHH release in *P. monodon* and *Macrobrachium malcolmsonii* (Kuo et al., 1995; Komali et al., 2005). While dopamine showed no effect when administered to *P. elegans*, the injection of serotonin into the same species induced a massive release of CHH from the sinus gland into the hemolymph (Lorenzon et al., 2005). Serotonin also mediated the release of CHH in *P. clarkii* (Lee et al., 2001; Escamilla-Chimal et al., 2002).

Evidence exists that enkephalins indirectly influence crustacean reproduction through mediating the release of neurohormones from the X-sinus gland (for review, see Nagabushanam et al., 1995). However, the injection of enkephalins shows conflicting results in different species. Both methionine and leucine-enkephalin inhibited the release of CHH in *Orconectes limosus* (Ollivaux et al., 2002). In addition, the authors found evidence for direct inputs from enkephalinergic neurons into dendrites of CHH-immunoreactive cells and hypothesized that enkephalins directly inhibit the release of CHH via synaptic contacts. Several other crustacean species also showed a hypoglycemic response when injected with leucine-enkephalin (Jaros, 1990; Lüschen et al., 1991; Rothe et al., 1991; Sarojini et al., 1995a; Lorenzon et al., 2004). On the other hand, leucine-enkephalin induced a hyperglycemic response in *Palaemon elegans, P. indicus, O. senex senex, Metapenaeus monocerus* and *Scylla serrata* (Lorenzon et al., 1999; Reddy and Basha, 2001; Reddy and Kishori, 2001; Kishori et al., 2001).

Crustacean hyperglycemic hormone levels also seem to be influenced by environmental stimuli. Different types of environmental stress have been shown to produce enhanced CHH concentrations (Webster, 1996). Chang et al. (1998) found that hypoxia and changes in temperature and salinity caused changes in CHH release into the hemolymph, while Lorenzon et al. (2000) detected an effect of heavy metal exposure on CHH hemolymph levels of *P. elegans*. In addition, it has been shown that CHH hemolymph concentrations show a daily and circadian rhythm in different species of crayfish (Kallen et al., 1990; Escamilla-Chimal et al., 2001).

Biogenic Amines

Richardson et al. (1991) proposed that biogenic amines synthesized in the central nervous system of crustaceans acted as neurotransmitters and were responsible for the synthesis and release of neurohormones such as GIH and GSH. A variety of these bioamines (e.g., serotonin or 5-hydroxytryptamine (5-HT), dopamine (DA) and octopamine (OA)) have been found to play a role in female reproduction in several crustacean species (Fingerman et al., 1994 (review); Sarojini et al., 1995b, c; Beltz, 1999 (review); Ragunathan and Arivazhagan, 1999; Vaca and Alfaro, 2000; Chen et al., 2003; Alfaro et al., 2004).

Localization studies have identified 5-HT from several tissues. Immunoreactivity was found in the medulla externa, medulla interna, medulla terminalis and brain of *Pacifastacus leniusculus* (Eloffson, 1983). Also, the central nervous systems of both *P. leniusculus* and *P. clarkii* were immunoreactive to 5-HT (Elofsson et al., 1982; Kulkarni and Fingerman,

1992). More recently, the expression of 5-HT mRNA and proteins was identified from the ovary, brain, thoracic ganglia, hepatopancreas, muscle, gill, stomach, lymphoid and eyestalk of *P. monodon* (Ongvarrasopone et al., 2006). While receptor expression in the eyestalks indicates that 5-HT might mediate the release of neurohormones such as CHH and MIH (Lee et al., 2001; Escamilla-Chimal et al., 2002), the expression in the ovary suggests that this neurotransmitter also plays an important role in ovarian development and spawning.

The use of anti-dopamine antibodies and High performance liquid chromatography (HPLC) have made it possible to determine the existence of DA in the last abdominal ganglion, the intestinal nerve and axons in the hindgut musculature of *Oronectes limosus* (Elekes et al., 1988). Dopamine was also detected in the optic ganglia and brain of *P. leniusculus* (Myhrberg et al., 1979) and the intestinal nerve of *P. clarkii* (Mercier et al., 1991). In *M. rosenbergii* DA was found in the brain and thoracic ganglia, though not in the eyestalks (Chen et al., 2003). This led to the proposition that DA exerts its inhibitory effect on vitellogenin synthesis indirectly by acting on the thoracic ganglia and thus inhibiting the release of GSH. Apart from DA, Elofsson et al. (1982) identified OA and norepinephrine (a derivative of DA) in the central nervous system of *P. leniusculus*.

Several studies have shown that the injection of 5-HT induces ovarian maturation and spawning in female crustaceans (e.g., Sarojini et al., 1995b; Vaca and Alfaro, 2000; Aktas and Kumlu, 2005). It has been suggested that ovarian maturation is not stimulated directly but through the release of GSH from the brain and the thoracic ganglia (Sarojini et al., 1995b), which in turn stimulates vitellogenesis. 5-HT stimulates hormone release by interacting with various receptor subtypes. Seven receptor families could be identified (5-HT_{1-7}), two of which (5-HT_1 and 5-HT_2) have recently been successfully cloned from crayfish, spiny lobster and several prawn species (Sosa et al., 2004; Clark et al., 2004; Ongvarrasopone et al., 2006). Observed effects of 5-HT administration vary in their intensity and expression and seem to be species- and dose-dependent.

Dose-dependent effects were found for the induction of ovarian maturation in both *M. rosenbergii* and *U. pugilator* (Richardson et al., 1991; Chen et al., 2003). Vaco and Alfaro (2000) found that when injected into female shrimp *P. vannamei*, 5-HT induced ovarian maturation and spawning but it did so at a later stage and a slower rate than eyestalk ablation. The authors hypothesized that other neurohormones such as GIH might have interacted with the injected 5-HT and therefore lowered the magnitude of its effect. The combined treatment with 5-HT and spiperone (a DA antagonist), however, induced maturation and spawning rates in *Litopenaeus stylirostris* and *Litopenaeus vannamei* similar to those found in eyestalk-ablated animals (Alfaro et al., 2004). No effect on maturation was found when 5-HT was injected into female *P. clarkii* (Kulkarni et al., 1992), which seems to be consistent with the hypothesis that 5-HT exerts its ovary-regulating action only indirectly through triggering GSH release.

In addition, the effect of 5-HT on maturation seemed to be dependent on receptor localization at different stages of the development. 5-HT receptors were found in the traberculae of stage 1 and 2 oocytes but on mature stage 3 and 4 oocytes they were found on and around the membranes of the oocytes (Ongvarrasopone et al., 2006). Moreover, Kirubagaran et al. (2005) found stage-dependent changes in the activity and distribution of 5-HT neurons in the brain and the thoracic ganglia of the spiny lobster *Panulirus homarus*. These changes coincided with changes in the ovary, thus indicating the regulation of ovarian development through inhibitory/stimulatory factors in the X-organ sinus gland of the eyestalk.

The role of DA in female crustacean reproduction is indirect, through its action on hormones or other neurotransmitters. Dopamine was shown to inhibit ovarian maturation in *P. clarkii* through the inhibition of 5-HT and consequent inhibition of GSH, or possible GIH release (Sarojini et al., 1995c, 1996; Fingerman et al., 1998). In *M. rosenbergii* Chen et al. (2003) found a dose-dependent effect on vitellogenesis for both 5-HT and DA. While 5-HT seemed to stimulate vitellogenesis, DA appeared to inhibit ovarian development. Dopamine was detected in the brain and the thoracic ganglia of *M. rosenbergii* and not in the eyestalks. This led to the proposition that it exerts its inhibitory effect on vitellogenin synthesis by acting on the thoracic ganglia and thus inhibiting the release of GSH.

Octopamine has been found to play a role in the ovarian maturation of several crustacean species (Sarojini et al., 1995b; Vaca and Alfaro, 2000; Chen et al., 2003; Alfaro et al., 2004; Tanboonteck et al., 2006). It inhibits the role of 5-HT by blocking GSH release (Richardson et al., 1991). Injected into *M. rosenbergii*, it showed no effect on ovarian maturation but stimulated spawning (Tanboonteck et al., 2006).

Apart from controlling the migration of pigment in the erythrophores of crustaceans, the red pigment concentrating hormone can also act as a neurotransmitter. It was proposed to stimulate ovarian maturation in *P. clarkii* through the release of GSH (Sarojini et al., 1995d).

Vaco and Alfaro (2000) proposed that these neurotransmitters stimulated the release of maturation-promoting pheromones into the water.

PERSPECTIVES ON HORMONAL MANIPULATIONS AND FUTURE STUDIES

The shortage of natural spawners and dwindling natural resources have made it necessary to develop new methods for artificial seed production (Wilder et al., 1999). To date, eyestalk ablation still represents the most commonly used manipulation to induce maturation and spawning in female crustaceans and has been successful in increasing reproductive output and spawning frequency and shortening the latency to first spawn (Racotta et al., 2003)

The research described in this chapter has shown that the endocrine regulation of both reproduction and molting is controlled by a complex interaction of neuropeptides, juvenoids and steroids. Although studies have confirmed that female gonadal development could successfully be induced through injections or transplants of these hormones and endocrine glands, the effects were highly variable depending on factors such as species, developmental stage of the animal and concentrations of hormones and neurotransmitters given.

For future applications of hormones in the manipulation of breeding stocks in captivity, it is first necessary to fully understand the interactions and functions of these endocrine systems. Only then can research be directed at finding a hormonal treatment that has clear advantages over eyestalk ablation and that is also effective in terms of reproductive output.

However, some promising advances have been made with the application of methyl farnesoate and 5-HT? Both approaches seemed to increase fecundity and hatching rates to rates similar to those found in eyestalk-ablated animals (Laufer et al., 1997; Vaca and Alfaro, 2000). Once we fully understand the dynamics of vitellogenin synthesis and the hormonal pathways involved, we can aim to develop hormone and neurotransmitter antagonists that will allow us to manipulate female reproduction in captivity more effectively and might also be useful in domesticating commercially important species and thus prevent the depletion of broodstocks in the wild.

Little attention has been paid to the effect of environmental stimuli on the hormonal regulation of reproductive development in Crustacea. For example, in cases of environmental sex determination, hormonal pathways are altered in response to cues from the environment that indicate whether male or female offspring would be desirable in order to maximize population stability and growth. On an endocrine level this is converted into a hormone translating the environmental signal such as a change in photoperiod or scarcity of food into a physiological response such as the production of monosex offspring (Olmstead and LeBlanc, 2003).

Studies have already shown that insecticides and xenobiotics interrupt the hormonal response of animals to environmental stimuli. Insecticides such as pyriproxyfen, fenoxycarb and metophrene were found to act as juvenile hormone or methyl farnesoate agonists and, depending on the species, resulted in the stimulation of the production of male progeny, decreased fecundity and inhibition of development (Olmstead and Le Blanc, 2002; Gagne et al., 2005; Tuberty and McKenney, 2005).

In species in which close links exist between environmental factors and reproduction the interruption of established metabolic pathways could have detrimental effects on the survival of the population and possibly also on their consumers. The resolution of connecting pathways between environmental stimuli and their hormonal responses would therefore be also beneficial for the conservation of populations in the wild.

References

Abdu, U., C. Davis, I. Khalaila and A. Sagi. 2002. The vitellogenin cDNA of *Cherax quadricarinatus* encodes a lipoprotein with calcium binding ability, and its expression is induced following the removal of the androgenic gland in a sexually plastic system. Gen. Comp. Endocrinol., 127: 263-272.

Adiyodi, R.G. 1978. Endocrine control of ovarian function in crustaceans. *In:* Comparative Endocrinology. P. Gaillard and H. Boer (eds.). Elsevier/North-Holland Biomedical Press, Amsterdam, pp. 25-28.

Adiyodi, R.G. 1985. Reproduction and its control. *In:* The Biology of Crustacea. D.E. Bliss and L.H. Mantel (eds.). Academic Press, New York, pp. 147-215.

Adiyodi, R.G., and T. Subramoniam. 1983. Arthropoda - Crustacea. *In:* Reproductive Biology of Invertebrates. K.G. Adiyodi and R.G. Adiyodi (eds.). John Wiley & Sons, New York, pp. 443-496.

Aguilar, M.B., L.S. Quackenbush, D.T. Hunt, J. Shabanowitz and A. Huberman. 1992. Identification, purification and initial characterization of the vitellogenesis-inhibiting hormone from the Mexican Crayfish *Procambarus bouvieri* (Ortmann). Comp. Biochem. Physiol., B 102: 491-498.

Aiken, D.E., and S.L. Waddy. 1980. Reproductive biology. *In:* The Biology and Management of Lobsters. J.S. Cobb and B.F. Phillips (eds.). Academic Press, New York, pp. 215-276.

Aktas, M., and M. Kumlu. 2005. Gonadal maturation and spawning in *Penaeus semisculus* (de Hann, 1984) by hormone injection. Turk. J. Zool., 29: 193-199.

Alfaro, J., G. Zuniga and J. Komen. 2004. Induction of ovarian maturation and spawning by combined treatment of serotonin and a dopamine antagonist, spiperone in *Litopenaeus stylirostris* and *Litopenaeus vannamei*. Aquaculture, 236: 511-522.

Auttarat, J., P. Phiriyangkul and P. Utarabhand. 2006. Characterization of vitellin from the ovaries of the banana shrimp *Litopenaeus merguiensis*. Comp. Biochem. Physiol., B 143: 27-36.

Avarre, J.-C., R. Michelis, A. Tietz and E. Lubzens. 2003. Relationship between vitellogenin and vitellin in a marine shrimp (*Penaeus semisculatus*) and molecular characterization of vitellogenin complementary DNAs. Biol. Reprod., 69: 355-364.

Basu, A.C., and E.A. Kravitz. 2003. Morphology and monoaminergic modulation of crustacean hyperglycemic hormone-like immunoreactive neurons in the lobster nervous system. J. Neurocytol., 32: 253-263.

Beltz, B.S. 1999. Distribution and functional anatomy of amine-containing neurons in decapod crustaceans. Micros. Res. Techniq., 44: 105-120.

Blanchet, M.F., H. Junéra and J.J. Meusy. 1975. La mue et la vitellogenèse d'*Orchestia gammarellus* (Crustacé Amphipode): étude de la synthèse de la fraction protéique femelle après implantation de cristaux d'ecdystérone. Experientia, 31: 865-867.

Böcking, D., H. Dircksen and R. Keller. 2002. The crustacean neuropeptides of the CHH/MIH/GIH family: structures and biological activities. *In:* The Crustacean Nervous System. K. Wiese (ed.). Springer-Verlag, Berlin-Heidelberg, pp. 84-97.

Borst, D.W., H. Laufer, M. Landau, E.S. Chang, W.A. Hertz, F.C. Baker and D.A. Schooley. 1987. Methyl farnesoate and its role in crustacean reproduction and development. Insect Biochem., 17: 1123-1127.

Borst, D.W., J. Ogan, B. Tsukimura, T. Claerhout and K.C. Holford. 2001. Regulation of the crustacean mandibular organ. Amer. Zool., 41: 430-441.

Byard, E.H., R.R. Shivers and D.E. Aiken. 1975. Mandibular organ of lobster, *Homarus americanus*. Cell Tiss. Res., 162: 13-22.

Carlisle, D.B., and C.G. Butler. 1956. The queen-substance of honeybees and the ovary-inhibiting hormone of crustaceans. Nature, 177.

Carlisle, D.B., and B.F. Knowles. 1959. Endocrine Control in Crustaceans. Chapter 6: Sex. Cambridge University Press, Cambridge.

Chaix, J.C., and M. DeReggi. 1982. Ecdysteroid levels during ovarian development and embryogenesis in the spider crab *Acanthonyx lunulatus*. Gen. Comp. Endocrinol., 47: 7-14.

Chan, S.M., P.L. Gu, K.H. Chu and S.S. Tobe. 2003. Crustacean neuropeptide genes of the CHH/MIH/GIH family: implications from molecular studies. Gen. Comp. Endocrinol., 134: 214-219.

Chang, E.S. 1984. Ecdysteroids in Crustacea: role in reproduction, molting, and larval development. *In:* Advances in Invertebrate Reproduction. W. Engels, W.H. Clark, A. Fisher, P.J.W. Olive and D.F. Went (eds.). Elsevier, Amsterdam, pp. 223-230.

Chang, E.S. 1992. Endocrinology. *In:* Marine Shrimp Culture: Principles and Practices. A.W. Fast and L.J. Lester (eds.). Elsevier Science Publishers, Amsterdam, pp. 53-91.

Chang, E.S. 1997. Chemistry of crustacean hormones that regulate growth and reproduction. *In:* Recent Advances in Marine Biotechnology. M. Fingerman, R. Nagabushanam and M.F. Thompson (eds.). Oxford and IBH Publishing Co., New Delhi, pp. 163-178.

Chang, E.S., and W.R. Kaufman. 2005. Endocrinology of Crustacea and Chelicerata. *In:* Comprehensive Molecular Insect Science. L.I. Gilbert, K. Iatrou and S.S. Gill (eds.). Elsevier B. V., Oxford, pp. 805-842.

Chang, E.S., G.D. Prestwich and M.J. Bruce. 1990. Amino-acid-sequence of a peptide with both molt-inhibiting and hyperglycemic activities in the lobster, *Homarus americanus*. Biochem. Biophys. Res. Co., 171: 818-826.

Chang, E.S., M.J. Bruce and S.L. Tamone. 1993. Regulation of crustacean molting – a multi-hormonal system. Amer. Zool., 33: 324-329.

Chang, E.S., R. Keller and S.A. Chang. 1998. Quantification of crustacean hyperglycemic hormone by ELISA in hemolymph of the lobster, *Homarus americanus*, following various stresses. Gen. Comp. Endocrinol., 111: 359-366.

Chang, E.S., S.A. Chang, B.S. Beltz and E.A. Kravitz. 1999a. Crustacean hyperglycemic hormone in the lobster nervous system: Localization and release from cells in the subesophageal ganglion and thoracic second roots. J. Comp. Neurol., 414: 50-56.

Chang, E.S., S.A. Chang, R. Keller, P.S. Reddy, M.J. Snyder and J.L. Spees. 1999b. Quantification of stress in lobsters: Crustacean hyperglycemic hormone, stress proteins, and gene expression. Amer. Zool., 39: 487-495.

Chang, E.S., S.A. Chang and E.P. Mulder. 2001. Hormones in the lives of crustaceans: an overview. Amer. Zool., 41: 1090-1097.

Charmantier, G., M. Charmantier-Daures and F. Van Herp. 1997. Hormonal control of growth and reproduction in crustaceans. *In:* Recent Advances in Marine Biotechnology. M. Fingerman, R. Nagabushanam and M.F. Thompson (eds.). Oxford and IBH Publishing Co., New Delhi, pp. 109-161.

Charniaux-Cotton, H. 1980. Experimental studies of reproduction in malacostraca Crustaceans. Description of vitellogenesis and of its endocrine control. *In:* Advances in Invertebrate Reproduction. J.W.H. Clark and T.S. Adams (eds.). Elsevier, Amsterdam, pp. 177-186.

Charniaux-Cotton, H. 1985. Vitellogenesis and its control in Malacostracan Crustacea. Amer. Zool., 25: 197-206.

Charniaux-Cotton, H., and G.G. Payen. 1988. Crustacean reproduction. *In:* Endocrinology of Selected Invertebrate Types. H. Laufer and R.G.H. Downer (eds.). Alan R. Liss Inc., New York, pp. 279-303.

Chen, S.H., C.Y. Lin and C.M. Kuo. 2005. In silico analysis of crustacean hyperglycemic hormone family. Mar. Biotechnol., 7: 193-206.

Chen, Y.N., D.Y. Tseng, P.Y. Ho and C.M. Kuo. 1999. Site of vitellogenin synthesis determined from a cDNA encoding a vitellogenin fragment in the freshwater giant prawn, *Macrobrachium rosenbergii*. Mol. Reprod. Dev., 54: 215-222.

Chen, Y.N., H.F. Fan, S.L. Hsieh and C.M. Kuo. 2003. Physiological involvement of DA in ovarian development of the freshwater giant prawn, *Macrobrachium rosenbergii*. Aquaculture, 228: 383-395.

Chung, J.S., M.C. Wilkinson and S.G. Webster. 1998. Amino acid sequences of both isoforms of crustacean hyperglycaemic hormone (CHH) and corresponding precursor-related peptide in *Cancer pagurus*. Regul. Peptides, 77: 17-24.

Clark, M.C., T.E. Dever, J.J. Dever, P. Xu, V. Rehder, M.A. Sosa and D.J. Baro. 2004. Arthropod 5-HT_2 receptors: A neurohormonal receptor in decapod crustaceans that displays agonist independent activity resulting from an evolutionary alteration to the DRY motif. J. Neurosci., 24: 3421-3435.

Couch, E F., N. Hagino and J.W. Lee. 1987. Changes in 17β-estardiol and progesterone immunoreactivity in tissues of the lobster, *Homarus americanus*, with developing and immature ovaries. Comp. Biochem. Physiol., A 85: 765-770.

De Kleijn, D.P.V., F. Sleutels, G.J.M. Martens and F. Vanherp. 1994. Cloning and expression of messenger-RNA encoding prepro-gonad-inhibiting hormone (GIH) in the lobster *Homarus americanus*. Febs Letters, 353: 255-258.

De Kleijn, D.P.V., and F. Van Herp. 1995. Molecular biology of neurohormone precursors in the eyestalk of Crustacea. Comp. Biochem. Physiol., B 112: 573-579.

De Kleijn, D.P.V., and F. Van Herp. 1998. Involvement of the hyperglycaemic neurohormone family in the control of reproduction in decapod crustaceans. Invert. Reprod. Dev., 33: 263-272.

Demeusy, N. 1962. Rôle de la gland de mue dans l'evolution ovarianne du crabe *Carcinus maenas*. Linn. Can. Biol. Mar., 3: 37-56.

Dircksen, H., D. Bocking, U. Heyn, C. Mandel, J.S. Chung, G. Baggerman, P. Verhaert, S. Daufeldt, T. Plosch, P.P. Jaros, E. Waelkens, R. Keller and S.G. Webster. 2001. Crustacean hyperglycaemic hormone (CHH)-like peptides and CHH-precursor-related peptides from pericardial organ neurosecretory cells in the shore crab, *Carcinus maenas*, are putatively spliced and modified products of multiple genes. Biochem. J., 356: 159-170.

Dircksen, H., and U. Heyn. 1998. Crustacean hyperglycaemic hormone-like peptides in crab and locust peripheral intrinsic neurosecretory cells. Ann. NY Acad. Sci. Trends Comp. Endocrinol. Neurobiol., 839: 392-394.

Eastman-Reks, S., and M. Fingerman. 1984. Effect of neuroendocrine tissue and cyclic AMP on ovarian growth in vivo and in vitro in the fiddler crab *Uca pugilator*. Comp. Biochem. Physiol., A 79: 679-684.

Elekes, K., E. Florey, M.A. Cahill, U. Hoeger and M. Geffard. 1988. Morphology, synaptic connections and neurotransmitters of the efferent neurons of the crayfish hindgut. Symp. Biol. Hungarica, 36.

Elofsson, R. 1983. 5-HT-like immunoreactivity in the central nervous system of the crayfish, *Pacifastacus leniusculus*. Cell Tiss. Res., 232: 221-236.

Elofsson, R., L. Laxmyr, E. Rosengren and C. Hansson. 1982. Identification and quantitative measurements of biogenic amines and DOPA in the central nervous system and hemolymph of the crayfish *Pacifastacus leniusculus*. Comp. Biochem. Physiol., C 71: 195-201.

Escamilla-Chimal, E.G., F. Van Herp and M.L. Fanjul-Moles. 2001. Daily variations in crustacean hyperglycaemic hormone and serotonin immunoreactivity during the development of crayfish. J. Exp. Biol., 204: 1073-1081.

Escamilla-Chimal, E.G., M. Hiriart, M.C. Sanchez-Soto and M.L. Fanjul-Moles. 2002. Serotonin modulation of CHH secretion by isolated cells of the crayfish retina and optic lobe. Gen. Comp. Endocrinol., 125: 283-290.

Fainzilber, M., M. Tom, S. Shafir, S.W. Applebaum and E. Lubzens. 1992. Is there extraovarian synthesis of vitellogenin in Penaeid shrimp? Biol. Bull., 183: 233-241.

Fairs, N.J., P.T. Quinlan and L.J. Goad. 1990. Changes in ovarian unconjugated and conjugated steroid titers during vitellogenesis in *Penaeus monodon*. Aquaculture, 89: 83-99.

Fanjul-Moles, M.L. 2006. Biochemical and functional aspects of crustacean hyperglycaemic hormone in decapod crustaceans: Review and update. Comp. Biochem. Physiol., C 142: 390-400.

Fingerman, M. 1987. The endocrine mechanisms in crustaceans. J. Crustacean Biol., 7: 1-24.

Fingerman, M. 1995. Endocrine mechansims in crayfish, with emphasis on reproduction and neurotransmitter regulation of hormone release. Amer. Zool., 35: 68-78.

Fingerman, M. 1997. Roles of neurotransmitters in regulating reproductive hormone release and gonadal maturation in decapod crustaceans. Invert. Reprod. Dev., 31: 47- 54.

Fingerman, M., R. Nagabhushanam, R. Sarojini and P.S. Reddy. 1994. Biogenic amines in crustaceans: identification, localization and roles. J. Crustacean Biol., 14: 413-437.

Fingerman, M., N.C. Jackson and R. Nagabhushanam. 1998. Hormonally-regulated functions in crustaceans as biomarkers of environmental pollution. Comp. Biochem. Physiol., C 120: 343-350.

Gagne, F., C. Blaise and J. Pellerin. 2005. Altered exoskeleton composition and vitellogenesis in the crustacean *Gammarus* sp. collected at polluted sites in the Saguenay Fjord, Quebec, Canada. Environ. Res., 98: 89-99.

Gohar, M., C. Soutygrosset, G. Martin and P. Juchault. 1984. Demonstration of an inhibition of vitellogenin synthesis by a neurohumoral factor (VIH) in a Crustacean Isopoda, *Porcellio dilatatus* Brandt. Cr. Acad. Sci. III-Vie., 299: 785-787.

Gosh, D., and A.K. Ray. 1993. Subcellular action of estradiol 17α in a freshwater prawn *Macrobrachium rosenbergii*. Gen. Comp. Endocrinol., 90: 274-281.

Greve, P., O. Sorokine, T. Berges, C. Lacombe, A. Van Dorsselaer and G. Martin. 1999. Isolation and amino acid sequence of a peptide with vitellogenesis inhibiting activity from the terrestrial isopod *Armadillidium vulgare* (Crustacea). Gen. Comp. Endocrinol., 115: 406-414.

Gunamalai, V., R. Kirubagaran and T. Subramoniam. 2004. Hormonal coordination of molting and female reproduction by ecdysteroids in the mole crab *Emerita asiatica* (Milne Edwards). Gen. Comp. Endocrinol., 138: 128-138.

Gunamalai, V., R. Kirubagaran and T. Subramoniam. 2006. Vertebrate steroids and the control of female reproduction in two decapod crustaceans, *Emerita asiatica* and *Macrobrachium rosenbergii*. Curr. Sci., 90: 119-123.

Hasegawa, Y., E. Hirose and Y. Katakura. 1993. Hormonal control of sexual differentiation and reproduction in Crustacea. Amer. Zool., 33: 403-411.

Hazel, C.M. 1986. Steroidogenesis in the female crab *Carcinus maenas*. PhD thesis, University of Liverpool.

Hinsch, G.W. 1980. Effect of mandibular organ implants upon the spider crab ovary. T. Am. Microsc. Soc., 99: 317-322.

Hinsch, G.W. 1981. The mandibular organ of the female spider crab, *Libinia emarginata*, in immature, mature, and ovigerous crabs. J. Morph., 168: 181-187.

Hinsch, G.W., and D.C. Bennet. 1979. Vitellogenesis stimulated by thoracic ganglion implants into detsalked immature spider crabs *Libinia emarginata*. Tiss. Cell, 11: 345-351.

Hsu, Y.W.A., D.I. Messinger, J.S. Chung, S.G. Webster, H.O. de la Iglesia and A.E. Christie. 2006. Members of the crustacean hyperglycaemic hormone (CHH) peptide family are differentially distributed both between and within the neuroendocrine organs of Cancer crabs: implications for differential release and pleiotropic function. J. Exp. Biol., 209: 3241-3256.

Huberman, A. 2000. Shrimp endocrinology. A review. Aquaculture, 191: 191-208.

Imai, K., T. Konno, Y. Nakazawa, T. Komiya, M. Isobe, K. Koga, T. Goto, T. Yaginuma, K. Sakakibara, K. Hasegawa and O. Yamashita. 1991. Isolation and structure of diapause hormone of the silkworm, *Bombyx mori*. P. Jpn. Acad. B. Phys., 67: 98-101.

Jaros, P.P. 1990. Enkephalins, biologically active neuropeptides in invertebrates, with special reference to crustaceans. *In:* Frontiers in Crustacean Neurobiology. (Advances in Life Sciences) K. Wiese, W.-D. Krenz, J. Tautz, H. Reichert and B. Mulloney (eds.). Birkhäuser, Basel, pp. 471-482.

Jaros, P.P., H. Dircksen and R. Keller. 1985. Occurrence of immunoreactive enkephalins in a neurohemal organ and other nervous structures in the eyestalk of the shore crab, *Carcinus maenas* (Crustacea, Decapoda). Cell Tiss. Res., 241: 111-117.

Jasmani, S., T. Ohira, V. Jayasankar, N. Tsutsui, K. Aida and M.N. Wilder. 2004. Localization of vitellogenin mRNA expression and vitellogenin uptake during ovarian maturation in the giant freshwater prawn *Macrobrachium rosenbergii*. J. Exp. Zool., A 301: 334-343.

Jo, Q.-T., H. Laufer, W.J. Biggers and H.S. Kang. 1999. Methyl farnesoate induced ovarian maturation in the spider crab, *Libinia emarginata*. Invert. Reprod. Dev., 36: 79-85.

Junéra, H., M. Zerbib, M. Martin and J.J. Meusy. 1977. Evidence for control of vitellogenin synthesis by an ovarian hormone in *Orchestia gammarellus* (Pallas), Crustacea Amphipoda. Gen. Comp. Endocrinol., 31: 457-462.

Kallen, J.L., S.L. Abrahamse and F. Van Herp. 1990. Circadian rhythmicity of the crustacean hyperglycaemic hormone (CHH) in the hemolymph of the Crayfish. Biol. Bull., 179: 351-357.

Keller, R. 1992. Crustacean neuropeptides: structures, functions and comparative aspects. Experientia, 48: 439-448.

Keller, R., P.P. Jaros and G. Kegel. 1985. Crustacean hyperglycaemic neuropeptides. Amer. Zool., 25: 207-221.

Khayat, M., E. Lubzens, A. Tietz and B. Funkenstein. 1994. Are vitellin and vitellogenin coded by one gene in the marine shrimp *Penaeus semisulcatus*? J. Mol. Endocrinol., 12: 251-254.

Khayat, M., W.J. Yang, K. Aida, H. Nagasawa, A. Tietz, B. Funkenstein and E. Lubzens. 1998. Hyperglycaemic hormones inhibit protein and mRNA synthesis in in vitro-incubated ovarian fragments of the marine shrimp *Penaeus semisulcatus*. Gen. Comp. Endocrinol., 110: 307-318.

Kirubagaran, R., D.M. Peter, G. Dharani, N.V. Vinithkumar, G. Sreeraj and M. Ravindran. 2005. Changes in vertebrate-type steroids and 5-hydroxytryptamine during ovarian recrudescence in the Indian spiny lobster, *Panulirus homarus*. Nzy. Mar. Fresh. Res., 39: 527-537.

Kishori, B., and P.S. Reddy. 2000. Antagonistic effects of opioid peptides in the regulation of ovarian growth of the Indian rice field crab, *Oziotelphusa senex senex* Fabricius. Invert. Reprod. Dev., 37: 107-111.

Kishori, B., and P.S. Reddy. 2003. Influence of leucine-enkephalin on moulting and vitellogenesis in the freshwater crab, *Oziotelphusa senex senex* (Fabricius, 1791) (Decapoda, Brachyura). Crustaceana, 76: 1281-1290.

Kishori, B., B. Premasheela, R. Ramamurthi and P.S. Reddy. 2001. Evidence for a hyperglycaemic effect of methionine-enkephalin in the prawns *Penaeus indicus* and *Metapenaeus monocerus*. Gen. Comp. Endocrinol., 123: 90-99.

Klek-Kawinska, E., and A. Bomirski. 1975. Ovary inhibiting hormone activity in shrimp (*Crangon crangon*) eyestalks during the annual reproductive cycle. Gen. Comp. Endocrinol., 25: 9-13.

Komali, M., V. Kalarani, C. Venkatrayulu and D.C.S. Reddy. 2005. Hyperglycaemic effects of 5-hydroxytryptamine and dopamine in the freshwater prawn, *Macrobrachium malcolmsonii*. J. Exp. Zool., 303A: 448-455.

Koskela, R.W., J.G. Greenwood and P.C. Rothlisberg. 1992. The influence of prostaglandin E_2 and steroid hormones, 17α -hydroxyprogesterone and 17 β-estradiol on molting and ovarian development in the tiger prawn *Penaeus esculentus*, Haswell (Crustacea: Decapoda). Comp. Biochem. Physiol., A 101: 295-299.

Kulkarni, G.K., and M. Fingerman. 1992. Quantitative analysis by Reverse Phase High-Performance Liquid-Chromatography of 5-Hydroxytryptamine in the central nervous system of the Red Swamp Crayfish, *Procambarus clarkii*. Biol. Bull., 182: 341-347.

Kulkarni, G.K., and R. Nagabhushanam. 1980. Role of ovary-inhibiting hormone from eyestalks of marine penaeid prawns (*Parapenaeopsis hardwickii*) during ovarian developmental cycle. Aquaculture, 19: 13-19.

Kulkarni, G.K., R. Nagabhushanam and P.K. Joshi. 1981. Neuroendocrine regulation of reproduction in the marine female prawn, *Parapenaeopsis hardwickii* (Miers). Indian J. Mar. Sci., 10: 350-352.

Kulkarni, G.K., R. Nagabhushanam, G. Amaldoss, R.G. Jaiswal and M. Fingerman. 1992. *In vivo* stimulation of ovarian development in the Red Swamp Crayfish, *Procambarus clarkii* (Girard), by 5-Hydroxytryptamine. Invert. Reprod. Dev., 21: 231-240.

Kuo, C.M., C.R. Hsu and C.Y. Lin. 1995. Hyperglycaemic effects of dopamine in tiger shrimp, *Penaeus monodon*. Aquaculture, 135: 161-172.

Lachaise, F., G. Carpentier, G. Somme and J. Colardeau. 1989. Ecdysteroid synthesis by crab Y-organs. J. Exp. Zool., 252: 283-292.

Landau, M., H. Laufer and E. Homola. 1989. Control of methyl farnesoate synthesis in the mandibular organ of the crayfish *Procambarus clarkii*: evidence for peptide neurohormones with dual functions. Invert. Reprod. Dev., 16: 165-168.

Laufer, H., and W.J. Biggers. 2001. Unifying concepts learned from methyl farnesoate for invertebrate reproduction and post-embryonic development. Amer. Zool., 41: 442-457.

Laufer, H., M. Landau, E. Homola and D.W. Borst. 1987. Methyl farnesoate - its site of synthesis and regulation of secretion in a juvenile crustacean. Insect Biochem., 17: 1129-1131.

Laufer, H., J.S.B. Ahl and A. Sagi. 1993. The role of juvenile hormones in crustacean reproduction. Amer. Zool., 33: 365-374.

Laufer, H., P. Takac, J.S.B. Ahl and M.R. Laufer. 1997. Methyl farnesoate and the effect of eyestalk ablation on the morphogenesis of the juvenile female spider crab *Libinia emarginata*. Invert. Reprod. Dev., 31: 63-68, and Erratum, 32: 187.

Laufer, H., W.J. Biggers and J.S.B. Ahl. 1998. Stimulation of ovarian maturation in the crayfish *Procambarus clarkii* by methyl farnesoate. Gen. Comp. Endocrinol., 111: 113-118.

Lee, C.Y., P.F. Yang and H.S. Zou. 2001. Serotonergic regulation of crustacean hyperglycaemic hormone secretion in the crayfish, *Procambarus clarkii*. Phys. Biochem. Zool., 74: 376-382.

Legrand, J.J., G. Martin, P. Juchault and G. Besse. 1982. Contrôle neuroendocrine de la reproduction chez les Crustacés. J. Physiol. Paris, 78: 543-552.

Li, H., and D.W. Borst. 1991. Characterization of a methyl farnesoate binding protein in hemolymph from *Libinia emarginata*. Gen. Comp. Endocrinol., 81: 335-342.

Li, K., L. Chen, Z. Zhou, E. Li, X. Zhao and H. Guo. 2006. The site of vitellogenin synthesis in Chinese mitten-hand crab *Eriocheir sinensis*. Comp. Biochem. Physiol., B 143: 453-458.

Linder, C.J., and B. Tsukimura. 1999. Ovarian development inihibition by methyl farnesoate (MF) in the tadpole shrimp, *Triops longicaudatus*. Amer. Zool., 39: 20A-21A.

Liu, L., and H. Laufer. 1996. Isolation and characterization of sinus gland neuropeptides with both mandibular organ inhibiting and hyperglyacemic effects from the spider crab *Libinia emarginata*. Arch. Insect Biochem. Physiol., 32: 375-385.

Liu, L., H. Laufer, Y. Wang and T. Hayes. 1997. A neurohormone regulating both methyl farnesoate synthesis and glucose metabolism in a crustacean. Biochem. Biophys. Res. Commun., 237: 694-701.

Lorenzon, S., P. Pasqual and E.A. Ferrero. 1999. Biogenic amines control blood glucose level in the shrimp *Palaemon elegans*. *In:* The Biodiversity Crisis and Crustacea. Custacean Issues. F.B. Schramm (eds.). Balkema, Rotterdam, pp. 471-480.

Lorenzon, S., M. Francese and E.A. Ferrero. 2000. Heavy metal toxicity and differential effects on the hyperglycaemic stress response in the shrimp *Palaemon elegans*. Arch. Environ. Con. Tox., 39: 167-176.

Lorenzon, S., S. Brezovec and E.A. Ferrero. 2004. Species-specific effects on hemolymph glucose control by serotonin, dopamine, and L-enkephalin and their inhibitors in *Squilla mantis* and *Astacus leptodactylus* (Crustacea). J. Exp. Zool., 301A: 727-736.

Lorenzon, S., P. Edomi, P.G. Giulianini, R. Mettulio and E.A. Ferrero. 2005. Role of biogenic amines and CHH in the crustacean hyperglycemic stress response. J. Exp. Biol., 208: 3341-3347.

Lüschen, W., F. Buck, A. Willig and P.P. Jaros. 1991. Isolation, sequence analysis, and physiological properties of enkephalins in the nervous tissue of the shore crab *Carcinus maenas*. Proc. Natl. Acad. Sci. USA, 88: 8671-8675.

Mak, A.S.C., C.L. Choi, S.H.K. Tiu, J.H.L. Hui, J.G. He, S.S. Tobe and S.M. Chan. 2005. Vitellogenesis in the red crab *Charybdis feriatus*: Hepatopancreas-specific expression and farnesoic acid stimulation of vitellogenin gene expression. Mol. Reprod. Dev., 70: 288-300.

Marco, H.G., J.C. Avarre, E. Lubzens and G. Gade. 2002. In search of a vitellogenesis-inhibiting hormone from the eyestalks of the South African spiny lobster, *Jasus lalandii*. Invert. Reprod. Dev., 41: 143-150.

Mercier, A.J., I. Orchard and A. Schmoeckel. 1991. Catecholaminergic neurons supplying the hindgut of the crayfish *Procambarus clarkii*. Can. J. Zool. 69: 2778-2785.

Meusy, J.J. 1980. Vitellogenenin, the extraovarian precursor of the protein yolk in Crustacea: A review. Reprod. Nutr. Dev., 20: 1-21.

Meusy, J.J., and H. Charniaux-Cotton. 1984. Endocrine control of vitellogenesis in malacostraca Crustaceans. *In:* Advances in Invertebrate Reproduction. W. Engels (ed.). Elsevier Science Publishers,

Meusy, J.J., and G.G. Payen. 1988. Female reproduction in malacostracan Crustacea. Zool. Sci., 5: 217-265.

Meusy, J.J., and D. Soyez. 1991. Immunological relationships between neuropeptides from the sinus gland of the lobster *Homarus americanus*, with special references to the vitellogenesis-inhibiting hormone and crustacean hyperglycaemic hormone. Gen. Comp. Endocrinol., 81: 410-418.

Myhrberg, H.E., R. Elofsson, R. Aramant, N. Klemm and L. Laxmyr. 1979. Selective uptake of exogenous catecholamines into nerve fibres in crustaceans. A fluorescence histochemical investigation. Comp. Biochem. Physiol., C 62: 141-150.

Nagabhushanam, R., P.K. Joshi and G.K. Kulkarni. 1980. Induced spawning in the prawn *Parapenaeopsis stylifera* (H. Milne Edwards) using a steroid hormone 17α-hydroxyprogesterone. Ind. J. Mar. Sci., 9: 227-227.

Nagabhushanam, R., R. Sarojini, P.S. Reddy, M. Devi and M. Fingerman. 1995. Opioid peptides in invertebrates - localization, distribution and possible functional roles. Curr. Sci., 69: 659-671.

Nagamine, C., and A.W. Knight. 1987. Induction of female breeding characteristics by ovarian tissue implants in androgenic gland ablated male freshwater prawns *Macrobrachium rosenbergii* (Deman) (Decapoda, Palaemonidae). Int. J. Invert. Rep. Dev., 11: 225-234.

Nagaraju, G.P.C., N.J. Suraj and P.S. Reddy. 2003. Methyl farnesoate stimulates gonad development in *Macrobrachium malcolmsonii* (H. Milne Edwards) (Decapoda, Palaemonidae). Crustaceana, 76: 1171-1178.

Nagaraju, G.P.C., P.R. Reddy and P.S. Reddy. 2006. In vitro methyl farnesoate secretion by mandibular organs isolated from different molt and reproductive stages of the crab *Oziotelphusa senex senex*. Fish. Sci., 72: 410-414.

Nithya, M., and N. Munuswamy. 2002. Immunocytochemical identification of crustacean hyperglycaemic hormone-producing cells in the brain of a freshwater fairy shrimp, *Streptocephalus dichotomus* Baird (Crustacea : Anostraca). Hydrobiologia, 486: 325-333.

Okumura, T. 2004. Perspectives on hormonal manipulation of shrimp reproduction. Jpn. Agr. Res. Q., 38: 49-54.

Okumura, T., and K. Aida. 2000. Fluctuations in hemolymph ecdysteroid levels during the reproductive and non-reproductive molt cycles in the giant freshwater prawn *Macrobrachium rosenbergii*. Fish. Sci., 66: 876-883.

Okumura, T., and K. Sakiyama. 2004. Hemolymph levels of vertebrate-type steroid hormones in female kuruma prawn, *Marsupenaeus japonicus* (Crustacea: Decapoda: Penaeidae) during natural reproductive cycle and induced ovarian development by eyestalk ablation. Fish. Sci.,

Okumura, T., C.H. Han, Y. Suzuki, K. Aida and I. Hanyu. 1992. Changes in hemolymph vitellogenenin and ecdysteroid levels during reproductive and non-reporductive molt cycles in the freshwater prawn *Macrobrachium nipponense*. Zool. Sci., 9: 37-45.

Okuno, A., W.J. Yang, V. Jayasankar, H. Saido-Sakanaka, D.T.T. Huong, S. Jasmani, M. Atmomarsono, T. Subramoniam, N. Tsutsui, T. Ohira, I. Kawazoe, K. Aida and M.N. Wilder. 2002. Deduced primary structure of vitellogenin in the giant freshwater prawn, *Macrobrachium rosenbergii*, and yolk processing during ovarian maturation. J. Exp. Zool., 292: 417-429.

Ollivaux, C., H. Dircksen, J.Y. Toullec and D. Soyez. 2002. Enkephalinergic control of the secretory activity of neurons producing stereoisomers of crustacean hyperglycaemic hormone in the eyestalk of the crayfish *Orconectes limosus*. J. Comp. Neurol., 444: 1-9.

Olmstead, A.W., and G.A. LeBlanc. 2002. Methyl farnesoate is a sex determinant in *Daphnia magna*. J. Exp. Zool., 293: 736-739.

Olmstead, A.W., and G.A. LeBlanc. 2003. Insecticidal juvenile hormone analogs stimulate the production of male offspring in the crustacean *Daphnia magna*. Environ. Health Persp., 111: 919-924.

Ongvarrasopone, C., Y. Roshorm, S. Somyong, C. Pothiratana, S. Petchdee, J. Tangkhabuanbutra, S. Sophasan and S. Panyim. 2006. Molecular cloning and functional expression of the *Penaeus monodon* 5-HT receptor. Biochim. Biophys. Acta - Gene Struct. Expr., 1759: 328-339.

Panouse, J.B. 1943. Influence d'ablation du pédoncule oculaire sur la croissance de l'ovaire chez la Crevette *Leander serratus*. C. R. Acad. Sci. Paris, 217: 553-555.

Phiriyangkul, P., and P. Utarabhand. 2006. Molecular characterization of a cDNA encoding vitellogenin in the banana shrimp, *Penaeus* (*Litopenaeus*) *merguiensis* and sites of vitellogenin mRNA expression. Mol. Reprod. Dev., 73: 410-423.

Picaud, J.L., and C. Souty. 1980. Démonstration par immunoautoradiographie de la synthèse de la vitellogénine par le tissu adipeux de *Porcellio dilatatus* Brandt (Crustace Isopode). C.R. Acad. Sci. Paris, 290: 1019-1021.

Pooyan, R., K.C. Holford and B. Tsukimura. 1999. Determination of the site of vitellogenin synthesis in the ridgeback shrimp, *Sicyonia ingentis*. Amer. Zool., 38: 187A.

Quackenbush, L.S. 1986. Crustacean endocrinology, A review. Can. J. Fish. Aquat. Sci., 43: 2271-2282.

Quackenbush, L.S. 1992. Yolk synthesis in the marine shrimp *Penaeus vannamei*. Comp. Biochem. Physiol., A 103: 711-714.

Quinitio, E.T., A. Hara, K. Yamauchi and S. Nakao. 1994. Changes in the steroid hormone and vitellogenin levels during the gametogenic cycle of the giant tiger shrimp, *Penaeus monodon*. Comp. Biochem. Physiol., C 109: 21-26.

Racotta, I.S., E. Palacios and A.M. Ibarra. 2003. Shrimp larval quality in relation to broodstock condition. Aquaculture, 227: 107-130.

Ragunathan, M.G., and A. Arivazhagan. 1999. Influence of eyestalk ablation and 5-hydroxytryptamine on the gonadal development of a female crab, *Paratelphusa hydrodromous* (Herbst). Curr. Sci., 76: 583-587.

Raina, A.K. 1993. Neuroendocrine control of sex-pheromone biosynthesis in Lepidoptera. Annu. Rev. Entomol., 38: 329-349.

Raviv, S., S. Parnes, C. Segall, C. Davis and A. Sagi. 2006. Complete sequence of *Litopenaeus vannamei* (Crustacea: Decapoda) vitellogenin cDNA and its expression in endocrinologically induced sub-adult females. Gen. Comp. Endocrinol., 145: 39-50.

Reddy, P.R., G.P.C. Nagaraju and P.S. Reddy. 2004. Involvement of methyl farnesoate in the regulation of molting and reproduction in the freshwater crab *Oziotelphusa senex senex*. J. Crustacean Biol., 24: 511-515.

Reddy, P.S. 2000. Involvement of opioid peptides in the regulation of reproduction in the prawn *Penaeus indicus*. Naturwissenschaften, 87: 535-538.

Reddy, P.S., and M.R. Basha. 2001. On the mode of action of methionine enkephalin, FK 33-824 and naloxone in regulating the hemolymph glucose level in the freshwater field crab *Oziotelphusa senex senex*. Z. Naturforsch. C, 56: 629-632.

Reddy, P.S., and B. Kishori. 2001. Methionine-enkephalin induces hyperglycaemic through eyestalk hormones in the estuarine crab *Scylla serrata*. Biol. Bull., 201: 17-25.

Reddy, P.S., and P. Ramamurthi. 1998. Methyl farnesoate stimulates ovarian maturation in the freshwater crab *Oziotelphusa senex senex* Fabricius. Curr. Sci., 74: 68-70.

Reddy, P.S., and P.R. Reddy. 2006. Purification of molt-inhibiting hormone-like peptides with hyperglycaemic activity from the eyestalks of the crab *Scylla serrata*. Fish. Sci., 72: 415-420.

Richardson, H.G., M. Deecaraman and M. Fingerman. 1991. The effect of biogenic amines on ovarian development in the fiddler crab *Uca pugilator*. Comp. Biochem. Physiol., C 99: 53-56.

Rodriguez, E.M., D.A. Medesani, L.S.L. Greco and M. Fingerman. 2002a. Effects of some steroids and other compounds on ovarian growth of the red swamp crayfish, *Procambarus clarkii*, during early vitellogenesis. J. Exp. Zool., 292: 82-87.

Rodriguez, E.M., L.S.L. Greco, D.A. Medesani, H. Laufer and M. Fingerman. 2002b. Effect of methyl farnesoate, alone and in combination with other hormones, on ovarian growth of the red swamp crayfish, *Procambarus clarkii*, during vitellogenesis. Gen. Comp. Endocrinol., 125: 34-40.

Rothe, H., W. Lüschen, A. Asken, A. Willig and P.P. Jaros. 1991. Purified crustacean enkephalin inhibits release of hyperglycaemic hormone in the crab *Carcinus maenas*. Comp. Biochem. Physiol., C 99: 57-62.

Rotllant, G., N. Pascual, F. Sarda, P. Takac and H. Laufer. 2001. Identification of methyl farnesoate in the hemolymph of the Mediterranean deep-sea species Norway lobster, *Nephrops norvegicus*. J. Crustacean Biol., 21: 328-333.

Sagi, A., E. Homola and H. Laufer. 1993. Distinct reproductive types of male spider crabs *Libinia emarginata* differ in circulating and synthesising methyl farnesoate. Biol. Bull., 185: 168-173.

Sarojini, R., K. Jayalakshmi and S. Sambashivarao. 1986. Effect of external steroids on ovarian development in the freshwater prawn *Macrobrachium lamerii*. J. Adv. Zool,. 7: 50-53.

Sarojini, R., R. Nagabhushanam and M. Fingerman. 1995a. Evidence for opioid involvement in the regulation of ovarian maturation of the fiddler crab, *Uca pugilator*. Comp. Biochem. Physiol., A 111: 279-282.

Sarojini, R., R. Nagabhushanam and M. Fingerman. 1995b. Mode of action of the neurotransmitter 5-Hydroxytryptamine in stimulating ovarian maturation in the Red Swamp Crayfish, *Procambarus clarkii* - an *in-vivo* and *in-vitro* study. J. Exp. Biol., 271: 395-400.

Sarojini, R., R. Nagabhushanam and M. Fingerman. 1995c. *In vivo* inhibition by DA of 5-hydroxytryptamine-stimulated ovarian maturation in the red swamp crayfish *Procambarus clarkii*. Experientia, 51: 156-158.

Sarojini, R., R. Nagabhushanam and M. Fingerman. 1995d. A neurotransmitter role for red-pigment-concentrating hormone (RPCH) in ovarian maturation in the red swamp crayfish *Procambarus clarkii*. J. Exp. Biol., 198: 1253-1257.

Sarojini, R., R. Nagabhushanam and M. Fingerman. 1996. *In vitro* inhibition by DA of 5-hydroxytryptamine-stimulated ovarian maturation in the red swamp crayfish *Procambarus clarkii*. Experientia, 52: 707-709.

Schoofs, L., G.M. Holman, T.K. Hayes, R.J. Nachman and A. Deloof. 1991. Isolation, primary structure, and synthesis of locustapyrokinin - a myotropic peptide of *Locusta migratoria*. Gen. Comp. Endocrinol., 81: 97-104.

Serrano-Pinto, V., C. Vazquez-Boucard and H. Villarreal-Colmenares. 2003. Yolk proteins during ovary and egg development of mature female freshwater crayfish (*Cherax quadricarinatus*). Comp. Biochem. Physiol., A 134: 35-45.

Serrano-Pinto, V., M.G. Carrisoza-Velenzuela and M. Ramírez-Orozco. 2005. Determination site of vitellogenin synthesis in freshwater crayfish *Cherax quadricarinatus* at different maturation stages in females. Invest. Mar. Valparaíso, 33: 195-200.

Shechter, A., E.D. Aflalo, C. Davis and A. Sagi. 2005. Expression of the reproductive female-specific vitellogenin gene in endocrinologically induced male and intersex *Cherax quadricarinatus* crayfish. Biol. Reprod., 73: 72-79.

Sosa, M.A., N. Spitzer, D.H. Edwards and D.J. Baro. 2004. A crustacean serotonin receptor: Cloning and distribution in the thoracic ganglia of crayfish and freshwater prawn. J. Comp. Neurol., 473: 526-537.

Soumoff, C., and D.M. Skinner. 1983. Ecdysteroid titers during the molt cycle of the blue crab resemble those of other crustacea. Biol. Bull., 165: 321-329.

Soyez, D., J.E. Van Deynen, M. Martin, P. Jugan and F. Van Herp. 1986. Isolation and characterization of the vitellogenesis inhibiting hormone from the lobster *Homarus americanus*. B. Soc. Zool. Fr., 111: 27-28.

Soyez, D., J.P. Lecaer, P.Y. Noel and J. Rossier. 1991. Primary structure of 2 isoforms of the vitellogenesis-inhibiting hormone from the lobster *Homarus americanus*. Neuropeptides, 20: 25-32.

Spaziani, E.P., K. DeSantis, B.D. O'Rourke, W.L. Wang and J.D. Weld. 1997. The clearance *in vivo* and metabolism of ecdysone and 3-dehydroxyecdysone in tissues of the crab *Cancer antennarius*. J. Exp. Zool., 278: 609-619.

Subramoniam, T. 2000. Crustacean ecdysteroids in reproduction and embryogenesis. Comp. Biochem. Physiol., C 125: 135-156.

Subramoniam, T., D. Sedlmeier and R. Keller. 1999. Recent advances in the endocrine regulation of reproduction and molting in Crustacea. *In:* Comparative Endocrinology and Reproduction. K.P. Joy, A. Krishna and C. Haldar (eds.). Narosa Publishing House, New Delhi, pp. 462-475.

Summavielle, T., P.R. Rocha Monteiro, M.A. Reis-Henriques and J. Coimbra. 2003. In vitro metabolism of steroid hormones by ovary and hepatopancreas of the crustacean Penaeid shrimp *Marsupenaeus japonicus*. Sci. Mar., 67: 299-306.

Swevers, L., J.G.D. Lambert and A. Deloof. 1991. Metabolism of vertebrate-type steroids by tissues of 3 crustacean species. Comp. Biochem. Physiol., B 99: 35-41.

Takayanagi, H., Y. Yamamoto and N. Takeda. 1986. An ovary-stimulating factor in the shrimp, *Paratya compressa*. J. Exp. Zool., 240: 203-209.

Taketomi, T., M. Motono and M. Miyawaki. 1989. On the biological function of the mandibular gland of Decapod Crustacea. Cell Biol. Int. Rep., 13: 463-469.

Tamone, S.L., and E.S. Chang. 1993. Methyl farnesoate stimulates ecdysteroid secretion from crab Y-organs *in vitro*. Gen. Comp. Endocrinol., 89: 425-432.

Tanboonteck, S., P. Damrongphol and W. Poolsanguan. 2006. Stimulation of ovarian development and spawning in the giant freshwater prawn, *Macrobrachium rosenbergii* (de Man). Aquacult. Res., 37: 1259-1261.

Tang, C.H., W.Q. Lu, G. Wainwright, S.G. Webster, H.H. Rees and P.C. Turner. 1999. Molecular characterization and expression of mandibular organ-inhibiting hormone, a recently discovered neuropeptide involved in the regulation of growth and reproduction in the crab *Cancer pagurus*. Biochem. J., 343: 355-360.

Tensen, C.P., K.P.C. Janssen and F. Vanherp. 1989. Isolation, characterization and physiological specificity of the crustacean hyperglycaemic factors from the sinus gland of the lobster, *Homarus americanus*. Gen. Comp. Endocrinol., 74: 262-262.

Tiu, S.H.K., J.H.L. Hui, J.G. He, S.S. Tobe and S.M. Chan. 2006. Characterization of vitellogenin in the shrimp *Metapenaeus ensis*: expression studies and hormonal regulation of MeVg1 transcription in vitro. Mol. Reprod. Dev., 73: 424-436.

Torfs, P., J. Nieto, A. Cerstiaens, D. Boon, G. Baggerman, C. Poulos, E. Waelkens, R. Derua, J. Calderon, A. De Loof and L. Schoofs. 2001. Pyrokinin neuropeptides in a crustacean - Isolation and identification in the white shrimp *Penaeus vannamei*. Eur. J. Biochem., 268: 149-154.

Tsang, W.S., L.S. Quackenbush, B.K. Chow, S.H. Tiu, J.G. He and S.M. Chan. 2003. Organization of the shrimp vitellogenin gene: evidence of multiple genes and tissue specific expression by the ovary and the hepatopancreas. Gene, 303: 99-109.

Tseng, D.Y., Y.N. Chen, G.H. Kou, C.F. Lo and C.M. Kuo. 2001. Hepatopancreas is the extraovarian site of vitellogenin synthesis in black tiger shrimp, *Penaeus monodon*. Comp. Biochem. Physiol., A 129: 909-917.

Tseng, D.Y., Y.N. Chen, K.F. Liu, G.H. Kou, C.F. Lo and C.M. Kuo. 2002. Hepatopancreas and ovary are sites of vitellogenin synthesis as determined from partial cDNA encoding of vitellogenin in the marine shrimp, *Penaeus vannamei*. Invert. Reprod. Dev., 42: 137-143.

Tsukimura, B. 2001. Crustacean vitellogenesis: Its role in oocyte development. Amer. Zool., 41: 465-476.

Tsukimura, B., and D.W. Borst. 1992. Regulation of methyl farnesoate in the hemolymph and mandibular organ of the lobster, *Homarus americanus*. Gen. Comp. Endocrinol., 86: 297-303.

Tsukimura, B., J.S. Bender and C.J. Linder. 2000. Developmental aspects of gonadal regulation in the ridgeback shrimp, *Sicyonia ingentis*. Comp. Biochem. Physiol., A 127: 215-224.

Tsutsui, N., I.T. Kawasoe, T. Ohira, S. Jasmani, W.J. Yang, M. Wilder and K. Aida. 2000. Molecular characterization of a cDNA encoding vitellogenin and its expression in the hepatopancreas and ovary during vitellogenesis in the kuruma prawn, *Penaeus japonicus*. Zool. Sci., 17: 651-660.

Tsutsui, N., H. Saido-Sakanaka, W.-J. Yang, V. Jayasankar, S. Jasmani, A. Okuno, T. Ohira, T. Okumura, K. Aida and M. Wilder. 2004. Molecular characterization of a cDNA encoding vitellogenin in the coonstriped shrimp, *Pandalus hypsinotus* and site of vitellogenin mRNA expression. Comp. Exp. Biol., A 301: 802-814.

Tuberty, S.R., and C.L. McKenney. 2005. Ecdysteroid responses of estuarine crustaceans exposed through complete larval development to juvenile hormone agonist insecticides. Int. Comp. Biol., 45: 106-117.

Vaca, A.A., and J. Alfaro. 2000. Ovarian maturation and spawning in the white shrimp, *Penaeus vannamei*, by serotonin injection. Aquaculture, 182: 373-385.

Van Herp, F., and G.G. Payen. 1991. Crustacean neuroendocrinology: Perspectives for the control of reproduction in aquaculture systems. Bull. Inst. Zool. Acad. Sinica, 16: 513-539.

Van Herp, F., and D. Soyez. 1997. Reproductive biology in invertebrates: Arthropoda-Crustacea. *In:* Progress in Reproductive Endocrinology, Part A. T.S. Adams (ed.). pp. 247-275.

Vincent, S.G.P., R. Keller and T. Subramoniam. 2001. Development of vitellogenin-ELISA, an in vivo bioassay, and identification of two vitellogenesis-inhibiting hormones of the tiger shrimp *Penaeus monodon*. Mar. Biotechnol., 3: 561-571.

Wainwright, G., S.G. Webster, M.C. Wilkinson, J.S. Chung and H.H. Rees. 1996. Structure and significance of mandibular organ-inhibiting hormone in the crab *Cancer pagurus*. Involvement in multihormonal regulation of growth and reproduction. J. Biol. Chem.. 271: 12749-12754.

Webster, S.G. 1996. Measurement of crustacean hyperglycaemic hormone levels in the edible crab *Cancer pagurus* during emersion stress. J. Exp. Biol., 199: 1579-1585.

Webster, S.G. 1998. Neuropeptides inhibiting growth and reproduction in crustaceans. *In:* Recent Advances in Arthropod Endocrinology. G.M. Coast and S.G. Webster (eds.). Cambridge University Press, Cambridge, pp. 33-52.

Webster, S.G., and R. Keller. 1986. Purification, characterization and amino acid composition of the putative molt-inhibiting hormone (MIH) of *Carcinus maenas* (Crustacea: Decapoda). J. Comp. Physiol., 156B: 617-624.

Wilder, M.N., S. Okada, N. Fusetani and K. Aida. 1995. Hemolymph profiles of juvenoid substances in the giant freshwater prawn *Macrobrachium rosenbergii* in relation to reproduction and molting. Fish. Sci., 61: 175-176.

Wilder, M.N., W.-J. Yang, D.T.T. Huong, M. Maeda, T.T.T. Hien, T.Q. Phu, N.T. Phuong and H.Y. Ogata. 1999. Reproductive mechanisms in the giant freshwater prawn *Macrobrachium rosenbergii* and cooperative research to improve seed production technology in the Mekong Delta region of Vietnam. 28, Japan International Research Center for Agricultural Sciences.

Wilder, M.N., T. Subramoniam and K. Aida. Yolk proteins of Crustacea. 2002. *In:* Reproductive Biology of Invertebrates. Progress in Vitellogenesis. A.S. Raikhel and T.W. Sappington (eds.). Science Publishers, Enfield, New Hampshire, pp. 131-174.

Yang, F., H.T. Xu, Z.M. Dai and W.J. Yang. 2005. Molecular characterization and expression analysis of vitellogenin in the marine crab *Portunus trituberculatus*. Comp. Biochem. Physiol., B 142: 456-464.

Yano, I. 1985. Induced ovarian maturation and spawning in greasyback shrimp *Metapenaeus ensis*, by progesterone. Aquaculture, 47: 223-229.

Yano, I. 1987. Effect of 17α-hydroxyprogesterone on vitellogenin secretion in kuruma prawn, *Penaeus japonicus*. Aquaculture, 61: 49-57.

Yano, I. 1993. Ultraintensive culture and maturation in captivity of penaeid shrimp. *In:* CRC Handbook of Mariculture: Crustacean Aquaculture. J.P. McVey (ed.). CRC Press, Boca Raton, Florida, pp. 289-313.

Yano, I., R.M. Krol, R.M. Overstreet and W.E. Hawkins. 1996. Route of egg yolk protein uptake in the oocytes of kuruma prawn, *Penaeus japonicus*. Mar. Biol., 125: 773-781.

Young, N.J., S.G. Webster and H.H. Rees. 1993. Ovarian and hemolymph ecdysteroid titers during vitellogenesis in *Macrobrachium rosenbergii*. Gen. Comp. Endocrinol., 90: 183-191.

Yudin, A.I., R.A. Diener, W.H. Clark and E.S. Chang. 1980. Mandibular gland of the blue crab, *Callinectes sapidus*. Biol. Bull., 159: 760-772.

Zerbib, C., and J.J. Meusy. 1983. Electronmicroscopic observations of the subepidermal fat body changes following ovariectomy in *Orchestia gammarellus* (Pallas) (Crustacea, Amphipoda). Int. J. Invert. Rep. Dev., 6: 123-127.

Zou, H.S., C.C. Juan, S.C. Chen, H.Y. Wang and C.Y. Lee. 2003. Dopaminergic regulation of crustacean hyperglycaemic hormone and glucose levels in the hemolymph of the crayfish *Procambarus clarkii*. J. Exp. Zool., 298A: 44-52.

7

Male Reproductive Hormones

E.S. Chang[1] and A. Sagi[2]

[1]Bodega Marine Laboratory, University of California-Davis, PO Box 247, Bodega Bay, CA 94923, USA, E-mail: eschang@ucdavis.edu

[2]Department of Life Sciences and the National Institute for Biotechnology in the Negev, Ben Gurion University, PO Box 653, Beer Sheva 84105, Israel. E-mail: sagia@bgu.ac.il

INTRODUCTION

Sex differentiation and reproduction in male vertebrates are regulated by several different peptide and steroid hormones from various endocrine organs. In contrast, insects are thought not to have sex hormones at all (Maas and Dorn, 2005). In the closely related crustaceans, the androgenic gland (AG) appears to be unique, since this single gland regulates both male sex differentiation and male reproductive physiology and since, unlike in vertebrates, the endocrine and gametogenic functions are clearly separated into distinct organs, the androgenic gland and the testis, respectively. As to the other invertebrate phyla, very little is known about the role (or even the existence) of hormones that regulate sex differentiation or male reproductive physiology.

This review is not intended to be comprehensive. It gives a brief historical overview of hormonal regulation of reproduction in male crustaceans, including the androgenic gland hormone (AGH), and then describes some of the more recent work on its isolation, characterization, and mode of action. We then summarize evidence for the role of other chemical mediators in the regulation of male reproduction (ecdysteroids, vertebrate-like steroids, and methyl farnesoate). A brief discussion of pheromones is included, since they are chemical mediators that influence reproduction and are mechanistically similar to hormones.

The reader is directed to several comprehensive reviews that discuss some aspects of the material covered in this review (Ginsburger-Vogel and Charniaux-Cotton, 1982; Adiyodi, 1985; Charniaux-Cotton, 1985; Charniaux-Cotton and Payen, 1988; Sagi, 1988; Sagi et al., 1997; Sagi and Khalaila, 2001).

THE CRUSTACEAN ANDROGENIC GLAND

Sex Differentiation and the Androgenic Gland

The regulation of reproduction in crustaceans is highly diverse; most species are dioecious (separate sexes), but there are also many hermaphroditic species. In the Malacostraca (e.g., isopods, amphipods, decapods), experiments have demonstrated that individuals of most species possess the genetic information for the development of both males and females. In genetic males, the AG develops and begins to secrete AGH. In the absence of a developed, active AG (i.e., in females), there is an absence of AGH and female structures develop.

The AG was initially discovered by Cronin (1947) in the blue crab (*Callinectes sapidus*) and later suggested by Charniaux-Cotton (1954a) to be the regulator of spermatogenesis and male differentiation in the amphipod *Orchestia gamarella*. The AG is located at the distal portion of the sperm duct. Charniaux-Cotton (1955) observed that the AG is the only male tissue that can mediate the transformation of the ovaries of an immature female into testes. An ovary transplanted into an andrectomized male remained an ovary (Charniaux-Cotton, 1954b, 1955, 1957). Following removal of the AG from a male, spermatogenesis waned and in some cases oocytes appeared (Charniaux-Cotton, 1964). These observations were confirmed in the isopod *Armadillidium vulgare* (Legrand, 1955). More recent experiments have been conducted in the giant freshwater prawn, *Macrobrachium rosenbergii*. Nagamine et al. (1980a) demonstrated complete sex reversal following the removal of the AGs at an early immature stage. This operation resulted in complete female differentiation, complete with ovaries and oviducts including reproductive behavior, successful mating and production of offspring (Sagi and Cohen, 1990; Sagi et al., 1990, 1997). Implantation of AGs into females resulted in the development of male copulatory organs with reported cases of functional sex reversal and progeny obtained when fertile sex-reversed animals were crossed with normal prawns (Nagamine et al., 1980a, b; Malecha et al., 1992).

Macrobrachium rosenbergii males progress through a succession of male morphotypes beginning with small males to orange-claw males to the dominant blue-claw males. Sagi et al. (1990) demonstrated that the AG was necessary for this morphotypic progression. Histological and biochemical supporting evidence for the relationship between morphotypic differentiation and the structure of the AG were presented by Sun et al. (2000). The necessity of the AG for complete maleness in decapods was demonstrated in other species, including the crayfish *Procambarus clarkii* (Nagamine and Knight, 1987; Taketomi and Nishikawa, 1996) and *Cherax destructor* (Fowler and Leonard, 1999). The wide array of effects for which this gland is responsible was demonstrated in *Cherax quadricarinatus* (Khalaila et al., 1999, 2001; Sagi et al., 2002; Manor et al., 2004), including the development of male secondary characteristics and inhibition of female secondary characteristics and vitellogenesis. It was, moreover, shown that the AG induces male-like reproductive and aggressive behavior (Barki et al., 2003; Karplus et al., 2003).

The Androgenic Gland and Biotechnology

Most economically important cultured crustaceans show sexually bimodal growth patterns in which males grow faster than females or vice versa. Monosex culture has been recently suggested as one of the most promising ways to improve production efficiencies of crustacean aquaculture. Nair et al. (2006) demonstrated in *M. rosenbergii* that profits from monosex culture could be higher by as much as 60% over a mixed population. An example of the application of endocrine research in crustaceans is the potential for the production of all male monosex *M. rosenbergii*. This latter example would be the result of AG removal from immature males, which results in sex reversal with complete and functional female differentiation (neo-female). In the prawn *M. rosenbergii*, neo-female animals are capable of mating with normal specimens to produce all-male offspring. The reason for the occurrence of all-male progeny in such a crossing is the homogametic (ZZ) nature of the males (Sagi and Aflalo, 2005; Aflalo et al., 2006).

Androgenic Gland Hormone

Much effort has been devoted to the purification and characterization of the active AGH belonging to the insulin-like super family. From the isopod *A. vulgare*, Hasegawa et al. (1987) purified two proteins, AGHI and AGHII, consisting of 157 and 166 amino acids, respectively. Molecular masses of 17.0 and 18.3 kDa were estimated. Independent work by Martin et al. (1990) on the same species led to similar results, though there were slight discrepancies in the amino acid compositional data. Bioassays involved masculinization of young female isopods. Protein blotting experiments with polyclonal antisera raised against AGH indicated the presence of a larger, biologically inactive protein. This indicates that AGH may be translated as a prohormone (Martin et al., 1990; Hasegawa et al., 1991). Recent work has resulted in the determination of the structure of the AGH and the gene that codes for it (Martin et al., 1999; Okuno et al., 1999) (Fig. 1). Recombinant isopod AGH has been produced (Okuno et al., 2002).

Signal Peptide	B Chain	KR	C Peptide	KR	A Chain
(21)	(44)		(46)		(29)

Fig. 1 Schematic structure of the androgenic gland hormone (AGH) from *Armadillidium vulgare*. As shown, Okuno et al. (1999) reported an additional signal peptide of 21 amino acid residues adjacent to the B-chain in the preprohormone form. Loss of the signal peptide results in the AGH prohormone. Martin et al. (1999) reported a glycan moiety (not shown) attached to one amino acid of the A-chain. The numbers in parentheses indicate the number of amino acid residues in each domain of AGH. The possible cleavage sites are indicated by KR (Lys-Arg). Removal of the C-peptide and formation of disulfide bonds between the A- and B-chains results in the formation of the mature hormone.

In a decapod, a subtractive cDNA library from the AG of the red-claw crayfish *C. quadricarinatus* has been established revealing an AG-specific gene, expressed exclusively in males, even at early stages of maturation. This gene is termed *Cq-IAG* (insulin-like AG factor from *C. quadricarinatus*). *In situ* hybridization of *Cq-IAG* confirmed the exclusive localization of its expression in the AG. Following cloning and complete sequencing of the gene, its cDNA was found to contain 1,445 nucleotides encoding a deduced translation product of 176 amino acids. The proposed protein sequence encompasses Cys residue and putative cleaved peptide patterns whose linear and 3D organization are similar to those of members of the insulin/insulin-like growth factor/relaxin family and their receptor recognition surface (Fig. 2). The peptide and its activity remain to be elucidated (Manor et al., 2007).

Other biologically active factors have been isolated from the green crab (*Carcinus maenas*) AG. The terpenes farnesylacetone and hexahydroxfarnesylacetone were isolated and inhibited the incorporation of radiolabeled leucine by ovaries *in vitro* (Ferezou et al., 1977). Farnesylacetone was also able to inhibit radiolabeled uridine incorporation, indicating that it inhibits transcription (Berreur-

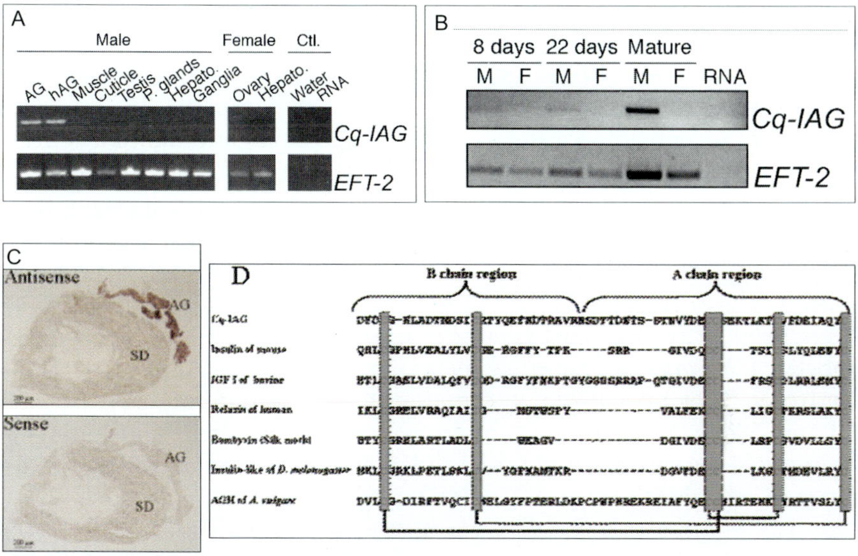

Fig. 2 Identification and characterization of *Cq-IAG* (insulin-like AG factor from *Cherax quadricarinatus*), an AG-specific expressed gene. A. RT-PCR (reverse transcriptase-polymerase chain reaction) showing expression of *Cq-IAG* only in the AG. hAG: hypertrophied androgenic gland; P. glands: peripheral glands; Hepato.: hepatopancreas; EFT-2: control for equal sample loading. B. RT-PCR showing expression of *Cq-IAG* in juvenile males (M) (8 and 22 days post-release from the mother) and in the base of the fifth walking leg (containing the AG) of a mature male but not in females (F). In the lane marked RNA, RNA was added as a template to the PCR in order to rule out genomic contamination. C. Localization of the expression of *Cq-IAG* was performed by RNA *in situ* hybridization. A strong, specific signal in the AG was detected by the antisense probe only. SD: sperm duct. D. Multiple sequence alignment of the putative mature Cq-IAG peptide with representative members of the insulin/insulin-like growth factor/relaxin family (based on SMART results and done by Clustal X; modified from Manor et al., 2007, with permission from Elsevier).

Bonnenfant and Lawrence, 1984). The roles of these terpene AG factors remain to be elucidated, though it is most likely that the decapod AGH is also a peptide as in the isopods (for review, see Sagi and Khalaila, 2001). Secretions from the eyestalk X-organ/sinus gland complex may regulate the AG (Khalaila et al., 2002).

STEROID HORMONES

Ecdysteroids

A number of studies indicate that the arthropod molting hormones (ecdysteroids) are important regulators of female reproduction in crustaceans (Subramoniam, 2000; Chang and Kaufman, 2005, for reviews). Much less is known about the involvement of ecdysteroids in male reproduction. While injections of ecdysteroids into crustaceans induce many physiological changes, it is not clear whether the hormone is directly responsible for the associated morphological and biochemical changes (for review see Adiyodi, 1985).

In vitro studies on the effects of ecdysteroids on crustaceans have demonstrated some direct effects of the hormone on male-related physiology. For example, DNA synthesis increased after 2-3 days in sheath cells of cultured isopod (*Idotea wosnesenskii*) testes incubated with 20-hydroxyecdysone (Fig. 3) (Matlock and Dornfeld, 1982). Exposure of lobster (*Homarus americanus*) testicular primary cell cultures to 20-hydroxyecdysone caused spermatogonia to proliferate (Brody and Chang, 1989). Addition of 20-hydroxyecdysone resulted in a significant increase in the incorporation of radiolabeled thymidine in cultured testes from *M. rosenbergii* (Sagi et al., 1991).

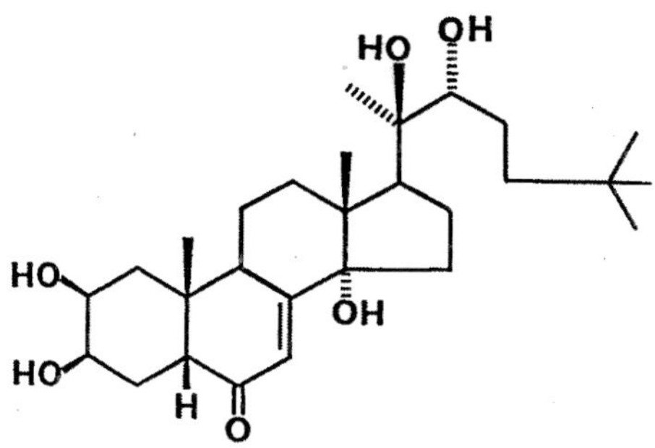

Fig. 3 Structure of 20-hydroxyecdysone.

Several reports circumstantially implicate ecdysteroids as being involved in male reproduction. In shore crabs (*Pachygrapsus crassipes*), the testes had the highest hydroxylating enzyme activity in converting the prohormone ecdysone to the active molting hormone 20-hydroxyecdysone (Chang et al., 1976). Expression of cytochrome P_{450} genes that code for this enzyme activity were also high in the testes of *C. maenas* (Styrishave et al., 2004). Recently, Parnes et al. (2006) discovered a molt-dependent mechanism by which old sperm is periodically removed from the reproductive system of male *Litopenaeus vannamei* shrimp. It was shown that male shrimp go through reproductive cycles that are strictly associated with their molt cycles, which, in turn, are hormonally regulated. Intact intermolt spermatophores disappeared about 12 h premolt, and a new pair of spermatophores appeared in the ampullae the day after the males had molted.

OTHER STEROIDS

There are a few reports of male vertebrate steroids (such as testosterone) being present (Burns et al., 1984) or capable of being metabolized in various species of male crustaceans (Gilgan and Idler, 1967; Teshima and Kanazawa, 1971; James and Shiverick; 1984, Baldwin and LeBlanc, 1994; LeBlanc and McLachlan, 2000; Verslycke et al., 2002). Histochemical evidence for possible steroidogenic activity through the distribution of lipids 3-alpha and 3-beta-hydroxysteroid dehydrogenase in the AG was reported in *M. rosenbergii* (Veith and Malecha, 1983). Nagabhushanam and Kulkarni (1981) suggested that exogenous testosterone affects the AG of a marine penaeid prawn, *Parapenaeopsis hardwickii*. Recent studies suggest that vertebrate steroid androgens may alter crustacean sex ratio causing the appearance of more males in prawn populations (Baghel et al., 2004; Ohs et al., 2006). The mere presence, exogenous effect and/or metabolism of a steroid, however, do not necessarily mean that the molecule has a physiological role in normal development. Whether testosterone has a physiological role in male reproduction in crustaceans remains to be seen.

METHYL FARNESOATE

Methyl farnesoate (MF; Fig. 4) is a sesquiterpenoid that was identified in crustacean hemolymph using gas chromatography/mass spectrometry with selected ion monitoring (Borst et al., 1987; Laufer et al., 1987). It is produced by the paired mandibular organs (Fig. 5). Methyl farnesoate is very similar in structure to insect juvenile hormone III, differing only by the absence of an epoxide group in MF. In insects, the juvenile hormones regulate metamorphosis and reproduction (Goodman and Granger, 2005). Crustaceans have many morphological, developmental, and physiological features in common with insects. Because of the striking similarities between these two groups of arthropods, biologists have speculated about the existence of crustacean juvenile hormone-like

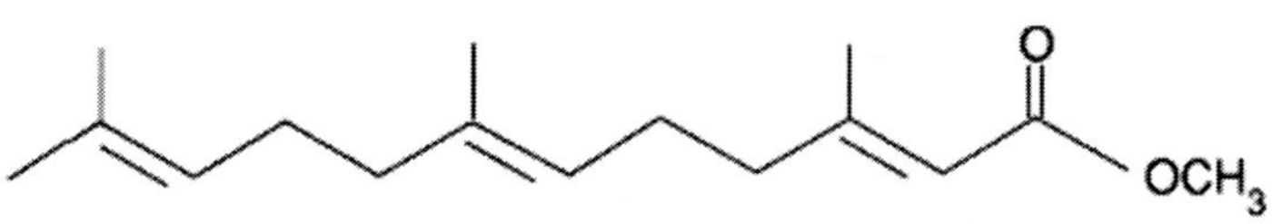

Fig. 4 Structure of methyl farnesoate.

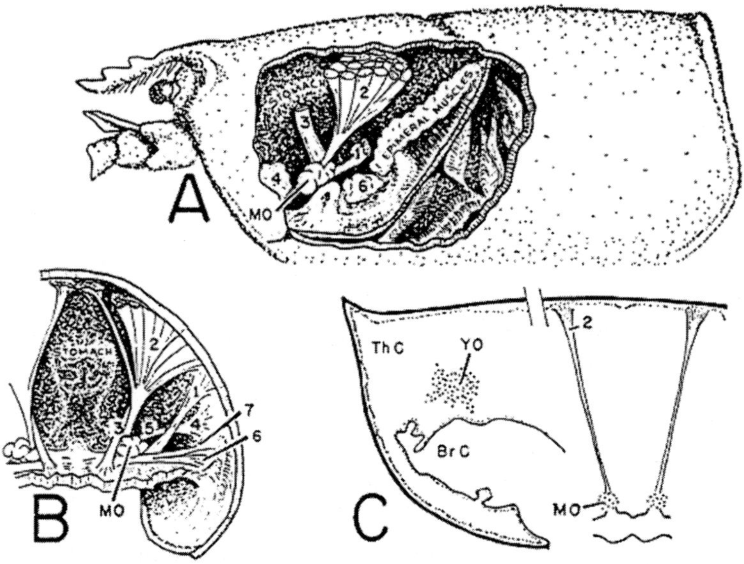

Fig. 5 Location of mandibular organs (MO) and Y-organs (YO, molting glands) in *Carcinus* sp. and *Homarus* sp. Cephalothorax of an adult *Homarus* sp. viewed from lateral (A) and posterior (B) aspects, showing the position of the MO. Organs are also shown in a diagrammatic transverse section of an adult *Carcinus* sp. (C). BrC and ThC are the branchial and thoracic cavities, respectively. Muscles and tendons are the following: (1) maxillulary abductor coxopoditis; (2) posterior mandibular adductor; (3) anterior dorsoventral; (4) lateral mandibular adductor; (5) major mandibular adductor; (6) posterior dorsoventral; (7) transverse mandibular tendon. Reprinted from Homola and Chang (1997), with permission from Elsevier; modified from Sochasky et al. (1972), with permission from National Research Council Canada.

molecules for at least 50 years (Chaudonneret, 1956). Support for this hypothesis was presented shortly thereafter (Schneiderman and Gilbert, 1958). A number of different effects on crustacean development and reproduction have been attributed to MF (for review, see Homola and Chang, 1997). The function of MF in crustaceans is still ripe for investigation. It is probable that MF has roles similar to those of juvenile hormones in insects, but it is also possible that it has novel functions. In many species of insects, juvenile hormones play several distinct roles in female reproduction (Raikhel et al., 2005, for review). Primarily, they stimulate the synthesis of the yolk protein precursor vitellogenin in the female. In some species of insects, juvenile hormones maintain the functional state of the male accessory sex organs. One effect of MF is on the duration of the crustacean molt interval (Yudin et al., 1980; Borst et al., 1987; Tamone and Chang, 1993). Methyl farnesoate may also be involved in mediating various stress responses (Lovett et al., 1997, 2001). Several lines of evidence indicate that it may have a permissive or stimulatory effect upon reproduction in both female and male crustaceans (see other chapters in this volume for discussions on the role of MF in female reproduction).

Methyl Farnesoate as a Gonadotropin

A number of studies implicate MF as a gonadotropin—a stimulator of gonadal development. Spider crabs (*Libinia emarginata*) can be divided into various morphotypes on the basis of overall size, claw size, and shell condition (abraded or unabraded). Homola et al. (1991) observed that the group of males with the largest reproductive index comprised the large, unabraded males with large claws. These male crabs had the highest rates of MF production when their mandibular organs were cultured *in vitro*. They also had the greatest reproductive activity (Sagi et al., 1994).

Injection of MF into crabs (*Oziotelphusa senex senex*) resulted in significant increases in the size of individual testicular follicles and the overall testicular index (Kalavathy et al., 1999; Reddy et al., 2004).

Methyl Farnesoate and the Regulation of Morphotypic Development

As mentioned in the previous section, differences in MF status are correlated with reproduction in male *L. emarginata* (Homola et al., 1991). Adult males of this species have alternative morphotypes, or distinct external features corresponding to differences in reproductive development and behavior (Laufer et al., 1994). In *L. emarginata*, male morphotypes can be distinguished by the relative size of the claws and the condition of the epicuticle. Males with large claws and worn epicuticles display mating behavior and have high MF titers (39-67 ng/ml) in the hemolymph. Males with small claws and intact epicuticles do not display mating behavior. They have hemolymph with significantly lower MF titers (3-30 ng/ml) (Sagi et al., 1993, 1994). This crab species offers the opportunity to predict differences in MF titer using external morphological markers. Male morphotypes of *L. emarginata* could be an important model for testing the reproductive function of MF when a technique for manipulating MF titer becomes available. Sex-specific differences in MF titer occur in several species of crustaceans. Male *L. emarginata* and *H. americanus* have higher levels of MF in the hemolymph than females (Borst et al., 1987; Laufer et al., 1987). In *H. americanus*, the relative size of the mandibular organ increases in males after sexual maturity, but not in females, suggesting that the mandibular organ may function differently in males and females (Waddy et al., 1995). Methyl farnesoate titers in females may be lower than in males because the mandibular organs are smaller or less active, or because MF is sequestered by the gonads.

Development of different morphotypes depends upon the differential growth of various body parts (allometric growth). Laufer et al. (2002) were able to modify ecdysteroid levels via eyestalk ablation. This operation removes the X-organ/sinus gland complex resulting in the disinhibition of ecdysteroid secretion by the molting gland. These authors were also able to modify MF levels via injections. They concluded that ecdysteroids in the presence of low MF concentrations promoted allometric claw growth, while ecdysteroids with relatively high concentrations of MF inhibited it (Laufer et al., 1997).

Methyl Farnesoate and Sexual Differentiation

Methyl farnesoate also acts as a sex determinant. In *Daphnia magna,* application of physiological amounts of exogenous MF (400 nM) to egg-maturing females resulted in all-male broods. There was a correlation with lower MF concentrations and increasing proportions of males in the broods (Olmstead and LeBlanc, 2002). Similar effects were seen in four other species of cladocerans (Kim et al., 2006). The production of male offspring following MF induction will be a powerful tool for the elucidation of the hormone's action at the molecular level (Rider et al., 2005).

PHEROMONES

For decapod crustaceans, pheromones may play an important role in both soft-shelled mating crab families (Eales, 1974) and hard-shelled mating crab families (Warner, 1977; Hardege et al., 2002). It has been suspected that during the days leading up to the actual molt, the female of some species releases a pheromone into the environment to attract a male (Ryan, 1966; Snow and Neilsen, 1966; Dunham, 1978; Ekerholm and Hallberg, 2005). Males in the vicinity of the premolt female begin a search behavior that is distinct from the behavior displayed upon the detection of either food or other males (Atema and Engstrom, 1971; Eales, 1974; Kamio et al., 2002). Upon location of the female, the male may initiate a courtship display. It may then attempt to grasp the female and carry her upside-down with their ventral surfaces in contact until she molts. The male protects the female from predators and other males. After the female molts, the male copulates and then maintains a post-mating embrace until her exoskeleton has partially hardened.

The source of these attractant pheromones is likely urine. There have been a number of studies focused on isolating and identifying the urinary pheromone, but with limited success (Dunham, 1978). Hormones such as the ecdysteroids have been proposed to be pheromones (Kittredge et al., 1971; Seifert, 1982; Gleeson et al., 1984). Those results have been inconclusive. It may be that a suite of compounds acts as a multi-component signal rather than a solitary chemical messenger (Warner, 1977; Wyatt, 2003). The attractant pheromone of the female crab *Erimacrus isenbeckii* has been identified as a novel ceramide (Asai et al., 2000).

Other research indicates that alternative sources of pheromones may exist. This was demonstrated by showing that post-molt female lobsters were still attractive to males, even though the female's urine was collected by catheters and not released into the environment (Snyder et al., 1993). In *C. maenas*, the pheromone may be contact mediated and not released in the urine (Ekerholm and Hallberg, 2005).

Males also likely produce pheromones. In *H. americanus* and crayfish (*Astacus leptodactylus*), release of urine is timed with agonistic interactions between males (Breithaupt and Atema, 2000; Breithaupt and Eger, 2002) and, along with visual cues, likely provides information about the status of the males. Establishment of dominance results in more successful mating by the winner of the contest.

CONCLUSIONS

This is an exciting time for research on crustacean endocrinology. There is increased interest in the topic due to aquaculture applications and the focus on crustaceans as keystone species in aquatic environments. Some key areas of future research on the hormonal control of male reproduction include the following:

1. The identification of the AGH in decapod crustaceans will enable a better understanding of the regulation of sexual differentiation and masculine reproductive physiology. In case the AG active factor is found to be a member of the insulin-like super family, it will support the notion that insulin may have evolved in the context of regulating sexual differentiation.
2. Except for the ecdysteroid receptor, research on crustacean hormone receptors has been minimal. The characterization of MF and AGH receptors will yield insight into the molecular action of these important regulatory molecules.
3. New functions are being attributed to members of the crustacean hyperglycemic hormone (CHH) peptide family. Mandibular organ-inhibiting hormone is a member of this family (see Chang

and Kaufman, 2005 for review). Structure-function studies of the peptides comprising this family will yield information about key amino acid positions and their resulting biological activities.

4. Further research on the biological activities and mode of action of MF will be of comparative interest in relation to juvenile hormone studies in insects. Both the developmental and reproductive aspects of MF need to be addressed.
5. There are indications of other chemical mediators involved in male reproduction. These factors include neurotransmitters and peptides such as gonad-inhibiting hormone (also known as vitellogenesis-inhibiting hormone). Most of the work on these factors has been focused on the female. More research needs to be conducted to determine whether they have a role in male reproduction.
6. There is growing interest in the field of invertebrate endocrine disrupters (deFur et al., 1999). Because of the biological importance of the Crustacea, there will probably be much more research devoted to the determination of the effects of exogenous chemicals upon various physiological processes, including male reproduction. Crustaceans will likely be useful indicator species for environmental contamination.
7. The isolation and characterization of crustacean pheromones will be important for basic research, development of novel methods of harvesting commercially important fisheries species, and the control of nuisance and alien crustaceans.

Acknowledgments

We thank Sharon A. Chang for editorial assistance. E.S.C. is grateful to Prof. Howard A. Bern for introducing him to the crab androgenic gland over three decades ago.

References

Adiyodi, R.G. 1985. Reproduction and its control. *In*: The Biology of Crustacea, Vol. 9. D.E. Bliss and L.H. Mantel (eds.). Academic Press, Orlando, USA, pp. 147-215.

Aflalo, E.D., T.T.T. Hoang, V.H. Nguyen, Q. Lam, D.M. Nguyen, Q.S. Trinh, S. Raviv and A. Sagi. 2006. A novel two-step procedure for mass production of all-male populations of the giant freshwater prawn *Macrobrachium rosenbergii*. Aquaculture, 256: 468-478.

Asai, N., N. Fusetani, S. Matsunaga and J. Sasaki. 2000. Sex pheromones of the hair crab *Erimacrus isenbeckii*. Part 1: Isolation and structures of novel ceramides. Tetrahedron, 56: 9895-9899.

Atema, J., and D. Engstrom. 1971. Sex pheromone in the lobster, *Homarus americanus*. Nature, 232: 261-263.

Baghel, D.S., W.S. Lakra and G.P.S. Rao. 2004. Altered sex ratio in giant fresh water prawn, *Macrobrachium rosenbergii* (de Man) using hormone bioencapsulated live *Artemia* feed. Aquacult. Res., 35: 943-947.

Baldwin, W.S., and G.A. LeBlanc. 1994. *In vivo* biotransformation of testosterone by phase I and II detoxication enzymes and their modulation by 20-hydroxyecdysone in *Daphnia magna*. Aquat. Toxicol., 29: 103-117.

Barki, A., I. Karplus, I. Khalaila, R. Manor and A. Sagi. 2003. Male-like behavioral patterns and physiological alterations induced by androgenic gland implantation in female crayfish. J. Exp. Biol., 206: 1791-1797.

Berreur-Bonnenfant, J., and F. Lawrence. 1984. Comparative effect of farnesylacetone on macromolecular synthesis in gonads of crustaceans. Gen. Comp. Endocrinol., 54: 462-468.

Borst, D.W., H. Laufer, M. Landau, E.S. Chang, W.A. Hertz, F.C. Baker and D.A. Schooley. 1987. Methyl farnesoate (MF) and its role in crustacean reproduction and development. Insect Biochem., 17: 1123-1127.

Breithaupt, T., and J. Atema. 2000. The timing of chemical signaling with urine in dominance fights of male lobsters (*Homarus americanus*). Behav. Ecol. Sociobiol., 49: 67-78.

Breithaupt, T., and P. Eger. 2002. Urine makes the difference: chemical communication in fighting crayfish made visible. J. Exp. Biol., 205: 1221-1231.

Brody, M.D., and E.S. Chang. 1989. Development and utilization of crustacean long-term primary cell cultures: Ecdysteroid effects *in vitro*. Invert. Reprod. Dev., 16: 141-147.

Burns, B.G., G.B. Sangalang, H.C. Freeman and M. McMenemy. 1984. Isolation and identification of testosterone from the serum and testes of the American lobster (*Homarus americanus*). Gen. Comp. Endocrinol., 54: 429-432.

Chang, E.S., and W.R. Kaufman. 2005. Endocrinology of Crustacea and Chelicerata. *In*: Comprehensive Molecular Insect Science, Vol. 3. L.I. Gilbert, K. Iatrou and S.S. Gill (eds.). Elsevier B.V., Oxford, pp. 805-842.

Chang, E.S., B.A. Sage and J.D. O'Connor. 1976. The qualitative and quantitative determinations of ecdysones in tissues of the crab, *Pachygrapsus crassipes*, following molt induction. Gen. Comp. Endocrinol., 30: 21-33.

Charniaux-Cotton, H. 1954a. Découverte chez un Crustacé Amphipode (*Orchestia gammarella*) glande endocrine responsable de la différenciation de caractères sexuels primaires et secondaires mâles. C.R. Acad. Sci. Paris, 239: 780-782.

Charniaux-Cotton, H. 1954b. Implantation de gonads de sexe oppose à des males et des femelles chez un Crustacé Amphipode (*Orchestia gammarella*). C.R. Acad. Sci. Paris, 238: 953-955.

Charniaux-Cotton, H. 1955. Le déterminisme hormonal des caractères sexuels d'*Orchestia gammarella* (Crustacé Amphipode). C.R. Acad. Sci. Paris, 240: 1487-1489.

Charniaux-Cotton, H. 1957. Croissance, régénération et déterminisme endocrinien des caractères sexuels d'*Orchestia gammaella* (Pallas) (Crustacé Amphipode). Ann. Sci. Nat., 19: 411-559.

Charniaux-Cotton, H. 1964. Endocrinologie et génétique du sexe chez les Crustacés Supérieurs. Ann. Endocrinol., 25: 36-42.

Charniaux-Cotton, H. 1985. Sexual differentiation. *In*: The Biology of Crustacea, Vol. 9. D.E. Bliss and L.H. Mantel (eds.). Academic Press, Orlando, USA, pp. 217-299.

Charniaux-Cotton, H., and G. Payen. 1988. Crustacean reproduction. *In*: Endocrinology of Selected Invertebrate Types. H. Laufer and R.G.H. Downer (eds.). Alan R. Liss, New York, pp. 279-303.

Chaudonneret, J. 1956. Le systeme nerveux de la région gnathale de l'écrevisse *Cambarus affinis* (Say). Ann. Sci. Nat. Zool. Biol. Anim., 18: 33-61.

Cronin, L.E. 1947. Anatomy and histology of the male reproductive system of *Callinectes sapidus* Rathbun. J. Morphol., 81: 209-239.

deFur, P.L., M. Crane, C. Ingersoll and L. Tattersfield. 1999. Endocrine Disruption in Invertebrates: Endocrinology, Testing, and Assessment. SETAC Press, Pensacola. USA.

Dunham, P. 1978. Sex pheromones in Crustacea. Biol. Rev., 53: 555-583.

Eales, A. 1974. Sex pheromone in the shore crab *Carcinus maenas*, and the site of its release from females. Mar. Behav. Physiol., 2: 345-355.

Ekerholm, M., and E. Hallberg. 2005. Primer and short-range releaser pheromone properties of premolt female urine from the shore crab *Carcinus maenas*. J. Chem. Ecol., 31: 1845-1863.

Ferezou, J.P., J. Berreur-Bonnenfant, A. Tekitek, M. Rojas, M. Barbier, S. Wipf, H.K. Wipf and J.J. Meusy. 1977. Biologically-active lipids from the androgenic gland of the crab. *In*: Marine Natural Products Chemistry. D.J. Faulkner and W.H. Fenical (eds.). Plenum Press, New York, pp. 361-366.

Fowler, J.F., and B.V. Leonard. 1999. The structure and function of the androgenic gland in *Cherax destructor* (Decapoda: Parastacidae). Aquaculture, 171: 135-148.

Gilgan, M., and D. Idler. 1967. The conversion of androstenedione to testosterone by some lobster (*Homarus americanus* Milne Edwards) tissues. Gen. Comp. Endocrinol., 9: 319-324.

Ginsburger-Vogel, T., and H. Charniaux-Cotton. 1982. Sex determination. *In*: The Biology of Crustacea, Vol. 2. L.G. Abele (ed.). Academic Press, New York, pp. 257-281.

Gleeson, R.A., M.A. Adams and A.B. Smith. 1984. Characterization of a sex pheromone in the blue crab, *Callinectes sapidus*: Crustecdysone studies. J. Chem. Ecol., 10: 913-921.

Goodman, W.G., and N.A. Granger. 2005. The juvenile hormones. *In*: Comprehensive Molecular Insect Science, Vol. 3. L.I. Gilbert, K. Iatrou and S.S. Gill (eds.). Elsevier B.V., Oxford, pp. 319-408.

Hardege, J.D., A. Jennings, D. Hayden, C.T. Muller, D. Pascoe, M.G. Bentley and A.S. Clare. 2002. Novel behavioural assay and partial purification of a female-derived sex pheromone in *Carcinus maenas*. Mar. Ecol. Progr. Ser., 244: 179-189.

Hasegawa, Y.K., K. Haino-Fukushima and Y. Katakura. 1987. Isolation and properties of androgenic gland hormone from the terrestrial isopod, *Armadillidium vulgare*. Gen. Comp. Endocrinol., 67: 101-110.

Hasegawa, Y.K., K. Haino-Fukushima and Y. Katakura. 1991. An immunoassay for the androgenic gland hormone of the terrestrial isopod, *Armadillidium vulgare*. Invert. Reprod. Dev., 20: 59-66.

Homola, E., and E.S. Chang. 1997. Methyl farnesoate: Crustacean juvenile hormone in search of functions. Comp. Biochem. Physiol., B 117: 347-356.

Homola, E., A. Sagi and H. Laufer. 1991. Relationship of claw form and exoskeleton condition to reproductive system size and methyl farnesoate in the male spider crab, *Libinia emarginata*. Invert. Reprod. Dev. 20: 219-225.

James, M.O., and K.T. Shiverick. 1984. Cytochrome P-450-dependent oxidation of progesterone, testosterone, and ecdysone in the spiny lobster, *Panulirus argus*. Arch. Biochem. Biophys., 233: 1-9.

Kalavathy, Y.P., P. Mamatha and P.S. Reddy. 1999. Methyl farnesoate stimulates testicular growth in the freshwater crab *Oziotelphusa senex senex* Fabricius. Naturwissenschaften, 86: 394-395.

Kamio, M., S. Matsunaga and N. Fusetani. 2002. Copulation pheromone in the crab *Telmessus cheiragonus* (Brachyura: Decapoda). Mar. Ecol. Progr. Ser., 234: 183-190.

Karplus, I., A. Sagi, I. Khalaila and A. Barki. 2003. The influence of androgenic gland implantation on the agonistic behavior of female crayfish (*Cherax quadricarinatus*) in interactions with males. Behaviour, 140: 649-663.

Khalaila, I., S. Weil and A. Sagi. 1999. Endocrine balance between male and female components of the reproductive system in intersex *Cherax quadricarinatus* (Decapoda: Parastacidae). J. Exp. Zool., 283: 286-294.

Khalaila, I., T. Katz, U. Abdu, G. Yehezkel and A. Sagi. 2001. Effects of implantation of hypertrophied androgenic glands on sexual characters and physiology of the reproductive system in the female red claw crayfish *Cherax quadricarinatus*. Gen. Comp. Endocrinol., 121: 242-249.

Khalaila, I., R. Manor, S. Weil, Y. Granot, R. Keller and A. Sagi. 2002. The eyestalk-androgenic gland-testis endocrine axis in the crayfish *Cherax quadricarinatus*. Gen. Comp. Endocrinol., 127: 147-156.

Kim, K., A.A. Kotov and D.J. Taylor. 2006. Hormonal induction of undescribed males resolves cryptic species of cladocerans. Proc. R. Soc., B 273: 141-147.

Kittredge, J., M. Terry and F. Takahashi. 1971. Sex pheromone activity of the molting hormone, crustecdysone, on male crabs. Fish. Bull., 69: 337-343.

Laufer, H., D. Borst, F.C. Baker, C. Carrasco, M. Sinkus, C.C. Reuter, L.W. Tsai and D.A. Schooley. 1987. Identification of a juvenile hormone-like compound in a crustacean. Science, 235: 202-205.

Laufer, H., A. Sagi and J.S.B. Ahl. 1994. Alternate mating strategies of polymorphic males of *Libinia emarginata* appear to depend on methyl farnesoate. Invert. Reprod. Dev., 26: 41-44.

Laufer, H., P. Takac, J.S.B. Ahl and M.R. Laufer. 1997. Methyl farnesoate and the effect of eyestalk ablation on the morphogenesis of the juvenile female spider crab *Libinia emarginata*. Invert. Reprod. Dev., 31: 63-68.

Laufer, H., J. Ahl, G. Rotllant and B. Baclaski. 2002. Evidence that ecdysteroids and methyl farnesoate control allometric growth and differentiation in a crustacean. Insect Biochem. Molec. Biol., 32: 205-210.

LeBlanc, G.A., and J.B. McLachlan. 2000. Changes in the metabolic elimination profile of testosterone following exposure of the crustacean *Daphnia magna* to tributyltin. Ecotoxicol. Environ. Saf., 45: 296-303.

Legrand, J.J. 1955. Rôle endocrinien de l'ovaire dans la differenciation des oostégites chez les Crustacés Isopodes terrestres. C.R. Acad. Sci. Paris, 241: 1083-1087.

Lovett, D.L., P.D. Clifford and D.W. Borst. 1997. Physiological stress elevates hemolymph levels of methyl farnesoate in the green crab *Carcinus maenas*. Biol. Bull., 193: 266-267.

Lovett, D.L., M.P. Verzi, P.D. Clifford and D.W. Borst. 2001. Hemolymph levels of methyl farnesoate increase in response to osmotic stress in the green crab, *Carcinus maenas*. Comp Biochem. Physiol., A 128: 299-306.

Maas, U., and A. Dorn. 2005. No evidence of androgenic hormone from the testes of the glowworm, *Lampyris noctiluca*. Gen. Comp. Endocrinol., 143: 40-50.

Malecha, S.R., P.A. Nevin, P. Ha, L.E. Barck, Y. Lamadrid-Rose, S. Masuno and D. Hedgecock. 1992. Sex-ratios and sex-determination in progeny from crosses of surgically sex-reversed freshwater prawns, *Macrobrachium rosenbergii*. Aquaculture, 105: 201-218.

Manor, R., D.E. Aflalo, C. Segall, S. Weill, D. Azulay and A. Sagi. 2004. Androgenic gland implantation promotes growth and inhibits vitellogenesis in *Cherax quadricarinatus* females grown in individual compartments. Invert. Reprod. Dev., 45: 151-159.

Manor, R., S. Weil, S. Oren, L. Glazer, E.D. Aflalo, T. Ventura, V. Chalifa-Caspi, M. Lapidot and A. Sagi. 2007. Insulin and gender: an insulin-like gene expressed exclusively in the androgenic gland of the male crayfish. Gen. Comp. Endocrinol., 150: 326-336.

Martin, G., P. Juchault, O. Sorokine and A. Van Dorsselaer. 1990. Purification and characterization of androgenic hormone from the terrestrial isopod *Armadillidium vulgare* Latr. (Crustacea, Oniscidea). Gen. Comp. Endocrinol., 80: 349-354.

Martin, G., O. Sorokine, M. Moniatte, P. Bulet, C. Hetru and A. Van Dorsselaer. 1999. The structure of a glycosylated protein hormone responsible for sex determination in the isopod, *Armadillidium vulgare*. Eur. J. Biochem., 262: 727-736.

Matlock, D.B., and E.J. Dornfeld. 1982. The effect of crustecdysone on DNA synthesis in polyploid somatic cells of an isopod. Comp. Biochem. Physiol., B 73: 603-605.

Nagabhushanam, R., and G.K. Kulkarni. 1981. Effect of exogenous testosterone on the androgenic gland and testis of a marine penaeid prawn, *Parapenaeopsis hardwickii* (Miers) (Crustacea, Decapoda, Penaeidae). Aquaculture, 23: 19-27.

Nagamine, C., and A.W. Knight. 1987. Induction of female breeding characteristics by ovarian tissue implants in androgenic gland ablated male freshwater prawns *Macrobrachium rosenbergii* (de Man) (Decapoda, Palaemonidae). Intl. J. Invert. Reprod. Dev., 11: 225-234.

Nagamine, C., A.W. Knight, A. Maggenti and G. Paxman. 1980a. Effects of androgenic gland ablation on male primary and secondary sexual characteristics in the Malaysian prawn, *Macrobrachium rosenbergii* (de Man) (Decapoda, Palaemonidae), with first evidence of induced feminization in a nonhermaphroditic decapod. Gen. Comp. Endocrinol., 41: 423-441.

Nagamine, C., A.W. Knight, A. Maggenti and G. Paxman. 1980b. Masculinization of female *Macrobrachium rosenbergii* (de Man) (Decapoda, Palaemonidae) by androgenic gland implantation. Gen. Comp. Endocrinol., 41: 442-457.

Nair, M.C., K.R. Salin, M.S. Raju and M. Sebastian. 2006. Economic analysis of monosex culture of giant freshwater prawn (*Macrobrachium rosenbergii* DeMan): a case study. Aquacult. Res., 37: 949-954.

Ohs, C.L., L.R. D'Abramo and A.M. Kelly. 2006. Effect of dietary administration of 17α-methyltestosterone on the sex ratio of postlarval freshwater prawn, *Macrobrachium rosenbergii*, during the nursery stage of culture. J. World Aquacult. Soc., 37: 328-333.

Okuno, A., Y. Hasegawa, T. Ohira, Y. Katakura and H. Nagasawa. 1999. Characterization and cDNA cloning of androgenic gland hormone of the terrestrial isopod *Armadillidium vulgare*. Biochem. Biophys. Res. Commun., 264: 419-423.

Okuno, A., Y. Hasegawa, M. Nishiyama, T. Ohira, R. Ko, M. Kurihara, S. Matsumoto and H. Nagasawa. 2002. Preparation of an active recombinant peptide of crustacean androgenic gland hormone. Peptides, 23: 567-572.

Olmstead, A.W., and G.A. LeBlanc. 2002. Juvenoid hormone methyl farnesoate is a sex determinant in the crustacean *Daphnia magna*. J. Exp. Zool., 293: 736-739.

Parnes, S., S. Raviv, A. Shechter and A. Sagi. 2006. Cyclic disposal of old spermatophores, timed by the molt cycle, in a marine shrimp. J. Exp. Biol., 209: 4974-4983.

Raikhel, A.S., M.R. Brown and X. Belles. 2005. Hormonal control of reproductive processes. *In*: Comprehensive Molecular Insect Science, Vol. 3. L.I. Gilbert, K. Iatrou and S.S. Gill. (eds.). Elsevier B.V., Oxford, pp. 433-491.

Reddy, P.R., G.P.C. Nagaraju and P.S. Reddy. 2004. Involvement of methyl farnesoate in the regulation of molting and reproduction in the freshwater crab *Oziotelphusa senex senex*. J. Crustacean Biol., 24: 511-515.

Rider, C.V., T.A. Gorr, A.W. Olmstead, B.A. Wasilak and G.A. LeBlanc. 2005. Stress signaling: Coregulation of hemoglobin and male sex determination through a terpenoid signaling pathway in a crustacean. J. Exp. Biol., 208: 15-23.

Ryan, E. 1966. Pheromone: Evidence in a decapod crustacean. Science, 151: 340-341.

Sagi, A. 1988. The androgenic gland in Crustacea—with emphasis on the cultured freshwater prawn *Macrobrachium rosenbergii*—a review. Israeli J. Aquacult., 40: 107-112.

Sagi, A., and E.D. Aflalo. 2005. The androgenic gland and monosex culture of freshwater prawn *Macrobrachium rosenbergii* (De Man): A biotechnological perspective. Aquacult. Res., 36: 231-237.

Sagi, A., and D. Cohen. 1990. Growth, maturation and progeny of sex-reversed *Macrobrachium rosenbergii* males. World Aquacult., 21: 87-90.

Sagi, A., and I. Khalaila. 2001. The crustacean androgen: A hormone in an isopod and androgenic activity in decapods. Amer. Zool., 41: 477-484.

Sagi, A., D. Cohen and Y. Milner. 1990. Effect of androgenic gland ablation on morphotypic differentiation and sexual characteristics of male freshwater prawns, *Macrobrachium rosenbergii*. Gen. Comp. Endocrinol., 77: 15-22.

Sagi, A., L. Karp, Y. Milner, D. Cohen, A.M. Kuris and E.S. Chang. 1991. Testicular thymidine incorporation in the prawn *Macrobrachium rosenbergii*: Molt cycle variation and ecdysteroid effects in vitro. J. Exp. Zool., 259: 229-237.

Sagi, A., E. Homola and H. Laufer. 1993. Distinct reproductive types of male spider crabs *Libinia emarginata* differ in circulating and synthesizing methyl farnesoate. Biol. Bull., 185: 168-173.

Sagi, A., J.S.B. Ahl, H. Danaee and H. Laufer. 1994. Methyl farnesoate levels in male spider crabs exhibiting active reproductive behavior. Horm. Behav., 28: 261-272.

Sagi, A., E. Snir and I. Khalaila. 1997. Sexual differentiation in decapod crustaceans: Role of the androgenic gland. Invert. Reprod. Dev., 31: 55-61.

Sagi, A., R. Manor, C. Segall, C. Davis and I. Khalaila. 2002. On intersexuality in the crayfish *Cherax quadricarinatus*: An inducible sexual plasticity model. Invert. Reprod. Dev., 41: 27-33.

Schneiderman, H., and L. Gilbert. 1958. Substances with juvenile hormone activity in Crustacea and other invertebrates. Biol. Bull., 115: 530-535.

Seifert, P. 1982. Studies on the sex pheromone of the shore crab, *Carcinus maenas*, with special regard to ecdysone excretion. Ophelia, 21: 147-158.

Snow, C.D., and J.R. Neilsen. 1966. Premating and mating behavior of the Dungeness crab (*Cancer magister* Dana). J. Fish. Res. Board Can., 23: 1319-1323.

Snyder, M.J., C. Ameyaw-Akumfi and E.S. Chang. 1993. Sex recognition and the role of urinary cues in the lobster, *Homarus americanus*. Mar. Behav. Physiol., 24: 101-116.

Sochasky, J.D., D.E. Aiken and N.H.F. Watson. 1972. Y organ, molting gland, and mandibular organ: A problem in decapod Crustacea. Can. J. Zool., 50: 993-997.

Styrishave, B., K. Rewitz, T. Lund and O. Andersen. 2004. Variations in ecdysteroid levels and cytochrome P_{450} expression during moult and reproduction in male shore crabs *Carcinus maenas*. Mar. Ecol. Progr. Ser., 274: 215-224.

Subramoniam, T. 2000. Crustacean ecdysteroids in reproduction and embryogenesis. Comp. Biochem. Physiol., C 125: 135-156.

Sun, P.S., T.M. Weatherby, M.F. Dunlap, K.L. Arakaki, D.T. Zacarias and S.R. Malecha. 2000. Developmental changes in structure and polypeptide profile of the androgenic gland of the freshwater prawn *Macrobrachium rosenbergii*. Aquacult. Intl., 8: 327-334.

Taketomi, Y., and S. Nishikawa. 1996. Implantation of androgenic glands into immature female crayfish, *Procambarus clarkii*, with masculinization of sexual characteristics. J. Crustacean Biol., 16: 232-239.

Tamone, S.L., and E.S. Chang. 1993. Methyl farnesoate stimulates ecdysteroid secretion from crab Y-organs *in vitro*. Gen. Comp. Endocrinol., 89: 425-432.

Teshima, S., and A. Kanazawa. 1971. In vitro bioconversion of progesterone to 17α-hydroxyprogesterone and testosterone by the sliced testes of crab, *Portunus trituberculatus*. Bull. Jpn. Soc. Sci. Fish., 37: 524-528.

Veith, W.J., and S.R. Malecha. 1983. Histochemical study of the distribution of lipids 3-alpha and 3-beta-hydroxysteroid dehydrogenase in the androgenic gland of the cultured prawn *Macrobrachium rosenbergii* (de Man) (Crustacea; Decapoda). S. African J. Sci., 79: 84-85.

Verslycke, T., K. De Wasch, H.F. De Brabancer and C.R. Janssen. 2002. Testosterone metabolism in the estuarine mysid *Neomysis integer* (Crustacea; Mysidacea): Identification of testosterone metabolites and endogenous vertebrate-type steroids. Gen. Comp. Endocrinol., 126: 190-199.

Waddy, S.L., D.E. Aiken and D.P.V. De Kleijn. 1995. Control of reproduction. *In*: Biology of the Lobster *Homarus americanus*. J.R. Factor (ed.). Academic Press, San Diego, pp. 217-266.

Warner, G.F. 1977. The Biology of Crabs. Paul Elek (Scientific Books), London.

Wyatt, T.D. 2003. Pheromones and Animal Behaviour: Communication by Smell and Taste. Cambridge University Press, Cambridge, UK.

Yudin, A.I., R.A. Diener, W.H. Clark and E.S. Chang. 1980. Mandibular gland of the blue crab, *Callinectes sapidus*. Biol. Bull., 159: 760-772.

8

Nutrition in Relation to Reproduction in Crustaceans

G. Cuzon[1], G. Gaxiola[2] and C. Rosas[2]

[1]Ifremer/COP, BP 7004 Taravao, Tahiti, French Polynesia. E-mail: gcuzon@ifremer.fr

[2]Unidad multidisciplinaria de docencia e investigation, Faculdad de ciencias UNAM, Porto de Abrigo s/n, Sisal, Yucatan

E-mails: mggc@hp.fciencias.unam.mx; cru@hp.fciencias.unam.mx

INTRODUCTION

Interest in studies on reproduction of penaeid Crustacea grew in the 1970s with the advent of mariculture. In the wild, this biological process is successful when a variety of foods (e.g., invertebrates, debris, decomposing organisms) are available in the benthos. Feeding habit is usually attributed to scavengers eating meiofauna and organic debris. In culture it is more appropriate to refer to feeding rather than nutrition given the significant difficulty encountered in replacing live or fresh foods with compounded feeds.

The issue of breeder nutrition *sensu stricto* became relevant in the early 1970s, when the first attempts were made to raise crustacean species in captivity for commercial purposes. For many years researchers preferred the use of live or fresh food as a means of guaranteeing good quality offspring. Indeed, some researchers continue to prefer the use of such foodstuffs.

Why was so little progress made in developing artificial feeds for crustacean reproduction?

There are a number of possible explanations. Breeding stock was expensive. For example, a female *Penaeus monodon* in Asia cost more than US$60. Farmers were often reluctant to provide animals for experimental purposes. Some researchers reported difficulties in obtaining animals for experimental use. Others considered the duration of breeding trials to be a limiting factor. Finally, the demand for a functional feed remained low for many years owing to the relative ease with which live or fresh food could be obtained. Today feeding schedules vary, but the quantity and quality of feeding is a prerequisite for good gonadal maturation.

The importance of quality food had already been identified (Aquacop, 1977). Many authors had observed maturation in captivity of various penaeid species that were fed *ad libitum* with marine foods such as mussels (*Mytilus* sp.), clams, oysters, squid, shrimp, crabs, worms, gastropod (*Troca*), at times used together with dry feed (Moore et al., 1974; Primavera et al., 1979; Middleditch et al., 1980; Beard and Wickins, 1980; Brown et al., 1979, 1980; Lawrence et al., 1980; Chamberlain and Lawrence, 1981). In Centre Océanologique du Pacifique (COP)/Tahiti, fresh foods and locally prepared dry feeds with 40% Crude Protein (CP) or commercial feeds with 68% CP allowed maturation. Many trials and errors were reported by Ricardez-Carrion (1985). For example, a 47% CP dry feed gave some signs of maturation followed by regression of the ovary (Aquacop, 1975). Similar results with a 60% CP diet were reported by Shigueno (1975). Aquacop (1977) found maturation signs with a diet high in protein levels combined with other components. Lipids, for example, clearly play a major role in successful maturation. However, lipids or substances such as vitamins, or native proteins, may be altered, for example, in case of a double extrusion of a Japanese feed (Aquacop, 1989). Maturation can be observed at a successful rate when fresh troca (*Trochus niloticus*) is fed to female animals. The feeding of troca has a clear positive effect on the last stages of gonadal maturation, egg quality and larva viability. It is presumed that the ingestion of gastropod gonad produced such a beneficial effect on shrimp females. This observation is confirmed by results found with several penaeid species fed on molluscs and polychaetes, such organisms producing ripe gonads that contain a favourable lipid composition (Middleditch et al., 1980). Moreover, Aquacop (1983) evidenced the importance of fresh food on the further quality of breeders, not only at maturation stage but also during the grow-out phase.

In 1972, a Japanese diet, Nippaï®, which served as a partial substitute for fresh food, became available. A "local" maturation feed used to make a reference diet was later formulated (Patrois et al., 1990, 1991).

The development of this approach taking into account a reference diet that should be upgraded as far as possible to obtain similar results as with a Japanese diet was aided with partial research on the nutritional requirements of penaeids. However, this type of study required an extensive investment in facilities and time, which explains the paucity of scientific data at the time.

Lipid is one of the major nutrients involved in the biochemical processes during penaeid shrimp maturation. Comprehensive data for the further improvement of ovary maturation is available (Teshima and Kanazawa, 1978). Data relating to the nutrition of male animals is also available (Rosas et al., 1993), although difficulties with male nutrition remained a minor concern compared to the nutrition of female animals.

In spite of the difficulties encountered in the early 1970s, these afforded an opportunity to rediscover what was known as the "Panouse effect".

Today, some information is available to enable aquaculturists to control reproduction in captivity with success. Several concepts were used: (1) Reproductive process in the scope of the culture of aquatic crustaceans without a catch of wild breeders. The broodstock in captivity is a necessity in areas where no indigenous species of commercial value are available. Tahiti is a case in point (Aquacop, 1975). Moreover, the scarcity of suitable lands for extensive culture previously recommended for such purposes gave rise to a high density culture system. (2) Criteria for candidate species to raise in absence of all advantages linked to one single species. For reproduction the choice may have hidden specificities from a physiological or nutritional point of view. It has been found that *P. indicus* is the easiest species to reproduce in the conditions prevalent in Tahiti, but unfortunately it does not grow well in ponds. This is in contrast to its performance in other places such as Madagascar or Saudi Arabia. Given that this species was easy to reproduce in captivity it has served as an experimental animal for a long time. (3) "Background of animals" with regard to reproduction: "background" is a term used to characterize the conditions of stocking density, food quality, variation in non-biotic factors, and other factors. The period during which shrimps were kept in earthen ponds would possibly interfere with gonadal development and later larva viability.

The present chapter compiles and analyses most research work on broodstock nutrition in captivity for different species (not only tropical species) and also the accumulation of knowledge since the early 1970s. Most examples refer to penaeids, but there are some paragraphs on *Macrobrachium rosenbergii*, a palaemonidae. Advances in research relating to shrimp nutrition will benefit species such as lobster, spiny lobster and crabs in the future.

EMPIRICAL APPROACH

It is clear that reproduction of penaeids in captivity started with breeders caught at sea. For example, kuruma shrimp (*P. japonicus*) or *P. kerathurus* along the Rio Ebro in Spain were caught by trawling on fishing grounds and the best females then selected to spawn. These females received natural food (worms, polychaetes, molluscs, debris). Aquaculturists followed suit, feeding breeders in captivity with similar natural food (worms, *Artemia* biomass, squid, troca). In France, *P. japonicus* juveniles imported from Japan received fresh mussel (*Mytilus edulis*) to induce maturation signs and successful spawning. This was extremely laborious, intensive and costly. Natural food was also used in tropical areas, such as Taiwan, French Polynesia (Aquacop, 1975), and the Philippines (Primavera et al., 1979). In Japan, where marine sources are numerous, a pioneering group successfully contributed to the development of one of the first "maturation diets" intended to complement or completely replace live food (Shigueno, 1975; Teshima and Kanazawa, 1980). Live food is used as a means of maturing penaeid females in captivity. Large ponds for grow-out operations in big commercial complexes exist in Latin America (Ecuador, Colombia, Mexico, Venezuela, Brazil, Belize) as well as in South East Asia (Thailand, Malaysia, Indonesia), India and Madagascar. Commercial maturation feeds were largely used. In Tahiti, Aquacop maintained a stock of *L. stylirostris* (28th generation) completely on dry feed, except for pre-breeders, which still received a meal of fresh food (squid) once a week. This may be a way to wait until a full artificial diet is efficient enough to bring about gonadal development on purpose.

Formulated feeds usually contain more n-6 fatty acids than natural food and a different ratio of n-3/n-6. There were similar proportions of highly unsaturated fatty acid content of the omega-3 series in marine food and two commercial feeds (Table 1).

Table 1 Fatty acids in bloodworms and formulated feeds (Lytle et al., 1990). Nippai® is a Japanese commercial feed for shrimp breeders; Ziegler® is also a commercial shrimp feed.

	Bloodworms	*Nippai®*	*Ziegler®*
total FA%	0.72	3.8	6.4
PUFA%	0.19	1.5	1.9
unsat/sat	3.6	1.9	3.1
n-6 ppm	260	4,700	8,000
n-3 ppm	1,600	10,000	11,000
n-3/n-6	6.6	2.1	1.4
EPA+DHA/n-3	75	87	77

FA, fatty acids; PUFA, polyunsaturated fatty acids; EPA, eicosapentanoic acid; DHA, docosahexanoic acid.

EPA (20:5n-3)+DHA (22:6n-3)/n-3 expressed in percentage gives similar values but the main difference between natural food and feeds lies in n-3/n-6 mainly because of the low linoleic level in natural food (table 1).

It is also interesting to compare the potential loss of essential nutrients during storage between fresh or live bloodworms and frozen or stored ones. Total fatty acids decreased from 0.79 to 0.7 after storage for 18 days and 0.42 after 8 months. Polyunsaturated fatty acids (PUFAs) also decreased from 0.12 to 0.09 and 0.075 and n-3 fatty acids from 1100 to 800 and 320 ppm respectively, n-3/n-6 decreased from 8.7 to 7.1 and 5.8 after 18 days, one month and 8 months respectively.

Generally, live food is more dense in essential nutrients than frozen food. This explains why shrimps are still fed with a mixture of dry feed and fresh or briefly frozen food in order to maintain shrimp with the best reproductive performances in breeding condition or to maintain strains with specific characteristics in captivity (Marsden et al., 1997; Lytle et al., 1990; Xu et al., 1994).

DIETARY REQUIREMENTS

A comprehensive model of the dynamics of nutrients during oocyte formation in decapod crustaceans is a prerequisite, or at least an excellent, tool in discussions relating to dietary requirements. This model is based on lipid and protein intake. It explained the route of dietary lipids (triglycerides, TG and phospholipids, Pls), protein, carbohydrates (glucose mainly) and carotenoïds. It assumed good digestibility of nutrients such as lipids, protein and carbohydrates (82-90, 92-95 and 80% respectively). Ingredients such as marine animal protein (squid; mussel; low temperature or LT-fishmeal; or hydrolysed fish protein concentrate, HFPC), as well as native protein (wheat gluten), wheat starch, or marine oil would be selected in the formulation (Table 5). According to these assumptions, metabolic routes were defined. At hepatopancreas (HP) level, absorption was facilitated in M-cells and R-cells and a formation of micelles occurred via hemolymph; different lipid classes reached the oocytes. In oocytes yolk received vitellogenin by endocytosis and storage of triglycerides (Harrison, 1990). This process was controlled by hormones (vitellogenin-inhibiting hormone, VIH). Hormones were produced by the sinus gland neurosecretory system located in the eyestalks. Oocytes build-up could be reversed extremely rapidly in case of stress during manipulation of breeding animals. The phenomenon of re-mobilization of nutrients was so rapid via hemolymph as to lead certain aquaculturists to assume the existence of a "canal" between the ovary and the HP. From a biochemical perspective it became clear that requirements for reproduction would focus on *P. indicus* with the effect of Pls (phosphatidyl-choline and phosphatidyl-ethanolamin), HUFAs (eicosapentanoic acid, EPA, docosahexanoic acid, DHA), neutral lipids, cholesterol, amino acids (basic ones), carotenoids and vitamins, especially vitamins E and C. Actual knowledge on existing maturation shrimp feed composition is obviously not available because of confidentiality linked with commercial feeds. Since then, a certain number of studies carried out on different penaeids resulted in nutrient requirement values and a synthesis of results on a single one given for *P. stylirostris* (Cuzon et al., 1995).

AN EFFICIENT COMPOUNDED COMMERCIAL FEED

Since 1983, Nippaï®, a commercial feed, has been used for breeding shrimps on a large scale with good results, which led to a number of studies to enable the production of a similar compounded feed. Two questions arise: First, is it possible to replace natural diets with a compounded feed? Second, is it possible to replace Nippaï® with an open-formula compounded feed?

First Step

The first step was to take an analytic approach using proximate analysis of available commercial feed and examine in particular different classes of lipids (Table 2) from a commercial maturation feed Nippaï®-83 (Galois, 1983).

Table 2 Composition of lipid (% dry weight) for a commercial maturation feed Nippaï®-83.

moisture %FW	*7.78*	*fatty acid*	*%TL*
total lipid	10.14	16:0	23.46
polar lipid	7.28	18:1	23.9
neutral lipid	2.61	18:2n-6	12.33
sterols	6.59	20:4n-6	1.12
TG	18.37	18:3n-3	3.42
others	0.82	20:5n-3	6.92
free FA	0.25	22:6n-3	6.55

FW, fresh weight; TG, triglycerides; FA, fatty acids expressed in %TL (TL total lipids).

The proportion of polar lipids is higher than that of neutral lipids. Sterols and TG part of neutral lipids were well represented and essential fatty acids were present in fair amounts to allow maturation.

Nippaï® vs Nippaï® + Fresh Food

Nippaï® was fed with and without fresh food, such as *Troca* meat (18% daily ration) and mussel (6% daily ration), and the reproduction performances of the shrimps were compared. Females were raised for 40 days and TG levels in HP were measured. The levels in HP decreased respectively from 130 to 1.9 µg/mg dw for females receiving dry feed or dry feed + fresh food. Arachidonic acid (AA) and also EPA levels were higher in shrimps fed on a composite diet than in those fed on dry feed alone. After spawning twice (rank 2), HP reserves of shrimp females were completely used and subsequently the ovaries could not supply TG in sufficient quantities to allow the production of eggs. The difference in the TG content of HP in females fed Nippaï® could be explained whether with a low TG content in the dry feed or as a result of eyestalk ablation that would produce a higher transfer of lipids from HP to ovary compared to what was found with females fed on Nippaï® + fresh food. For example, in *P. notialis*, 4 days after eyestalk ablation the lipid level in the ovaries of females is approximately twice as high as the lipid level in the ovaries of unablated females (Ceccaldi and André, 1971). So far as AA is concerned, females fed dry feed + mussel + *Troca* showed a higher content of fatty acids than females on dry feed alone. *Troca* content of lipids is 9% vs 1.4% with dry feed. It is, therefore, possible that *Troca* is responsible for a good hatch rate and nauplii count (NA). Finally, EPA measured on larvae at Zoea stage while starved or fed could explain a high demand for EPA. Some PUFAs were known to play a key role in metabolism. Arachidonic acid was among the precursors for prostaglandins (PGs) such as 20:3n-6, EPA or DHA. Prostaglandin synthesized from EPA had an effect on the hatching rates of barnacles (Hill et al., 1993).

Development of a Practical Maturation Feed as a Substitute for Nippaï®

Three aspects were considered: lipid content in feeds, fatty acid composition of eggs and formulation of a specific maturation diet.

Low efficiency and poor reproductive performances were reported for breeding animals fed on locally made dry feeds even with the addition of fresh food to the daily ration. Therefore, a set of experiments were conducted on *Litopenaeus stylirostris, Litopenaus vannamei* and *P. monodon.*

A deficit in moults was regularly recorded and a correction of the mineral balance of the feed led to a significant improvement. Finally, a similar performance between Nippaï® and Monogal^{++} (a formulated feed from Aquacop) was achieved with the incorporation of around 15% squid meal.

This type of dry feed was used successfully to feed pre-breeding large *P. monodon* shrimps in Tahiti and New Caledonia and as a finisher-feed in super-intensive shrimp tanks (Aquacop, 1989). Fatty acids analysis and proximate composition analysis was conducted. At analytical level, feeds did not differ except in unsaponifiable fraction (simple lipids such as free sterols, terpens), which had a lower value for Monogal feed than the two others; this seemed unrelated to a putative lack of reproductive performances. In total lipids, Pls or TG formed the main source of fatty acids that were examined for the purpose of reproduction.

Fatty acids from dietary TGs are major fraction and sterols are essential for maturation and their levels can be increased for maturation purposes. There existed a good correspondence in composition between a commercial feed and an experimental diet; both would perform better if supplemented with squid (Table 3). The FA profile of the eggs is significantly improved when Monogal is enriched with squid meal. This produces a significant increase in the n-3/n-6 ratio and this ratio is important for the eggs so far as further viability of the larvae is concerned (Table 4).

Table 3 Lipid composition of three maturation feeds.

Feed	*lipid (%)*	*FA/TL (%)*	*unsap/TL (%)*
Nippaï®	8	81	3.4
Monogal*	9	82	2.5
Monogal + squid	9	83	3.6

FA, fatty acids; TL, total lipids; unsap, unsaponifiable fraction.
*Formulated feed from Aquacop.

Table 4 Fatty acids composition of eggs (% total FA) from spawns obtained from shrimps fed three compounded feeds.

	Nippai®	*Monogal*	*Monogal + squid*
Sat	31	26	27
n-7	14	10	14
n-9	21	23	19
n-6	8	10	6
n-3	22	24	27
n-3/n-6	2.7	2.4	4.5

n = three spawns per type of food.

The formulations of the breeder diets (Table 5 and Table 21) were tested on several shrimp species and their reproduction performances were as good as the performances of breeding animals fed on Nippaï®. Such formulations combined quality protein sources, leading to a balance in amino acids that did not always take into account arginine (ARG) level, which could be a limiting factor. Digestibility of protein from all ingredients fell in the range of 82-93%. Triglycerides and phospholipids, cholesterol, as well as carotenoids were present. Binder helps to keep the food stable during ingestion by the breeding animals.

Table 5 Breeders diet Monogal^{+} (left) and maturation diet Monogal^{++} (right) (Aquacop, 1989)

	%		%
squid meal	14	sequid meal	35
shrimp meal	23	shrimp meal	15
Fish meal+HFPC	11	mussel slurry	6
Bel yeast	11	lactic yeast	11
native wheat gluten	20	wheat gluten	15
soy protein	5	soy protein	
Px_1	2	Px^1	1
Fish oil	3	fish oil	3
lecithin+Generol[122]	1	soy lecithin	1
astaxanthin	0	astaxanthin	0, 1
vit. mix	3	vit. mix	3
min mix	4	min mix	4
binder	0, 7	Na alginate	2
qsp starch	3	qsp starch	
	100		100

However, such formulation, based on weekly preparation, did not compete with a similar formulation of feed produced in a pilot plant or a factory. The main inconvenience of an experimental feed was water contamination, which led to a low quality batch resulting in reduced reproduction performances.

SPECIES–TO-SPECIES APPROACH OF NUTRITION AND REPRODUCTION

This section is the most important of the chapter because it explores specific aspects of nutrition and reproduction. Penaeids in particular have never exhibited the same behavioural patters in captivity. There are closed-thelycum and open-thelycum species. However, it was widely accepted that no single species has ever possessed all desirable traits for culture in captivity (in terms of acceptance of feed, ease of handling, routine mating, good spawning, high hatchability, high growth rate). Thus, this chapter is in several parts, each one dedicated to a single species.

Nutrition and Reproduction of Several Species of Penaeids

Since 1974, Aquacop has conducted research on penaeid reproduction in captivity in Ifremer Centre Océanologique du Pacifique (COP), Tahiti. The food was distributed twice a day. Fresh food based on frozen squid, live mussel (*Perna viridis*) and troca (*Trochus niloticus*) was fed in the morning. A compounded feed with 60% protein content was distributed during the afternoon. The daily feeding rate was 3-5% of the biomass (Aquacop, 1982). By and large, a schedule for the feeding of breeding animals was set up (Table 6).

Table 6 Feeding sequences applied for preparing breeders in tank or in ponds.

	initial stage	*final weight*	*feeding regimes*	
	PLs 30/sq m	20 g		
Indoor tanks	20 g 14/sq m >	50	gjapanese diet+squid	5%
	PLs	20 g	PGN (COP diet)	10%
	1.5-2/sq m		6 d/w	
Earthen ponds	20 g >	50 g	1) squid	5%
	0.6/sq m		2) PGN	
			3) special maturation diet	

% biomass daily amount of feed

The percentage of feed varied according to rearing conditions and stage of development. However, the dominant factors in obtaining breeder size in the minimum of time and with predominantly fresh feed remained stock density (Table 6). This schedule applied to various species studied at COP and led to a consistent level of successful reproduction. Reproduction performances helped sustain all other research activities on penaeid shrimp.

Nutrition and Reproduction of *Penaeus japonicus*

Penaeus japonicus is one of the most extensively studied species from which the Nippai® formulae derived. Several research programmes were initiated in the early 1970s including research into lipid and protein metabolism in reproduction.

Laubier-Bonichon (1978) reported a single case of control of reproductive process without eyestalk ablation. Feeding had a high nutritive value owing to live mussel given daily throughout the reproductive period. With manipulation of two abiotic factors, temperature and photoperiod, the initiation of ovary development was successful.

Concentration of lipoproteins and vitellogenin varies according to the gametogenesis stage. Very low density lipoprotein is the molecule that transports TG and high density lipoprotein (HDL) is the transporter of structural lipids (phospholipid). These metabolites tended to increase during reproduction. Incorporation of fatty acids EPA and DHA were reduced in lipoproteins when the food was inadequate or insufficient. High density lipoprotein can contain up to 30% DHA and decreases in vitellogenin (10-19%) to 23-25% according to the dietary status. In trout, it was found that deprivation of food triggered a mobilization of reserves in essential fatty acids to the egg compartments (Fremont et al., 1984).

The serum lipoproteins of *P. japonicus* were separated by the ultracentrifugation method. Most of the hemolymph lipids of this prawn, regardless of the sex, were associated with high density lipoproteins (HDL_2 and HDL_3 and very high density lipoproteins or VHDL), and only rarely with low density lipoprotein (LDL). No lipid was detected in the chylomicron. As lipid moieties, HDL_3 and VHDL contained an abundance of polar lipids (PL) (65-85% of lipids); however, HDL_2 involved substantial amounts of free sterols and diglycerides, and free fatty acids apart from PL accounted for 40-47%. In every lipoprotein, the lipids contained palmitic, palmitoleic, stearic, oleic, eicosamonoenoic, eicosapentaenoic, and docosahexaenoic acids as the prominent fatty acids. These data were discussed in relation to the lipid transport mechanism in crustaceans by Teshima and Kanazawa (1980).

To clarify the transport mechanism of dietary lipids, the labelling pattern of lipid classes was investigated on the lipids extracted from the intestinal content, hind-gut, hepatopancreas, muscle, and hemolymph after oral administration of ^{14}C tri-palmitin to *P. japonicus*. In addition, the incorporation of radioactivity into the serum lipoproteins was examined. Most of the radioactive serum lipids were associated with high density lipoproteins (HDL_2 and HDL_3 and VHDL), but not with the low density lipoprotein, chylomicron, and very low density lipoprotein. The lipids of HDL_2, HDL_3, and VHDL contained polar lipids as the major radioactive lipid classes. The above data is discussed in relation to the lipid transport mechanism in the prawn *P. japonicus* (Teshima and Kanazawa, 1980).

Nutrition and Reproduction of *Penaeus indicus*

Research on *P. indicus* nutrition and reproduction contains the maximum information obtained during the 1980s. The reason is that this species was one of the easiest to reproduce in captivity to the point where some natural spawning in earthen ponds, without eyestalk ablation or particular feeding of the females, was observed. However, because of the species' poor growth rate in ponds, it did not create the interest that aquaculturists had in other penaeid species. However, it remains an excellent species for research purposes. The first step was to test the freshness of the food on reproductive performances and then compare the quality of fresh and frozen mussel.

Fresh vs Frozen Mussel Fed on P. indicus

The daily preparation and feeding of fresh food was time consuming. Then it appeared useful to improve the productive process through a comparison of the range of difference using fresh or frozen mussel (Table 7). Potential loss of nutrients in frozen natural food as compared to fresh was examined.

Table. 7 Ascorbic acid content (μg/g dry matter) of eggs and nauplii (NA) count.

	fresh mussel	*frozen mussel*
ascorbic acid	689	194
eggs number	$8.4\ 10^6$	$4.4\ 10^6$
NA number	$3.8\ 10^6$	$1.8\ 10^6$

In comparison to fresh mussel, the loss of ascorbic acid content in frozen mussel produced a 50% decrease in egg numbers and in nauplii numbers (Table 7). Moreover, the ascorbic acid content of eggs decreased drastically from 700 to 200 µg/g DM. Ascorbic acid is implicated in biochemical reactions to maintain a low level of cellular oxidation and it acts in a vicariance effect with vitamin E. In frozen mussel, it is probable that insufficient vitamins remain or that the components that affect the tissue build-up at oocytes level is degraded. Ascorbic acid is essential for egg formation at the stage of intense hyperplasia in order to protect new cells from oxidation and contribute globally to the cell integrity.

α-tocopherol as a Biochemical Indicator of Quality Feed and Eggs

The dietary vitamin E requirement was evaluated using practical diets (Alvares del Castillo Cueto, 1988). Females were fed three different diets:

one deficient in vitamin E (RD, at 50 ppm), another supplemented with vitamin E at 350 ppm (RS) and a control diet (RT fresh mussel). The results indicated that hatching rate (HR expressed in percentage calculated from the number of nauplii, NA divided by the number of fertilized eggs placed to incubate) was negatively affected by the absence of vitamin E ($p < 0.05$, Fig. 1).

The RD diet allowed the maintenance of an average HR around 40% until rank #4 (Fig. 1), after which it dropped progressively to 10%. The importance of vitamin E is clearly demonstrated in this experiment, as is the relationship between nauplii viability and vitamin E supplied from the feed.

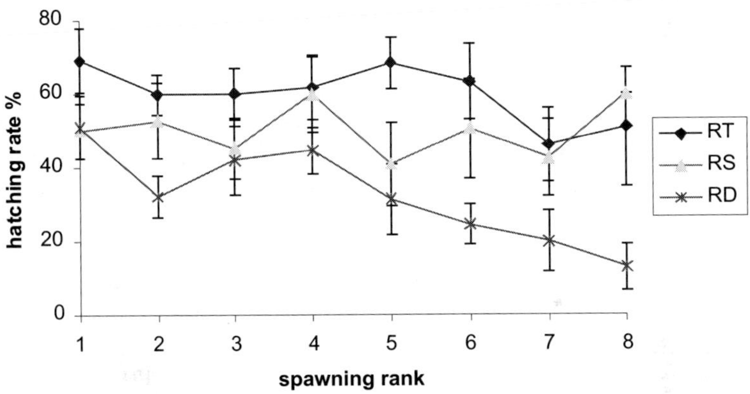

Fig. 1 Influence of diet composition (vitamin E levels *vs* fresh mussel) given to *P. indicus* breeders on hatching rate percentage (RD in ×; RS in triangle; RT in diamond).

Role of α-tocopherol in P. indicus Reproduction

Trials were conducted in order to examine nutrition and reproduction in a given species that was found easily maintained in captivity. The comparative effect of fresh food and compounded feeds on reproductive performances was evaluated.

Experiments with Marine Food. The mussel species *Mytilus edulis* was the chosen fresh food. It is common along the coast of Brittany and has been previously used successfully in reproduction experiments on kuruma shrimp (*P. japonicus*) in the laboratory of Ifremer, Brest (Laubier-Bonichon, 1978). At that time the use of fresh mussel was found to be labour intensive, and therefore the purpose of the experiment was to compare fresh and frozen mussels in order to simplify the daily procedure. In fact the experiments conducted on *P. indicus* made it possible to verify analytically the reasons for the differences between fresh and frozen mussels, in both male and female. Fresh mussel significantly increased the vitamin E content in males HP ($p < 0.05$); vitamin E content in female ovary also was higher with fresh mussel than with frozen ($p > 0.05$), evidencing a remarkable advantage in using fresh mussel (Table 8).

Table 8 Influence of fresh and frozen mussel on tissue content in α-tocopherol (μg/g dw) after 40 d feeding 10 g average weight, male and female *P. indicus*. Superscript letters indicate significant difference ($p < 0.05$).

	MF		*MC*	
	females	males	females	males
ovary	108.7 +/-16.8^{a}		95.8 +/ -18.2^{a}	
HP	16.7 +/ -1.1^{a}	326 +/ -37.4^{b}	41.4 +/ -9.9^{c}	16.7 +/ -17.7^{d}
muscle	11.5 +/ -0.8^{a}	33.62 +/ -1.7^{b}	19.8 +/ -2.0^{a}	31.7 +/ 2.2^{b}

The presence of high vitamin E level in HP of males was beneficial in the reproductive process. A positive effect of vitamin E on male fecundity was evidenced (Draper, 1980). The vitamin content of fresh mussels is intact and they contain no peroxides, whereas it is probable that the vitamin E in frozen mussels is destroyed when fatty acids oxidize. The transfer of vitamin E from HP to ovary and from the food to the tissues is clearly demonstrated, and the fact that the food composition is reflected by and large in body tissue composition is confirmed.

Experiments with Compounded Feeds. α-tocopherol concentration in eggs was analysed from spawns of successive ranks (female re-maturation, Fakhfakh, 1989). Dietary vitamin E concentration directly influenced the reproductive performances of *P. indicus* and improved egg hatchability and larvae viability in the early developmental stages. Shrimps fed a compounded feed with 40 ppm vitamin E spawned with decreasing hatching rate during several weeks (Fig. 2).

Spawns that reached a rank equal to or above 8 did not produced viable eggs. Shrimps fed a diet with 350 ppm yielded consistent hatching rates (Fig. 2) similar to those obtained for animals fed fresh mussels (control diet). Analytical data from eggs and shrimp tissues showed that a concentration of 350 ppm vitamin E in the feed did not limit hatching rate in the experimental conditions described. But, shrimps fed a diet low (40 ppm) or high (350 ppm) in vitamin E produced only half the number of eggs produced by shrimps fed the control diet (Fakhfakh, 1989).

Previous results (Alvares del Castillo Cueto, 1988) demonstrated that egg yield number was, among other parameters, linked to dietary Pls content. The incidence of successive spawning decreased in accordance

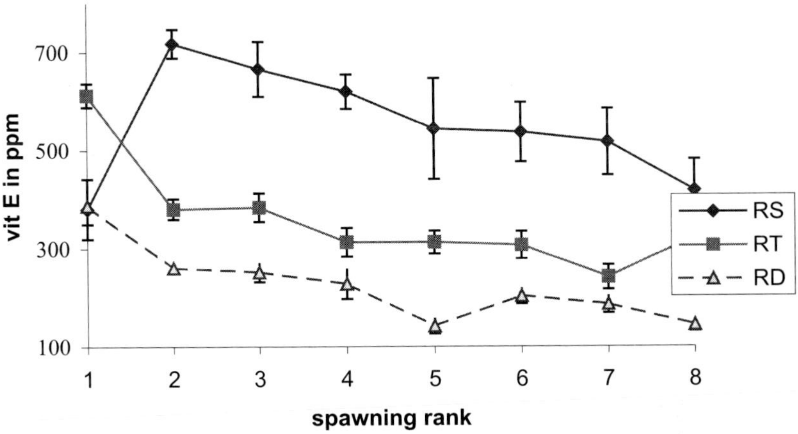

Fig. 2 Egg concentration in D-Lα-tocopherol from spawns of successive ranks (Fakhfakh, 1989).
RT = fresh mussels; RS = 600 ppm vitamin E; RD = basal diet without vitamin E.

with the spawning rank and when the vitamin E content in eggs decreased from 305 to 98 μg/g DM. Successive spawning can induce an exhaustion of lipid reserves in the HP (Boucard et al., 2004). However, as described above, a decrease in lipid reserves can be counterbalanced with dietary supplements.

Requirement in Phospholipids

Phospholipids and vitamin E requirements regarding reproductive performances were investigated by Villette (1990) using three dietary treatments: PHO (Pl without vitamin E supplement; PHO + vitamin E and NEUT (neutral lipid) + vitamin E). The HP of animals obtained from the experiment contained large quantities of lipids (primarily neutral lipids). Neutral lipids contributed to "fatty liver syndrome", which, unlike in fish, is uncommon in shrimp. However, shrimps fed bivalves had an HP containing twice as few lipids.

The HP is a metabolic crossroads where lipids vary in quantity and in quality according to absorption, synthesis, catabolism and transfer to other tissues. Lipid content in eggs was the same in all treatments, showing that the treatments did not affect lipid storage in the eggs. However, in spite of low Pls content in NEUT + E treatment (32%) compared to treatments PHO and PHO + E (40%), eggs contained the same amount of phospholipids (45%) in the three treatments. Phospholipids are an essential component of vitellin reserves; in case of a slight deficiency in Pls, their synthesis by HP cells is increased and it is done from neutral lipids hydrolysed in di- and monoglycerides that will contribute to Pls synthesis. Phospholipids are then carried out via haemolymph to different organs, among them ovary, particularly during secondary vitellogenesis. Phosphatidyl-choline and phosphatidyl-ethanolamin are the main flowing lipids in hemolymph (Lee and Pupione, 1978). However, in cases of severe deficiency in dietary Pls, HP cannot sustain its Pls level and these can fall to 13% TL in females entering secondary vitellogenesis. Nor can hepatopancreas maintain egg concentration, which fell from 45 to 41%. The HP composition of fatty acids was very much related to the feeding regime and to HUFAs in particular. The HP of shrimp fed PHO and PHO + E contained much more EPA and DHA than NEUT + E diet (Table 9). In contrast, such a difference does not appear in eggs where the HUFA content remained high whatever the feed composition. This proves the importance of such fatty acids in eggs. The HUFAs played a metabolic role as well as a structural role in membrane. It should be noted that 80% of muscle lipids consisted of Pls, which contributed to stabilize tissue membranes.

A large amount of HUFAs coming from dietary origin are deposited in tissue membranes during vitellogenesis. This accumulation can be done efficiently with the presence of antioxidant substances, some of which are vitamins. Vitamins E and C influenced the quality of a spawn (hatching rate, larva viability), unlike phospholipids, which would act upon a quantitative aspect (number of eggs in a spawn). It was also clear that a combination of phospholipids and vitamin E was related with performances leading up to eight spawnings per female with 94,000 NA per spawning (Table 9).

Effect of Some Biochemical Components of Broodstock Diet on Egg Composition in P. indicus

Biochemical components that presented the most significant variations with broodstock diet composition were polyunsaturated fatty acids and vitamins, in particular E and C.

Table 9 Reproduction performances and phospholipids, vitamin E requirement.

	PHO	*PHO + E (vit.E)*	*NEUT + E (vit.E)*
% f. spawning	94	100	94
max # spawning per female	6	8	8
# eggs/spawning	89000	94000	92000
# NA/spawning	16000	21000	24000
fertilized rate (%)	45	49	52
hatching rate (%)	45	42	34

HUFA. Experiments were conducted by feeding *Penaeus indicus* broodstock with compounded diets (high HUFA, HH; medium HUFA, MH; low HUFA, LH) containing different concentrations of HUFA. Fresh mussels were used as the control diet (MU). The number of spawn was not affected but the number of eggs per spawn was significantly higher with MU (112,000). Hatching rate was also significantly higher with mussel compared to other treatments (HH, MH, LH), but HUFAs gradually improved their hatching rate from a low to a high level (Table 10; Cahu et al., 1994).

Table 10 Egg number and hatching rate with breeders fed on HUFAs, (HH, MH, LH) *vs* mussel (MU).

	MU	*HH*	*MH*	*LH*
spawn #no.	20	24	21	18
#no. eggs per spawn ($\times 10^3$)	112 ± 17^a	94 ± 10^b	93 ± 11^b	98 ± 11^b
hatching rate (%)	61 ± 11^a	42 ± 9^b	34 ± 7^c	8 ± 3^d

This shows that HUFAs play a major role from a qualitative viewpoint (hatching rate) rather than from a quantitative one (number of eggs). The hatching rate difference between MU and HH can be explained at FA level by a decrease in EPA from 15 to 10% total FA with MU and HH respectively. At an equivalent level of n-3HUFAs (33.9 and 31.2 with MU and HH respectively), EPA decreased with HH and could be responsible for this drop in hatching rate from 61 to 42% (Table 10).

Vitamins. Vitamin concentration in tissues was found to be directly related to their concentration in the diet. This observation was emphasized by the concept of "live food", whether fresh or frozen, with a better success when breeders received fresh food.

α-tocopherol and ascorbic acid were studied in relation to reproductive performances with practical diets that vary in FA content (Table 11).

The HUFA concentration 3.8% of dry matter in eggs was maintained at a constant value promoting embryonic development. Eggs with poor HUFA concentration (2% dry matter) were obtained by feeding broodstock with low dietary HUFA content, resulting in low ability of eggs to develop. These results suggest that this concentration of 2% HUFAs was insufficient to sustain membrane development during embryo genesis.

Vitamin E has a potential role as an antioxidant able to prevent free-radical-mediated tissue damage. It has a strong antioxidant capacity. Both gamma- and delta-tocopherol may be necessary for preventing lipid peroxidation and in counteracting the pro-oxidant effect of α-tocopherol.

Table 11 Full description of a practical experimental maturation feed (Cahu et al., 1994).

HUFA-rich		*HUFA-deficient*	*Phospholipid-deficient*
Casein	6.25		
$CPSP^{80}$	17.5		
shrimp meal	16.2		
lactic yeast	12.5		
wheat gluten	15		
sugars	19.6		
vit mix+stayC	2.6		
$Na_2HPO_4+CaCO_3$	4.4		
cholesterol	0.3		
hen PL	1.9		
cod liver oil	3.8		
soy lecithin		1.9	
sunflower oil		3.8	
hen TG			1.9
sunflower oil			3.8

Table 12 α-tocopherol concentration (μg/g dry matter^{-1}) in feeds, eggs, and tissues.

	MU	*HE*	*LE*
feed	63 ± 10^b	352 ± 11^a	82 ± 3^b
eggs	290 ± 26^b	587 ± 66^a	177 ± 17^c
male HP	327 ± 38^a	355 ± 67^a	58 ± 20^b
female HP	62 ± 16^b	363 ± 61^a	30 ± 4^b
male muscle	52 ± 2^c	105 ± 4^a	85 ± 3^b
female muscle	45 ± 3^b	103 ± 7^a	50 ± 4^b

HP, hepatopancreas; MU, mussel; HE, high vitamin E; LE, low vitamin E.

The HP of males maintained vitamin reserves, more than 300 μg dm^{-1} with MU or HE and only 58 with LE (Table 12). In contrast, female HP accumulate vitamin E with feed rich in vitamin E (HE) at a level significantly higher than in the HP of females fed MU, and an expected higher level than the level of vitamin E in LE diet. The α-tocopherol and ascorbic acid concentrations in the broodstock diet affected their concentration in eggs and hatchability. The action of these two vitamins in preventing HUFA oxidation in developing eggs is discussed. Moreau et al. (1998) had already demonstrated the rapid accumulation of ascorbic acid in the HP of juveniles according to the concentration of vitamin C in the feed. There is a similar accumulation during the reproduction period as shown with *P. indicus* (Cahu et al., 1995). Vitamin C was identified with essential function to get shrimp developing ovary and the presence in tissues at levels indicated above could be a reference for maintaining good reproductive performances in captivity using compounded feeds of fresh food.

There is a correspondence between the vitamin C concentration in the compounded feed and the concentration in eggs. Interestingly, HC feed provided a vitamin C level as high as with MU feed (Table 13).

Table 13 Ascorbic acid concentration (μg/g dry matter) in feeds.

	MU	*HC*	*LC*
eggs	484 ± 52.7[a]	523 ± 83.8[a]	197 ± 20.5[b]

MU, mussel; HC, high vitamin C; LC; low vitamin C content and eggs.

In summary, *P. indicus* was an excellent model to work on shrimp requirements for reproduction. It was possible to show the importance of fresh food rather than frozen food, demonstrate the role of vitamin E for quality spawning, obtain preliminary results on vitamin C and its possible association with vitamin E, and obtain an idea of the essentiality of phospholipids in the reproductive period. It was not possible to differentiate the nutritional requirements of the males and females of the species. This aspect is further studied with *Litopenaeus setiferus*.

Nutrition and Male Reproduction: The Case of L. setiferus

The possible role of PUFAs in ovarian maturation was discussed by Middleditch et al. (1980). These authors observed that, in the same way as other species, PUFAs were needed for *L. setiferus* female ovarian development. *Litopenaeus setiferus* grown in the laboratory spawned successfully after being fed an annelid rich in lipids; major fatty acids in lipids from mature ovaries were C20 and C22-PUFAs. Lilly and Bottino (1981) identified the level of arachidonic acid in relation to the reproduction period.

Why dissociate male and female feeding or nutrition requirements in terms of reproduction? Males and females, for example, with *L. stylirostris* or *L. vannamei,* were conventionally separated into two different tanks and fed *ad libitum*. However, some attempts were made, using mussels as the main food, to feed male mussels to male shrimps and male mussels (with red gonads) to female shrimps. No apparent differences were found and the project was abandoned. Normally (Aquacop, 1975), shrimps enter the maturation unit for 4 weeks, with the males and females separated into two tanks. The females received squid, frozen mussels in abundance and dry feed. The males received squid, restricted quantities of frozen mussels and dry feed. Such practices are more or less empirical.

Male reproduction was more deeply studied with *L. setiferus* because of certain specific problems linked to spermatophore formation.

Several problems remain to be solved regarding *L. setiferus* culture. One problem is related to the nutritional requirements required during sexual maturation in order to obtain healthy broodstocks. In fact, Alfaro (1990), Rosas et al. (1995), Sanchez et al. (2001), and Goimier et al. (2006) have shown that nutritional conditions affect spermatic quality in *L. setiferus*.

In captive conditions, the reproductive performances of *L. setiferus* males are limited. Massive nauplii production, by means of natural mating, has not been reported. The main reason is the decrease of spermatophore attachment to the female thelycum, which has been associated with two syndromes: Male Reproductive Tract Degenerative Syndrome (MRTDS) and Male Reproductive System Melanization (MRSM). Both lead to male sterility.

These two syndromes are characterized by a reduction in the number of spermatic cells and an increase in the percentage of abnormal and dead cells. The MRSM is due to melanin production by the immune system of the shrimp after a chitinolytic bacterial infection (Alfaro, 1990). These syndromes are related to captivity and stress management affecting the physiological state and the immune system of shrimps (Pascual et al., 2003a, b). Sanchez et al. (2002) demonstrated that the effect of temperature on nutritional conditions could affect the immune response and alter the molecular mechanisms associated with melanin production, explaining the MRSM.

Evidence exists to show that male reproductive performance increases with age of shrimp. Ramos et al. (1996) showed that 7-month-old *L. schmitti* could be mature but that 9- to 10-month-old shrimps had a better sperm count than the younger shrimps. On the other hand, Alfaro (1990) showed that, during the maturation process, the spermatic quality of *L. stylirostris* changes with age with a higher proportion of abnormal cells in recently mature shrimps (7 months old) than in 9-month-old or older shrimps. Although the influence of age and size on male reproductive performance has been studied in several shrimp species (Alfaro, 1990; Cevallos-Vázquez et al., 2003; Crocos and Coman, 1997; Peixoto et al., 2004; Rosas et al., 1993), recent studies demonstrated that the influence of age on male reproduction was independent of male live weight (Cevallos-Vázquez et al., 2003; Peixoto et al., 2004). A previous study (Rosas et al., 1993) showed that maturation of *L. setiferus* adult males increases with the age of shrimp and shrimps weighing 10 g are suitable as sperm donors for artificial insemination. Similarly, Peixoto et al. (2004) observed that 10-month-old *F. paulensis* of 10 g weight have a similar sperm count to that observed in 16-month-old shrimps weighing 25 g.

Other studies (Cevallos-Vázquez et al., 2003; Sánchez et al., 2001, 2002) have suggested that food proteins, rather than lipids and carbohydrates, are responsible for the quality of spermatophores and sperm in *L. vannamei* and *L. setiferus*. In particular, these proteins play a crucial role in the formation of the sperm spike, taking place in the vas deferens. Besides, spikes that have a protein base have an important function during the acrosomal reaction with the ovule. Sperm without spikes cause most of the abnormalities observed in males, leading most often to abortion of the acrosomal reaction. In a first attempt to define the protein requirement for adult males, the relationship between dietary protein level and spermatic quality, immune condition and nutritional condition has been demonstrated (Goimier et al., 2006).

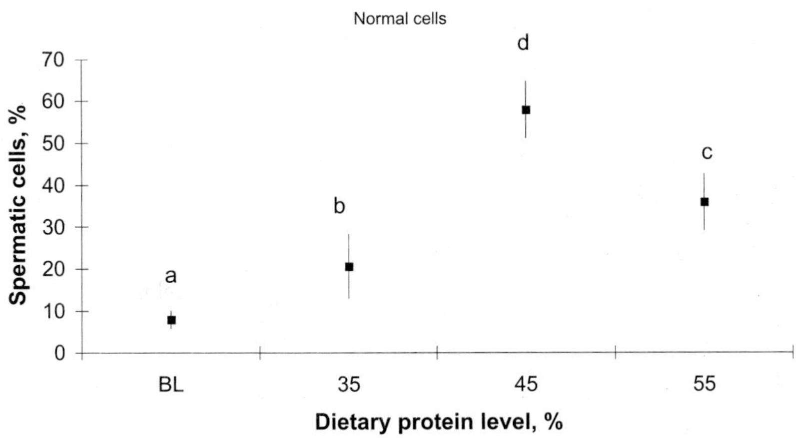

Fig. 3 Effect of dietary protein level on normal sperm proportion of *L. setiferus* adult male. mean ± S.E. Different letter means statistical differences ($p < 0.05$). BL = base line.

As can be seen, a high normal sperm proportion was observed in males fed 45% CP, with intermediate values in shrimps fed 55% CP and low values in shrimps fed 35% CP. Why did a greater amount of dietary protein level cause a reduction in sperm quality? Pascual et al. (2003b) hypothesized that *L. setiferus* sperm quality is related to the loss of immune control. With that hypothesis a lesser sperm quality was related with differences in haemocyte concentrations and pro phenol oxidase activity that affected the sperm quality through melanization. Observations made using electron microscopy showed that sperm cells can accumulate melanin and as a consequence haemocytes attack the vas deferens when the shrimps are stressed (Dougherty and Dougherty,

1990). In the study of Goimier et al. (2006), excess dietary protein seemed to produce stress with a loss of immune control, melanization and reduction in sperm quality. In previous studies, protein in excess of 45% CP reduced weight gain for several shrimp species (see review from Guillaume, 1999). Protein catabolism and stress induced changes in internal ammonia concentrations related to growth rate reduction due to excess protein (Rosas et al., 1995; Taboada et al., 1998). High blood protein content was observed in shrimps fed 55% CP indicating that greater protein metabolism occurred in these animals owing to increased internal ammonia concentration. High dietary protein levels also enhanced haemocyte synthesis, preparing shrimps for an immune defence. If shrimps were stressed by internal ammonia an immune reaction could be activated as a response. Stress affects several physiological functions including sperm production and quality, as was shown for high temperature in males of *L. setiferus* (Pascual et al., 2003b).

Nutrition of *Penaeus chinensis* Breeders

The effect of feeding four semi-purified diets containing different lipid sources (anchovy oil, linseed oil, corn oil and pork lard) on fecundity, egg hatchability and egg and spent gonadal tissue fatty acid composition of Chinese prawn (*Penaeus chinensis*) broodstock was compared with that of a fresh clam diet in a 60-day feeding trial. Broodstock prawn fed the diet containing pork lard showed poor fecundity and low egg hatchability. Broodstock fed the diets containing linseed or corn oil showed improved egg production ($p < 0.05$); however, no significant improvement in hatchability was observed.

When broodstock were fed the diet containing anchovy oil, both fecundity and egg hatchability significantly improved (Table 14).

Table 14 Effect of dietary lipids on female reproduction (Xu et al., 1994).

oils	*ω per female* ($x10^4$)	*NA/50 ml*	*hatch rate*
pork lard	39	17	22
corn	80	45	28
linseed	90	53	30
anchovy	110	174	80
clam	113	187	83

Eggs from broodstock fed anchovy oil as the sole dietary lipid had a higher n-3 HUFA content (27.6%) than those from broodstock fed prawn diets containing linseed oil (19.5%), corn oil (14.0%) or pork lard (12.8%). Good correlations between the 20:5n-3 content of the egg lipid and fecundity and between 22:6n-3 content and hatchability were observed. The results suggest that each of these n-3 HUFAs may play different specific roles in crustacean reproduction and that either or both must be included in the broodstock diet. In a recent study, EPA was related to egg lipid and fecundity, DHA content and hatch ability and fecundity, and hatchability and n-3/n-6 ratio (Wen et al., 2002).

Effect of vitamin A supplementation in broodstock fed on reproductive performances in *P. chinensis* (Mengqing et al., 2004) showed a correlation between fecundity, hatching rate and vitamin A content in eggs. The dose of 60 ppm vitamin A acetate in a semi-purified diet produced the best results.

Nutrition and Reproduction of *P. monodon*

Several investigations were reported (Aquacop, 1975; Alava and Kanazawa, 1995; Millamena, 1996).

Penaeus monodon can reach a very large size and, in Indonesia and other countries, breeding animals are sold by piece; a single wild female could spawn enough eggs to run a whole larva production cycle in a backyard hatchery.

Penaeus monodon gave different reproductive performances according to its origin. For example, comparison between animals from Indonesia, Malaysia, Taiwan and Fiji identified differences, with the best-performing animals originating in Indonesia. In Indonesia it was found that a female from the wild (250-300 g) after unilateral eyestalk ablation could spawn 14 times in a period of 2 months. The same female could spawn 2 days in a row, producing 1 million eggs and about 90% fertilization. It showed the potential of reproduction for one given species with a rate of 7 spawnings per month.

Penaeus monodon from Malaysia was not so well adapted to captivity and its performance tended to decline after the first spawning. *Penaeus monodon* from Fiji spawned at an average weight of 90 g. Generally three or four spawnings per hatchery female per month was considered acceptable. But in fact such performances were well below the real potential of the species. Unfortunately, academic research did not develop along this interesting theme (Goguenheim, 2006; pers.comm.).

The effect of nutrition on crustacean gonadal development and reproduction was described by several authors in South East Asia (Alava and Kanazawa, 1995; Millamena, 1996) and reproductive performance of captive *P. monodon* fed various sources of carotenoids (Pangantihon-Kuhlmann et al., 1998) highlighted a number of essential requirements for females. For example, in the Philippines a whole unit was dedicated for controlled reproduction in captivity of breeding animals fed a mixture of fresh food and compounded feeds with carotenoid-containing natural food: mussel + shrimp broodstock pellets (MBP), crab + BP (CBP), and *Artemia* + BP (ABP). After eyestalk ablation, MBP-fed shrimps initially spawned 20 days later and at the end of 4 months maturation and spawning rates did not differ significantly among treatments (Quinitio et al., 1996). Dietary carotenoids were provided in sufficient amount from all diets.

The study of fatty acid content of maturation feed and male performances was based on the statement that reproductive success was due entirely to female spawners (Meunpol et al., 2005). However, prawn larval quality can decline, indicating that the quality of male spawners is contributing to reproductive success (Aquacop, 1983). Authors significantly enriched an experimental feed with 20:4n-6 (AA) and found an accumulation (18% TFA) above the level found in wild animals but the percentage of abnormal spermatozoid (sptz) from animals fed enriched AA was high (53%) compared to 24-39% in the control animals. Arachidonic acid is known to play a role in fertilization success in fish. With shrimp, a large supplementation of AA in feed resulted in reasonable male reproductive performances (total sperm and percentage live sperm) except for the presence of a higher incidence of abnormal sperm than in the control animals. The authors conclude that there is an optimum AA level. Special attention was to be given for n-3/n-6 ratio that could easily reach a value out of range such as 0.4. Accumulation of high proportions of HUFAs in the tissue membranes exists in fish. However, the quantity of HUFAs used for prostaglandins, for example, in upper vertebrates, is about 1/100 of HUFAs deposited in Pls (membranes mainly). Then AA leading to prostaglandins could facilitate reproductive performances. Recently, feeding bio-encapsulated prey to broodstock was described (Kian et al., 2004).

Nutrition and Reproduction of *L. stylirostris*

Reports from several teams referred to the blue shrimp (Aquacop, 1975; Ottogalli et al., 1988; Cuzon and Aquacop, 1998).

Under normal conditions of rearing starting with an average weight of 30 g, following a given feed chart based on percentage of biomass per day, and adding a meal of squid meat once a week, it takes about a year to rear breeders of a given size (50 g) that achieve good spawning and a good percentage of viable larvae. It is commonly achieved at COP in Tahiti through the use of low stocking density at juvenile stages (1-4 shrimp/m^2), supply of a water-stable balanced feed (43% CP, 6% lipid, 2% fibre), and an established feeding chart. Recently, *L. stylirostris* pre-breeder juveniles were maintained in an earthen pond (500 m^2) and raised with algal bloom at low density (2 shrimp/m^2). Rather than adhering to the regular feeding chart it was decided to add 100 g compounded feed to the normal daily ration. A number of year-old animals reaching 60-70 g, which performed excellently at reproduction level, were observed. In addition, a highly inbred strain of shrimp such as *L. stylirostris* demonstrated the potential to grow significantly above the 25 g normally obtained in a super-intensive farm (Sopomer, Tahiti) and inbreeding effect on reproductive performances was not shown even after 28 generations obtained in captivity. Moreover, the results show that the rationing does not need be so strict (Aquacop, 1975). A commercial formulated feed conveniently balanced in protein and energy addresses not only growth performances but also maturation needs. In fact, this operation is not run on a commercial scale; it simply aims to maintain a virus-resistant strain of *L. stylirostris* IHHN. Therefore, there is a significant link between the quality and quantity of feed and hardy large breeders capable of spawning large numbers of eggs. This is an example in which the preparation of breeders was optimized: large animals were available in one year instead of the previously required 14-15 months.

Shrimps of up to 50 g are raised in earthen ponds fed with extruded commercial feed (40/8). Sometimes shrimps of 70 g can be obtained by this method but the manipulation of the breeders seems more difficult. In tropical areas such as in Tahiti, a hot season of 28-29°C from December to May contrasted with a cold season of 25-26°C from June to September. It was preferred to remove the breeders from the earthen ponds in June in order to obtain males with good spermatophores. Otherwise, males that were kept in the earthen ponds during the hot season showed traces of necrosis on spermatophores and low fertilization (down to 2%) after artificial insemination. Breeding animals harvested from earthen ponds were transferred to scobalit® (fibreglass) indoor tanks (13 m^3). Males and females were kept separate. Photoperiod could be reversed to facilitate manipulation. The daily feeding schedule consisted of extrusion dry feed in the morning, squid at noon and frozen mussels at night. Once females matured from a whitish colour to a light brown, each mature female was inseminated with the sperm of two or three males and placed in the spawning tank. The females' eggs were collected a few hours later (sequential spawning or full spawning). Each female spawned about 200,000 eggs.

Some Requirements

Different species of shrimp were analysed periodically on an annual basis in the Gulf of Mexico to follow the alteration of FA in the shrimps.

It suggested that the changes were occurring through the food chain (Bottino et al., 1980) rather than by endogenous adjustment to an alteration in water temperature. Water temperature decreased in austral winter time (New Caledonia) and, as suggested for a species such as *L. stylirostris* under laboratory conditions (Chim et al., 2003), this change would lead to a change in fatty composition in Pls. The importance of food from a quantitative or qualitative point of view is evident. Also, the FA composition of a wide variety of invertebrates that had been used for successful maturation was described (Middleditch et al., 1980; Galois, 1983).

Some fundamentals of breeder nutrition with an emphasis on *P. stylirostris* were described in a short review (Cuzon et Aquacop, 1998; Cuzon et al., 2004). Reproduction with eyestalk ablation of *P. stylirostris* fed various levels of total dietary lipid was reported (Bray et al., 1990). The effects of total dietary lipid on reproduction of *P. stylirostris* were compared using a series of diets ranging from 8% to 14% total dietary lipid. In addition, a commercial feed was compared to the experimental feeds, and a single fresh diet component (squid as 40% of diet) was compared to a multiple fresh component portion composed of squid, bloodworm, shrimp, and brine shrimp as 40% of diet. This demonstrated the effectiveness of multiple sources of fresh food.

Energetics

Very little information is available on this aspect of energy expenditure. Energy for breeders was calculated taking into account maintenance and metabolizable energy dedicated to activity and egg formation. At 28 weeks, females would consume 354 kJ per shrimp per day. This value can be applied between weeks 24 and 34.

Squid Factor

Squid factor was important for growth promotion in juveniles; probably based on active peptides (not purified by HPLC to a single peak, unfortunately), it could also be relevant to gonad maturation (Cruz-Suarez et al., 1992). However, squid meal as a whole would actively act upon gonad maturation because it contained among lipid content an especially good proportion of Pls (Mendoza et al., 1997).

Breeder Quality and "Moulinettes" Rearing System

"Moulinettes" is a system of rearing shrimp in high densities with low water renewal per day, which creates a biomass sufficient to start and maintain the development of bacterial "floc" composed principally of the bacteria required for nitrification. The moulinettes rearing system was developed in 1975 for *L. stylirostris* and *L. vannamei* during a collaborative work between AQUACOP and Ralston Purina. Experiments were conducted simultaneously in Crystal River, Panama, and Tahiti. The principle was to minimize water inflow in a round tank 12 m^2 up to 1000 m^2, adjusting oxygen intake with aeration devices that moreover constantly put the water in movement. This movement in 12 m^2 round tanks was organized in multiple convection cells between the wall and outlet or a circular movement in a 1000 m^2 tank. Shrimp could stay near the bottom of the tank and receive food solely through the movement of the water. Sea water was rich in algae, bacteria, protozoans, ciliates, artificial feed (marine ration MR20 or a locally made feed), organic matter microballs and faeces.

A comparative strategy between "moulinettes" and earthen ponds is described (Table 15), demonstrating once again the importance of stocking density for breeding animals.

Table 15 Comparative growth in two systems to produce breeders fed dry feed+squid.

	area (m^2)	*no.*	*shrimp/m^2*	*biomass (g m^{-1})*	*wt (g)*	*duration (mo)*
"moulinettes"	30	300	10	500	59	5
earthen pond	500	600	1	60	70	8

Earthen ponds produced larger animals, while "moulinettes" produced very healthy shrimp. Both systems have advantages and drawbacks (Table 15) and are better suited to one purpose (growth rate, for example) rather than another (biomass, gonad development, healthy animals).

Analysis of the bacterial "floc" in terms of fatty acids, growth factors, vitamins, and other nutrients partly explained the effectiveness of the "floc" (Table 16).

Table 16 Fatty acids (%) of total lipid (TL) from turbid water ("floc"), food and feed (Galois, 1983).

	"floc" walls	*"floc" water*	*mussel*	*Nippaï*
saturated	34			34
mono	31			33
di, tri	9	8	17	17
AA, EPA, DHA	20	50	16	16
n-3/n-6	1.4	5.3	1.3	1.3

AA, arachidonic acid; EPA, eicosapentanoic acid; DHA, docosahexanoic acid.

During this recirculation process, and as soon as nitrification bacteria started to stabilize, the medium (sequence NO_2, NO_3, NH_3) shrimp grew rapidly (25 g in 3 months with a survival rate of up to 98%), and the cleanliness was excellent, with no accumulated sludge at the tank bottom. After a set of experiments for rapid growth, shrimp produced in this way were hardy, healthy animals. The production of breeding animals was attempted and both species produced consistent results in terms of reproduction performances.

Table 17 *Litopenaeus vannamei* performance in turbid water (with bacterial "floc") compared to clear water with black or white sand bottom (Cahu et al., 1987).

	"floc"	*black*	*white*
#no. females	10	10	10
spawning	++	+	+
fertilized eggs	+++	+	+
hatching rate	+++	+	++
larva viability	+++	++	++

The fatty acids profile and lipid classes indicated that the inclusion of "bacterial floc" in maturation feed could be sufficient to provide essential nutrients for pre-breeders. The reproductive performances of shrimps in "floc" water and those in clear water (black or white sand bottom) were compared and an improvement in the performance of the former group was shown (Table 17). For example, the number of spawns per female would increase up to 6 per month compared to 4-5 per month with breeding animals coming from earthen ponds (Goguenheim, pers. comm.).

When *L. vannamei* and *L. stylirostris* were reproduced in captivity, the issue of whether to add substrate to the tanks of the breeding animals or place them in tanks without substrate arose. Some authors recommended the addition of sand to the bottom of the tanks, while others preferred not to add a substrate. A round fibreglass tank was filled with clear water and a piece of coral was added to the bottom of the tank (for pH control). In addition, green light was used because the wavelength produced by green light has been shown to be particularly suitable for penaeids.

Since then, it has been proved that the absence of a substrate, clear water and a regular light (sunlight spectra) is suitable for sustaining a reproductive process in captivity.

A tank filled with turbid water seems to give the best eco-physiological conditions to obtain maturation of *L. vannamei* in captivity. Shrimp are able to eat the phytoplankton and periods of phytoplanktonic blooms correspond to the spawning times of wild animals (Wurts and Stickney, 1984).

Table 18 Comparison of the spawning rates of *L. vannamei* in three experimental conditions (Cahu et al., 1987).

	"moulinette"	*black sand bottom*	*white sand bottom*
#no. females	17	17	17
no. #spawns	42	19	15
#d after eyestalk ablation	6	9	17
#no. spawn per tank	2.4	1.1	0.9

Moreover, suspended organic matter induced bacteria development, which constitutes an additional food source. "Bacterial flock" contributed significantly to the diet of *L. vannamei* (Aquacop, 1985) and "moulinette" tanks produced excellent breeders. Spawn per females were 1.4 more numerous in clear water tanks than in earthen ponds in 5 weeks (Table 20).

Nutritional deficiency was identified with the change in structure leading to defection of cortical rods. Cortical rods are the outer membrane of eggs. They form after the ova receive the sperm but before egg fertilization. In New Caledonia, a malformation of cortical rods leading to an absence of the membrane and the egg remaining soft was reported. In that case, the composition of ova reflected a presence of high lipid level: as a percentage of dry weight, the values for a wild specimen of *P. stylirostris* were 37.3 ± 1 for protein, 3.0 ± 0.1 for carbohydrate, and 16.1 ± 1.8 for lipid.

	protein(1)	CHO(1)	lipid(1)
P. stylirostris	37.3+/-1	3.0+/-0.1	16.1+-1.8

The problem was attributed to a nutritional deficiency. Analysis of the lipids found in females and the comparison in HP, ovary and muscle tissues of the FA profiles, especially of those of Pls, indicated a possible malfunction of lipid metabolism. It affected 90% of the spawn of the strain *L. stylirostris*. Since it was impossible to identify the putative nutritional origin of this symptom, the strain was simply eradicated. The origin of this malformation is still unknown.

Nutrition and Reproduction of *L. vannamei*

The broodstock in captivity is a necessity in places where no indigenous species of commercial value are available, as is the case in Tahiti

(Aquacop, 1975). Moreover, the scarcity of suitable lands for extensive culture previously recommended for such purposes led to high density culture in small 100 m^2 tanks, which subsequently increased in size to 1000 m^2.

Breeding *P. vannamei* are first imported. Selection is carried out after a growing period of 6 months from post-larval stage at a density of 50-100/m^2. Animals ranging from 22 to 30 g are then cultured at a stocking density of 20 animals/m^2 for a further 6 months until a reproductive size of 40 g or more is achieved. Such broodstock can spawn consistently over a 3 month period when placed in 12 m^3 maturation tanks. After unilateral eyestalk ablation, the number of spawn per female depends on the composition of the bottom of the tank (bare or sandy) and varies with the feeding regime. The fertilization rate using artificial insemination is dependent on the quality of the sperm and oocytes. The number of spermatozoids (sptz) varies in quantity and quality during the spermatophore life. The quality of oocytes is highly correlated with the fresh food given during the very rapid secondary oogenesis period. Manipulation of photoperiod allows control of the time of spawning, facilitating artificial insemination. These procedures are now used routinely in commercial production in Tahiti (Goguenheim et al., 1987).

Studies on the feeding of penaeids during the reproduction period should also consider the effect of the feeding of breeding animals on larva development. It is possible that hatching rate and larva viability at least at zoea stages depends on the "reserves" an embryo can rely on, bearing in mind that accumulation of such reserves by the female occurs in oocytes during vitellogenesis (Galgani and Aquacop, 1989): the influence of diet on reproduction in captivity was also part of a whole set of investigations on *P. vannamei* and *L. stylirostris*.

A study on natural diets for *L. vannamei* found the same amino acids and FA patterns between several natural diets (squid, worms, clam, oyster). There was a protein level > 50% in sources such as squid. Amino acids such as Arg, Lys, His were also found. However, differences were also found in the concentrations of protein, total lipids, cholesterol, vitamin E and C. The addition of squid to the diet resulted in good overall consumption consistent with ovary development while the addition of worms tended to trigger maturation due to their HUFA composition (Du et al., 2005). Dietary fatty acid composition on *P. vannamei* reproduction was previously described for its effect on egg quality (Cahu et al., 1986) and FA identified in relation to their incidence on n-3/n-6 ratio (Galois, 1980) with a special attention to arachidonic acid (Lilly and Bottino, 1980). Requirement of cholesterol led some researchers to hypothesize that up to about 5% dietary cholesterol was required to improve maturation rates.

Influence of Pls and HUFAs on spawning rate and egg and tissue composition in *L. vannamei* fed semi-purified diets (Cahu et al., 1994) was investigated. Egg lipid composition varies according to the type of feed given to breeders. Fatty acid composition of eggs reflected that of feeds. However, females fed low HUFA tended to maintain a tissue HUFA level in both neutral and polar lipids higher than in dietary lipids.

Table 19 Food and feed composition influence on spawning results (Cahu et al., 1994).

	mussel	*HR*	*HD*	*PLD*
initial weight (g)	36	37	37	36
final weight (g)	38	37	38	36
#no. of spawn	22	22	23	8
no. of #spawn per fem. per d	0.06	0.06	0.06	0.02
no. of #eggs per spawn x 10^3	98 ± 43	72 ± 25	64 ± 30	69 ± 25

HR = HUFA rich; HD = HUFA deficient; PLD = Pl deficient.

Muscle and HP HUFA levels of these females decreased first in neutral lipids and then in polar lipids. Phospholipids proved to be essential for spawning and high levels of HUFAs were required for the production of eggs and deposited as reserves. The study conducted with few fertilized eggs did not demonstrate the influence of dietary lipid on hatching rate (Table 18).

The effect of FA composition of foods fed to *L. vannamei* broodstock on egg quality is clearly demonstrated. Compounded collets (meaning compounded feed processed through a cooker-extruder) and fresh food (mussels) were tested for their effect on gonadal maturation and spawning performance. The average number of eggs per spawn was 105,000 and frequency of spawning two per female per month. The deficient qualities of collets as feed is reflected in the hatching rates: 15% for eggs spawned by females fed collets and 75% for those fed mussels. Eggs spawned by females fed on collets had lower HUFAs (AA, EPA, DHA) than eggs spawned by females fed on mussels. The n-3/n-6 ratio was about 2 for the group fed on collets and 4.3 for the other group. The HUFA level of eggs cannot be the only reason for such a difference in hatching rates (Cahu et al., 1986). Then, lipid metabolism was largely studied (Kanazawa et al., 1981) for bioconversion capability of FAs. However, since it is already known that the biochemical balance of n-6 and n-3 FAs is extremely important for the amount of prostaglandins formation, FA profile was used as a comparative tool in assessing maturation diets (Lytle et al., 1990).

Hormone-like factors were isolated from squid (Mendoza, 1997). During the main steps, the following points were examined: (1) study of unablated females (*P. vannamei* and *P. stylirostris*) as control; (2) identification of a squid factor for ovary development, which came after the identification of a squid factor for somatic growth; (3) enhancement of vitellogenin concentration in blood that was examined in relation with ovary development; (4) identification of an excellent model that fit between intact females supplemented with squid extracts and eyestalk-ablated females used as a control; and (4) squid factor incorporated in a formulated dry diet for reproduction (varep, Table 19).

Table 20 Trials conducted to ascertain whether there is a positive response in the presumably final phase of maturation (Mendoza, 1997).

varep	*test-*	*varep*
varep + lipid fraction		varep + HAS
varep + HAS fraction		varep + HAS-polar frac.
varep + structural protein		varep + HAS-apolar frac.
varep + soluble protein		varep + lipid
varep + eyestalk ablated		varep + lipid- polar frac.
varep + squid		varep + lipid-apolar frac.
fresh squid	test+	varep (eyestalk ablated)

Varep is the basal diet. HAS: hydrosoluble fraction.

Finally, the components that determined the start of vitellogenesis of females were molecules such as steroids, terpens, and carotenoids with the vitellogenin molecule and from squid extracts tested one could consider polar and non-polar fractions that were common for both HAS and lipid fractions.

Vegetable protein sources and wheat starch produced good growth during larva development; Pls also produced good growth but tended to stabilize at juvenile stages and breeders appeared to refrain from developing ovaries with very little spawning performance. There was probably already a deficiency in nutrients at juvenile stages that hampered further development of ovaries, despite fresh food being available in the tanks. Further research is required to confirm an apparent dysfunction during maturation process possibly in the absence of enough animal protein bearing growth factor or maturation factor. Such factors were not available during juvenile stages and pre-adult phase.

The role of the supply of basic amino acids from marine proteins as well as substances such as muco-polysaccharides to enhance weight gain had long been suspected. On the other hand, the presence of inhibitors in vegetable proteins, such as saponins, may disturb cholesterol metabolism. In addition, a certain level of the moulting hormone or ecdysterone seems necessary to maintain vitellogenin synthesis, thus leading to secondary vitellogenesis (Payen, 1980). There is a possible lack of essential FA accumulation in tissues, particularly in the ovaries, leading to subsequent imbalances in the n-3/n-6 ratio. In the absence of analytical data on eggs, tissues such as ovary, HP and muscle, investigations on this issue are still pending. Pigments play an essential role in reproduction because of their impact on larva viability. Studies have been carried out on red spots appearing on penaeid eggs, related to a good possibility of obtaining viable zoea. Several supplements to enrich the feed were subsequently proposed, including synthetic astaxanthin and ingredients rich in carotenoids such as paprika (Wyban et al., 1997). Pigments such as astaxanthin derived from paprika seemed to increase larva viability.

Maturation performances of *P. vannamei* were described unsing co-fed *Artemia* biomass (Naessens et al., 1997). Maturation diets for prawn spawners are generally composed of a mix of maine organisms and a well-balanced, high protein dry feed. Marine organisms are rich in essential fatty acids and the n-3/n-6 ratio (in polychaetes, for example) contributes to ovary development (Gelabert et al., 2003). *Artemia* biomass plays a similar role in enhancing the reproductive performances of species such as *L. vannamei* (Naessens et al., 1997) or its larval quality (Wouters et al., 1999). The maturation performances of *L. schmitti* and *L. setiferus* were enhanced when 11% of total broodstock diet consisted of *Artemia franciscana* adults. Higher numbers of eggs per spawn and of spawn per females were obtained with both species compared to the control (Gelabert et al., 2003).

Effect of Maturation Diets on *P. vannamei* Male Performance

Dietary requirements for reproduction may be sex-specific. In a study on the effect of maturation diets on *P. vannamei* (Ogle, 1994; Tirado et al., 1998), the pelleted diet contained lower protein, cholesterol, phosphatidylcholine and less arachidonic acid and DHA than the two composite diets (pelleted diet + squid and pelleted diet + squid + polychaetes). A higher content of essential nutrients may contribute to enhanced performance in males. This result contrasts with the actual management of breeders in captivity where both males and females received the same diet up until the reproductive stage since their performance tended to indicate an absence of deficiencies if performance was measured by the quality and quantity of spawn per female, or the degree of fertilization. The results seemed to suggest the benefit of giving and equal amount of feed + food during the entire period when preparing pre-adults for breeders.

In summary, from the species-specific studies on penaeid nutrition and reproduction, there are differences according to species in the best mode of handling animals in captivity (hardiness of animals). For example, penaeibs results on the requirements of in captivity show that it is easier to work with than any other species. The requirements of *P. indicus*, by and large, also apply to other species. This assertion is corroborated by the following observations: (1) the proprietary feed Nippaï is fed in many different aquaculture centres or reproduction units and (2) the choice of natural food for shrimp broodstock largely focused on a few invertebrates (basically worms, mussels, and squid). However, the formulated feed seems insufficient alone and natural food continues to contribute to diets.

At present, there is no known alternative to eyestalk ablation and, with the exception of unablated *P. japonicus* females spawning for a year (Laubier-Bonichon, 1978), most maturation units around the tropical zone continue to use ablated females.

There was a difference between open- and closed-thelycum species in terms of nutrition and reproduction under laboratory conditions, but the difference could disappear in the future, as has happened with fish reproduction. There is an advantage in despermiation and convenience of artificial insemination for open-thelycum species. However, the most important factor is good health for pre-breeders, which means good nutrition from the beginning, i.e., early Pls during seeding into earthen ponds. Quality feeds and fresh food provided the main essential nutrients (Pls, HUFAs, sterols, carotenoids, vitamins, native protein sources). Such feeding remained an important source of nutrients for successful reproduction of animals under laboratory conditions. In male reproduction, necrosis of spermatophores, ovary regression, and low hatching rates are the consequences of poor management of pre-breeders, poor environmental conditions (e.g., pH, salinity, light intensity, temperature) and, above all, poor quality of feeds, as clearly demonstrated in this chapter, particularly in the case of *P. indicus*.

Macrobrachium and Other Crustaceans

Several authors conducted studies on freshwater prawn (d'Abramo and Sheen, 1994), but little information on breeding animals was available. Male performances of *Macrobrachium* were described by Samuel et al. (1998). Cavalli et al. (2000) conducted experiments first with females weighing 26.2 ± 5.5 g, and second with females weighing 21.4 ± 4.2 g. Although the females were a little smaller in the second experiment, fecundity was estimated in terms of eggs per female weight (g) in an attempt to minimize differences caused by female size. Is there a relationship between linoleic FA and egg production in *Macrobrachium*? In the first experiment (1999), the increase in dietary levels of linoleic acid (from 3 to 12.6) resulted in a significant increase in fecundity (eggs/g female), which is an impressive result because it is fairly difficult to obtain significant differences in the fecundity of crustaceans given the large variation in spawn sizes. When the data is interpreted from the perspective of "reproductive effort" (g eggs/g female), n-3 HUFA may also have played a role in increasing fecundity, but data from only one experiment is insufficient to make that finding. Interestingly, d'Abramo and Sheen (1994) worked with prawn juveniles and suggested that the response to dietary linoleic acid and linolenic acids might require a minimum level of unsaturated fatty acids > 20 carbons. That finding may also apply to this case. To summarize, linoleic acid in combination with long chain unsaturated fatty acids (20:5n-3) EPA and (22:6n-3) DHA may have a role in prawn fecundity, but this cannot be proven solely with the data from the 1999 experiment.

Effect of dietary supplementation with vitamins C and E on maternal performance and larval quality of the prawn *Macrobrachium rosenbergii* (Cavalli et al., 2003) confirmed the results obtained with penaeids. The content of ascorbic acid and tocopherol in the tissues and eggs of wild *M. rosenbergii* during maturation (Cavalli et al., 2001) probably decreased after multiple spawnings. Therefore, a reasonable level could only be maintained by selecting food rich in such vitamins or supplementing the diet with fresh food.

Sonia et al. (2006) explained the effect of low molecular weight components (LMWC) from healthy juvenile and adult *M. rosenbergii* haemolymph on lectin activity and oxidative burst OB in haemocytes. In an attempt to identify the LMWC that affect the lectin's hemagglutinating activity or oxidative burst, the haemolymph carbohydrates and free amino acid (FAA) concentration were determined. The LMWC (< 2000 Da) were obtained after dialysis of the hemolymph. Results showed that LMWC from juveniles exerted a greater inhibition on lectin than LMWC from adult haemolymph. Production of superoxide radicals by haemocytes was lower in the presence of juvenile ($p < 0.05$) as compared to adult LMWC. Free amino acid composition of the haemolymph from adults showed higher proportions of alanine, which corresponded to 25% of total FAA and proline (> 20%); in juveniles, the main FAA identified were glycine (> 40%) and alanine (26%). N-acetyl-D-glucosamine

(GlcNAc) was the main sugar residue in the haemolymph and LMWC from juveniles; its concentration was 2.4 times that of glucose (Glc), whereas, in adults, Glc was the main free sugar residue. Results suggest that the proportion of FAA and carbohydrates in the haemolymph of *M. rosenbergii* seems to be correlated with the maturation process; furthermore, in the juvenile stage the high proportion of free GlcNAc and glycine regulated lectin activity and cellular oxidative mechanisms, respectively.

Other Species

The reproductive performances of mud crab and their offspring fed diets of different quality have been studied (Djunaidah et al., 2003). The FA compositions of the crab *Eriocheir sinensis* during different physiological stages (maturation, spawning and miscarriage) were determined with capillary gas chromatography (Ying et al., 2004). The results showed that 22 types of FA were present in the HP of *E. sinensis*. The major FAs in each group were C18:1 (35-41%), C16:0 (18-22%) and C16:1(12-15%). Saturated fatty acids (SFA), monounsaturated fatty acids (MUFA) and PUFA represented 23-29%, 53-59% and 9-21% respectively of total FAs in HP at the different physiological stages of *E. sinensis*. From maturation to spawning, ratios of SFA, MUFA and n-3HUFA to the total FAs increased significantly, but ratios of n-6HUFA and PUFA to the total FAs significantly decreased. Nevertheless, it is noteworthy that contents of n-3HUFA, n-6HUFA and PUFA all increased during miscarriage, the DHA + EPA content also being much higher than that during the maturation and spawning stage. Comparative research on the Wenzhou and Taihu spawning crab revealed significant differences in the contents of FA, SFA, MUFA, PUFA and n-3/n-6. The content of FAs C18:2, C18:3, PUFA and n-6HUFA in the HP of the Taihu spawning crab were much higher than that of the Wenzhou spawning crab. However, the content of DHA was much lower in the Taihu spawning crab.

In summary, the results indicate that C18:2 might form egg membranes and stalks, and EPA and DHA might play an important role in the growth and development of *E. sinensis* embryos. The main sterol is cholesterol in both sexes (82 to 93%). The highest cholesterol contents are found in the structural lipids of both specimens and also in the eggs of *Cancer pagurus*. Depot sterols in comparison with structural sterols in *C. pagurus* and *E. sinensis* were reported (Zandee and Kruitwagen, 1975).

Lobster nutrition in early juveniles has been studied fairly extensively. Working with the spiny lobster, Lee and Pupione (1978) described an increase in flowing lipids and noticeably Pls, largely dominant in hemolymph of *Palinurus interruptus*. More recently, the influence of diet on broodstock lipid and FA composition and larval competency was shown in the spiny lobster *Jasus lalandii.*

In the shrimp *Pleoticus muelleri*, Jeckel et al. (1989) reported that lipids in spermatophores and testis are between 1.3 and 2% of the wet weight, while triacylglycerides varied between 0.2 and 0.5% and Pls between 1.1 and 1.7%.

CONCLUSION

Knowledge of nutrition relative to the reproduction of Crustaceans was sparse in the early 1970s and it was thus more correct to refer to their feeding. Indeed, prior to the 1970s, when aquaculture was in its infancy, nutrition was not a consideration because shrimp breeders were captured from fishing grounds in the wild, where shrimp found sufficient quantities of quality food.

The evolution of the knowledge of nutrition is difficult to retrace; however, some experiments led to significant improvement in knowledge of shrimp requirements (fatty acids, vitamins, protein). Nutritional requirement in relation to reproduction took into account not only the nutrition of breeding animals but also the control of larva development (Racotta et al., 2003) and the importance of nutrition appeared clearly through successive spawning under normal environmental conditions. Penaeid reproduction has now been mastered and knowledge on nutrition has been properly integrated in feed formulations. Many research projects concentrated on ascertaining the feed required for successful maturation. There were still few attempts to completely replace fresh food with quality dry feed for maturation purposes. One recent example concerned a quarantine trial that was implemented in order to prepare the import of a new strain of *L. stylirostris* from Hawaii to New Caledonia so as to increase the level of polymorphism to improve resistance to pathogenic bacteria and to maintain the reproduction potential. Another example was that of marine fish, the reproduction of which was dependent on fresh food in the past. However, the development of formulations has led to the feeding of a complete dry feed that has allowed the reproduction of several generations of seabream and seabass in captivity.

Nevertheless, for most species, reproduction in captivity still relied on eyestalk ablation, and nutritional solutions (including additives) that would enable the industry to maintain shrimp in captivity and induce ovary development without ablation in the future are still required. The growing industry of shrimp production expected an improvement in the

Table 21 Formulation of maturation feed (CPSP for fish protein concentrate).

	%	*CP*	*lipids*	*n-3*	*n-6*	*n-3/n-6*	*cholesterol*	*20:4n-6*
squid meal NZ	35	75	3.5	1.4	0.2	8.2	0.42	0.04
Mussel	10	70	0.6					0.06
lactic yeast	15	50	0.7					
native wheat gluten	12	80						
CPSP[90]	7	90						
PX_1	1	50						
wheat starch	4							
fish oil	5		5	1.5	0.3	7.5	0.1	0.05
aqualipid	4		4	0.4	2.4			0.04
cholesterol	1		1			0.2	1	
vit mix (1)	2							
min mix (2)	3.2							
trace elements (3)	0.8							
ethoxyquin	0.02							
	100	***57***	***15***	*3*	*2.4*	***1***	*2*	*1.4*
% CP	57							
% lipids	14							
% HUFA	3			3				
% cholesterol	2							
% astaxanthin	60 ppm							
% LYS	2.5							
% ARG	2.5							
% HIS	1							
energy MJ/kg	22							
DE MJ/kg	19							
Cbh %	7							
EAA/prot. %	>40%							
A IU/kg	.6M							
D IU/kg	.12M							
E g/kg	8							
B1 mg/kg	1425							
B2 mg/kg	4000							
B6 mg/kg	800							
B_{12} mg/kg								
biotin mg/kg	120							
folic acid mg/kg	300							
inositol g/kg	80							
choline g/kg	240							
vit C APP g/kg	40							
menadione g/kg	0.8							

Contd.

Contd.

PAB g/kg	80
Se ppm	2
Cu ppm	30
Ca %	1
P %	0.8
K %	1
Mg %	0.5
Ca/P	1.0:1.0

areas of maintenance and nutrition of breeding animals for the management of several families under selection for different characters (Goyard et al., 2004). Maintaining animals in good health represented a constant concern at research level and of course at farming level. Scientific progress in the understanding of the nutritional requirements of penaeids should benefit other species such as crabs, *Squilla*, and spiny lobster. Furthermore, the predictions of an expansion of crustacean culture in the coming years would be based on continuing research efforts and the objective of satisfying the growing demand for seafood. This review was not exhaustive. Species such as *P. semisulcatus, Metapenaeus monoceros, P. schmitti, P. notialis* and *P. merguiensis* were not described in order to allow the authors to concentrate on species of commercial importance.

Acknowledgements

Dr J. Guillaume advised us to write this review and gave us continuous support. We would like to express our sincere gratitude to him. Also, we would like to thank Viviane Le Berre for her prodigious help in maintaining the highest standard in relation to the references.

References

Ai, C., L. Chen, Z. Zhou and X. Wen. 2000. Proceedings of the Third World Fisheries Congress: Feeding the World with Fish in the Next Millenium—the Balance between Production and Environment. pp. 223-229. Am. Fish. Soc. Symp., Vol. 38.

Alava, V.R., and A. Kanazawa. 1995. The effect of nutrition on crustacean gonadal development and reproduction. *In:* Abstracts of the 5th Intl Working Group on Crustacean Nutrition Symposium, Kagoshima University, Japan.

Alfaro, J. 1990. A contribution to the understanding and control of the male reproductive system melanization disease of broodtsock *Penaeus setiferus*. Texas A & M University, 1-65 pp.

Alfaro, J., A.L. Lawrence and D. Lewis. 1993. Interaction of bacteria and male reproductive system blackening disease of captive *Penaeus setiferus*. Aquaculture, 117: 1-8.

Alvares del Castillo Cueto, M. 1988. Comparaison des performances de reproduction de *P. indicus* nourrie avec des moules fraîches et des moules congelées. Etude de α-tocopherol comme indicateur biochimique de la qualité de la nourriture et des œufs. Mém. DEA Univ. Aix-Marseille, Fac. Sci. Luminy, 42 pp.

Aquacop. 1975. Maturation and spawning in captivity of penaeid shrimps: *Penaeus merguiensis* de Man, *P. japonicus* Bate, *P. aztecus* Ives, *Metapenaeus ensis* de Man and *P. semisulcatus* de Man. Proc. World Maricult. Soc., 6: 123-132.

Aquacop. 1977. Observations on the maturation and reproduction of penaeid shrimp in captivity in a tropical medium. 3rd Aquaculture workshop, ICES, 10-13 May, Brest, France. Actes de colloques du CNEXO, 4: 157-178.

Aquacop 1982. Constitution of broodstock, maturation, spawning and hatching systems for penaeids shrimp in COP. In: Handbook of mariculture vol. I. Crustacean Aquaculture, CRC Press, Inc. Boca Raton, Florida, USA 1983. 12pp.

Aquacop. 1983. Constitution of broodstock, maturation, spawning and hatching systems for penaeid shrimps in the Center Océanologique du Pacifique. *In:* Handbook of Aquaculture, CRC Press Inc., Boca Raton, Florida, pp. 105-121.

Aquacop. 1989. Selected ingredients for shrimp feed. Advances in tropical aquaculture, Tahiti, 20 Feb-4 March 1989. Aquacop, Ifremer Actes colloque 9 pp. 405-412.

Beard, TW., and J.F. Wickins. 1980. Breeding of *Penaeus monodon* Fabricius in laboratory recirculation systems. Aquaculture, 20(2): 79-89.

Bottino, N.R., J. Gennity, M.L. Lilly, E. Simmons and G. Finne. 1980. Seasonal and nutritional effects on the fatty acids of three species of shrimp, *P. setiferus, P. aztecus* and *P. duorarum*. Aquaculture, 19: 139-148.

Boucard, C., J. Patrois and H.J. Ceccaldi. 2004. Exhaustion of lipid reserves in the HP of F. indicus in relation to successive spawnings. Aquaculture, 236(1): 523-537.

Bray, W.A., A.L. Lawrence and L.J. Lester. 1990. Reproduction of eyestalk-ablated *Penaeus stylirostris* fed various levels of total dietary lipid. J. World Aquacult. Soc., 21(1): 41-52.

Brown, A., Jr., J. Macvey, B.S. Middleditch and A.L. Lawrence. 1979. Maturation of white shrimp (*Penaeus setiferus*) in captivity. World Maricult. Soc., 10: 435-444.

Brown, A., J. McVey, B.M. Scott, T.D. Williams and A.L. Middleditch. 1980. The maturation and spawning of white shrimp *Penaeus stylirostris* under controlled laboratory conditions. Proc. World Maricult. Soc., 11: 488-499.

Cahu, C., C. Fauvel and Aquacop. 1986. Effect of food fatty acids composition of *P. vannamei* broodstock on egg quality. ICES Maricult. Committee CM.F:28, 8 pp.

Cahu, C., et al. 1987. Comparison of the spawning rates of *L. vannamei* in 3 experimental conditions with different substrates. ICES Maricult. Committee CM.F publication, 5 pp.

Cahu, C., J. Guillaume, G. Stephan and L. Chim. 1988. Essentiality of phospholipids and highly unsaturated fatty acids in *P. vannamei* spawners fed purified diets. J. World Aquacult. Soc., 2: 25-29.

Cahu, C., J.C. Guillaume, G. Stéphan and L. Chim. 1994. Influence of phospholipid and highly unsaturated fatty acids on spawning rate and egg and tissue composition in *P. vannamei* fed semi-purified diets. Aquaculture, 126: 159-170.

Cahu, C., G. Cuzon and P. Quazuguel. 1995. Effect of some biochemical components of broodstock diet on egg composition in *P. indicus*. Comp. Biochem. Physiol., A, 112(3-4): 417-424.

Cavalli, RO., G. Menschaert, P. Lavens and P. Sorgeloos. 2000. Maturation performance, offspring quality and lipid composition of *Macrobrachium rosenbergii* females fed increasing levels of dietary phospholipids. Aquacult. Intl., 8(1): 41-58.

Cavalli, R.O., M. Tamtin, P. Lavens, P. Sorgeloos, H.J. Nelis, and A. De Leenheer. 2001. The content of ascorbic acid and tocopherol in the tissues and eggs of wild Macrobrachium rosenbergii during maturation. J. Shellfish Res., 20(3): 939-943.

Cavalli, R.O., F.M. Batista, P. Lavens, P. Sorgeloos, H.J. Nelis and A. De Leenheer. 2003. Effect of dietary supplementation of vitamins C and E on maternal performance and larval quality of the prawn *Macrobrachium rosenbergii*. Aquaculture, 227(1-4) : 131-146.

Ceccaldi, H.J., and J.P. Andre. 1971. Note sur l'ovogenese chez le peneide *P. Kerathurus*. Rapp. Comm. Int. Mer Méditer., 20(3): 263.

Cevallos-Vázquez, P., C. Rosas and I. Racotta. 2003. Sperm quality in relation to age and weight of white shrimp *Litopenaeus vannamei*. Aquaculture, 229: 141-151.

Chamberlain, G.W., and A.L. Lawrence. 1981. Maturation, reproduction and growth of *Penaeus vannamei* and *Penaeus stylirostris* fed natural diets. Proc. World Maricult. Soc., 12: 209-224.

Chim, L., R. Galois, J.L.M. Martin, P. Lemaire, N. Wabete, J.C. Massabuau and G. Cuzon. 2003. Temperature effects on the nutrition of *Litopenaeus stylirostris*. Consequences on the formulation and distribution of food in function of rearing seasons. Styli 2003. Thirty years of shrimp farming in New Caledonia. Proc. Symp., Noumea-Kone, 2-6 June 2003. Actes Colloq. Ifremer. n° 38, pp. 99-105.

Crocos, P.J., and G.J. Coman. 1997. Seasonal and age variability in the reproductive performance of *Penaeus semisulcatus* broodstock: optimizing broodstock selection. Aquaculture, 155: 57-69.

Cruz-Suarez, L.E., D. Ricque and Aquacop. 1992. Effect of squid meal on growth of *Penaeus monodon* juveniles reared in pond pens and tanks. Aquaculture, 106(3-4): 293-299.

Cuzon, G., and Aquacop. 1998. Nutritional review of *Penaeus stylirostris*. Rev. Fish. Sci., 6: 129-141.

Cuzon, G., J. Patrois, C. Cahu and Aquacop. 1995. Some fundamentals of penaeids breeders nutrition. *In:* Proc. III Congreso Ecuatoriano de Acuicultura, Guayaquil, Oct.-Nov. 1995.

Cuzon, G., L. Arena, J. Goguenheim, E. Goyard and Aquacop. 2004. Is it possible to raise offspring of the 25th generation of *Litopenaeus vannamei* (Boone) and 18th generation *Litopenaeus stylirostris* (Stimpson) in clear water to 40 g? Aquacult. Res., 1-9.

D'Abramo, L.R., and S.S. Sheen. 1994. Nutritional requirements, feed formulation, and feeding practices for intensive culture of the freshwater prawn *Macrobrachium rosenbergii*. Rev. Fish. Sci., 2(1): 1-21.

Diaz, A.C., and J.L. Fenucci. 2004. Effect of prepared diet on the induction of precocious maturation in *Pleoticus muelleri* Bate (Crustacea, Penaeoidea) Aquacult. Res., 35(12): 1166-1171.

Djunaidah, I.S., M. Wille, E.K. Kontara and P. Sorgeloos. 2003. Reproductive performance and offspring quality in mud crab (Scylla paramamosain) broodstock fed different diets. Aquacult. Intl., 11(1-2): 3-15.

Dougherty, W.J., and M.M. Dougherty. 1990. Ultrastructural observations on melanized sperm in developing and fully formed spermatophores of male shrimp, *Penaeus vannamei*. Academic Press, London, pp. 387-394.

Draper, H. 1980. Role of vitamin E in plants, microbes, invertebrates and fish. *In:* Vitamin E: A Comprehensive Treatise. L. Lachlin (ed.). New York, pp. 391-396. Machlin L.P. (editor), E. Deuuer, New York, NY.

Du, S., C. Hu, Qi Shen and S. Zheng. 2005. Comparative study on biochemical composition of main natural diets for broodstock *Litopenaeus vannamei*. J. Trop. Oceanogr./Redai Haiyang Xuebao., 24(1): 50-59.

Fakhfakh, M. 1989. Role of α-tocopherol on reproduction of *P. indicus;* experiments with marine food and compounded feeds. Mém. ISPA option aquaculture, 92 pp.

Fremont, L., C. Leger, B. Petridou and M.T. Gozzelino. 1984. Effects of a (n-3) polyunsaturated fatty acid-deficient diet on profiles of serum vitellogenin and lipoprotein in vitellogenic trout (*Salmo gairdneri*). Lipids, 19(7): 522-528.

Galgani, M.L., and Aquacop. 1989. Influence du régime alimentaire sur la reproduction en captivité de *P. vannamei* et *L. stylirostris*. Aquaculture, 80: 97-109.

Galois, R. 1983. Aspects du métabolisme lipidique chez qq crustacés décapodes natantia. Thèse doctorat état, Univ. Aix-Marseille II, 273 pp.

Gelabert, R., L. Ramos, M. Oreyana, M. Mascaró, A. Sánchez, L.A. Soto, C. Re and C. Rosas. 2003. The effect of including *Artemia franciscana* biomass in the diet of *Litopenaeus setiferus* and *Litopenaeus schmitti* Broodstock. Rev. Invest. Mar., 24: 29-40.

Goguenheim, J., J. Barret, J. Patrois, C. Cahu, C. Fauvel and Aquacop. 1987. *P. vannamei* broodstock constitution, maturation and artificial insemination. Proc. WAS Guayaquil, p. 63.

Goimier, Y., C. Pascual, A. Sanchez, G. Taboada, G. Gaxiola, A. Sanchez and C. Rosas. 2006. Relation between reproductive, physiological and immunological conditions of *L. setiferus* males fed dietary protein levels. Anim. Reprod. Sci., 92: 193-208.

Goyard, E., E. Bedier, J. Patrois, V. Vonau, D. Pham, G. Cuzon, L. Chim, L. Penet, T. Dao and P. Boudry. 2004. Synthesis of the "shrimp" genetic programme in Tahiti: What profit for Caledonian shrimp industries? Styli 2003. Thirty years of shrimp farming in New Caledonia. Proc. Symp., Noumea-Kone, 2-6 June 2003. Actes Colloq. Ifremer. no. 38: 126-133.

Guillaume, J. 1999. Généralités. *In:* Nutrition et Alimentation des Poissons et Crustacés. J. Guillaume, S. Kaushik, P. Bergot and R Métailler (eds.). INRA, Brest, France, pp. 25-39.

Harrison, K.E. 1990. The role of nutrition in maturation, reproduction and embryonic development of decapod crustaceans: a review. J. Shellfish Res., 9(1): 1-28.

Hill, E.M., D.L. Holland and J. East. 1993. Egg hatching activity of trihydroxylated eicosanoids in the barnacle *Balanus balanoides*. Biochim. Biophys. Acta., 1157(3): 297-303.

Jeckel, W.H., J.E. Aizpun and V.J. Moreno. 1989. Biochemical composition, lipid classes and fatty acids in the male reproductive system of *Pleoticus muelleri* Bate. Comp. Biochem. Physiol., 93B(4): 807-811.

Kanazawa, A., 1981. Penaeid nutrition. Proceedings of the 2nd Intl. Conf. on Aquaculture nutrition. Biochemical and physiological approaches to Shellfish nutrition. Special Pub. N°2 WMS, America, p. 87-105.

Kian, A.Y.S., S. Mustafa and R.A. Rahman. 2004. Broodstock condition and egg quality in tiger prawn, *Penaeus monodon* resulting from feeding bioencapsulated live prey. Aquacult. Intl., 12(4-5): 423-433.

Laubier-Bonichon, A. 1978. Ecophysiology of reproduction in *P. japonicus*. Three years of experiments in control conditions. Oceanol. Acta, 1: 135-150.

Lawrence, A., Y. Akamine, B. Middelditch, G. Chamberlain and D. Huchins. 1980. Maturation and reproduction of *P. setiferus* in captivity. Proc. World Maricult. Soc., 11: 481-487.

Lee, R.F., and D.L. Pupione. 1978. Serum lipoproteins in the spiny lobster *Palinurus interruptus*. Comp. Biochem. Physiol., 59B: 239-243.

Lilly, M.L., and N.R. Bottino. 1981. Identification of arachidonic acid in Gulf of Mexico and degree of biosynthesis in *P. setiferus*. Lipids, 16: 871-875.

Lytle J.S., T.S. Lytle and J.T. Ogle. 1990. Polyunsaturated fatty acids profile as a comparative tool in assessing maturation diets of *P. vannamei*. Aquaculture, 89: 287-299.

Marsden, G.E., J.J. McGuren, S.W. Hansford and M.J. Burke. 1997. A moist artificial diet for prawn broodstock: Its effect on the variable reproductive performance of wild caught *Penaeus monodon*. Aquaculture, 149: 145-156.

Mendoza R., A. Revol, C. Fouve, J. Patrois and G.S. Guillaume. 1997. Influence of squid extract on the triggering of secondary vitellogenesis in *Penaeus vannamei*. Aquacult. Nutr., 3(1): 55-63.

Mengqing, L., J. Wenjuan, C. Qing and W. Jialin. 2004. The effect of vitamin A supplementation in broodstock feed on reproductive performance and larval quality in *Penaeus chinensis* Aquacult. Nutr., 10(5): 295-300.

Meunpol, O., P. Meejing and S. Piyatiratitivorakul. 2005. Maturation diets based on fatty acids content for male *P. monodon* (Fabricius) broodstock. Aquacult. Res., 32: 1216-1225.

Middleditch, B.S., S.R. Missler, H.B. Hines, J.P. McVey, A. Brown, D.G. Ward and A.L. Lawrence. 1980. Metabolic profiles of penaeid shrimp: dietary lipids and ovarian maturation. J. Chromatogr., 195(3): 359-368.

Millamena, M. 1996. Reproduction of *Penaeus monodon* and nutrition constraints. Aquaculture, 102(2): 98-101.

Moore, D.W., R.W. Sherry and F. Montanez. 1974. Maturation of *Penaeus californiensis* in captivity. Proc. World Maricult. Soc., 5: 445-450.

Moreau, R., G. Cuzon and J. Gabaudan. 1998. Efficacy of silicone-coated ascorbic acid and ascorbyl-2-polyphosphate to fast-growing tiger shrimp (*Penaeus monodon*) Aquacult. Nutr., 4(1): 23-29.

Naessens, E., P. Lavens, L. Gomez, C.L. Browdy, K. McGovern-Hopkins, A.W. Spencer, D. Kawahigashi and P. Sorgeloos. 1997. Maturation performance of *Penaeus vannamei* co-fed *Artemia* biomass preparations. Aquaculture, 155(1-4): 89-103.

Ogle, J.T. 1994. The effect of diet on mating in *Penaeus vannamei. In:* Book of Abstracts, World Aquaculture, '94, WAS Baton Rouge, Louisiana, pp. 315 (abstract only).

Ottogalli, L., C. Galinie and D. Goxe. 1988. Reproduction in captivity of *Penaeus stylirostris* over ten generations in New Caledonia. J. Aquacult. Trop., 3: 111-125.

Pangantihon-Kuhlmann, M.P., O. Millamena and Y. Chern. 1998. Effect of dietary astaxanthin and vitamin A on the reproductive performance of *Penaeus monodon* broodstock. Aquat. Living Resour., 11: 403-409.

Pascual, C., G. Gaxiola and C. Rosas. 2003a. Blood metabolites and hemocyanin of the white shrimp *Litopenaeus vannamei*: the effect of culture conditions and a comparison with other crustacean species. Mar. Biol., 142: 735-745.

Pascual, C., A. Sánchez, A. Sánchez, F. Vargas-Albores, G. LeMoullac and C. Rosas. 2003b. Haemolymph metabolic variables and immune response in *Litopenaeus setiferus* adult males: the effect of an extreme temperature. Aquaculture, 218: 637-650.

Patrois J., G. Cuzon and Aquacop. 1990. Mise au point d'un aliment maturation, I. Comparaison de 3 aliments, cas de *P. vannamei*. Rapport Interne IFREMER COP Tahiti, 13 pp.

Patrois, J., P. Levy, G. Cuzon and Aquacop. 1991. Mise au point d'un aliment maturation, II. Effet de la supplementation du Monogal en vitamine E, cas de *P. vannamei*. Rapport Interne IFREMER, 10 pp.

Payen, G. 1980. Aspects fondamentaux de l'endocrinologie de la reproduction chez les crustacés marins. Oceanis, 5(3): 309-339.

Peixoto, S., R.O. Cavalli, W. Wasielesky, F. d'Incao, D. Krummenaeur and A.M. Milach 2004. Effects of age and size on reproductive performance of captive *F. paulensis* broodstock. Aquaculture, 238: 173-182.

Primavera, J., C. Lim and E. Borlongan. 1979. Feeding regimes in relation to reproduction and survival of ablated *P. monodon*. Kalikasan Philipp. J. Biol., 8: 227-235.

Quinitio, E.T., F.D. Parado-Estepa, O.M. Millamena, and H. Biona. 1996. Reproductive performance of captive *Penaeus monodon* fed various sources of carotenoids feeds for small-scale aquaculture. Proc. National Seminar-Workshop on Fish Nutrition and Feeds. Tigbauan, Iloilo, Philippines, 1-2 June 1994, pp. 74-82.

Racotta, I.S., E. Palacios and A.M. Ibarra. 2003. Shrimp larval quality in relation to broodstock condition. Aquaculture, 227(1-4): 107-130.

Ramos, L., C. Vazquez Boucard, M. Oliva, E. Garcia and I. Fernandez. 1996. Variations in the total lipids content, lipids and fatty acids class during the ovarian maturation in the *Penaeus schmitti* and *P. notialis* Rev. Invest. Mar., 17(2-3) : 157-165.

Ricardez-Carrion, R. 1985. Analyse des techniques d'élevage des crevettes pénéides dans la région du Pacifique et du SEA. Thèse Université Bretagne Occidentale, 355 pp.

Rosas, C., A. Sánchez, M.E. Chimal, G. Saldaña, L. Ramos and L.A. Soto. 1993. Effect of electrical stimulation on spermatophore regeneration in white shrimp *Penaeus setiferus*. Aquat. Living Resour., 8: 139-144.

Rosas, C., A. Bolongaro-Crevenna, A. Sánchez, G. Gaxiola, L. Soto and E. Escobar. 1995. Role of the digestive gland in the energy metabolism of *L. setiferus*. Biol. Bull., 189: 168-174.

Samuel, M.J., T. Kannupandi and P. Soundarapandian. 1998. Fatty acid profile during embryonic development of cultivable freshwater prawn *Macrobrachium malcolmsonii* (H. Milne Edwards). Indian J. Fish., 45(2): 141-148.

Sanchez, A., C. Pascual, A. Sanchez, F. Vargas-Albores, G. LeMoullac and C. Rosas. 2001. Hemolymph metabolic variables and immune response in *L. setiferus* males, effect of acclimation. Aquaculture, 198: 13-28.

Sánchez, A., C. Pascual, A. Sánchez, F. Vargas-Albores, G. LeMoullac and C. Rosas. 2002. Acclimation of adult males of *Litopenaeus setiferus* exposed at 27°C and 31°C: Bioenergetic balance. *In*: Modern Approaches to the Study of Crustacea. E.G. Escobar-Briones and F. Alvarez (eds.). Kluwer Academic Press, New York, pp. 45-52.

Shigueno, K. 1975. Shrimp culture in Japan. AITP, Tokyo (Japan). 153 p.

Smith, G.G., A.J. Ritar, D. Johnston and G.A. Dunstan. 2004. Influence of diet on broodstock lipid and fatty acid composition and larval competency in the spiny lobster, *Jasus edwardsii*. Aquaculture, 233(1-4): 451-475.

Sonia, F., C. Sierra, S. Bouquelet, C. Brassart, C. Agundis, E. Zenteno and L. Vazquez. 2006. The effect of sugars and free amino acids from the freshwater prawn *Macrobrachium rosenbergii* hemolymph on lectin activity and on oxidative burst. Comp. Biochem. Physiol. C Toxicol. Pharm., 142(3-4): 212-219.

Taboada, G., G. Gaxiola, T. Garcia, R. Pedroza, A. Sanchez, L.A. Soto and C. Rosas. 1998. Oxygen consumption and ammonia-N excretion related to protein requirements for growth of white shrimp, *P. setiferus* (L.) juveniles. Aquacult. Res., 29: 823-833.

Teshima, S. and A. Kanazawa, 1978. Release and transport of lipids in the prawn. Bull. Jap. Soc. Sci. Fish., 44(11): 1269-1274.

Teshima, S., and A. Kanazawa. 1980. Transport of dietary lipids and role of serum lipoproteins in the prawn. Bull. Jap. Soc. Sci. Fish., 46(1): 51-55.

Tirado, M.C., J.M. Lotz and W.D. Youngs. 1998. Effect of maturation diets on *Penaeus vannamei* male performance. Aquaculture '98, Book of Abstracts. p. 543.

Villette, M. 1990. Influence des phospholipides et de l'α-tocopherol sur les performances de reproduction de *Penaeus indicus*. Mémoire DAA ENSAT, Brest, 49 pp.

Wen, X.B., L.Q. Chen, Z.L. Zhou, C.X. Ai and G.Y. Deng. 2002. Reproduction response of chinese mitten crab mitten handed crab (*Eriocheir sinensis*) fed different sources of dietary lipid. Comp. Biochem. Physiol., A, 131: 675-681.

Wounters, R.S. 1999. Feeding enriched *Artemia* biomass to *P. vannamei*: effect on performance and larval quality. J. Shellfish Res., 18(2): 651-656.

Wurtz, W., and R. Stickney. 1984. An hypothesis on the light requirements for pawning penaeid shrimp, with emphasis on *Penaeus setiferus*. Aquaculture, 1: 93-98.

Wyban, J., G. Martinez and J. Sweeney. 1997. Adding paprika to *P. vannamei* maturation diet improves nauplii quality. J. World Aquacult. Soc., 28(2): 59-62.

Xu, X.L., W.J. Ji, J.D. Castell and R.K. O'Dor. 1994. Influence of dietary lipid sources on fecundity, egg hatchability and fatty acid composition of Chinese prawn (*Penaeus chinensis*) broodstock. Aquaculture, 119(4): 359-370.

Ying, X., Y. Zhang and W. Yang. 2004. Comparative study on fatty acids composition in crab *Eriocheir sinensis* during maturation, spawning and miscarriage. Oceanol. Limnol., 35(2): 141-148.

Zandee, D.I., and E.C.J. Kruitwagen. 1975. Depot sterols in comparison with structural sterols in *Cancer pagurus* and *Eriocheir sinensis*. Neth. J. Sea Res., l. 9(2): 214-221.

9

Coordination of Reproduction and Molt in Decapods

S. Raviv[1], S. Parnes[1] and A. Sagi[1,2]

[1]Department of Life Sciences and [2]National Institute for Biotechnology in the Negev, Ben-Gurion University of the Negev, P.O. Box 653, Beer-Sheva 84105, Israel

INTRODUCTION

Factors Affecting the Reproductive Cycle

The molt cycle is a constitutive phenomenon in crustaceans, with molting being a complex and energy-demanding process. Therefore, most aspects of the life history of crustaceans are, at least to some extent, synchronized with the molt cycle (Chang, 1995). Reproduction and metabolism in crustaceans depend on the molt stage, and all three phenomena—molt, reproduction and metabolism—are correlated with the seasons (Aiken, 1969a; Conan, 1985; Bouchon et al., 1992). Breeding cycles are correlated with seasonal changes in such a way that the offspring are produced at a time most favorable to their survival. The relative importance of particular environmental factors may vary among different species and indeed among different environments (Sastry, 1983; Dall et al., 1990; Bauer, 1992; Bouchon et al., 1992). Generally, the frequency and amplitude of changes in temperature and day length are the proximate factors controlling the temporal patterns of reproduction, especially in species inhabiting the subtropical and temperate latitudes, i.e., in areas with clear differences between seasons (Stephens, 1952; Aiken, 1969b; Rice and Armitage, 1974; Penn, 1980; Lipcius and Herrnkind, 1987; Nelson et al., 1988; Justo et al., 1991; Ituarte et al., 2004). In some species, the reproduction cycle is synchronized with the circatidal and circalunar cycles (Nascimento et al., 1991; Morgan and Christy, 1994; Morgan, 1996; Yamahira, 2004; Skov et al., 2005).

The interrelationships between reproductive and molt cycles during the reproductive season are discussed in this chapter. Environmental factors are dealt with only as they pertain to reproduction. Aspects of energy allocation for growth versus reproduction and the apparent competition between these two processes are addressed in brief. A large number of studies describe reproduction in decapod crustaceans (Meusy and Payen, 1988; Wilder et al., 2002), but only a few studies relate reproduction to the molt cycle (reviewed previously by Adiyodi and Adiyodi, 1970; Hartnoll, 1985; Nelson, 1991). The reproductive cycle comprises gonad development and copulation plus spawning and, in Pleocyemata, also brooding (this latter feature is common to this suborder and distinguishes it from the other decapod suborder, Dendrobranchiata). A variety of strategies for coordinating the reproductive and molt cycles are found in the different decapod species; in some, each of the above-mentioned reproductive stages is synchronized with the molt cycle, while in others there is no apparent synchronization between molt and reproduction.

Integumental Machinery Essential for Mating and Brooding

As a rule, in organisms with exoskeletons, any morphological change necessary for reproductive activity must involve a molting event. Successful brooding may depend on overt integumental modifications (e.g., abdominal broadening and changes in setal length and in the morphology of the seminal receptacle). The changes may be abrupt, occurring in a single puberty molt, or they may develop gradually with age and size or disappear and reappear according to the breeding season (common in non-continuously reproductive species). In crab species, for example, an abdominosternal chamber is formed only at the terminal molt in *Ebalia tuberosa* and *Chionoecetes opilio* (Brachyura) (Schembri, 1982; Alunno-Bruscia and Sainte-Marie, 1998), and in *Maja squinado* (Brachyura) the spermathecae (an enlarged portion of the genital duct in which sperm is stored and, in species with indeterminate growth, retained over molting) develop only after the terminal molt (Gonzalez-Gurriaran et al., 1998). In the lobster *Homarus americanus* (Astacidea), the development of ovigerous setae in the female parallels the gradual broadening of the second abdominal segment, rather than occurring suddenly at a maturity molt (Aiken and Waddy, 1980). In the spiny lobster *Panulirus argus* (Palinura), in addition to a gradual change, ovigerous setae lengthen and shorten in pre- and post-reproductive molt cycles, respectively (Lipcius and Herrnkind, 1987). Loss of ovigerous setae during the non-reproductive season has been observed in *Crangon crangon* (Caridea: Bomirski and Klek, 1974), *Pandalus borealis* (Caridea: Bergstrom, 2000) and *Jasus edwardsii* (Palinura: MacDiarmid, 1989), whereas ovigerous setae, once formed, remain permanent structures that are retained throughout molt in *Panulirus homarus* and *Palinurus delagoae* (Palinura: Berry, 1971a; Berry, 1971b). In *Jasus lalandii* the nature of ovigerous setae has not been established unequivocally (Palinura: Fielder, 1964; Berry, 1971b; Lipcius, 1985). Palaemonidae species (Caridea) have extra setae for egg bearing during the breeding season, and in these species the periodic appearance of chromatophores is believed to play a camouflage role in otherwise transparent animals (Antheunisse et al., 1968). In the female *Penaeus latisulcatus*, the thelycum (external organ into which the female receives and stores spermatophores from the male during copulation) develops gradually, with development being completed by the time the carapace length is about 25 mm, but mated females have rarely been observed with a carapace length of less than 28 mm (Penn, 1980). An exception to the general requirement for molting to precede reproductive activity is the decalcification of the opercula (lids covering the female vulvae) prior to mating in brachyuran species (Hartnoll, 1969; Fukui, 1993; Brockerhoff and McLay, 2005).

Strategies for Coordination of Molt and Reproductive Cycles in Decapods

In decapods, molt and reproduction are coordinated in a wide variety of ways. Some reproductive strategies can be related to a particular taxon, and others may be found in a number of taxa. Alternatively, various species within the same taxon might exhibit different strategies. Moreover, a single species might use more than one reproductive strategy.

The reproductive cycle is regulated by internal cues that are related to the physiological nature of a species (e.g., molt cycle interval, integumental reproductive-related machinery, energy allocation, maternal care, and embryonic development of the offspring) and by external cues related to the environment (e.g., habitat and season). Different reproductive events, i.e., mating and oviposition, may be coupled to different molt stages (i.e., A-E). Since mating may indicate the

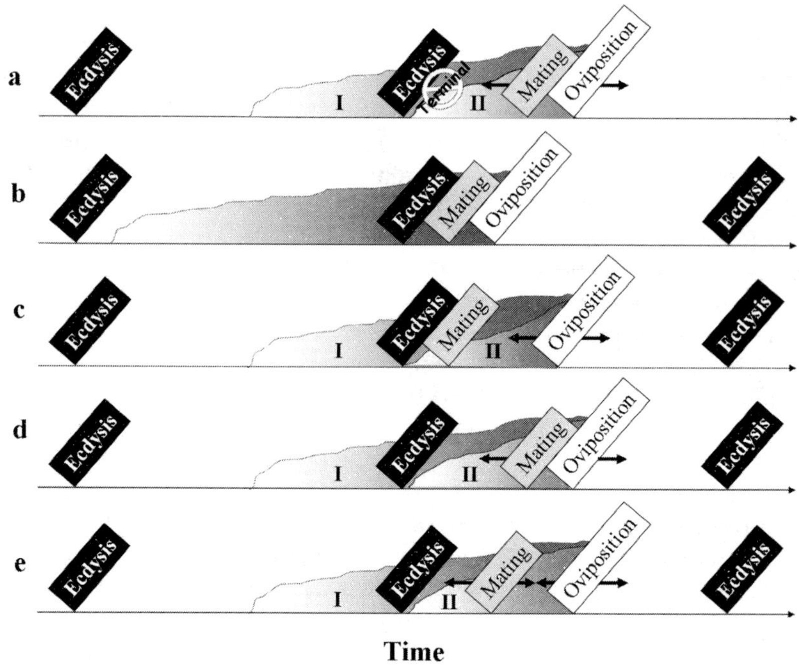

Fig. 1 Five strategies describing the different ways in which the reproductive cycle is coordinated with the molt cycle. To cover an entire reproductive cycle, two consecutive molt cycles are drawn for which the first cycle does not include spawning. a. Species that cease to grow in a pubertal molt, and mating/s followed by oviposition/s occur when the female exoskeleton is hard. b. Species in which the ovary develops in a parturial molt and that mate and oviposit immediately after ecdysis. c. Species in which mating is coupled to ecdysis; mating occurs when the female is soft-shelled. Oviposition/s take/s place during intermolt and in some species also in premolt. d. Species that oviposit in concert with mating during intermolt or premolt, while the female is hard-shelled. e. Species in which ecdysis, mating, and oviposition are not coupled and are relatively independent events. Two possible alternatives for the vitellogenic periods are marked by gradient shading I and II. I relates to species whose ovaries begin to develop during the previous molt cycle, and II applies to those species that do not carry developing ovaries through ecdysis. Black arrows indicate how timing may vary according to the vitellogenesis rate.

onset, acceleration (Fig. 1c), or completion (Fig. 1a, b and d) of vitellogenesis and since oviposition indicates the start of egg incubation (in Pleocyemata), the sequence of mating and oviposition combined with the occurrence of ecdysis comprise different reproductive strategies, which we have classified into the five categories shown in Fig. 1. The patterns shown in Fig. 1 have been generalized and are not scaled to time: A particular strategy can thus describe the reproductive cycle found in species belonging to different taxa, even though the time scale may vary considerably among the different species. Specific examples of the five categories of reproductive strategy are given in Table 1.

Table 1. Examples for the five categories described in Fig. 1 in which the reproductive cycle is coordinated with the molt cycle.

Strategy	*Species*	*Infraorder or Superfamily*	*Reference*
a	*Ebalia tuberosa*	Brachyura	Schembri, 1982
a	*Inachus dorsettensis*	Brachyura	Jones and Hartnoll, 1997
a	*Libinia emarginata*	Brachyura	Hinsch, 1972
a	*Maja brachydactyla*	Brachyura	Gonzalez-Gurriaran et al., 1998
a	*Maja squinado*	Brachyura	Corgos et al., 2006
b	*Alpheus normanni*	Caridea	Bauer, 1989
b	*Callianassa japonica*	Thalassinidea	Tamaki et al., 1996
b	*Crangon crangon*	Caridea	Oh and Hartnoll, 2004
b	*Emerita asiatica*	Anomura	Gunamalai and Subramoniam, 2002
b	*Hippolyte curacaoensis*	Caridea	Bauer, 1989
b	*Jasus edwardsii*	Palinura	MacDiarmid, 1989
b	*Latreutes fucorum*	Caridea	Bauer, 1989
b	*Latreutes parvulus*	Caridea	Bauer, 1989
b	*Lysmata californica*	Caridea	Bauer and Newman, 2004
b	*Macrobrachium rosenbergii*	Caridea	O'Donovan, 1984
b	*Metanephrops andamanicus*	Astacidea	Aiken and Waddy, 1980
b	*Nephrops norvegicus*	Astacidea	Farmer, 1974
b	*Palaemon gravieri*	Caridea	Kim and Hong, 2004
b	*Palaemonetes pugio*	Caridea	Bauer and Abdalla, 2000
b	*Pandalus hypsinotus*	Caridea	Bergstrom, 2000
b	*Paralithodes brevipes*	Anomura	Wada et al., 1997
b	*Periclimenes americanus*	Caridea	Bauer, 1989
b	*Philocheras trispinosus*	Caridea	Oh and Hartnoll, 1999
b	*Stenopus hispidus*	Stenopodidea	Zhang et al., 1998
b	*Thor manningi*	Caridea	Bauer, 1989
First cycle b, thereafter a	*Chionoecetes bairdi*	Brachyura	Adams and Paul, 1983; Paul, 1984
First cycle b, thereafter a	*Chionoecetes opilio*	Brachyura	Elner and Beninger, 1995; Sainte-Marie et al., 1999
c	*Cancer magister*	Brachyura	Shields, 1991
c	*Cancer pagurus*	Brachyura	Shields, 1991

Contd.

Contd.

c	*Fenneropenaeus indicus*	Penaeoidea	Emmerson, 1980; Peeters and Diter, 1994
c	*Fenneropenaeus merguiensis*	Penaeoidea	Crocos and Kerr, 1983
c	*Homarus americanus*	Astacidea	Aiken and Waddy, 1976; Cowan and Atema, 1990
c	*Panulirus interruptus*	Palinura	Aiken and Waddy, 1980
c	*Paratelphusa hydrodromous*	Brachyura	Gupta et al., 1987; Adiyodi, 1968
c	*Penaeus monodon*	Penaeoidea	Primavera, 1979; Beard and Wickins, 1980; Emmerson, 1983; Quinitio, 1993
c	*Sicyonia ingentis*	Penaeoidea	Anderson et al., 1985
First cycle c, thereafter a	*Callinectes sapidus*	Brachyura	Van Engel, 1958
d	*Austropotamobius pallipes*	Astacidea	Holdich, 2002
d	*Cherax destructor*	Astacidea	Sokol, 1988
d	*Cherax quadricarinatus*	Astacidea	Barki et al., 1997; Holdich, 2002
d	*Gaetice depressus*	Brachyura	Fukui, 1993
d	*Helice crassa*	Brachyura	Nye, 1977; Brockerhoff and McLay, 2005
d	*Litopenaeus vannamei*	Penaeoidea	Yano, 1995; Raviv et al., 2006; Parnes et al., 2007
d	*Metopograpsus messor*	Brachyura	Sudha and Anilkumar, 1996; Anilkumar et al., 1999
d	*Palinurus delagoae*	Palinura	Berry, 1971b; Aiken and Waddy, 1980
d	*Panulirus argus*	Palinura	Lipcius, 1985
d	*Panulirus homarus*	Palinura	Berry, 1971a
d	*Pleoticus muelleri*	Penaeoidea	Diaz et al., 2003
e	*Candidiopotamon rathbunae*	Brachyura	Liu and Li, 2000
d or e (temperature dependent)	*Astacus astacus*	Astacidea	Cukerzis, 1988

The most complete separation between molt and reproduction, such that none of the reproductive stages are linked to any particular molt stage, is found in some crab species (Fig. 1 and Table 1, strategy a). These species cease to grow after the terminal molt, which is also the pubertal molt. In this group, copulation may take place immediately after the terminal molt when the females are soft-shelled (Van Engel, 1958), during the intermolt when the females are in the hard-shelled condition (Hinsch,

1972; Schembri, 1982; Jones and Hartnoll, 1997; Gonzalez-Gurriaran et al., 1998; Corgos et al., 2006), or during both stages (Adams and Paul, 1983; Paul, 1984; Bryant and Hartnoll, 1995; Elner and Beninger, 1995). In these species, the ovary matures after the terminal molt, and mating marks the completion of the maturation process. Mating is followed by oviposition, and the brooding of eggs may be accompanied by redevelopment of the ovary. Sperm deposited in a single mating in the female's spermathecae may be used to fertilize a few successive broods, without the need for re-mating. All stages of the reproductive cycle (i.e., ovarian development, mating, spawning and egg brooding) in species adopting this strategy are thus independent of the molt cycle and are not restricted to a particular molt stage.

In sharp contrast, in caridean species, each reproductive stage is restricted to a particular molt stage (Hartnoll, 1985, Fig. 1 and Table 1, strategy b). A complete reproductive cycle spans two molt cycles: In the first, the ovary matures to become fully vitellogenic at ecdysis, and immediately thereafter, during the second molt cycle, mating and spawning take place, followed by brooding of eggs during post- and intermolt, which may last during premolt (Hartnoll, 1985; also see section on Molting as an integral part of the reproductive cycle, below). Decapod taxa other than caridean species follow similar patterns (Table 1, strategy b). In the lobster species *Nephrops norvegicus* and *Metanephrops andamanicus* (Astacidea; Farmer, 1974; Aiken and Waddy, 1980), the females mate while their exoskeleton is soft and oviposit within one month thereafter. The spiny lobster *J. edwardsii* may also be included in the group of decapods for which the reproductive stage is related to a molt stage (MacDiarmid, 1989).

In yet a different strategy, mating is coupled only to ecdysis and not to the completion of ovarian maturation (Fig. 1 and Table 1, strategy c). For example, in the crab *Paratelphusa hydrodromous* (Brachyura), mating occurs in the soft-shelled period immediately after ecdysis, and spawning takes place several months later in the intermolt period. Ovarian maturation, which has progressed slowly during the previous molt cycle, simultaneously with the development of the previous brood, accelerates in the intermolt period (Adiyodi, 1968; Gupta et al., 1987). *Paratelphusa hydrodromous* may spawn up to three times per molt cycle (Adiyodi, 1968). An example from Astacidea is *Homarus americanus*. Females of this species may molt at different vitellogenic stages (Aiken and Waddy, 1976; Cowan and Atema, 1990) but rarely mate while hard-shelled (Waddy and Aiken, 1990).

Chionoecetes bairdi and *Chionoecetes opilio* use a combination of the first and second strategies described above: Copulation takes place in both hard- and soft-shelled conditions. The first reproductive cycle comprises the following stages: premolt ovary maturation, which is coupled to the terminal molt; mating while the female is still in a soft-shelled condition; and oviposition (Fig. 1b). In all the subsequent reproductive cycles, mating occurs with the female in the hard-shelled condition (Fig. 1a) (Adams and Paul, 1983; Paul, 1984; Elner and Beninger, 1995; Sainte-Marie et al., 1999). The blue crab *Callinectes sapidus* (Brachyura) uses a combination of the first and third strategies described above: Females copulate only immediately after the pubertal molt, which is also the terminal molt, while in the soft-shelled condition and while the ovary is immature (Van Engel, 1958). Thus, in this case, an individual's fitness depends solely on the timing of a single event—copulation. Subsequently, the remainder of the sperm stored in the spermathecae may be used to fertilize additional broods of eggs, and thus fertilization of the eggs is not dependent on the molt cycle. This combination of strategies is not a general rule for the entire genus, since other *Callinectes* species continue to molt after the puberty molt (Costa and Negreiros-Fransozo, 1998).

Mating may be coupled solely to ovarian maturation (Fig. 1 and Table 1, strategy d), rather than to ecdysis (as in Fig. 1c), or to both ecdysis and ovarian maturation (as in Fig. 1b) (Greenspan, 1982; Henmi and Murai, 1999; Murai et al., 2002). In the crab *Gaetice depressus* (Brachyura), the ovary is immature during molting (Fukui, 1993). Mating, followed by oviposition, takes place when the ovary is ripe; at this stage, the females are hard-shelled. Oviposition is usually followed by another vitellogenic cycle, and the females are capable of producing more than one brood per molt cycle.

Mating may be independent of both ecdysis and the developmental stage of the ovary (Fig. 1 and Table 1, strategy e). Mating in the freshwater crab *Candidiopotamon rathbunae* can occur all year round, independently of the developmental stage of the ovary, although this species breeds only in a particular season (Liu and Li, 2000). Cases of mating of soft-shelled females (Liu and Li, 2000) and of ovigerous females (Fukui 1993; Liu and Li, 2000), although not common, have been recorded.

In the following sections (on Molt-cycle lengthening by reproductive activity and Inhibition of reproduction by molt-related activity), the reciprocal interrelationships between molt and reproduction are discussed. More detailed discussion is devoted to the second strategy (Fig. 1b), with special reference to taxa of the Caridea (section on Molting as an integral part of the reproductive cycle), and to the third and fourth strategies (Fig. 1c and d), with special reference to taxa of the Penaeoidea (section on Molting as an antagonistic process to the reproductive cycle). The reason for this choice is dictated by the availability of molecular data on the coordination of molt and reproduction in these strategies (discussed in the last sections in this chapter, on Regulation and programming of vitellogenesis and the molt cycle and Male and female reproduction and molt).

MOLT-CYCLE LENGTHENING BY REPRODUCTIVE ACTIVITY

Beyond the question of energy allocation (Nelson, 1991; Hartnoll, 2006), a clash of molting with reproduction could result in inhibition of either one of these processes. Some authors have indeed described inhibition of molting during incubation (Caridea: Hess, 1941; Astacidea: Tack, 1941; Scudamore, 1948; Brachyura: Cheung, 1969). The removal of eggs from the pleopods of egg-bearing females resulted in an earlier onset of molting in some species (Hess, 1941; Scudamore, 1948) but failed to cause significant differences in others (Caridea: Pandian and Balasundaram, 1982). Instances of egg-bearing *Cambarus immunis* females found with small gastroliths suggest that the reproductive cycle may overlap the premolt stage (Scudamore, 1948). Premolt development was also found to be unaffected by attached eggs in *Homarus americanus* and in several species of spiny lobster (Aiken and Waddy, 1980). In the spiny lobster *Panulirus argus*, intermolt duration was significantly increased, probably due to ovarian development (Lipcius, 1985). In the crayfish *Cherax quadricarinatus* (Barki et al., 1997) and in the crab *Macrophthalmus boteltobagoe* (Brachyura: Kosuge, 1993), a few successive spawns lengthened the molt cycle and postponed the molt event. In the freshwater prawn *Macrobrachium rosenbergii* (Caridea), the molt cycle was inhibited by ovarian development rather than by egg incubation (Wickins and Beard, 1974). A similar situation was recorded for the shrimp *Palaemonetes pugio* (Caridea: Bauer and Abdalla, 2000). Inhibition of the molt cycle in this species might be due to release of larvae when the ovary is immature, thus delaying the parturial molt while the ovary matures or, alternatively, a non-parturial molt may intervene immediately after release of the larvae, followed by a relatively fast parturial molt cycle. The interspawn time for the two scenarios is similar (Bauer and Abdalla, 2000). In *Homarus americanus*, brooding does not generally protract premolt, and, moreover, instances of brood loss as a result of ecdysis have been observed (Aiken and Waddy, 1976). However, the females almost always oviposit in molt stage C_4; thus transition to D_0 is apparently suspended until oviposition occurs (Aiken and Waddy, 1980). Two instances of molting ovigerous *Panulirus homarus* that stripped off fertilized premature eggs have been recorded (Berry, 1971a). In Penaeidae, only a few studies suggest that reproductive molt cycles are significantly longer than non-reproductive cycles (Peeters and Diter, 1994; Diaz et al., 2003).

INHIBITION OF REPRODUCTION BY MOLT-RELATED ACTIVITY

Although it is commonly accepted that in decapod crustaceans ovarian development is associated with intermolt, there are relatively few studies that actually describe the down-regulation and synchronization of vitellogenesis with the molt cycle. In *Homarus americanus* females, the ovary is resorbed under certain conditions in which molt is timed to occur immediately after oviposition (Aiken and Waddy, 1976). Resorption of the ovary during premolt has also been recorded in *N. norvegicus* (Conan, 1985) and *Gecarcinus lateralis* (Brachyura: Weitzman, 1964). A few other examples of the inhibitory effect of molt-related activity on reproduction have also appeared in the literature. In *P. hydrodromous*, resorption of oocytes was reported in a female just beginning its second vitellogenic cycle when molt stage D was initiated (Adiyodi, 1968). In *Rhithropanopeus harrisii*, *Metopograpsus messor* and *Chasmagnathus granulatus*, vitellogenesis was inhibited during premolt, and the ovaries were shown to be immature at ecdysis (Brachyura: Bomirski and Klek, 1974; Sudha and Anilkumar, 1996; Ituarte et al., 2004). A similar phenomenon was found for the shore crab *Carcinus maenas*, with the exception of a report of a female crab molting twice with a full-sized mature ovary (Cheung, 1966). The phenomenon of ovary resorption in the Penaeoidea, a small taxon within the Decapoda, is well documented mainly for species of *Penaeus* (Nelson, 1991; see also section on Molting as an antagonistic process to the reproductive cycle and the following sections). In some species, molting at the stage of ripe ovaries constitutes part of the reproductive strategy; in these species, ovaries mature during premolt or earlier and are not resorbed but rather retained through late premolt and ecdysis.

MOLTING AS AN INTEGRAL PART OF THE REPRODUCTIVE CYCLE

In some decapod taxa, interrelationships between the molt and reproductive cycles are strictly controlled (Fig. 1b), with almost all phases of the reproductive cycle (i.e., gonad maturation, mating, spawning and embryo brooding) being confined to particular stages along the molt cycle. As a rule, females that employ this strategy undergo a pre-mating (i.e., parturial) molt cycle during which the ovary matures, and mating takes place immediately after ecdysis, followed by spawning within minutes or hours or at most a day or two (Fig. 1b). Thereafter, the females incubate the fertilized eggs, even during the premolt stage (Bauer, 2004).

Examples of this strategy are to be found in Anomura species (Wada et al., 1997; Gunamalai and Subramoniam, 2002) and in the following genera from different superfamilies of the Caridea: *Macrobrachium*, *Palaemonetes*, *Palaemon* and *Periclimenes* (O'Donovan et al., 1984; Bauer, 1989; Bauer and Abdalla, 2000; Kim and Hong, 2004); *Crangon* (Oh and Hartnoll, 1999, 2004); the protandric hermaphrodite *Pandalus* (Bergstrom, 2000); *Alpheus*, *Hippolyte* and the simultaneous hermaphrodite *Lysmata* (Bauer, 1989; Bauer and Newman, 2004).

As mentioned above, breeding is controlled by environmental factors that can vary with season, latitude and water depth (Bauer, 1992). At the beginning of the breeding season, after gonad maturation during the parturial molt cycle, the female integument undergoes several modifications that become evident after this parturial molt, such as the presence of extra setae for the purpose of egg bearing (Antheunisse et al., 1968; Bomirski and Klek, 1974; Hartnoll, 1985; Bergstrom, 2000; Bauer, 2004). During the reproductive season, females can spawn several times in consecutive molt cycles, but all species adopting such a strategy will spawn only once per molt cycle. In this pattern of successive brood production, new cohorts of oocytes may mature simultaneously with the development of incubated embryos from the previous spawning (Pandian and Balasundaram, 1982; Bauer, 1989; Oh and Hartnoll, 1999, 2004; Bauer and Abdalla, 2000; Kim and Hong, 2004). The synchronic nature of vitellogenesis and embryogenesis has been demonstrated in *M. rosenbergii* in a study that recorded oocyte diameter with reference to color changes of the developing embryos, zoea hatching, and parturial molt of the next cycle (O'Donovan et al., 1984).

A reproductive molt cycle can be followed by a non-reproductive molt cycle, meaning that after releasing the larvae the female with immature ovaries undergoes a regular molt. The number and sequence of these two molt types vary among species and reproductive seasons (Pandian and Balasundaram, 1982; Damrongphol et al., 1991; Justo et al., 1991; Oh and Hartnoll, 2004). Embryonic development spans almost the entire molt cycle (stages A to D) in many caridean species (Bauer and Abdalla, 2000; Oh and Hartnoll, 2004).

The importance of timing of mating in this strategy has been demonstrated in the king crab species *Paralithodes camtschatica* and *Paralithodes brevipes* (Anomura: McMullen, 1969; Sato et al., 2005). As the number of days from the parturial molt to mating increased, the fertilization rate decreased significantly. The spawning time of unmated *M. rosenbergii* (Caridea) females is delayed by 40 h on average after the parturial molt (Damrongphol et al., 1991), and the non-fertilized eggs are usually discarded within one to two days.

Let us now consider some adaptations and fitness aspects of this strategy. Separation of vitellogenesis and egg incubation into two successive molt cycles and oviposition early in the molt cycle could perhaps constitute an adaptation of caridean species to their relatively short molt cycle. Oviposition early in the molt cycle may confer the advantage of enabling the use of the renewed ovigerous setae for the new brood, and mating immediately after molting grants the female protection by the male in the very vulnerable stage of ecdysis. Further synchronization of the reproductive cycle with the molt cycle is required before mating so as to recruit the male in a particular molt stage. It is likely that communication to coordinate mating is mediated by contact pheromones (Bauer, 2004) and chemotactile signals (Correa and Thiel, 2003; Bauer, 2004).

MOLTING AS AN ANTAGONISTIC PROCESS TO THE REPRODUCTIVE CYCLE

The reproductive cycle in Penaeoidea females is completed within a single molt cycle (Fig. 1cII and 1dII). Females of different taxa within the Penaeoidea exhibit a variety of seasonal patterns in relation to spawning (Anderson et al., 1985; Courtney and Dredge, 1988; Dall et al., 1990; Crocos, 1991; Demestre, 1995; Diaz et al., 2003), but only the interrelation of molt and reproductive cycles will be addressed here. As opposed to brooding decapod species, especially caridean species, free-spawning penaeids seem to express "looser" relationships between the different reproductive stages (i.e., mating, gonad maturation and spawning) and particular molt stages. In general, Penaeoidea have relatively short molt cycle intervals, with the premolt stage occupying most of the molt cycle (Robertson et al., 1987; Dall et al., 1990). Reproductive activity spans almost the entire molt cycle, which means that the ovary matures simultaneously with the synthesis of the new cuticle. For example, ripe ovaries could be seen from the postmolt stage B to premolt D_2 in *Fenneropenaeus* and *Sicyonia* (Emmerson, 1980; Crocos and Kerr, 1983; Anderson et al., 1985) and from C to D_0 in *Artemesia longinaris* and *Pleoticus muelleri* (Petriella and Bridi, 1992; Diaz et al., 2003). In some species, females may spawn more than once during a single molt cycle, completing successive cycles of gonad maturation followed by spawning (Emmerson, 1980, 1983; Beard and Wickins, 1980). Thus, the rate of ovary development does not depend on the progress of the molt cycle (in sharp contrast to Caridea).

With this strategy, it appears that the reproductive cycle will not delay the molt, and if the female does not spawn, the oocytes will be resorbed before molting (Emmerson, 1980, 1983; Penn, 1980; Courtney and Dredge, 1988; Quinitio et al., 1993; Raviv et al., 2006). However, two studies have recorded somewhat longer intermolt periods during the breeding season in which the oocytes were nevertheless resorbed before spawning (Peeters and Diter, 1994; Diaz et al., 2003). It has further been suggested that impregnation of *Fenneropenaeus indicus* females acts as a stimulating factor for lengthening the molt cycle (Peeters and Diter, 1994). In species belonging to other taxa, such as Astacidea and Brachyura, molt inhibition seems more common (see section on Molt-cycle lengthening by reproductive activity, above).

In closed-thelycum species, mating is synchronized with ecdysis, which takes place soon after molting when the female cuticle is still soft (Fig. 1c) (Hudinaga, 1942; Shlagman et al., 1986; Dall et al., 1990). In these species, mating is independent of ovary maturation and is not coupled to spawning, in contrast to the parturial molt of Caridea. However, there is a report in the literature that the ovaries of *Penaeus semisulcatus* do not mature in females with no sperm in the thelycum (Shlagman et al., 1986). In *F. indicus*, impregnation was indeed found to act as a stimulatory factor for vitellogenesis (Peeters and Diter, 1994). Females may use the sperm deposited in the thelycum for several spawns within a single molt cycle. On the other hand, in open-thelycum species, re-mating is required before each spawn, which takes place between molt stages C and D_2 and marks the completion of vitellogenesis (Fig. 1d) (Yano, 1995; group of Sagi, personal observations). Thus, it appears that the stimulus for vitellogenesis differs between open- and closed-thelycum species but takes place in the same molt stages in both. The completion of the female reproductive cycle depends on the timing of the recruitment of a male. In *Penaeus monodon* and *Sicyonia dorsalis*, courtship and copulation increase dramatically after the female molts (Primavera, 1979; Bauer, 1996), but *S. dorsalis* females that were introduced to a pair of males as long as three weeks after ecdysis were also found to be receptive (Bauer, 1996). In the latter species, the attractiveness of a female is associated with insemination rather than with chemical or tactile stimuli, which should diminish as the cuticle of the recently molted female hardens. As opposed to caridean species, there are no reports of precopulatory guarding of females by males in closed-thelycum species such as *S. dorsalis*.

During the reproductive season, the interspawn time with an intervening molt can be very short in some species, for example, less than 10 days in eyestalk-ablated *P. monodon* (Beard and Wickins, 1980), in intact *Penaeus indicus* (Emmerson, 1980), and in eyestalk-ablated *Litopenaeus vannamei* (Parnes et al., 2007). During the molt cycle, the non-mated female may carry its mature ovary for up to 10 days (Emmerson,

1980; group of Sagi, personal observation). Thus, females are capable of carrying mature ovaries for relatively long periods of time, and it is the approaching molt that causes their resorption. These findings ultimately raise the following questions: Why are ovaries resorbed before molt? What is the logic, in terms of energy, underlying the resorption and re-maturation of the ovary just before and just after the molt as opposed to retaining a fully vitellogenic ovary throughout ecdysis? The resorption may be caused by the direction of the energy amassed within the yolk protein to the needs of molting or it may express an adaptation caused by a physiological constraint related to the molt. It has been suggested that water uptake just prior to ecdysis might inhibit ovarian development and cause ovarian resorption (Emmerson, 1980).

REGULATION AND PROGRAMMING OF VITELLOGENESIS AND THE MOLT CYCLE

In some studies of the interrelation between reproduction and molt, data were collected on the number of spawns and molts in the whole population during a designated period of time. In other studies, the reproductive parameters studied were indicative of the precise state of particular vitellogenic process stages (e.g., gonadosomatic index, average oocyte diameter and ovary pigmentation). However, in most of the studies, information reported on molt stage and vitellogenic state was not based on continuous observation of the changes in a single individual.

Since the 1990s, cDNA sequences of genes related to both vitellogenic and molt cycles (e.g., vitellogenin and ecdysone receptor) and their regulatory machinery (e.g., molt-inhibiting hormone or MIH, and gonad-inhibiting hormone) have become available. At present, however, there are only very few studies on the functional significance and differential expression of these genes along the molt cycle. This lack of information is regrettable, since the precise timing of a physiological process could be detected through the recording of differential expression of pivotal genes related to each process. During premolt, genes that encode essential proteins for the replacement of the old cuticle by a new one (El Haj et al., 1997; Koenders et al., 2002; Gao and Wheatly, 2004; Buda and Shafer, 2005) and genes related to other metabolic processes (Klein et al., 1996; Le Boulay et al., 1996) are differentially expressed.

One of the most direct, and therefore precise and reliable, indicators of vitellogenesis is the expression of the vitellogenin gene. It has been observed, for example, in eyestalk-ablated *M. rosenbergii* that vitellogenin mRNA levels increased as the molt cycle progressed and decreased sharply prior to the molt event (Jayasankar et al., 2002). There is also a report of vitellogenin levels in *M. rosenbergii* hemolymph starting to decrease at molt stage D_2 and remaining low until oviposition after ecdysis (Okumura and Aida, 2000a). In the penaeids *P. semisulcatus, L. vannamei* and *Metapenaeus ensis*, vitellogenin gene expression was down-regulated in proximity to ecdysis (Avarre et al., 2003; Raviv et al., 2006; Tiu et al., 2006). Apparently, even in species that molt with a fully vitellogenic ovary as part of their reproductive strategy (i.e., caridean species), vitellogenin gene expression is inhibited in proximity to ecdysis. The reason for this inhibition in Caridea could be that the ovaries are completely ripe just prior to ecdysis. However, in light of the vitellogenin gene expression pattern in penaeids and the apparent inhibition of vitellogenesis in other well-studied decapod infraorders (i.e., Astacidea, Brachyura and Palinura), it seems that the inhibition of vitellogenin expression in proximity to ecdysis constitutes the general antagonistic pattern for all decapod species.

Dall (1986) showed that metabolic rates during the molt cycle of *Penaeus esculentus* did not change until late premolt (three days prior to ecdysis), when they increased by 55%; the rates then fell to previous levels one day after ecdysis. During the same period of time, an increase in oxygen consumption was recorded in *Xiphopenaeus kroyeri* (Carvalho and Phan, 1998), and higher oxygen consumption in proximity to ecdysis was also recorded in *M. rosenbergii* (Stern and Cohen, 1982). A drop in lipid concentration was reported in the hepatopancreas of *P. esculentus* during late premolt (D_3) (Barclay et al., 1983). Presumably, there is a transition point at D_2 D_3 at which the hormonal balance causes an extreme change in the physiological programming of the animal. During this change, expression of a set of genes involved in ecdysis is up-regulated, and expression of another set of genes, among them the vitellogenin gene and probably other vitellogenesis-related genes, is down-regulated.

The currently accepted model of molt control in decapods is based on a hypothesis suggesting that molting can proceed only following a decline in MIH titers. Studies on the expression patterns of the neuropeptides controlling this programming along the molt cycle have been performed for a number of species but have not yielded decisive results that could be applied to decapods in general (Lee et al., 1998; Nakatsuji et al., 2000; Chung and Webster, 2003, 2005; Nakatsuji and Sonobe, 2004; Chen et al., 2007). The pattern of the molting hormone, ecdysone, which is governed by MIH and may also have direct effects on reproduction (Nelson, 1991), has been recorded in a number of species. The peak of ecdysteroid titer measured in decapod hemolymph corresponds to the D_2 D_3 molt stage (Skinner, 1985). In *Macrobrachium nipponense and M. rosenbergii*, ecdysteroids reach a peak at D_3, with the opposite trend occurring for the vitellogenin titer at that point of time (Okumura et al., 1992; Okumura and Aida, 2000b). In *Emerita asiatica*, ecdysteroid titer in the hemolymph peaks at D_2 (Gunamalai et al., 2004). A recent microarray study in the crayfish *C. quadricarinatus* showed that a set of hepatopancreatic genes are up-regulated while others are down-regulated under the influence of ecdysteroids (Shechter et al., 2007). The results of the above studies stress the necessity for further molecular research on patterns of expression of genes under the control of eyestalk neuropeptides in both the molt and reproductive cycles.

MALE AND FEMALE REPRODUCTION AND MOLT—THE CASE OF *LITOPENAEUS VANNAMEI*

A comprehensive study of resorption of the ovary in *L. vannamei* females one to three days before ecdysis (at the approximate molt stage of D_2 D_3) and its redevelopment starting approximately one day after ecdysis (a time point that corresponds to molt stage B) showed that these phenomena correlated with the quantities of vitellogenin mRNA at those times (Fig. 2a) (Raviv et al., 2006). Male reproduction and its relation to molt have been far less well researched. In contrast to the case of the crab *G. depressus*, in which semen is stored in the sperm ducts irrespective of the stage of the molt cycle (Fukui, 1993), a recent study of male *L. vannamei* shrimp revealed that old sperm is periodically removed from the reproductive system (Fig. 2b) (Parnes et al., 2006). In the latter study, it was demonstrated that intact intermolt spermatophores disappeared about 12 h premolt, and a new pair of spermatophores appeared in the ampoules the day after the males had molted. It is well known that males do not copulate at times close to the molt event because of decalcification of their integument and their vulnerable physiological state after the molt, but to the best of our knowledge this is the first report of males of any animal species exhibiting endogenous reproductive cycles scheduled by the molt cycle, as is the case for females of many species. The above study demonstrates similar antagonistic patterns between ecdysis and reproduction in both male and female *L. vannamei* shrimp. The periodical changes in both females and males of the species might have important ramifications at the population level regarding the synchronization of reproductive and molting peaks of the different sexes in the natural population with respect to the fitness of this shrimp species. Whether the above-described antagonistic mechanism is unique to *L. vannamei* or whether it is a broader phenomenon must await further research on other species.

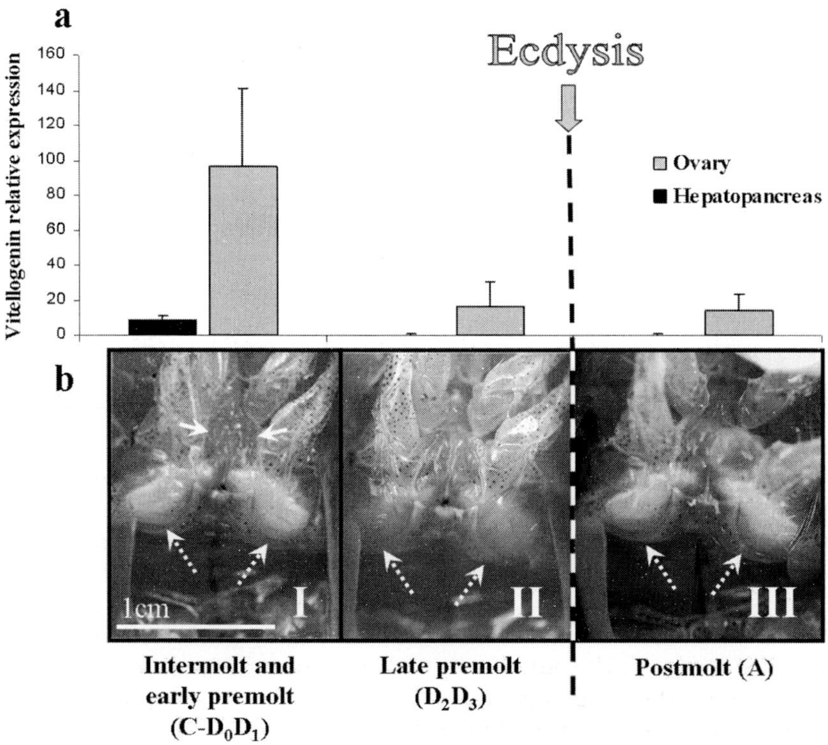

Fig. 2 Cyclic reproductive phenomena along the molt cycle in male and female penaeid shrimp (*L. vannamei*). a. Relative vitellogenin gene expression levels at different molt stages in females (determined by real-time RT-PCR). In I intermolt and early premolt (C-D_0 D_1), there were significantly higher levels of vitellogenin transcripts in the ovary and hepatopancreas. In II late premolt (D_2 D_3) and III postmolt (A), vitellogenin expression was down-regulated (adapted from Raviv et al., 2006). b. Disappearance of spermatophores in male shrimp. Broken arrows point to the ampoules. I – Intermolt with intact spermatophores. Solid arrows point to the genital papillae that house the sexual openings. II – At ~12 h before ecdysis, in the late-premolt (D_3) the ampoules are empty but are visible as two lumps beneath the cuticle. III – At ~ 36 hours postmolt (A-B), a new pair of spermatophores are visible through the cuticle (adapted from Parnes et al., 2006).

References

Adams, A.E., and A.J. Paul. 1983. Male parent size, sperm storage and egg production in the crab *Chionoecetes bairdi* (Decapoda, Majidae). Intl. J. Invert. Reprod., 6: 81-87.

Adiyodi, R.G. 1968. On reproduction and molting in the crab *Paratelphusa hydrodromous*. Physiol. Zool., 41: 204-209.

Adiyodi, K.G., and R.G. Adiyodi. 1970. Endocrine control of reproduction in decapod crustacea. Biol. Rev. Cambridge Phil. Soc., 45: 121-165.

Aiken, D.E. 1969a. Photoperiod, endocrinology and the crustacean molt cycle. Science, 164: 149-155.

Aiken, D.E. 1969b. Ovarian maturation and egg laying in the crayfish *Orconectes virilis*: Influence of temperature and photoperiod. Canad. J. Zool., 47: 931-935.

Aiken, D.E., and S.L. Waddy. 1976. Controlling growth and reproduction in the American lobster. Proc. Ann. Meeting World Maricult. Soc., 7: 415-430.

Aiken, D.E., and S.L. Waddy. 1980. The biology and management of lobster. Vol. 1. *In:* Reproductive Biology. J.S. Cobb and B.F. Philips (eds.). Academic Press, New York, pp. 215-276.

Alunno-Bruscia, M., and B. Sainte-Marie. 1998. Abdomen allometry, ovary development, and growth of female snow crab, *Chionoecetes opilio* (Brachyura, Majidae), in the northwestern Gulf of St. Lawrence. Canad. J. Fish. Aquat. Sci., 55: 459-477.

Anderson, S.L., W.H. Clark and E.S. Chang. 1985. Multiple spawning and molt synchrony in a free spawning shrimp (*Sicyonia ingentis*: Penaeoidea). Biol. Bull., 168: 377-394.

Anilkumar, G., K. Sudha and T. Subramoniam. 1999. Spermatophore transfer and sperm structure in the brachyuran crab *Metopograpsus messor* (Decapoda: Grapsidae). J. Crustacean Biol., 19: 361-370.

Antheunisse, L.J., N.P. Van Den Hoven and D.J. Jefferies. 1968. The breeding characters of *Palaemonetes varians* (Leach) Decapoda Palaemonidae. Crustaceana, 14: 259-270.

Avarre, J.C., R. Michelis, A. Tietz and E. Lubzens. 2003. Relationship between vitellogenin and vitellin in a marine shrimp (*Penaeus semisulcatus*) and molecular characterization of vitellogenin complementary DNAs. Biol. Reprod., 69: 355-364.

Barclay, M.C., W. Dall and D.M. Smith. 1983. Changes in lipid and protein during starvation and the moulting cycle in the tiger prawn, *Penaeus esculentus* Haswell. J. Exp. Mar. Biol. Ecol., 68: 229-244.

Barki, A., T. Levi, G. Hulata and I. Karplus. 1997. Annual cycle of spawning and molting in the red-claw crayfish, *Cherax quadricarinatus*, under laboratory conditions. Aquaculture, 157: 239-249.

Bauer, R.T. 1989. Continuous reproduction and episodic recruitment in 9 shrimp species inhabiting a tropical seagrass meadow. J. Exp. Mar. Biol. Ecol., 127: 175-187.

Bauer, R.T. 1992. Testing generalizations about latitudinal variation in reproduction and recruitment patterns with sicyoniid and caridean shrimp species. Invert. Reprod. Dev., 22: 193-202.

Bauer, R.T. 1996. A test of hypotheses on male mating systems and female molting in decapod shrimp, using *Sicyonia dorsalis* (Decapoda: Penaeoidea). J. Crustacean Biol., 16: 429-436.

Bauer, R.T. 2004. Remarkable Shrimps – Adaptations and Natural History of the Carideans. University of Oklahoma Press, Norman.

Bauer, R.T., and J.H. Abdalla. 2000. Patterns of brood production in the grass shrimp *Palaemonetes pugio* (Decapoda: Caridea). Invert. Reprod. Dev., 38: 107-113.

Bauer, R.T., and W.A. Newman. 2004. Protandric simultaneous hermaphroditism in the marine shrimp *Lysmata californica* (Caridea: Hippolytidae). J. Crustacean Biol., 24: 131-139.

Beard, T.W., and J.F. Wickins. 1980. Breeding of *Penaeus monodon* Fabricius in laboratory recirculation systems. Aquaculture, 20: 79-89.

Bergstrom, B.I. 2000. The biology of Pandalus. Adv. Mar. Biol., 38: 55-245.

Berry, P.F. 1971a. The biology of the spiny lobster *Panulirus homarus* (Linnaeus) off the east coast of southern Africa. Oceanogr. Res. Inst. Invest. Rep., 28: 1-75.

Berry, P.F. 1971b. The biology of the spiny lobster *Palinurus delagoae* Bernard, off the east coast of Natal, South Africa. Oceanogr. Res. Inst. Invest. Rep., 31: 1-27.

Bomirski, A., and E. Klek. 1974. Action of eyestalk on the ovary in *Rhithropanopeus harrisii* and *Crangon crangon* (Crustacea: Decapoda). Mar. Biol., 24: 329-337.

Bouchon, D., C. Souty-Grosset and J.P. Mocquard. 1992. Photoperiodism and seasonal breeding in aquatic and terrestrial Eumalacostraca. Invert. Reprod. Dev., 22: 203-212.

Brockerhoff, A.M., and C.L. McLay. 2005. Comparative analysis of the mating strategies in grapsid crabs with special references to the intertidal crabs *Cyclograpsus lavauxi* and *Helice crassa* (Decapoda: Grapsidae) from New Zealand. J. Crustacean Biol., 25: 507-520.

Bryant, A.D., and R.G. Hartnoll. 1995. Reproductive investment in two spider crabs with different breeding strategies. J. Exp. Mar. Biol. Ecol., 188: 261-275.

Buda, E.S., and T.H. Shafer. 2005. Expression of a serine proteinase homolog prophenoloxidase-activating factor from the blue crab, *Callinectes sapidus*. Comp. Biochem. Physiol., B 140: 521-531.

Carvalho, P.S.M., and V.N. Phan. 1998. Oxygen consumption and ammonia excretion during the moulting cycle in the shrimp *Xiphopenaeus kroyeri*. Comp. Biochem. Physiol., A 119: 839-844.

Chang, E.S. 1995. Physiological and biochemical changes during the molt cycle in decapod crustacean: an overview. J. Exp. Mar. Biol. Ecol., 193: 1-14.

Chen, H.Y., Watson, R.D., Chen. J.C., Lee, C.Y. 2007. Molecular characterization and gene expression pattern of two putative molt-inhibiting hormones from *Litopenaeus vannamei*. Gen. Comp. Endocrinol. 151: 72-81.

Cheung, T.S. 1966. The interrelations among three hormonal-controlled characters in the adult female shore crab, *Carcinus maenas* (L). Biol. Bull., 130: 59-66.

Cheung, T.S. 1969. The environmental and hormonal control of growth and reproduction in the adult female stone crab, *Menippe mercenaria* (Say). Biol. Bull., 136: 327-346.

Chung, J.S., and S.G. Webster. 2003. Moult cycle-related changes in biological activity of moult-inhibiting hormone (MIH) and crustacean hyperglycaemic hormone (CHH) in the crab, *Carcinus maenas*—From target to transcript. Eur. J. Biochem., 270: 3280-3288.

Chung, J.S., and S.G. Webster. 2005. Dynamics of *in vivo* release of molt-inhibiting hormone and crustacean hyperglycemic hormone in the shore crab, *Carcinus maenas*. Endocrinology, 146: 5545-5551.

Conan, G.Y. 1985. Periodicity and phasing of molting. *In:* Crustacean Issues, Vol. 3, Factors in Adult Growth. A.M. Wenner (ed.). Balkema Press, Rotterdam, The Netherlands, pp. 73-99.

Corgos, A., P. Verisimo and J. Freire. 2006. Timing and seasonality of the terminal molt and mating migration in the spider crab, *Maja brachydactyla*: Evidence of alternative mating strategies. J. Shellfish Res., 25: 577-587.

Correa, C., and M. Thiel. 2003. Mating system in caridean shrimp (Decapoda: Caridea) and their evolutionary consequences for sexual dimorphism and reproductive biology. Rev. Chilena Hist. Natur., 76: 187-203.

Costa, T.M., and M.L. Negreiros-Fransozo. 1998. The reproductive cycle of *Callinectes danae* Smith, 1869 (Decapoda, Portunidae) in the Ubatuba region, Brazil. Crustaceana, 71: 615-627.

Courtney, A.J., and M.C.L. Dredge. 1988. Female reproductive biology and spawning periodicity of two species of king prawns, *Penaeus longistylus* Kubo and *Penaeus latisulcatus* Kishinouye, from Queensland's east coast fishery. Austral. J. Mar. Freshwater Res., 39: 729-741.

Cowan, D.F., and J. Atema. 1990. Moult staggering and serial monogamy in American lobsters, *Homarus americanus*. Anim. Behav., 39: 1199-1206.

Crocos, J.P. 1991. Reproductive dynamics of three species of penaeidae in tropical Australia, and the role of reproductive studies in fisheries management. *In:* Crustacean Issues, Vol. 7, Crustacean Egg Production. A.M. Wenner and A. Kuris (eds.). Balkema Press, Rotterdam, The Netherlands, pp. 317-331.

Crocos, P.J., and J.D. Kerr. 1983. Maturation and spawning of the banana prawn *Penaeus merguiensis* de Man (Crustacea: Penaeidae) in the Gulf of Carpentaria, Australia. J. Exp. Mar. Biol. Ecol., 69: 37-59.

Cukerzis, J.M. 1988. Astacus astacus in Europe. *In:* Freshwater Crayfish. D.M. Holdich and I.D. Reeve (eds.). Croom Helm, London, pp. 309-340.

Dall, W. 1986. Estimation of routine metabolic rate in a penaeid prawn, *Penaeus esculentus* Haswell. J. Exp. Mar. Biol. Ecol., 96: 57-74.

Dall, W., B.J. Hill, P.C. Rothlisberg and D.J. Staples. 1990. The biology of the penaeidae. Adv. Mar. Biol., 27: 1-461.

Damrongphol, P., N. Eangchuan and B. Poolsanguan. 1991. Spawning cycle and oocyte maturation in laboratory-maintained giant freshwater prawns (*Macrobrachium rosenbergii*). Aquaculture, 95: 347-357.

Demestre, M. 1995. Moult activity related spawning success in the Mediterranean deep water shrimp *Aristeus antennatus* (Decapoda: Dendrobranchiata). Mar. Ecol. Prog. Ser., 127: 57-64.

Diaz, A.C., A.M. Petriella and J.L. Fenucci. 2003. Molting cycle and reproduction in the population of the shrimp *Pleoticus muelleri* (Crustacea, Penaeoidea) from Mar del Plata. Ciencias Marinas, 29: 343-355.

El Haj, A.J., S.L. Tamone, M. Peake, P.S. Reddy and E.S. Chang. 1997. An ecdysteroid-responsive gene in a lobster - a potential crustacean member of the steroid hormone receptor superfamily. Gene, 201: 127-135.

Elner, R.W., and P.G. Beninger. 1995. Multiple reproductive strategies in snow crab, *Chionoecetes opilio*: Physiological pathways and behavioral plasticity. J. Exp. Mar. Biol. Ecol., 193: 93-112.

Emmerson, W.D. 1980. Induced maturation of prawn *Penaeus indicus*. Mar. Ecol. Prog. Ser., 2: 121-131.

Emmerson, W.D. 1983. Maturation and growth of ablated and unablated *Penaeus monodon* Fabricius. Aquaculture, 32: 235-241.

Farmer, A.S.D. 1974. Reproduction in *Nephrops norvegicus* (Decapoda: Nephropidae). J. Zool., 174: 161-183.

Fielder, D.R. 1964. The spiny lobster, *Jasus lalandei* (H. Milne-Edwards), in south Australia. Austral. J. Mar. Freshwater Res., 15: 133-144.

Fukui, Y. 1993. Timing of copulation in the molting and reproductive cycles in a grapsid crab, *Gaetice depressus* (Crustacea: Brachyura). Mar. Biol., 117: 221-226.

Gao, Y., and M.G. Wheatly. 2004. Characterization and expression of plasma membrane Ca^{2+} ATPase (PMCA3) in the crayfish *Procambarus clarkii* antennal gland during molting. J. Exp. Biol., 207: 2991-3002.

Gonzalez-Gurriaran, E., L. Fernandez, J. Freire and R. Muino. 1998. Mating and role of seminal receptacles in the reproductive biology of the spider crab *Maja squinado* (Decapoda, Majidae). J. Exp. Mar. Biol. Ecol., 220: 269-285.

Greenspan, B.N. 1982. Semi-monthly reproductive cycles in male and female fiddler crabs, *Uca pugnax*. Anim. Behav., 30: 1084-1092.

Gunamalai, V., and T. Subramoniam. 2002. Synchronization of molting and oogenic cycles in a continuously breeding population of the sand crab *Emerita asiatica* on the Madras Coast, South India. J. Crustacean Biol., 22: 398-410.

Gunamalai, V., R. Kirubagaran and T. Subramoniam. 2004. Hormonal coordination of molting and female reproduction by ecdysteroids in the mole crab *Emerita asiatica* (Milne Edwards). Gen. Comp. Endocrinol., 138: 128-138.

Gupta, N.V.S., K.N.P. Kurup, R.G. Adiyodi and K.G. Adiyodi. 1987. The antagonism between somatic growth and ovarian growth during different phases in intermoult (stage C4) in sexually mature freshwater crab, *Paratelphusa hydrodromous*. Intl. J. Invert. Reprod. Dev., 12: 307-318.

Hartnoll, R.G. 1969. Mating in the Brachyura. Crustaceana, 16: 161-181.

Hartnoll, R.G. 1985. Growth, sexual maturity and reproductive output. *In:* Crustacean Issues, Vol. 3, Factors in Adult Growth. A.M. Wenner (ed.). Balkema Press, Rotterdam, The Netherlands, pp. 101-128.

Hartnoll, R.G. 2006. Reproductive investment in Brachyura. Hydrobiologia, 557: 31-40.

Henmi, Y., and M. Murai. 1999. Decalcification of vulvar operculum and mating in the ocypodid crab *Ilyoplax pusilla*. J. Zool., 247: 133-137.

Hess, W.N. 1941. Factors influencing moulting in the crustacean, *Crangon armillatus*. Biol. Bull., 81: 215-220.

Hinsch, G.W. 1972. Some factors controlling reproduction in the spider crab, *Libinia emarginata*. Biol. Bull., 143: 358-366.

Holdich, D.M. 2002. Background and functional morphology. *In:* Biology of the Freshwater Crayfish. D.M. Holdich (ed.). Blackwell Science, Oxford, pp. 3-29.

Hudinaga, M. 1942. Reproduction, development and rearing of *Penaeus japonicus* bate. Japan. J. Zool., pp. 305-393.

Ituarte, R.B., E.D. Spivak and T.A. Luppi. 2004. Female reproductive cycle of the southwestern Atlantic estuarine crab *Chasmagnathus granulatus* (Brachyura: Grapsoidea: Varunidae). Sci. Mar., 68: 127-137.

Jayasankar, V., N. Tsutsui, S. Jasmani, H. Saido-Sakanaka, W.J. Yang, A. Okuno, T.T.T. Hien, K. Aida and M.N. Wilder. 2002. Dynamics of vitellogenin mRNA expression and changes in hemolymph vitellogenin levels during ovarian maturation in the giant freshwater prawn *Macrobrachium rosenbergii*. J. Exp. Zool., 293: 675-682.

Jones, D.R., and R.G. Hartnoll. 1997. Mate selection and mating behaviour in spider crabs. Estuar. Coast. Shelf Sci., 44:185-193.

Justo, C.C., K. Aida and I. Hanyu. 1991. Effects of photoperiod and temperature on molting, reproduction and growth of the freshwater prawn *Macrobrachium rosenbergii*. Nippon Suisan Gakkaishi, 57: 209-217.

Kim, S.H., and S.Y. Hong. 2004. Reproductive biology of *Palaemon gravieri* (Decapoda: Caridea: Palaemonidae). J. Crustacean Biol., 24: 121-130.

Klein, B., G. LeMoullac, D. Sellos and A. Van Wormhoudt. 1996. Molecular cloning and sequencing of trypsin cDNAs from *Penaeus vannamei* (Crustacea, Decapoda). Use in assessing gene expression during the moult cycle. Intl. J. Biochem. Cell Biol., 28: 551-563.

Koenders, A., X. Yu, E.S. Chang and D.L. Mykles. 2002. Ubiquitin and actin expression in claw muscles of land crab *Gecarcinus lateralis* and American lobster *Homarus americanus*: Differential expression of ubiquitin in two slow muscle fiber types during molt-induced atrophy. J. Exp. Zool., 292: 618-632.

Kosuge, T. 1993. Molting and breeding cycles of the rock-dwelling ocypodid crab *Macrophthalmus boteltobagoe* (Sakai, 1939) (Decapoda, Brachyura). Crustaceana, 64: 56-65.

Le Boulay, C., A. Van Wormhoudt and D. Sellos. 1996. Cloning and expression of cathepsin L-like proteinases in the hepatopancreas of the shrimp *Penaeus vannamei* during the intermolt cycle. J. Comp. Physiol. B: Biochem. System. Environ. Physiol., 166: 310-318.

Lee, K.J., R.D. Watson and R.D. Roer. 1998. Molt-inhibiting-hormone mRNA levels and ecdysteroid titer during a molt cycle of the blue crab, *Callinectes sapidus*. Biochem. Biophys. Res. Comm., 249: 624-627.

Lipcius, R.N. 1985. Size-dependent reproduction and molting in spiny lobsters and other long-lived decapods. *In:* Crustacean Issues, Vol. 3, Factors in Adult Growth. A.M. Wenner (ed.). Balkema Press, Rotterdam, The Netherlands, pp. 129-148.

Lipcius, R.N., and W.F. Herrnkind. 1987. Control and coordination of reproduction and molting in the spiny lobster *Panulirus argus*. Mar. Biol., 96: 207-214.

Liu, H.C., and C.W. Li. 2000. Reproduction in the freshwater crab *Candidiopotamon rathbunae* (Brachyura: Potamidae) in Taiwan. J. Crustacean Biol., 20: 89-99.

MacDiarmid, A.B. 1989. Moulting and reproduction of spiny lobster *Jasus edwardsii* (Decapoda: Palinuridae) in northern New Zealand. Mar. Biol., 103: 303-310.

McMullen, J.C. 1969. Effects of delayed mating on the reproduction of king crab, *Paralithodes camtschatica*. J. Fish. Res. Board Canada, 26: 2737-2740.

Meusy, J.J., and G.G. Payen. 1988. Female reproduction in malacostracan Crustacea. Zool. Sci., 5: 217-265.

Morgan, S.G. 1996. Influence of tidal variation on reproductive timing. J. Exp. Mar. Biol. Ecol., 206: 237-251.

Morgan, S.G., and J.H. Christy. 1994. Plasticity, constraint, and optimality in reproductive timing. Ecology, 75: 2185-2203.

Murai, M., T. Koga and H.S. Yong. 2002. The assessment of female reproductive state during courtship and scramble competition in the fiddler crab, *Uca paradussumieri*. Behav. Ecol. Sociobiol., 52: 137-142.

Nakatsuji, T., and H. Sonobe. 2004. Regulation of ecdysteroid secretion from the Y-organ by molt-inhibiting hormone in the American crayfish, *Procambarus clarkii*. Gen. Comp. Endocrinol., 135: 358-364.

Nakatsuji, T., H. Keino, K. Tamura, S. Yoshimura, T. Kawakami, S. Aimoto and H. Sonobe. 2000. Changes in the amounts of the molt-inhibiting hormone in sinus glands during the molt cycle of the American crayfish, *Procambarus clarkii*. Zool. Sci., 17: 1129-1136.

Nascimento, I.A., W.A. Bray, J.R.L. Trujillo and A. Lawrence. 1991. Reproduction of ablated and unablated *Penaeus schmitti* in captivity using diets consisting of fresh-frozen natural and dried formulated feeds. Aquaculture, 99: 387-398.

Nelson, K. 1991. Scheduling of reproduction in relation to molting and growth in malacostracan crustaceans. *In:* Crustacean Issues, Vol. 7, Crustacean Egg Production. A.M. Wenner and A. Kuris (eds.). Balkema Press, Rotterdam, The Netherlands, pp. 77-113.

Nelson, K., D. Hedgecock and W. Borgeson. 1988. Effects of reproduction upon molting and growth in female American lobsters (*Homarus americanus*). Canad. J. Fish. Aquat. Sci., 45: 805-821.

Nye, P.A. 1977. Reproduction, growth and distribution of the grapsid crab *Helice crassa* (Dana, 1851) in the southern part of New Zealand. Crustaceana, 33: 75-89.

O'Donovan, P., M. Abraham and D. Cohen. 1984. The ovarian cycle during the intermoult in ovigerous *Macrobrachium rosenbergii*. Aquaculture, 36: 347-358.

Oh, C.W., and R.G. Hartnoll. 1999. Size at sexual maturity, reproductive output, and seasonal reproduction of *Philocheras trispinosus* (Decapoda) in Port Erin Bay, Isle of Man. J. Crustacean Biol., 19: 252-259.

Oh, C.W., and R.G. Hartnoll. 2004. Reproductive biology of the common shrimp *Crangon crangon* (Decapoda: Crangonidae) in the central Irish Sea. Mar. Biol., 144: 303-316.

Okumura, T., and K. Aida. 2000a. Hemolymph vitellogenin levels and ovarian development during the reproductive and non reproductive molt cycles in the giant freshwater prawn *Macrobrachium rosenbergii*. Fish. Sci., 66: 678-685.

Okumura, T., and K. Aida. 2000b. Fluctuations in hemolymph ecdysteroid levels during the reproductive and non-reproductive molt cycles in the giant freshwater prawn *Macrobrachium rosenbergii*. Fish. Sci., 66: 876-883.

Okumura, T., C.H. Han, Y. Suzuki, K. Aida and I. Hanyu. 1992. Changes in hemolymph vitellogenin and ecdysteroid levels during the reproductive and non-reproductive molt cycles in the freshwater prawn *Macrobrachium nipponense*. Zool. Sci., 9: 37-45.

Pandian, T.J., and C. Balasundaram. 1982. Moulting and spawning cycles in *Macrobrachium nobilii* (Henderson and Mathai). Mar. Ecol. Progr. Ser., 127: 57-64.

Parnes, S., S. Raviv, A. Shechter and A. Sagi. 2006. Males also have their time of the month! Cyclic disposal of old spermatophores, timed by the molt cycle, in a marine shrimp. J. Exp. Biol., 209: 4974-4983.

Parnes, S., Raviv, S., Azulay, D. and A. Sagi. 2007. Dynamics of reproduction in a captive shrimp broodstock: Unequal contribution of the female shrimp and a hidden shortage in competent males. Invert. Reprod. Dev. 50: 21-29.

Paul, A.J. 1984. Mating frequency and viability of stored sperm in the tanner crab *Chionoecetes bairdi* (Decapoda, Majidae). J. Crustacean Biol., 4: 375-381.

Peeters, L., and A. Diter. 1994. Effects of impregnation on maturation, spawning, and ecdysis of female shrimp *Penaeus indicus*. J. Exp. Zool., 269: 522-530.

Penn, J.W. 1980. Spawning and fecundity of the western king prawn, *Penaeus latisulcatus* Kishinouye, in Western Australian waters. Austral. J. Mar. Freshwater Res., 31: 21-35.

Petriella, A.M., and R.J. Bridi. 1992. Variaciones estacionales del ciclo de muda y la maduracion overica del camaron (*Artemesia longinaris*). Frente Maritimo, 11: 85-92.

Primavera, J.H. 1979. Notes on the courtship and mating behavior in *Penaeus monodon* Fabricius (Decapoda, Natantia). Crustaceana, 37: 287-292.

Quinitio, E.T., R.M. Caballero and L. Gustilo. 1993. Ovarian development in relation to changes in the external genitalia in captive *Penaeus-monodon*. Aquaculture, 114: 71-81.

Raviv, S., S. Parnes, C. Segall, C. Davis and A. Sagi. 2006. Complete sequence of *Litopenaeus vannamei* (Crustacea: Decapoda) vitellogenin cDNA and its expression in endocrinologically induced sub-adult females. Gen. Comp. Endocrinol., 145: 39-50.

Rice, P.R., and K.B. Armitage. 1974. The influence of photoperiod on processes associated with molting and reproduction in the crayfish *Orconectes nais* (Faxon). Comp. Biochem. Physiol., A 47: 243-259.

Robertson, L., W. Bray, J. Leung-Trujillo and A. Lawrence. 1987. Practical molt staging of *Penaeus setiferus* and *Penaeus stylirostris*. J. World Aquacult. Soc., 18: 180-185.

Sainte-Marie, B., N. Urbani, J.M. Sevigny, F. Hazel and U. Kuhnlein. 1999. Multiple choice criteria and the dynamics of assortative mating during first breeding season of female snow crab *Chionoecetes opilio* (Brachyura, Majidae). Mar. Ecol. Progr. Ser., 181: 141-153.

Sastry, A.N. 1983. Ecological aspects of reproduction. *In:* The Biology of Crustacea, Vol. 8. D.E. Bliss and W.B. Vemberg (eds.). Academic Press, New York, pp. 179-270.

Sato, T., M. Ashidate and S. Goshima. 2005. Negative effects of delayed mating on the reproductive success of female spiny king crab, *Paralithodes brevipes*. J. Crustacean Biol., 25: 105-109.

Schembri, P.J. 1982. The biology of population of *Ebalia tuberosa* (Crustacea: Decapoda: Leucosiidae) from the Clyde Sea area. J. Mar. Biol. Assoc. UK, 62: 101-115.

Scudamore, H.H. 1948. Factors influencing molting and the sexual cycles in the crayfish. Biol. Bull., 95: 229-237.

Shechter, A., Tom, M., Yudkovski, Y., Weil, S., Chang, S.A., Chang, E.S., Chalifa-Caspi, V., Berman, A., Sagi, A. 2007. Search for hepatopancreatic ecdysteroid-responsive genes during the crayfish molt cycle: from a single gene to multigenicity. J. Exp. Biol., 2007 210: 3525-3537.

Shields, J.D. 1991. The reproductive ecology and fecundity of *Cancer* crabs. *In:* Crustacean Issues, Vol. 7, Crustacean Egg Production. A.M. Wenner and A. Kuris (eds.). Balkema Press, Rotterdam, The Netherlands, pp. 193-213.

Shlagman, A., C. Lewinsohn and M. Tom. 1986. Aspects of the reproductive activity of *Penaeus-semisulcatus* De Haan along the southeastern coast of the Mediterranean. Mar. Ecol., 7: 15-22.

Skinner, D.M. 1985. Molting and regeneration. *In:* The Biology of Crustacea. D.E. Bliss (ed.). Vol. 9. Integument, Pigments, and Hormonal Processes. D.E. Bliss and L.H. Mantel (eds.). Academic Press, New York, pp. 147-215.

Skov, M.W., R.G. Hartnoll, R.K. Ruwa, J.P. Shunula, M. Vannini and S. Cannicci. 2005. Marching to a different drummer: Crabs synchronize reproduction to a 14-month lunar-tidal cycle. Ecology, 86: 1164-1171.

Sokol, A. 1988. The Australian Yabby. *In:* Freshwater Crayfish. D.M. Holdich and I.D. Reeve (eds.). Croom Helm, London, pp. 401-425.

Stephens, G.J. 1952. Mechanisms regulating the reproductive cycle in the crayfish, *Cambarus*. I. The female cycle. Physiol. Zool., 25: 70-83.

Stern, S., and D. Cohen. 1982. Oxygen consumption and ammonia excretion during the molt cycle of the freshwater prawn *Macrobrachium rosenbergii* (de Man). Comp. Biochem. Physiol., A 73: 417-419.

Sudha, K., and G. Anilkumar. 1996. Seasonal growth and reproduction in a highly fecund brachyuran crab, *Metopograpsus messor* (Forskal) (Grapsidae). Hydrobiologia, 319: 15-21.

Tack, P.I. 1941. The life history and ecology of the crayfish, *Cambarus immunis* Hagen. Amer. Midl. Natural., 25: 420-446.

Tamaki, A., H. Tanoue, J. Itoh and Y. Fukuda. 1996. Brooding and larval developmental periods of the callianassid ghost shrimp, *Callianassa japonica* (Decapoda: Thalassinidea). J. Mar. Biol. Assoc. UK, 76: 675-689.

Tiu, S.H.K., J.H.L. Hui, J.G. He, S.S. Tobe and S.M. Chan. 2006. Characterization of vitellogenin in the shrimp *Metapenaeus ensis*: Expression studies and hormonal regulation of MeVg1 transcription *in vitro*. Mol. Reprod. Dev., 73: 424-436.

Van Engel, W.A. 1958. The blue crab and its fishery in Chesapeake Bay: reproduction, early development, growth and migration. Comm. Fish. Rev., 20: 6-16.

Wada, S., M. Ashidate and S. Goshima. 1997. Observations on the reproductive behavior of the spiny king crab *Paralithodes brevipes* (Anomura: Lithodidae). Crustacean Res., 26: 56-61.

Waddy, S.L., and D.E. Aiken. 1990. Intermolt insemination, an alternative mating strategy for the American lobster (*Homarus americanus*). Canad. J. Fish. Aquat. Sci., 47: 2402-2406.

Weitzman, M.C. 1964. Ovarian development and molting in the tropical land crab, *Gecarcinus lateralis* (Freminville). Amer. Zool., 4: 329-330.

Wickins, J.F., and T.W. Beard. 1974. Observation on the breeding and growth of the giant freshwater prawn *Macrobrachium rosenbergii* (de Man) in the laboratory. Aquaculture, 3: 159-174.

Wilder, M.N., T. Subramoniam and K. Aida. 2002. Yolk proteins of Crustacea. *In:* Reproductive Biology of Invertebrates, Vol. 12. K.G. Adiyodi and R.G. Adiyodi (eds.). Progress in Vitellogenesis. A.S. Raikhel and T.W. Sappington (eds.). Science Publishers, Inc. Enfield, USA, pp. 131-174.

Yamahira, K. 2004. How do multiple environmental cycles in combination determine reproductive timing in marine organisms? A model and test. Funct. Ecol., 18: 4-15.

Yano, I. 1995. Final oocyte maturation, spawning and mating in penaeid shrimp. J. Exp. Mar. Biol. Ecol., 193: 113-118.

Zhang, D., J. Lin and R.L. Creswell. 1998. Mating behavior and spawning of the banded coral shrimp *Stenopus hispidus* in the laboratory. J. Crustacean Biol., 18: 511-518.

10

Reproductive Biology and Growth of Marine Lobsters

A.J. Ritar and G.G. Smith

Tasmanian Aquaculture and Fisheries Institute, Marine Research Laboratories, University of Tasmania

Nubeena Crescent, Taroona, Tasmania, 7053

E-mail: Arthur.Ritar@utas.edu.au,

E-mail: ggsmith@postoffice.utas.edu.au

OVERVIEW OF THE WORLD'S MARINE LOBSTER FISHERIES AND AQUACULTURE

Lobsters are found throughout the world's temperate and tropical oceans and are among the most highly esteemed and valuable seafood products (Butler, 2001). Of the many species of lobsters recorded world-wide, only a few are regarded as commercially important.

Two of the four families comprise the bulk of the world's lobster fisheries (Phillips et al., 1980) and will be the focus of discussion in this chapter. The first is the Nephropidae, or clawed lobsters, which include the American *Homarus americanus* and European *H. gammarus* species, both of which are fished and farmed in large quantities, and the Norway lobster *Nephrops norvegicus*. The second family is the Palinuridae, the spiny (or rock) lobsters, from temperate and tropical regions; and of the more than 47 species, 33 are fished commercially (Williams, 1988). The two remaining families are the Scyllaridae, or slipper lobsters, and the

Synaxidae, or coral lobsters. This review will examine the reproductive biology and growth of lobsters in the first two families, for which there is now considerable literature. The proceedings of the international conferences and workshops on lobster biology and management, held every three or four years and most recently (2004) in Hobart, Australia, provide the latest research findings across all commercial lobster species. The second edition on the biology of spiny lobsters by Phillips and Kittaka (2000) also provides recent information on the commercially important species for fisheries and aquaculture. The American clawed lobster, *H. americanus*, is probably the most extensively studied of the lobster species, being the subject of a detailed book edited by Factor (1995) with contributions by numerous researchers on all aspects of its biology. These publications are cited extensively in this chapter.

Lobsters are of considerable scientific interest as they are widely distributed, large, and long-lived, abundant and ecologically of great consequence (Butler, 2001). Therefore, it is important to understand their biology, which is different from that of vertebrates or sedentary invertebrates. They also offer a unique model for fisheries because their abundance, lower mobility and hardiness make them more amenable to detailed research.

Chetrick (2006) summarized international lobster production based on data from the United Nations Food and Agriculture Organization (FAO) (Fig. 1). World production increased steadily from 157,000 metric tonnes (MT) in 1980 to more than 233,000 MT in 1997 before stabilizing at levels near 230,000 MT through 2003 and rising to about 239,000 MT in 2004. Catches of American lobster (*H. americanus*) and spiny lobster of the genus *Panulirus* accounted for 67% of world lobster production in 2004. Other important species included the European lobster (*H. gammarus*) and *Jasus* spp. The United States and Canada are the largest lobster-producing countries, accounting for 37% of global production in 2004. Wild lobster is by far the main source of total production, but aquaculture, although minimal, is growing.

World exports and imports of lobster grew steadily over the decade to 2004. World lobster exports rose 87% from US$1.2 billion in 1992 to US$2.2 billion in 2004, primarily because of increased sales of frozen and fresh or chilled products. Canada was the major exporter of live lobster products with US$371 million in 2004, followed by the United States with $300 million. World lobster imports increased 63% from US$1.3 billion in 1992 to US$2.1 billion in 2004, mainly because of increased demand for live and, to a lesser extent, frozen products in the hotel and restaurant sector.

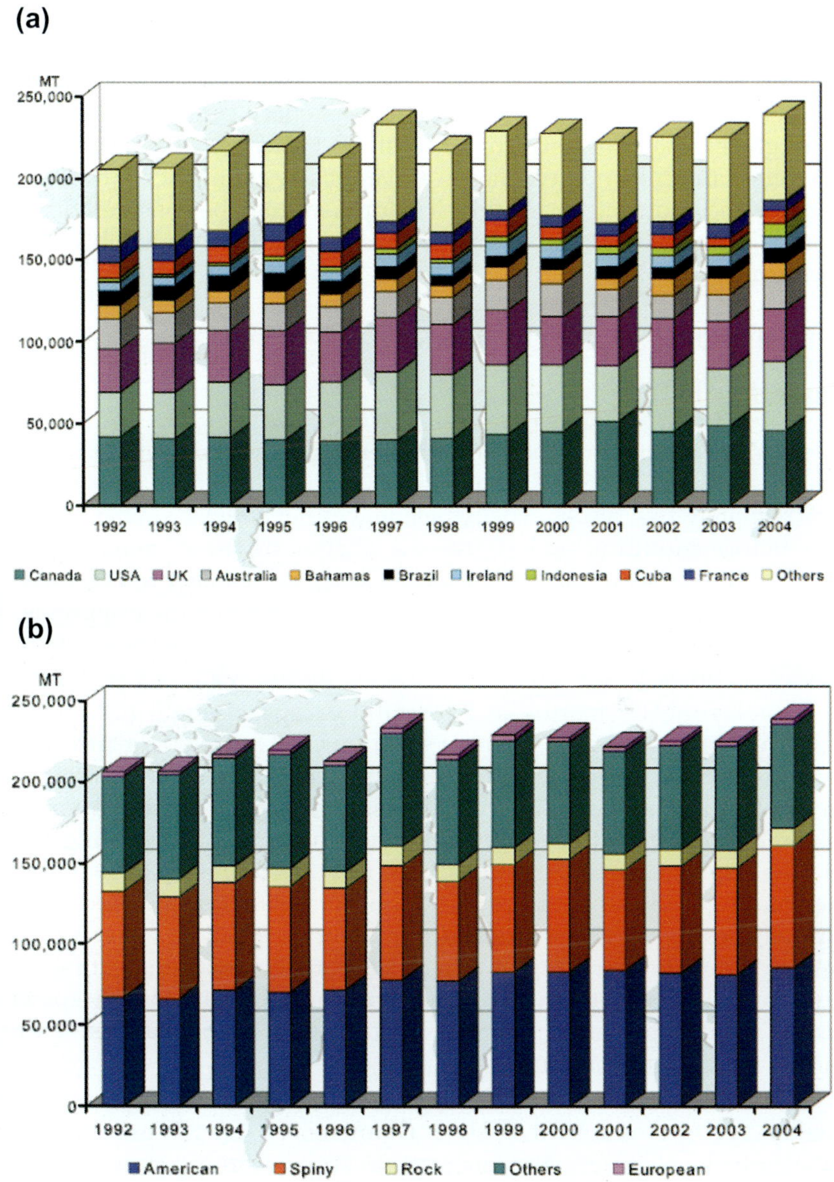

Fig. 1 World lobster production by (a) country and (b) species. From Chetrick (2006).

The intense interest in the culture of clawed lobster, especially *H. americanus*, peaked in the 1970s, which led to a need for biological information on the regulation of development, growth and reproduction

(Waddy et al., 1995). Summaries of the research were published in 1980 (Aiken, 1980; Aiken and Waddy, 1980; Van Olst et al., 1980) but much has been discovered since then. There is now also considerable interest in the commercial farming of spiny lobsters from hatchery-produced seedstock, although this has not yet proved feasible. However, the capture and growout of pueruli and juveniles to a market size commenced recently in Vietnam and India, although this development may be considered risky because it has the potential to threaten the long-term sustainability of the wild fishery.

LIFE HISTORIES OF LOBSTERS

Clawed Lobsters

Homarus americanus may be used to exemplify the life cycle of clawed lobsters (Factor, 1995). At spawning, mature lobsters pair up for approximately two weeks. The female leaves her solitary shelter about 7 days before moulting and shares the shelter of the dominant male. Although lobsters are normally cannibalistic, the female produces a sex pheromone around the time of moulting that suppresses male aggression and induces courtship displays to achieve successful pairing. Mating usually occurs within 40 min of the female moult, while she is still soft.

Females mostly mature at 70-90 mm carapace length (CL), while males may produce viable sperm at < 45 mm CL. However, for successful mating, the male is usually larger than the female. The male protects the female from predation and cannibalism for up to 7 days after the vulnerable moult, but also prevents other males from copulating with her. The female normally extrudes 3,000 to 115,000 eggs up to 13 months after mating and the eggs are fertilized by spermatozoa stored in the seminal receptacle. Oviposition occurs every second year even if the female carries no spermatozoa. There is strong evidence indicating that eggs are fertilized externally as they pass over the seminal receptacle during spawning. Females may fertilize two or more successive egg batches from a single mating.

Immediately after extrusion, eggs are dark green and irregular but quickly become spherical with a diameter of 1.5-1.7 mm. During embryonic development, they increase in size, becoming elongated and lighter in colour. After extrusion, fertilized eggs attach firmly to pleopods, with embryonic development taking about 11 months, the duration mainly determined by water temperature. Eggs hatch from May to October, reaching a peak in June to early July at 20°C. All eggs in a brood hatch in 2-3 days at 20°C but take 10-14 days at 15°C. All larvae are released quickly at night with the female vigorously beating her pleopods. The female then normally moults and mates within one month after hatch. Most pre-larvae moult into the first larval stage before release by the female. There are then four stages of free-swimming larvae before lobsters settle on the benthos, burrow into substrate shelters and moult to the juvenile stage. The larval stages are normally completed within 35 days from hatch but the duration is dependent on temperature.

The factors affecting larval abundance and distribution include the distribution of spawning females, the velocity and direction of surface currents, temperature, salinity, light intensity, hydrostatic pressure, and larval mortality. Larvae are planktonic from late May to October, appearing earlier in southern New England than in the Gulf of Maine. During the day, larvae are near the surface but they are in deeper water at night. The high densities along windward coasts suggest that larvae are transported by prevailing winds and surface currents.

Juvenile and adult lobsters prefer natural shelters such as sand substrate with overlying flattened rocks. They seek the dark in shelters where they can maintain contact with the roof and walls. Inshore lobsters in-shore prefer to be solitary, rarely sharing shelters. Offshore lobsters often share shelters or depressions in the substrate, possibly because of the scarcity of shelters. Juveniles less than 35 mm CL rarely leave their shelters, whereas juveniles of 35-40 mm CL forage nocturnally outside their shelters. Adults and juveniles are highly territorial, having a limited home range and seldom travelling more than 300 m from their shelter when foraging.

Lobsters moult approximately 10 times in their first year, 3-4 times in the second and third years, twice in the fourth year, and once or less annually thereafter. Temperature is the dominant factor influencing growth, although salinity, dissolved oxygen, food availability, and crowding may also be important. Mature females grow more slowly than males because they moult less frequently but the size increments between moults are similar.

Offshore lobsters grow faster and moult more often than inshore lobsters, possibly because of the differences in migratory habits. Offshore lobsters migrate annually from between the outer continental shelf and upper slope (8 to 12°C) to the shallower continental shelf (10-17.5°C) seeking the optimum water temperature for growth and breeding, whereas inshore lobsters are non-migratory and subject to low winter temperatures of < 5°C that inhibit growth. Lobsters suffer high mortality rates around the moult before shell hardening. They are vulnerable to general predation as well as territorial attacks and cannibalism from other lobsters, probably more so where there is a lack of shelter, such as rock crevices or sediments in which they can burrow.

Spiny Lobsters

The generalized life cycle of spiny lobsters was described by Lipcius and Eggleston (2000) and includes five major phases: egg, phyllosoma (larval), puerulus (post-larval), juvenile and adult (Phillips et al., 1980). Some of these life stages are markedly different from those of clawed lobsters, while others are similar.

Shortly before reaching adulthood, animals move from their inshore nursery habitat to deeper water reefs where they breed (Herrnkind, 1980; Lipcius, 1985). The male deposits the spermatophoric mass externally on the female's sternum, and she rasps the spermatophore shortly before spawning, releasing the sperm for fertilization of the eggs as she extrudes them onto her abdomen and pleopods (Cobb and Wang, 1985; Lipcius and Herrnkind, 1987; MacDiarmid, 1988).

Spawning usually occurs in spring-summer in females located in offshore reefs (Lipcius and Eggleston, 2000). After embryonic development, which lasts up to 5 months in some rock lobsters, larvae are hatched as naupliosoma with a pelagic swimming habit. This phase may last for less than an hour before larvae unfold to the characteristic phyllosoma (leaf-like body) form. The early-stage phyllosoma are then swept off-shore by surface currents into oceanic habitats (Phillips and McWilliam, 1986). The planktonic, free-swimming and transparent phyllosoma undergo a lengthy oceanic larval phase and, depending on the species, this may last between a few months and two years, during which they moult 20 or more times before metamorphosis to the puerulus stage (Phillips and Sastry, 1980).

Metamorphosis appears to be dictated by the development and nutritional state of the phyllosoma rather than by environmental cues (McWilliam and Phillips, 1997). The newly metamorphosed and transparent pueruli move in-shore by both active swimming and transportation via currents to settle on the benthos, predominantly on rocky reefs (Phillips and Sastry, 1980). The pueruli moult to juveniles, which remain in their benthic habitat until adulthood.

Growth and Moulting

Lobster growth, which is the accumulation of new body tissue, is almost continuous but can only be expressed at ecdysis, the moult, when the exoskeleton is shed (Phillips et al., 1980). Moulting and growth are influenced by numerous factors, individually and synergistically, including food, temperature, light, genetics, health, stress, water quality and seasonality (Waddy et al., 1995). In the wild, animals also suffer predation, whereas with aquaculture, animals often grow much faster although intensive conditions may precipitate the rapid development of disease and death.

The long intermoult period allows the lobster to recover from its last moult and to accumulate body resources for growth at the next moult (Phillips et al., 1980). As with reproduction, growth and ecdysis are controlled by the endocrine system, mostly influenced by temperature and photoperiod, and there is a close inter-relationship between the hormones involved in somatic growth and reproduction. Quackenbush (1994) and Waddy et al. (1995) summarized some of the findings in lobster endocrinology. The moult-inhibiting hormone, secreted from the X organ sinus gland complex of the eyestalk in *H. americanus*, is a large peptide that suppresses biosynthetic activity of the moulting glands (the Y organs). It blocks ecdysteroid biosynthesis in the Y organs, reduces ecdysteroid levels in the haemolymph, and delays the moult cycle. Vitellogenesis-inhibiting hormone, which blocks yolk biosynthesis, and crustacean hyperglycaemic hormone, the different forms of which have hyperglycaemic or gonad-stimulating activity as well as a role in regulating the moult, are also peptide hormones and secreted from the X organ sinus gland, but from different cells. The diverse functions of these hormones, which include regulation of blood glucose, moulting and vitellogenesis, are thought to be due to the variety in the C-terminal region of the peptide molecules (Webster, 1991). The moulting hormones, or ecdysteroids that induce the physiological and biochemical changes that lead to moulting, follow a predictable pattern throughout the moult cycle, being barely detectable immediately after ecdysis and reaching a peak when the pre-exuvial cuticle is being formed (Waddy et al., 1995). Methyl farnesoate, a terpenoid hormone, stimulates ecdysteroid secretion (Tamone and Chang, 1993) and is implicated in lobster reproduction as it occurs at levels in the haemolymph that are highest when there is no ovarian maturation but lowest during ovarian maturation (Quackenbush, 1994). With eyestalk ablation, levels of methyl farnesoate are considerably elevated. Thus, there is a complex interplay between the eyestalk neuroendocrine system, the steroid endocrine system and terpenoid endocrine system regulating moulting and gonadal maturation that is yet to be fully elucidated.

Size is usually expressed as body weight or carapace length, while growth rate depends on the time between successive moults and the increase in size at each moult (Phillips et al., 1980). Following metamorphosis from larvae, juveniles moult frequently in their first year with incremental size increases of 50% or more at each moult, but when they become adults, they may moult only once annually with a size increment of 10% or less. Rapidly growing juveniles become adults sooner and thus will be sexually mature earlier. Under aquaculture conditions, juveniles become precocious, maturing and breeding at much smaller sizes than is typically found in the wild. Our own results indicate that *Jasus edwardsii* will mature at < 65 mm CL and 150 g body weight when grown in captivity, considerably smaller than animals captured from the wild. Once sexual maturity is reached, much less energy is partitioned to growth in females than in males. Zoutendyk (1990) found that once *Jasus lalandii* reached sexual maturity, and especially in large animals, the ratio of apportionment of production into somatic growth and reproduction differed markedly between the sexes, reducing from 99:1 for males to 50:50 for females, i.e., considerably more energy was diverted to reproduction in females than in males.

Other than food availability, the most important environmental factor affecting growth is temperature. Holding animals constantly at their optimum temperature will result in rapid continuous growth. This is unlike the animals' natural environment, in which seasonal low temperatures usually reduce or prevent growth. For example, Hughes et al. (1972) found that holding *H. americanus* at a constant high 22°C could shorten the time to reach 450 g from 6 years to 2 years. There were similar findings in spiny lobsters, and the increased growth rate was due to reduced intermoult intervals while the percentage increase at each moult remained approximately the same (Chittleborough, 1975). Growth is adversely affected by the loss of appendages, which require regeneration at the next moult. On the other hand, eyestalk ablation removes the source of moult-inhibiting hormone, leading to precocious moulting and increased growth (Phillips et al., 1980).

The moult includes the physiological, morphological, biochemical, and behavioural changes involved in the preparation for and recovery from ecdysis (Ennis, 1995). Lobsters spend much of their lives preparing for and recovering from ecdysis. The preparation for ecdysis is controlled by the endocrine system, which in turn responds to internal cues (nutritional state, health) and external cues (temperature, photoperiod).

Premoult activities include limb regeneration, resorption and storage of cuticular components, deposition of new cuticle, histolysis of somatic muscle, selective water and ion absorption, and shifts in biochemical pathways (Waddy et al., 1995). These activities culminate in the shedding of the exoskeleton, ecdysis. After ecdysis, the lobster has usually increased in mass, volume and length by actively absorbing water. The epidermis regresses, additional cuticle is secreted, mineralization occurs, water is replaced by new tissue, and metabolic reserves are replenished. The lobster may then proceed to the preparation for the next moult or pause for an extended period (Carlisle and Dohrn, 1953, cited by Waddy et al., 1995).

Stages of the Moult Cycle

Waddy et al. (1995), citing Herrick (1909), described the three stages of the moult cycle for *H. americanus*:

1. *preparing to moult,* the growth of the new shell under the old;
2. *shedding,* the shedding of the outgrown shell and a sudden expansion in size; and
3. *recently moulted,* the gradual hardening of the new shell.

A detailed classification was developed by Knowles and Carlisle (1956) for crustaceans, dividing the intermoult cycle into five periods (Stages A to E) with numerous subdivisions, and this has been adapted for use with lobsters (see Waddy et al., 1995). Moult staging may be based on either shell hardness or changes in developing setae (see Aiken, 1980).

Larvae

Growth and development in larval lobsters is markedly influenced by temperature. For example, Bermudes (2002) found that the intermoult interval in *J. edwardsii* phyllosoma larvae from Stage I to II increased almost four-fold when culture temperature was reduced from 18°C to 10°C (Fig. 2) and this was associated with a 15% smaller length increment and 75% reduced daily feed intake at the cooler temperature. At warm temperatures, there was a marked increase in feed intake and metabolic activity (oxygen consumption, nitrogen excretion) for each stage up to Stage III (Bermudes and Ritar, 2004). However, there were only slight effects of photoperiod on intermoult interval, size and feeding rate (Bermudes, 2002).

In *H. americanus*, larval development in the wild occurs over a broad range of temperatures but ceases below 10°C where survival is poor (Waddy et al., 1995). Optimum development appears to occur at 15-18°C, although this is highly variable because many factors other than temperature influence larval size, including seasonality, geographical and depth locations, even differences between the first and last larvae to hatch from a brood.

Metamorphosis from the larval to the post-larval or juvenile stage is accompanied by many major changes in the morphology, ethology and physiological characteristics. This moult at metamorphosis appears to be under similar endocrine control, particularly from the eyestalk, as are moults in larvae, juveniles and adults (Charmantier and Charmantier-

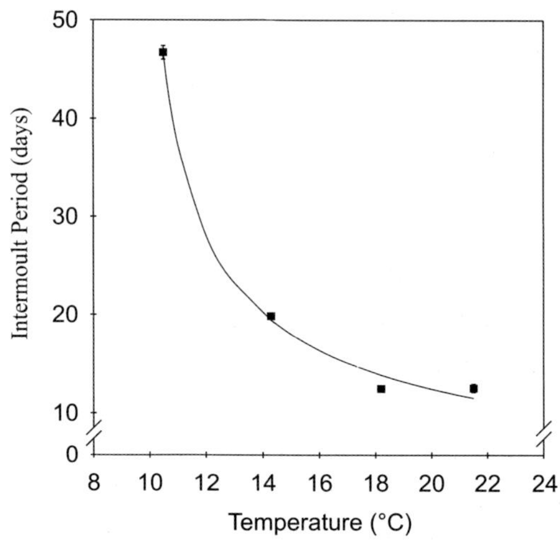

Fig. 2 Mean (± SE) intermoult period of stage I *Jasus edwardsii* phyllosoma larvae reared at four temperatures. The Bělehrádek expression was fitted to the raw time data (from Bermudes, 2002).

Daures, 1998). Moulting and morphogenesis are thought to be independent processes that are synchronized under normal conditions (McConaugha, 1985). Additional factors, such as the stresses of adverse environmental conditions or malnourishment, may influence or delay metamorphosis, especially in culture, resulting in intermediate stages or instars (Snyder and Chang, 1986). Larvae may need to reach a threshold size before they can undergo metamorphosis.

Juveniles and Adults

Temperature is the most important influence on growth in juvenile and adult clawed and spiny lobsters (Waddy et al., 1995; Thomas et al., 2000). In clawed lobsters, there is an acceleration in metabolic processes and growth rate proportional to the increase in temperature between 8°C to 25°C, while moulting is blocked at 5°C (Aiken, 1977). The normal seasonal moult inhibition is disengaged or overridden at 15-25°C, enabling animals to moult at any time of year (Aiken, 1980). Similarly, when lobsters were reared from egg at a constant 20°C, there was no distinct winter refractory period even though seasonal differences persisted (Waddy et al., 1995). The normal pattern sees a seasonal reduction in moulting in summer. The eyestalk is the neurosecretory centre controlling the circannual pattern of moulting. The seasonal refractory period is eliminated with eyestalk ablation.

Light, or more correctly photoperiod, controls many physiological processes including moulting in lobsters via the central nervous system (Waddy et al., 1995). By contrast, the intensity of light has other effects: high light intensity depresses activity in lobsters, whereas constant darkness may cause hyperactivity, antagonistic behaviour and stress (Aiken, 1980).

Social factors such as crowding increase the interaction between animals and have a negative effect on survival, moult frequency and increment in culture (Van Olst et al., 1980) and in the wild (Ennis, 1991). Growth rates increase in proportion to the amount of space available irrespective of the shape or depth of the container (Sastry et al., 1975). This negative effect of high density may be confounded by a short-lived, density-dependent, growth-inhibiting chemical factor that reduces growth of juveniles downstream (Nelson et al., 1980, 1983). This is exemplified in the case of lobsters that suffered reduced growth rates when held downstream from larger individuals (Nelson and Hedgecock, 1983; Rayns, 1991). In closed-water aquaculture systems, holding different-sized animals may contribute to growth inhibition and adverse aggressive encounters, partly mediated by chemical factors (Zeitlin-Hale and Sastry, 1978).

Reproduction

The ability of any lobster species to produce larvae is dictated by its reproductive characteristics, particularly the mode of gamete development and embryonic maturation. These will determine the quantity of gametes and embryos that individuals can produce and the frequency with which they become available. Reproduction in lobsters is extremely varied between and even within lobster families (Phillips et al., 1980). Copulation is generally head to head and sternum to sternum but sperm deposition may be external or internal. Embryonic development may last from 1 to 13 months with the eggs attached to the pleopods under the tail, and fecundity may vary from a few hundred to many hundreds of thousands of eggs per female. Some palinurids spawn up to four times each year, whereas *Homarus* females usually spawn every second year. Even within a species, size at sexual maturity varies greatly with geographic location.

Seasonal Effects on Reproduction

Crustaceans living in the deep sea, and possibly those living in caves, experience no seasonally related variations in their physical environment, except possibly for food supply, whereas those living closer in-shore and especially at higher latitudes experience pronounced seasonal variations (Sastry, 1983).

For lobsters, temperature and photoperiod are the principal environmental cues controlling the reproductive cycle. For example, gonadal development and spawning frequency in *Panulirus argus* are enhanced by long days (Lipcius and Herrnkind, 1987). Indeed, most species of spiny lobster are highly seasonal in their reproductive patterns. In *H. americanus*, females typically moult and mate one summer, spawn the following summer, and carry the eggs on their pleopods until the third summer, when they hatch (Waddy et al., 1995). The reproductive cycle in this species changes with increasing size, since they moult less frequently (less than annually) and take the opportunity to increase their fecundity by spawning twice between moults, thereby markedly increasing their fecundity.

Mating Behaviour

Copulation is usually characterized by a frontal approach between male and female (Phillips et al., 1980) and has been described in detail in palinurids by MacDiarmid and Kittaka (2000). The pair rear up on their fifth walking legs (pereiopods), hold each other sternum to sternum and gripping each other with the other pereiopods. In this tight embrace, the pair overbalances with the female usually on top, and the male rapidly deposits a spermatophore on the sternum of the female from his gonopores at the base of his fifth walking legs. The pair then quickly separates. Lobsters may copulate repeatedly, up to four times at night, most likely to achieve successful ejaculation and attachment of the spermatophore to the sternum of the female. Larger females seek multiple copulations from one or more males because more sperm may be required for the fertilization of eggs.

Sperm Release

In spiny lobsters, the spermatophore (Fig. 3) consists of a basal adhesive layer, an internal gelatinous matrix containing spermatozoa, and in most species an outer protective crust that may discolour to brown or black after several days' exposure to sea water (Berry and Heydorn, 1970; Radha and Subramoniam, 1985). *Jasus* spp. do not have the outer crust and the spermatophore is short-lived, disintegrating within a few hours (Paterson, 1969; Berry and Heydorn, 1970; Kittaka, 1987), and egg extrusion occurs immediately after successful copulation (MacDiarmid, 1988). By contrast, in other genera in which the spermatophore is longer-lived and protected by the hard outer layer, egg extrusion may occur two

Fig. 3 Spermatophore tarspot of *Panulirus penicillatus*.

to three times and up to two months after copulation (Chittleborough, 1976). The female scrapes the hard protective layer to release spermatozoa during oviposition.

Egg Extrusion and Fertilization

Egg extrusion in all palinurids appears to be similar (MacDiarmid and Kittaka, 2000). The female clings in a vertical head-up posture and forms a brood chamber by flexing her abdomen forwards beneath her body, extending the exopodites of the pleopods, and spreading her telson and uropods into a fan covering the gonopores located at the base of the third walking legs for *Jasus* (McKoy, 1979) or the posterior part of the spermatophoric mass as in *Panulirus homarus* (Berry, 1970). The eggs are then extruded from the gonopores, drawn by gravity, and a current is induced by beating of the endopodites of the posterior pleopods, over the spermatophore and down into the brood chamber. Fertilization occurs as the eggs pass over the spermatophore or in the brood chamber as eggs and spermatozoa are swirled by the beating of the pleopods (MacDiarmid and Kittaka, 2000). After fertilization, eggs attach to the setae on the endopodites of the pleopods; depending on the species, the process takes between 16 min (McKoy, 1979; Aiken and Waddy, 1980; MacDiarmid, 1987) and 4 h (Silberbauer, 1971).

Egg Incubation

During egg bearing, the pleopods beat slowly, probably to maintain oxygen supply to the eggs, and the female intermittently inspects or cleans the eggs using the chelate fifth walking legs (MacDiarmid and Kittaka, 2000). Any unfertilized, infected or parasitized eggs may be removed, occasionally leading to entire bundles being stripped. The duration of egg bearing may be as short as 7-9 days at 29°C for the tropical *P. homarus* (Radhakrishnan, 1977; Nair et al., 1981) to as long as 150 days at 11°C for cool temperate *J. edwardsii* (Paterson, 1969; MacDiarmid, 1989). However, even in *J. edwardsii*, embryonic development can either be extended considerably by holding at 10°C so that animals hatch up to 190 days after egg extrusion or reduced markedly by holding at 18°C so that animals hatch as soon as 90 days after extrusion (Fig. 4; Smith et al., 2002).

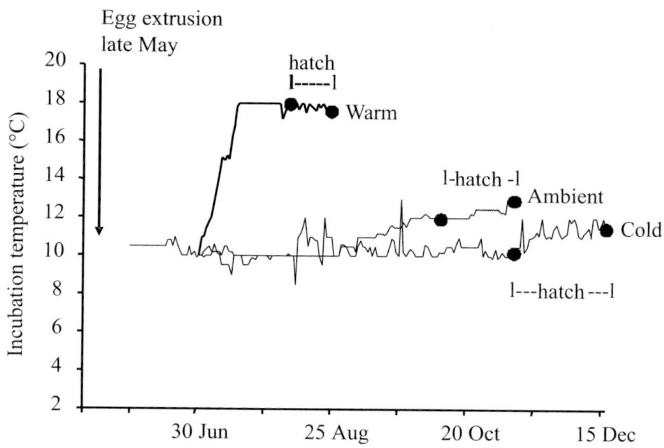

Fig. 4 Effect of temperature during egg incubation on date of larval hatch and hatch duration. Adapted from Smith et al. (2002)

Fecundity

The most important factor affecting the number of eggs extruded is body size (MacDiarmid and Kittaka, 2000) (Fig. 5). In any single population of spiny lobsters, the relationship between total egg weight (or fecundity) and body weight is curvilinear for most species, and is fit by a power relationship (Annala, 1991). For example, in *Panulirus japonicus*, the number of eggs increased from 50,000 in females of 40 mm CL to 800,000 at 95 mm CL. In *J. edwardsii*, the fecundity also increases markedly with

Fig. 5 Fecundity varies with size of *Jasus edwardsii* broodstock.

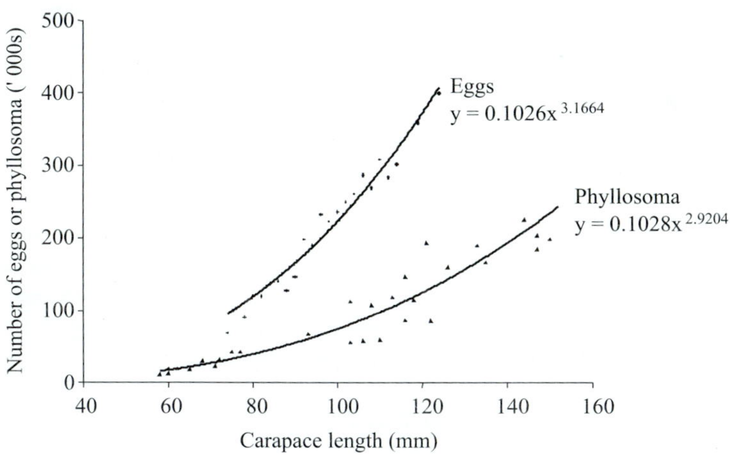

Fig. 6 Relationship between carapace length (mm) of female *Jasus edwardsii* and the number of eggs (Hickman, 1945) or newly hatched phyllosoma (Smith and Ritar, unpublished data). From Smith and Ritar (unpublished data).

animal size (Fig. 6). *Sagmariasus verreauxi* (formerly *Jasus verreauxi*; Booth et al., 2002) appears to be one of the most prolific and largest palinurids with fecundities reaching 2 million eggs or more (Kensler, 1967). Where species have multiple broods, the fecundity may decrease at successive spawnings (Creaser, 1950; Juinio, 1987). Fecundity may also be influenced by the amount of sperm extruded on to females (MacDiarmid and Butler, 1999), presumably related to the size of the male mated or the number of matings.

When discussing fecundity in spiny lobsters, it is important to recognize that some species have several spawnings each year. Thus, the term "brood size" becomes meaningful for such animals (Pollock and Goosen, 1991). While all *Jasus* spp. produce only a single brood each breeding season (Pollock, 1991), various *Panilurus* species produce up to four broods each year, although it is usual for them to produce only a single brood in their first year (Chubb, 2000). The repeat spawnings within the same season and after a single mating may have a somewhat smaller size in some species. Although very large females may undergo some senescence, this would represent only a minor impact on the egg production of highly exploited spiny lobster stocks because few of these animals would remain in the population.

Egg Loss

For *J. edwardsii*, the number of larvae hatched out, as measured by Smith and Ritar (unpublished data), may be considerably less than the number of eggs carried, as estimated by Hickman (1945) (Fig. 6), possibly because of some embryonic mortality and grooming. Tong et al. (2000) found < 5% loss of eggs during embryonic development, whereas Annala and Bycroft (1987) estimated the loss at about 20%, while MacDairmid and Kittaka (2000) found significant and sometimes total loss of eggs. Perkins (1971) reported egg loss in *H. americanus* during 9 months egg incubation and suggested that the extent of loss may be related to incubation duration. Stress during captivity can trigger over-grooming, causing losses associated with fouling (*Leucothrix mucor*), fungus (*Lagenidium*) or parasitic nemertean infestations (*Pseudocarcinonemertes homari*) (Silberbauer, 1971; Shields and Kuris, 1989; Kuris, 1991).

Development of Eggs and Embryos

Egg and embryonic development in palinurid species was described by MacDiarmid and Kittaka (2000). Immediately after extrusion and attachment, eggs are spherical and red (Silberbauer, 1971). The morphological changes during embryonic development in *P. japonicus* are presented in Table 1 and the progression from fertilization to hatch is similar but on different time frames for the other spiny lobsters. For *Jasus* spp., embryos hatch as short-lived naupliosoma, whereas other genera hatch as Stage I phyllosoma.

Table 1 Embryonic development of *Panulirus japonicus* at 24°C. Adapted from Shiino (1950), as cited by MacDiarmid and Kittaka (2000).

Stage	Stage of development	Time after egg extrusion
I	Blastula	0-72 hr
II	Morula	72-130 hr
III	Blastodisc formation	5.5-6.5 d
IV	Pre-nauplius	6.5-8 d
V	Nauplius	8-10 d
VI	Post-nauplius	10-13 d
VII	7 pairs of appendages	13-15 d
VIII	11 pairs of appendages	15-17 d
IX	Pigmented ocellus	17-20 d
X	Pigmented compound eye	20-25 d
XI	Pre-hatching	25-37 d

The egg diameter may vary considerably between genera, from 0.5 mm for *P. japonicus* to 0.8 mm for *Jasus* spp. (Silberbauer, 1971), although large differences also occur within species depending on size of the female (MacDiarmid and Kittaka, 2000). Egg size and weight increase during development, affecting the number of eggs carried per gram of egg mass (Annala, 1991). There also appears to be an increase in egg size with female body size in some spiny lobsters (Annala, 1991). Recently, MacDiarmid and Kittaka (2000) and Smith et al. (2003b) found that larger females produce larger larvae. These larger Stage I larvae have higher viability and become larger Stage II larvae (Smith et al., 2003b), which may have longer-term implications for their survival and recruitment to the fishery.

In *H. americanus*, embryonic development lasts 9-12 months, during which eggs are protected and maintained by the female parent, the eggs hatch and larvae are released into the water column (Ennis, 1995). Ovigerous females may migrate extensively from deep to shallow water and *vice versa* to take advantage of warmer conditions, which accelerate moulting and embryonic development. Hatching occurs over a 4 month period in summer in north-west Atlantic waters at temperatures from as low as 4.2°C to 20.0°C (Scarratt, 1964; Hughes and Matthieesen, 1962). *Homarus americanus* has a 6-8 week planktonic larval phase, in which there are three stages, followed by the post-larval benthic stage. Understanding the factors influencing the distribution, growth and survival of the planktonic stages is fundamental to understanding recruitment mechanisms and establishing a practical relationship between parental stock and recruitment to the population (Ennis, 1995).

In both clawed and spiny lobsters, the rate of embryological development is influenced by many factors (Phillips et al., 1980), of which temperature appears to be the most important and is easily manipulated experimentally (Perkins, 1972; Chittleborough, 1976; Aiken and Waddy, 1985; Beard et al., 1985; Charmantier and Mounet-Guillame, 1992). When the date of egg extrusion in clawed lobsters is known, the time to hatch can be predicted accurately (Waddy and Aiken, 1984). When the date is known, the appearance of distinct features such as eyes and chromatophores in the embryo can be used to predict hatch times (Tong et al., 2000; Fig. 7).

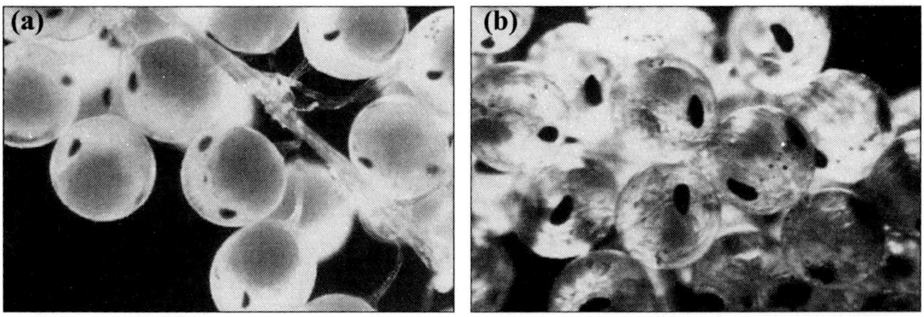

Fig. 7 *Jasus edwardsii* eggs (a) showing eye spots when they first appear (~40 days after fertilization) and (b) close to hatch when eye spots are enlarged and there are prominent chromatophores along the appendages. From Tong et al. (2000).

An eye index developed for clawed lobsters (Perkins, 1972) was adapted for *Jasus* spp. (Tong et al., 2000; Moss et al., 2004). For example, the following formula (from Tong et al., 2000) accurately predicts the number of days to hatch for *J. edwardsii* from the size of the developing eye spots (mean of length plus width):

$$\text{Days to hatch} = \frac{\text{EAT at hatch (699) - EAT on time index measured}}{\text{Rearing temperature - biological zero (7.53°C)}}$$

Egg Hatch

In many species of spiny lobster, the females migrate to deep water and on to a sandy bottom where they may aggregate in preparation for larval hatch (Ansell and Robb, 1977; Herrnkind, 1980; MacDiarmid, 1991; Kelly et al., 1999). MacDiarmid and Kittaka (2000) describe their observations of larval hatch in captive females. Broodstock briefly and vigorously beat their pleopods to release a "swarm" of larvae that are strongly photopositive and immediately swim towards bright light. Hatching usually occurs at dawn (MacDiarmid, 1985).

Ennis (1995) described the hatch and release of larvae in clawed lobster *H. americanus*. Eggs hatch in summer and this involves the escape of the larvae from egg membranes and then the release from the ovigerous female (Templeman, 1937). The outer egg membrane ruptures under internal pressure generated as water is absorbed by the developing embryo (Davis, 1964). The non-swimming pre-larval pre-zoea stage is enveloped in a cuticle, which is torn by the weight of the embryo and swimmeret movements by the female. The emerging larvae remain attached to their cuticles until released as free-swimming stage I larvae by the female's vigorous beating of the pleopods, usually within one minute (Templeman, 1937; Ennis, 1975) and after sunset (Ennis, 1995). Hatching and pre-larval moulting continue daily for 15-31 days, with the release of anywhere from just one larva to 2,000 larvae each day. Newly released larvae swim to the water surface, where they swarm (Herrick, 1895, as cited by Ennis, 1995; Templeman, 1937).

Population Egg Production

The high value of commercial lobsters means that they are mostly heavily exploited and require sound management to retain adequate breeding stocks for egg production and future recruitment to the fisheries (Chubb, 2000). Therefore, it is important to understand the fundamental relationship between parent stock and recruitment.

The biological and environmental factors that may influence egg production in a population include individual fecundity, size at onset of maturity (SOM), growth rate, natural mortality rate, water temperature and food availability (Annala, 1991). Hobday and Ryan (1997) also found a strong correlation between size of female *J. edwardsii* and egg production (Fig. 8).

In a fished population, exploitation rate, minimum legal size and mortality of discards also affect egg production. Exploitation has a bigger impact on population egg production where the SOM is high, possibly leading to only a small percentage of the total population producing eggs, and these being larger than the minimum legal size, compared to unfished populations.

Understanding the impact of fishing on egg production from a lobster population is crucial in maintaining a viable fishery (Chubb, 2000). Increases in fishing efficiency lead to increases in fishing mortality, which must be taken into account when calculating annual population fecundity from catch-rate-based data. Otherwise, the egg production will be increasingly over-estimated and, if incorrectly used for management, this would possibly threaten recruitment and the viability of the fishery.

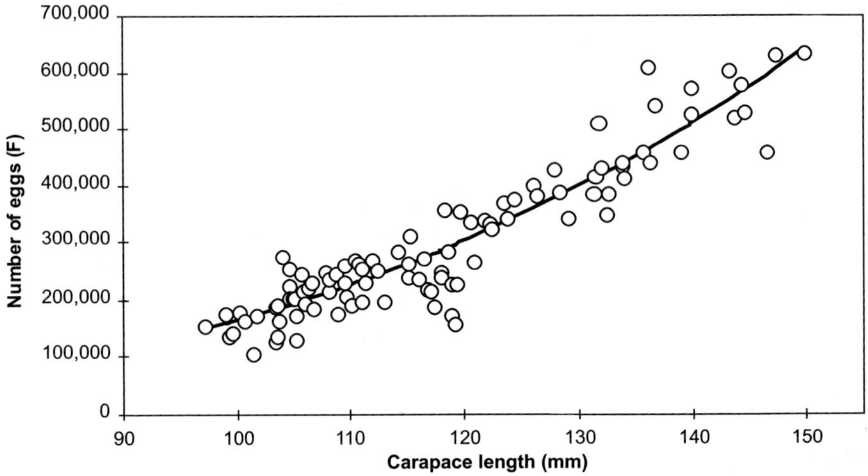

Fig. 8 Relationship between size of female *Jasus edwardsii* and number of eggs carried. From Hobday and Ryan (1997).

If egg production were reduced to low levels and recruitment were threatened, then managers would need to make difficult decisions to stabilize the fishery and maintain its viability. In addition, the inter-annual fluctuations in catches from the spiny lobster fisheries due to environmental influences need to be considered. For example, the correlation between El Nino Southern Oscillation and changes in the Leeuwin Current influence puerulus settlement of *P. cygnus* along the coast of Western Australia (Pearce and Phillips, 1988).

Some species of spiny lobsters undertake complex pre-spawning or spawning migrations, moving into deep water, generally against the prevailing current, and into their breeding areas to ensure the highest survival of larvae and then to settlement of pueruli (Herrnkind, 1980). Some species migrate on a broad front, such as *P. cygnus* and *P. argus*, while others such as *P. ornatus* and *S. verreauxi* follow distinct pathways. These routes need to be managed carefully to avoid over-exploiting the migrating lobsters or reducing the recruitment to the fishery.

Size at Onset of Maturity for Fisheries Management

Female spiny lobsters become mature when they are capable of extruding eggs; for fisheries assessment, this may be determined from secondary sexual characteristics (Aiken and Waddy, 1980) such as well-developed (ovigerous) pleopodal setae in *Jasus* spp. (Chubb, 2000), although these indirect measures should be exercised with caution as they do not guarantee maturity (Lyons et al., 1981). The data from Booth (1984) indicates that the presence of pleopodal setae or spermatophores overestimates the presence of egg-bearing females of *P. cygnus* and *S. verreauxi*.

Functional sexual maturity, defined as the capability to actively breed (Aiken and Waddy, 1980), occurs in male spiny lobsters when they not only become physiologically capable of producing sperm but also are able to copulate successfully (MacDiarmid and Kittaka, 2000). When assessing populations of lobsters, functional maturity may be estimated indirectly for male spiny lobsters on the basis of morphological changes such as the ratio of leg length to carapace length, which begins to change at maturity (George and Morgan, 1979).

For fisheries management of spiny lobsters, the SOM should not be taken when the smallest female becomes ovigerous, since this is misleading as to what is happening in the population (Fielder, 1964). Rather, the CL or weight at which 50% of females have ovigerous setae and are mated (indicated by unused or eroded spermatophore), combined with those that are berried, is the most practical indicator of maturity (Chubb, 2000). An indirect estimate of sexual maturation is the growth of the second walking leg (Evans et al., 1995), although this appears to be considerably more variable than the direct measures mentioned above, whereas estimates based on reproductive appendages such as the pleopods are more accurate than leg measurements (Minagawa and Higuchi, 1997).

The SOM is highly variable between species of spiny lobsters (Annala, 1991). It ranges from 56 mm CL for *J. lalandii* from Namibia to 120 mm CL for *J. edwardsii* from Otago in New Zealand, although even within *J. edwardsii* this may be as low as 64 mm CL, corresponding to less than 150 g body weight, in other locations. For *S. verreauxi*, which mature at a large size, SOM ranged between 160 and 180 mm from different areas around New Zealand (Booth, 1984). Our own unpublished data on captive *S. verreauxi* indicates that they need to reach at least 150 mm CL and 1.3 kg body weight to produce viable eggs, and that there is no indication of sexual precociousness as in captive young and small animals, unlike *J. edwardsii*, which may bear eggs at about 2 years, < 65 mm CL and 150 g body weight when cultured.

In southern Australian waters, SOM in *J. edwardsii* varies geographically from 114 mm CL in the northern, western and eastern extremes where there appears to be abundant food and animals grow rapidly, to a low 41 mm CL in the cooler south (Hobday and Ryan, 1997), where there may be limited food availability and animals grow slowly (Punt et al., 1997). Growth rate may also affect fecundity with females that grow more rapidly having a lower fecundity at the same size than females from areas with slower growth (Annala, 1991). This author suggested that such geographical differences in SOM as well as fecundity may be based solely on differences in water temperature in *J. edwardsii* but not in *J. lalandii*. Although there are few data for males, maturation and the onset of sperm production is also likely to be influenced by the above factors.

When Hobday and Ryan (1997) examined female *J. edwardsii* in the Victorian fishery in southern Australia, they found distinct differences in SOM between animals from different locations (Fig. 9). The smallest size classes of animals in the Eastern and Western Zones where 50% of females were mature had CL of 90 mm and 112 mm, respectively. This indicated that separate management strategies needed to be considered for the minimum legal size limits for the two zones.

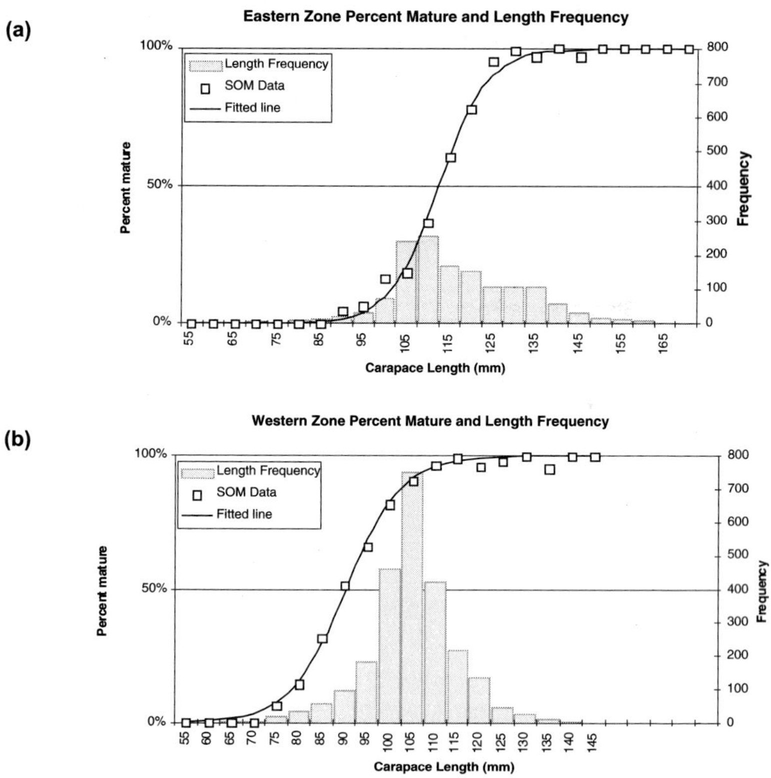

Fig. 9 Size of onset of maturity based on the relationship between carapace length and the presence of eggs or ovigerous setae on pleopods in *Jasus edwardsii* females caught in the (a) Eastern Zone and (b) Western Zone of the commercial fishery in Victoria, Australia. From Hobday and Ryan (1997).

The relationship between SOM and protection of the reproductive capacity of lobster stocks in the fishery are beyond the scope of this chapter. However, readers who wish to examine this issue in detail are referred to the review by Chubb (2000), who discusses the interrelationship between size limits in the spiny lobster fishery and the effects on recruitment. This author also analyses annual closures for the protection of females during their critical breeding period until egg hatch, which are effective in temperate areas where there is usually well-defined seasonality.

MANIPULATION OF REPRODUCTION IN LOBSTERS IN CAPTIVITY

Environmental Effects on Reproduction

Embryonic development to larval hatch in *J. edwardsii* is markedly influenced by water temperature and may be reduced or extended in ovigerous broodstock held in warm and cool water, respectively (Tong et al., 2000; MacDiarmid and Kittaka, 2000; Smith et al., 2002). When Smith et al. (2003a) compressed the seasonal changes in both temperature and photoperiod, mating and egg extrusion in *J. edwardsii* were delayed but embryonic development was greatly accelerated compared to the normal seasonality, resulting in much earlier hatch of larvae (Fig. 10).

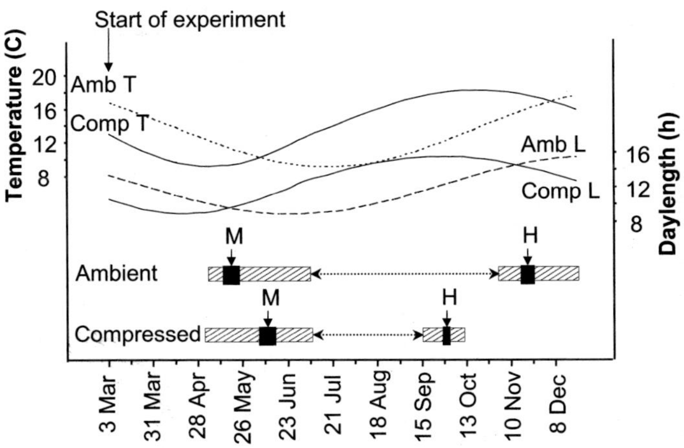

Fig. 10 Temperature (T) and daylength (L) patterns in ambient (Amb) and compressed (Comp) phototherm treatments. Mean of moult (M) and mean of hatch (H) are indicated by vertical arrows, standard errors by solid bars, and ranges by stippled bars (start and finish) (from Smith et al., 2003a).

A gradual advancement in the breeding season over three years was achieved in *J. edwardsii* broodstock by manipulating both temperature and photoperiod, as outlined in Fig. 11, resulting in mating, extrusion of eggs and hatching of viable larvae up to 6 months out of season compared to the natural pattern (Ritar and Smith, unpublished data). This allowed research to be undertaken on hatchery rearing of larvae at virtually any time of year and not constrained to the natural short seasonal hatch of larvae in spring. Holding *J. edwardsii* broodstock at constant 12°C or 15°C while manipulating photoperiod alone also allowed animals to be bred out of season, indicating that in this temperate species, light is the principal environmental cue for gonadal development, moulting and egg extrusion, but not for embryonic development. However, Kittaka (1987) found that abrupt reversal (within one year) of breeding and production of viable larvae of several palinurids was achieved when animals were transported from the southern hemisphere to Japan and placed immediately on northern hemisphere photoperiod and temperature. To obtain controlled hatching of individual *J. edwardsii* broodstock, a simple system was developed (Fig. 12) that ensured females did not suffer a change in water temperature or other environmental factors and yet could easily collect all larvae hatched each day (Smith et al., 2003a).

Early larval survival may be reduced when embryonic development is accelerated in warm water in clawed lobsters (Aiken et al., 1982) and in *J. edwardsii* (Smith et al., 2003b). In the latter case, there were other adverse effects of warm incubation temperature, including smaller size and reduced viability of larvae, but no obvious effects on their

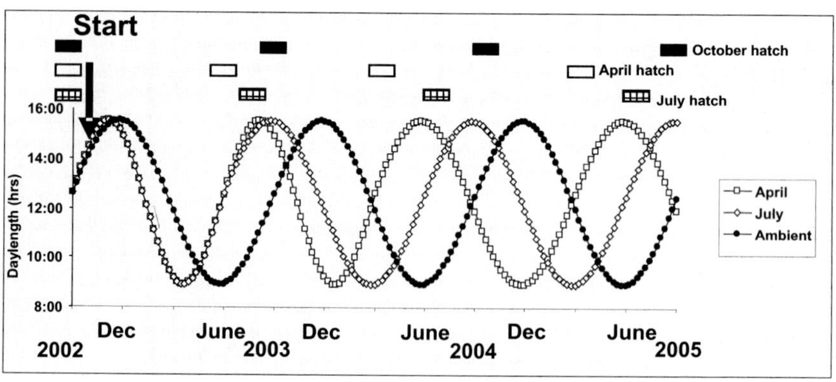

Fig. 11 The change in time of larval hatch (indicated by horizontal bars) in groups of *Jasus edwardsii* females after manipulation of photoperiod (daylength, hr). Females were placed on these regimes in November 2002. For clarity, changes in temperature are not shown but followed photoperiod by approximately two months.

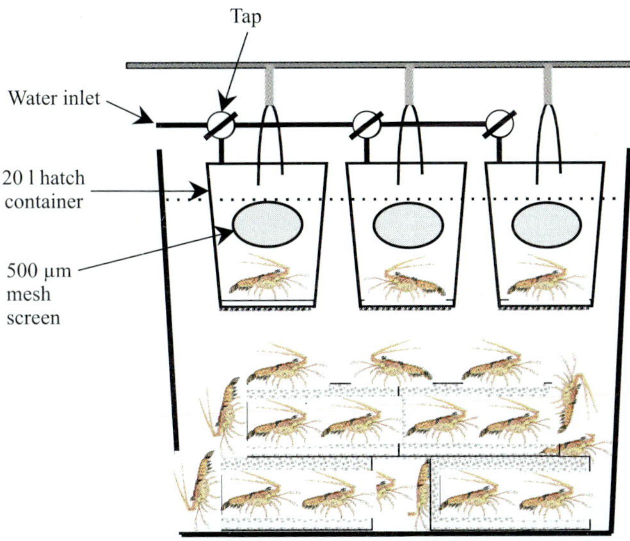

Fig. 12 Broodstock of *Jasus edwardsii* held in hatching containers suspended in 4000 l holding tanks. The 20 l containers were provided with constant flow-through water while a 500 μm screen on the side of the container retained newly hatched phyllosoma and afforded water escape. From Smith et al. (2003a).

biochemical composition. Further, for captive *S. verreauxi* broodstock held at warmer temperatures, larvae were significantly smaller, which may have implications for larval viability because there was a positive correlation between larval size and survival of larvae (Moss et al., 2004). By contrast, for *S. verreauxi* broodstock held at cooler temperatures, the duration of hatch was prolonged (Moss et al., 2004) and this was also seen previously in *J. edwardsii* (Smith et al., 2002) (Fig. 13).

Diet

Harrison (1997) described the three fundamental aspects of importance for crustacean broodstock nutrition. First, adequate nutrient and energy status appears to be necessary for the onset of gonadal maturation. Second, maturation after eyestalk ablation to induce rapid and precocious gonadal development occurs despite the nutritional status of the animals, so it is especially critical to ensure adequate nutrition to maintain the composition of the ovaries and the eggs. Third, as the larvae are lecithotrophic, they rely exclusively on nutrients supplied by the egg.

The optimum diet is usually one containing items from their natural habitat, and in palinurid lobsters, this includes live shellfish such as

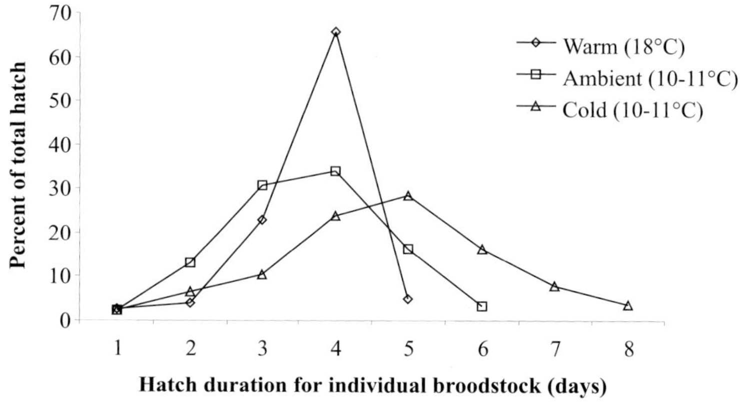

Fig. 13 The percentage and duration (mode) of total hatch of phyllosoma from individual *Jasus edwardsii* broodstock in three temperature treatments. Adapted from Smith et al. (2002).

molluscs (mussels and barnacles), coralline algae and crustaceans (Barkai et al., 1996; Frusher et al., 1999) but often reflects availability where food is scarce (Fielder, 1965; Edgar, 1990). For captive broodstock, fresh shellfish can be fed whole and stay alive in tanks for several days (Smith et al., 2004). Alternatives to fresh diet, such as frozen squid and fish, may be substituted with no apparent detrimental effect (MacDiarmid and Butler, 1999). Even on a diet completely devoid of seafood, adult *J. edwardsii* females still survived and produced viable larvae because they apparently had the ability to selectively sequester the required lipids and essential fatty acids (Smith et al., 2004). However, feeding animals on a poor diet in the long term may have adverse effects on reproduction and larval quality. Also, animals should not be underfed as they rapidly lose breeding condition especially at tropical temperatures (MacDiarmid and Kittaka, 2000).

Broodstock Handling and Stress

The adverse effects of frequent handling on *J. edwardsii* broodstock in captivity include immediate and long-term stressors that impact on breeding capacity (Smith and Ritar, 2005). This is manifested as decreased activity at dusk, possibly resulting in reduced foraging and less efficient sequestering of nutrients during ovarian recrudescence. The most important problem of frequently disturbing ovigerous females is that many have partial or total loss of eggs, possibly as a result of poor initial attachment (Talbot and Harper, 1984), with further sequential loss during incubation in the disturbed females due to over-grooming of the egg bundles (MacDiarmid, pers. comm.). The longer-term consequence of broodstock handling was that their newly hatched larvae performed poorly when subjected to a short-term low-salinity activity test (Smith and Ritar, 2005). Further, this was correlated with a lower survival in culture (Fig. 14) and larvae were smaller, indicating that the stress of handling produced less competent larvae.

Eyestalk ablation may improve growth and reproduction in lobsters. In *P. homarus*, eyestalk ablation accelerated gonadal growth in both males and females (Radhakrishnan and Vijayakumaran, 1984). However, this had the unfortunate negative effect of causing males to indiscriminately deposit spermatophores on other males and immature females, while ablated females sometimes moulted during the brooding of eggs, resulting in their loss.

Holding Conditions and Transportation

Most species of spiny lobsters can usually tolerate holding at high densities and biomass (Crear and Forteath, 1998). In contrast, clawed lobsters are highly cannibalistic under intensive conditions and are best cultured individually (Aiken and Waddy, 1995). The size of the tank in which spiny lobsters are held needs to be appropriate, so that for 300 g adult *P. guttatus*, a 40 l tank is suitable but much larger tanks are required

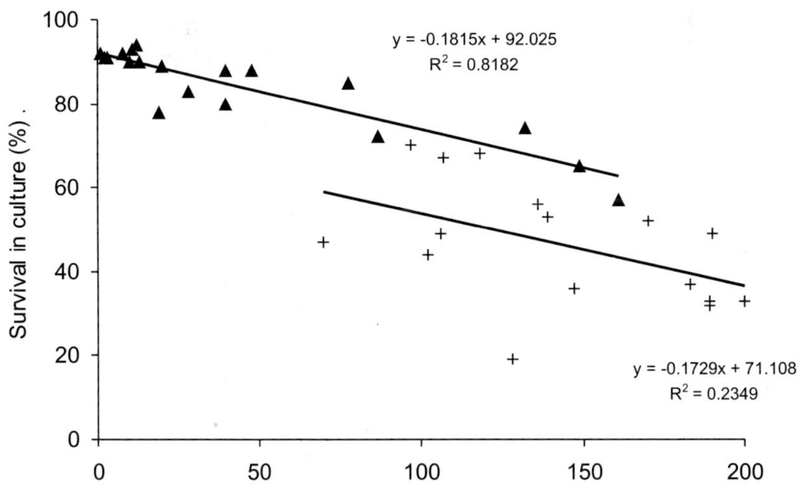

Fig. 14 The survival in culture of *Jasus edwardsii* phyllosoma larvae from hatch to Stage II of development related to the performance of newly hatched larvae in a short low-salinity activity test. Larvae were obtained from broodstock that had remained undisturbed or were handled frequently. From Smith and Ritar (2005).

for 4-5 kg *S. verreauxi* (MacDiarmid and Kittaka, 2000). Shelter offers a surface on which animals can retreat from perceived threats and to which females can cling during egg release. Shading and low light are generally preferred by all animals. Animals can be transported long distances in similar conditions to the live local and international trade for human consumption. Mature male and female spiny lobsters can withstand up to 3 days' road and air transportation without water when held at no more than 5°C for temperate species or 10°C for tropical species, and then still reproduce in captivity, although ovigerous animals need extra care to avoid dehydration or significant temperature changes of eggs (MacDiarmid and Kittaka, 2000).

Water Quality

Water quality should pose no problem for broodstock in flow-through systems as long as the lobsters are held at their appropriate temperature but careful attention is needed in recirculating systems to avoid hypoxia and deterioration due to waste build-up, including uneaten food, which may lead to microbial contamination. Disease is likely to occur in animals held sub-optimally (Handlinger et al., 2001). Ammonia is the main nitrogenous waste in lobster recirculation systems and can be fatal at high levels (Crear et al., 2003). Alkaline conditions, high temperature and low oxygen concentration exacerbate ammonia toxicity (Aiken and Waddy, 1995). All marine lobster species appear to suffer when salinities fall below normal oceanic conditions (Booth and Kittaka, 2000; Aiken and Waddy, 1995).

References

Aiken, D.E. 1977. Molting and growth in decapod crustaceans with particular reference to the lobster *Homarus americanus*. Division of Fisheries and Oceanography Circular (Australia, CSIRO) No. 7, pp. 41-73.

Aiken, D.E. 1980. Molting and growth. *In:* The Biology and Management of Lobsters, Vol. 1. J.S. Cobb and B.F. Phillips (eds.). Academic Press, New York, pp. 91-162.

Aiken, D.E., and S.L. Waddy. 1980. Reproductive biology. *In:* 1980. The Biology and Management of Lobsters, Vol. 1. J.S. Cobb and B.F. Phillips (eds.). Academic Press, New York, pp. 215-276.

Aiken, D.E., and S.L. Waddy. 1985. Production of seed stock for lobster culture. Aquaculture, 44: 103-114.

Aiken, D.E., and S.L. Waddy. 1995. Aquaculture. *In*: Biology of the Lobster *Homarus americanus*. J.R. Factor (ed.). Academic Press, San Diego, pp. 153-175.

Aiken, D.E., W.J. Rowe, D.J. Martin-Robiochaud and S.E. Waddy. 1982. Seasonal differences in the effect of photoperiod on survival and development of larval American lobsters (*Homarus americanus*). J. World Maricult. Soc., 13: 287-293.

Annala, J. 1991. Factors influencing fecundity and population egg production of *Jasus edwardsii*. *In:* Crustacean Egg Production, Vol. 7. A. Wenner and A. Kuris (eds.). A.A. Balkema, Rotterdam, The Netherlands, pp. 301-315.

Annala, J.H., and B.L. Bycroft. 1987. Fecundity of the New Zealand red rock lobsters, *Jasus edwardsii*. N.Z. J. Mar. Freshwater Res., 21: 591-597.

Ansell, A.D., and L. Robb. 1977. The spiny lobster *Palinurus elephas* in Scottish waters. Mar. Biol., 43: 63-70.

Barkai, A., C.L. Davis and S. Tugwell. 1996. Prey selection by the South African Cape rock lobster *Jasus lalandii*: ecological and physiological approaches. Bull. Mar. Sci., 58: 1-8.

Beard, T.W., P.R. Richards and J.F. Wickins. 1985. The techniques and practicability of year-round production of lobsters, *Homarus gammarus* (L), in laboratory recirculation systems. MAFF (UK) Fish. Tech. Rep. No. 79, 22 pp.

Bermudes, M. 2002. Environmental physiology of cultured early-stage southern rock lobster (*Jasus edwardsii* Hutton, 1875) larvae. PhD thesis, University of Tasmania, Australia.

Bermudes, M., and A.J. Ritar. 2004. The ontogeny of physiological response to temperature in early stage spiny lobster (*Jasus edwardsii*) larvae. Comp. Biochem. Physiol., Part A 138: 161-168.

Berry, P.F., and A.E.F. Heydorn. 1970. A comparison of the spermatophoric masses and mechanisms of fertilization in southern African spiny lobsters (Palinuridae). South African Association for Marine Biological Research, Investigational Report 25.

Berry, P.F. 1970. Mating behaviour, oviposition and fertilization in the spiny lobster *Panulirus homarus* (Linneaus). Oceanographic Research Institute (Durban) Investigational Report, 24, pp. 1-16.

Booth, J.D. 1984. Size at onset of breeding in female *Jasus verreauxi* (Decapoda: Palinuridae) in New Zealand. N.Z. J. Mar. Freshwater Res., 18: 159-169.

Booth, J.D., and J. Kittaka. 2000. Spiny lobster growout. *In:* Spiny Lobsters: Fisheries and Culture. B.F. Phillips and J. Kittaka (eds.). Blackwell Science Ltd (Fishing News Books), Oxford, pp. 556-585.

Booth, J.D., R. Webber, J. Kittaka and J. Ovenden. 2002. *Jasus* (*Sagmariasus*) *verreauxi* has a name change. Lobster Newslett., 15(1): 17-18.

Butler, M.J. IV. 2001. The 6th International Conference and Workshop on Lobster Biology and Management: an introduction. Mar. Freshwater Res., 52: 1033-1035.

Charmantier, G., and M. Charmantier-Daures. 1998. Endocrine and neuroendocrine regulations in embryos and larvae of crustaceans. Invert. Reprod. Dev., 33: 273-287.

Charmantier, G., and R. Mounet-Guillame. 1992. Temperature specific rates of embryonic development of the European lobster *Homarus gammarus* (L). Journal of Experimental Marine Biology and Ecology, 160: 61-66.

Chetrick, J. 2006. Trends in U.S. and world lobster production, imports and exports. Available at: http://www.fas.usda.gov/ffpd/Fishery_Products_

Presentations/Lobster_2006/lobster_2006.pdf#search=%22world%20lobster%20production%20by%20species%20FAO%22, interpreted from capture production 1950-2002 dataset available through Fishery Information, Data and Statistics Unit, Food and Agriculture Organization of the United Nations (FAO, Rome). 2004. FISHSTAT Plus: http://www.fao.org/fi/statist/FISOFT/FISHPLUS.asp.

Chittleborough, R.G. 1975. Environmental factors affecting growth and survival of juvenile western rock lobsters *Panulirus longipes* (Milne-Edwards). Austral. J. Mar. Freshwater Res., 26: 177-196.

Chittleborough, R.G. 1976. Breeding of *Panulirus longipes cygnus* George under natural and controlled conditions. Austral. J. Mar. Freshwater Res., 27: 499-516.

Chubb, C.F. 2000. Reproductive biology: issues for management. *In:* Spiny Lobsters: Fisheries and Culture. B.F. Phillips and J. Kittaka (eds.). Blackwell Science Ltd (Fishing News Books), Oxford, pp. 245-275.

Cobb, J.S., and D. Wang. 1985. Fisheries biology of lobsters and crayfishes. *In:* The Biology of Crustacea, Vol. 10. A.J. Provenzano, Jr., (ed.). Academic Press, New York, pp. 167-247.

Crear, B., J.M. Cobcroft and S.C. Battaglene. 2003. Guide for the rock lobster industry No. 2. Recirculating systems for holding rock lobsters. Tasmanian Aquaculture and Fisheries Institute, University of Tasmania, Australia, pp. 29.

Crear, B.J., and G.N.R. Forteath. 1998. A physiological investigation into methods of improving post-capture survival of both the southern rock lobster, *Jasus edwardsii*, and the western rock lobster, *Panulirus cygnus*. Fisheries Research and Development Corporation, Final Report for Project 94/134, University of Tasmania, Australia, 165 pp.

Creaser, E. P. 1950. Repetition of egg laying and number of eggs of the Bermuda spiny lobster. Proceedings of the Gulf Caribbean Fisheries Institute 2: 30–31.

Davis, C.C. 1964. A study of the hatching process in aquatic invertebrates. XIII. Events of eclosion in the American lobster, *Homarus americanus* Milne-Edwards (Astacura, Homaridae). Am. Midl. Nat., 72: 203-210.

Edgar, G.J. 1990. Predator-prey interactions in seagrass beds. I. The influence of macrofaunal abundance and size-structure on the diet and growth of the western rock lobster *Panulirus cygnus* George. J. Exp. Mar. Biol. Ecol., 139: 1-22.

Ennis, G.P. 1975. Observations on hatching and larval release in the lobster *Homarus americanus*. J. Fish. Res. Board Can., 32: 2210-2213.

Ennis, G.P. 1991. Annual variation in egg production in a Newfoundland population of the American lobster, *Homarus americanus*. *In:* Crustacean Issues, Vol. 7, Crustacean Egg Production. A.M. Wenner and A. Kuris (eds.). A.A. Balkema, Rotterdam, The Netherlands, pp. 291-299.

Ennis, G.P. 1995. Larval and postlarval ecology. *In:* Biology of the Lobster *Homarus americanus*. J.R. Factor (ed.). Academic Press, San Diego, pp. 23-46.

Evans, C.R., A.P.M. Lockwood, A.J. Evans and E. Free. 1995. Field studies of the reproductive biology of the spiny lobsters *Panulirus argus* (Latreille) and *P. guttatus* (Latreille) at Bermuda. Journal of Shellfish Research, 14: 371–381.

Factor, J.R. 1995. Introduction, anatomy, and life history. *In*: Biology of the Lobster *Homarus americanus*. J.R. Factor (ed.). Academic Press, San Diego, pp. 1-11.

Fielder, D.R. 1964. The spiny lobster, *Jasus lalandei* (H. Milne-Edwards), in South Australia. II. Reproduction. Austral. J. Mar. Freshwater Res., 15: 133-144.

Fielder, D.R. 1965. The spiny lobster, *Jasus lalandei* (H. Milne-Edwards), in South Australia. III. Food, feeding and locomotor activity. Austral. J. Mar. Freshwater Res., 16: 351-367.

Frusher, S., J. Prescott and M. Edmunds. 1999. Southern rock lobsters. *In:* Under Southern Seas - The Ecology of Australia's Rocky Reefs. N. Andrew (ed.). University of NSW Press, Sydney, pp. 106-113.

George, R.W., and G.R. Morgan. 1979. Linear growth stages in the rock lobster (*Panulirus versicolor*) as a method for determining size at first physical maturity. Rapp. P.-v. Reum. Cons. int. Explor. Mer., 175: 182-185.

Handlinger, J.H., J. Carson, A.J. Ritar, B.J. Crear, D.P. Taylor and D.J. Johnson. 2001. Disease conditions of cultured phyllosoma larvae and juveniles of the southern rock lobster (*Jasus edwardsii*, Decapoda; Palinuridae). *In:* Proceedings of the International Symposium on Lobster Health Management. L.H. Evans and J.B. Jones (eds.). pp. 75-87. Curtin University Publication. Website (http://www.muresk.curtin.edu.au/research/otherpublications/lhm/10_handlinger.pdf)

Harrison, K.E. 1997. Broodstock nutrition and maturation diets. *In:* Crustacean Nutrition: Advances in World Aquaculture, Vol. 6. L.R. D'Abramo, D.E. Conklin and D.M. Akiyama (eds.). The World Aquaculture Society, Baton Rouge, Louisiana, USA, pp. 390-408.

Herrick, F.H. 1909. Natural history of the American lobster. Bull. Bur. Fish., 29: 149-408 + 20 plates. As cited by Waddy et al. (1995).

Herrnkind, W.F. 1980. Spiny lobsters: patterns of movement. *In:* The Biology and Management of Lobsters, Vol. 1. J.S. Cobb and B.F. Phillips (eds.). Academic Press, New York, pp. 349-407.

Hickman, V.V. 1945. Notes on the Tasmanian marine crayfish, *Jasus lalandii* (Milne Edwards). Papers and Proceedings of the Royal Society of Tasmania, pp. 27-38.

Hobday, D.K., and T.J. Ryan. 1997. Contrasting sizes at sexual maturity of southern rock lobsters (*Jasus edwardsii*) in the two Victorian fishing zones: implications for total egg production and management. Mar. Freshwater Res., 48: 1009-1014.

Hughes, J.T., and G.C. Matthieesen. 1962. Observations on the biology of the American lobster, *Homarus americanus*. Limnol. Oceanogr., 7: 414-421.

Hughes, J.T., J.J. Sullivan and R. Shleser. 1972. Enhancement of lobster growth. Science, 177: 1110-1111.

Kelly, S., A.B. MacDiarmid and R.C. Babcock. 1999. Characteristics of the spiny lobster, *Jasus edwardsii*, aggregations in exposed reef and sandy areas. Mar. Freshwater Res., 50: 409-416.

Kensler, C.B. 1967. Fecundity in the marine spiny lobster *Jasus verreauxi* (H. Milne Edwards) (Crustacea: Decapoda: Palinuridae). N.Z. J. Mar. Freshwater Res., 2: 143-155.

Kittaka, J. 1987. Ecological survey on rock lobster *Jasus* in southern hemisphere: Ecology and distribution of *Jasus* along the coasts of Australia and New Zealand. Report to the Ministry of Education, Culture and Science, Japan, pp. 1-232.

Knowles, F.G.W., and D.B. Carlisle. 1956. Endocrine control in the Crustacea. Biol. Rev. Cambridge Phil. Soc., 31: 396-473.

Kuris, A.M. 1991. A review of patterns and causes of crustacean brood mortality. *In:* Crustacean Egg Production, Vol. 7. A. Wenner and A. Kuris (eds.). A.A. Balkema, Rotterdam, The Netherlands, pp. 117-141.

Lipcius, R.N. 1985. Size-dependent reproduction and molting in spiny lobsters and other long-lived Decapods. *In:* Factors in Adult Growth, Vol. 3. A. Wenner and A. Kuris (eds.). A.A. Balkema, Rotterdam, The Netherlands, pp. 129-148.

Lipcius, R.N., and D.B. Eggleston. 2000. Introduction: ecology and fishery biology of spiny lobsters. *In:* Spiny Lobsters: Fisheries and Culture. B.F. Phillips and J. Kittaka (eds.). Blackwell Sciences Ltd (Fishing News Books), Oxford, pp. 1-41.

Lipcius, R.N., and W.F. Herrnkind. 1987. Control and coordination of reproduction and molting in the spiny lobster *Panulirus argus*. Mar. Biol., 96: 207-214.

Lyons, W.G., D.G. Barber, S.M. Foster, F.S. Kennedy and G.R. Milano. 1981. The spiny lobster, *Panulirus argus*, in the middle and upper Florida Keys: population structure, seasonal dynamics, and reproduction. Florida Marine Research Publication 38: 1-38.

McConaugha, J.R. 1985. Nutrition and larval growth. *In:* Crustacean Egg Production, Vol. 2. F.R. Schramm (ed.). A.A. Balkema, Rotterdam, The Netherlands, pp. 127-154.

McKoy, J.L. 1979. Mating behaviour and egg laying in captive rock lobster, *Jasus edwardsii* (Crustacea: Decapoda: Palinuridae). New Zealand Journal of Marine and Freshwater Research, 13: 407–413

MacDiarmid, A.B. 1985. Sunrise release of larvae from the palinurid lobster *Jasus edwardsii*. Mar. Ecol. Prog. Ser. 21: 313-315.

MacDiarmid, A.B. 1988. Experimental confirmation of external fertilization in the southern rock lobster *Jasus edwardsii* (Hutton) (Decapoda: Palinuridae). J. Exp. Mar. Biol. Ecol., 120: 277-285.

MacDiarmid, A.B. 1989. Moulting and reproduction of the spiny lobster *Jasus edwardsii* (Decapoda: Palinuridae) in northern New Zealand. Mar. Biol., 103: 303-310.

MacDiarmid, A.B. 1991. Seasonal changes in depth distribution, sex ratio, and size frequency of spiny lobster *Jasus edwardsii* on a coastal reef in northern New Zealand. Mar. Ecol. Prog. Ser., 70: 129-141.

MacDiarmid, A.B., and M.J. Butler. 1999. Sperm economy and limitation in spiny lobsters. Behav. Ecol. Sociobiol., 46: 14-24.

MacDiarmid, A.B., and J. Kittaka. Breeding. *In:* Spiny Lobsters: Fisheries and Culture. B.F. Phillips and J. Kittaka (eds.). Blackwell Science Ltd (Fishing News Books), Oxford, pp. 485-507.

McWilliam, P.S., and B.F. Phillips. 1997. Metamorphosis of the final phyllosoma and secondary lecithotrophy in the puerulus of *Panulirus cygnus* George: a review. Mar. Freshwater Res., 48: 783-790.

Minagawa, M., and S. Higuchi. 1997. Analysis of size, gonadal maturation, and functional maturity in the spiny lobster *Panulirus japonicus* (Decapoda: Palinuridae). J. Crustacean Biol., 17: 70-80.

Moss, G.A., P.J. James, S.E. Allen and M.P. Bruce. 2004. Temperature effects on the embryo development and hatching of the spiny lobster *Sagmariasus verreauxi*. N.Z. J. Mar. Freshwater Res., 38: 795-801.

Nelson, K., and D. Hedgecock. 1983. Size-dependence of growth inhibition among juvenile lobsters (*Homarus*). J. Exp. Mar. Biol. Ecol., 66: 125-134.

Nelson, K., D. Hedgecock, W. Borgeson, E. Johnson, R. Daggett and D. Aronstein. 1980. Density-dependent growth inhibition in lobsters, *Homarus* (Decapoda: Nephropidae). Biol. Bull. (Woods Hole, Massachusetts), 159: 162-176.

Nelson, K., D. Hedgecock, B. Heyer and T. Nunn. 1983. On the nature of short-range growth inhibition in juvenile lobsters (*Homarus*). J. Exp. Mar. Biol. Ecol., 72: 83-89.

Paterson, N.F. 1969. Behaviour of captive cape rock lobsters, *Jasus lalandii* (H. Milne Edwards). Ann. S. Afr. Mus., 52: 225-264.

Pearce, A.F., and B.F. Phillips. 1988. ENSO events, the Leeuwin Current, and larval recruitment of the western rock lobster. J. Conseil (ICES), 45: 13-21.

Perkins, H.C. 1971. Egg loss during incubation from offshore northern lobsters (Decapoda: Homaridae). U.S. National Marine Fisheries Service. Fisheries Bulletin, 69: 452-453.

Perkins, H.C. 1972. Developmental rates at various temperatures of embryos of the northern lobster (*Homarus americanus* Milne-Edwards). Fish. Bull., 70: 95-99.

Phillips, B.F., and P.S. McWilliam. 1986. The pelagic phase of spiny lobster development. Canad. J. Fish. Aquat. Sci., 43: 2153-2163.

Phillips, B.F., and A.N. Sastry. 1980. Larval ecology. *In:* The Biology and Management of Lobsters, Vol. 2. J.S. Cobb and B.F. Phillips (eds.). Academic Press, New York, pp. 11-57.

Phillips, B.F., J.S. Cobb and R.W. George. 1980. General biology. *In:* The Biology and Management of Lobsters, Vol. 1. J.S. Cobb and B.F. Phillips (eds.). Academic Press, New York, pp. 1-82.

Phillies, B.F., and J. Kittaka 2000. Spiny Lobsters: Fisheries and Culture. Blackwell Science Ltd. (Fishing News Books), Oxford.

Pollock, D.E. 1991. Population regulation and stock-recruitment relationships in some crayfish and lobster populations. *In:* Crustacean Egg Production, Vol. 7. A. Wenner and A. Kuris (eds.). A.A. Balkema, Rotterdam, The Netherlands, pp. 247-266.

Pollock, D.E., and P.C. Goosen. 1991. Reproductive dynamics of two *Jasus* species in the south Atlantic region. S. Afr. J. Mar. Sci., 10: 141-147.

Punt, A.E., R.B. Kennedy and S.D. Frusher. 1997. Estimating the size-transition matrix for Tasmanian rock lobster, *Jasus edwardsii*. Mar. Freshwater Res., 48: 981-992.

Radha, T., and T. Subramonian, 1985. Origin and nature of spermatophoric mass of the spiny lobster *Panulirus homarus*. Marine Biology, 86: 13-19.

Radhakrishnan, E.V. 1977. Breeding of laboratory reared spiny lobster *Panulirus homarus* (Linnaeus) under controlled conditions. Indian Journal of Fisheries. 24: 269-270.

Radhakrishnan, E.V. and M. Vijayakumaran, 1984. Effect of eyestalk ablation in spiny lobster *Panulirus homarus* (Linnaeus). 1. On moulting and growth. Indian Journal of Fisheries, 31: 130-147.

Rayns, N. 1991. The growth and survival of juvenile rock lobster *Jasus edwardsii* held in captivity. PhD thesis, University of Otago, Dunedin, New Zealand, 225 pp.

Quackenbush, L.S. 1994. Lobster reproduction: A review. Crustaceana, 67: 82-94.

Sastry, A.N. 1983. Ecological aspects of reproduction. *In:* The Biology of Crustacea, Vol. 8, Environmental Adaptations. F.J. Vernberg and W.B. Vernberg (eds.). Academic Press, New York, pp. 179-269.

Sastry, A.N., D. Wilcox and J. Laczak. 1975. Effects of space on growth and survival of lobsters, *Homarus americanus. In:* Recent Advances in Lobster Aquaculture. J.S. Cobb (ed.). Sea Grant Lobster Aquaculture Workshop, University of Rhode Island, Kingston, Rhode Island, USA, p. 36.

Scarratt, D.J. 1964. Abundance and distribution of lobster larvae (*Homarus americanus*) in Northumberland Strait. J. Fish. Res. Board Can., 21: 661-680.

Shields, J.D., and A.M. Kuris. 1989. *Carcinonemertes wickhami* n-sp. (*Nemertea*), a symbiotic egg predator from the spiny lobster *Panulirus interruptus* in southern California, with remarks on symbiont-host adaptations. Fish. Bull; 88: 279-287.

Silberbauer, B.I. 1971. The biology of the South African rock lobster *Jasus lalandii* (H. Milne Edwards) 1. Development. South African Division Sea Fisheries Investigational Report 92: 1-70.

Smith, E.G., A.J. Ritar, C.G. Carter, G.A. Dunstan and M.R. Brown. 2003a. Morphological and biochemical characteristics of phyllosoma after photothermal manipulation of reproduction in broodstock of the spiny lobster, *Jasus edwardsii.* Aquaculture, 220: 299-311.

Smith, G.G., and A.J. Ritar. 2005. Effect of physical disturbance on reproductive performance in the spiny lobster, *Jasus edwardsii.* N.Z. J. Mar. Freshwater Res., 39: 317-324.

Smith, G.G., A.J. Ritar, P.A. Thompson, G.A. Dunstan and M.R. Brown. 2002. The effect of embryo incubation temperature on indicators of larval viability in Stage I phyllosoma of the spiny lobster, *Jasus edwardsii.* Aquaculture, 209: 157-167.

Smith, G.G., A.J. Ritar and G.A. Dunstan. 2003. An activity test to evaluate larval competency in spiny lobsters (*Jasus edwardsii*) from wild and captive ovigerous broodstock held under different environmental conditions. Aquaculture, 218: 293-307.

Smith, G.G., A.J. Ritar, D. Johnston and G.A. Dunstan. 2004. Influence of diet on broodstock lipid and fatty acid composition and larval competency in the spiny lobster, *Jasus edwardsii.* Aquaculture, 233: 451-475.

Snyder, M.J., and E.S. Chang. 1986. Effects of eyestalk ablation on larval molting and morphological development of the American lobster, *Homarus americanus*. Biol. Bull. (Woods Hole, Massachusetts), 170: 232-243.

Talbot, P., and R. Harper. 1984. Abnormal eyestalk morphology is correlated with clutch attrition in laboratory maintained lobsters. Biol. Bull., 166: 349-356.

Tamone, S.L., and E.S. Chang. 1993. Methyl farnesoate stimulates ecdysteroid secretion from crab Y-organs *in vitro*. Gen. Comp. Endocrinol., 89: 425-432.

Templeman, W. 1937. Egg-laying and hatching postures and habits of the American lobster (*Homarus americanus*). J. Biol. Board Can., 3: 339-342.

Thomas, C.W., B.J. Crear and P.R. Hart. 2000. The effect of elevated temperature on survival, growth, feeding and metabolic activity of early juvenile *Jasus edwardsii*. Aquaculture, 185: 73-84.

Tong, L.J., G.A. Moss, T.D. Pickering and M.P. Paewai. 2000. Temperature effects on embryo and early larval development of the spiny lobster *Jasus edwardsii*, and description of a method to predict larval hatch times. Mar. Freshwater Res., 51: 243-248.

Tsukimura, B., S. Waddy, J.M. Vogel and D.W. Borst. 1992. Regulation of vitellogenesis in the lobster, *Homarus americanus*. Amer. Zool., 32: 28a.

Van Olst, J.C., J.M. Carlberg and J.T. Hughes. 1980. Aquaculture. *In:* The Biology and Management of Lobsters, Vol. 2. J.S. Cobb and B.F. Phillips (eds.). Academic Press, New York, pp. 333-384.

Waddy, S.L., and D.E. Aiken. 1984. Broodstock management for year-round production of larvae for culture of the American lobster. Canad. Tech. Rep. Fish. Aquat. Sci., 1272: 1-18.

Waddy, S.L., D.E. Aiken and D.P.V. De Kleijn. 1995. Control of growth and reproduction. *In:* Biology of the lobster *Homarus americanus*. J.R. Factor (ed.). Academic Press, San Diego, pp. 217-266.

Webster, S.G. 1991. Amino acid sequence of putative moult-inhibiting hormone from the crab *Carcinus maenas*. Proc. Roy. Soc. London Ser. B-Biol. Sci., 244: 247-252.

Williams, A.B. 1988. Lobsters of the World—an Illustrated Guide. Osprey Books, Huntington, New York, 186 pp.

Zeitlin-Hale, L., and A.N. Sastry. 1978. Effects of environmental manipulation on the locomotory activity and agonistic behaviour of cultured American lobsters, *Homarus americanus*. Mar. Biol., 47: 369-379.

Zoutendyk, P. 1990. Gonad output in terms of carbon and nitrogen by the Cape Lobster *Jasus lalandii* (H. Milne Edwards, 1837) (Decapoda, Palinuridae). Crustaceana, 59: 180-192.

11

Male and Female Reproduction in Penaeid Shrimps

S. Parnes,[1] S. Raviv[1] and A. Sagi[1,2]

[1]Department of Life Sciences and [2]National Institute for Biotechnology in the Negev, Ben-Gurion University of the Negev, P.O. Box 653, Beer-Sheva 84105, Israel

INTRODUCTION

Penaeid shrimp are highly valued by people and other animals as a marine food source. In the last decade, there has been a formidable rise in aquaculture production of shrimp: the quantity of shrimp grown in captivity has more than doubled from 1994 to 2003, rising from 882,000 mt to more than 1,800,000 mt in 2003 (FAO, 2006), the latter weight being valued at US$9.3 billion. This is partly due to the tremendous amount of research that has been carried out on penaeid shrimp biology, most of which has focused on the genus *Penaeus*, which includes the largest and the most economically important species. According to the most recent taxonomy, however, all the former sub-genus taxa within *Penaeus* have been raised to genus level (Perez-Farfante and Kensley, 1997). Both the old and new species names are now used in the literature. In this chapter, the term "shrimp" is synonymous with "penaeid shrimp" and the new species names will be used even when referring to older literature.

The aforementioned economic importance of shrimp is the driving force behind all of the captive breeding programs involving these organisms. Therefore, most of the knowledge on reproduction in shrimp derives from such programs and many of the unanswered questions in this field are usually treated in the literature from the breeder's point of view. This chapter tries to look at the subject from the more basic biological direction but nevertheless concentrates on these unanswered questions. Thus, it will focus on only part of the captive shrimp's life-cycle, starting at the point when the shrimp is acquired by the breeder as an adult brooder and ending when the captive female shrimp is ready to spawn. The unanswered questions are summarized below in the form of intriguing facts that are essentially the most stubborn, long-standing puzzles in the field of penaeid shrimp reproduction (Browdy, 1992, 1998): (1) Vitellogenesis does not take place spontaneously in captivity: maturation of the ovaries and spawning has to be induced endocrinologically by unilateral eyestalk ablation together with the provision of a rich maturation diet (Wyban and Sweeney, 1991; Bray and Lawrence, 1992; Ogle, 1992; Robertson et al., 1993; Treece and Fox, 1993; Wickins and Lee, 2002). (2) Only a small proportion of the induced females contribute the larger part of total post-larvae production, while the ovaries of many induced females never mature or mature long after ablation (Wyban and Sweeney, 1991; Bray and Lawrence, 1992; Ogle, 1992; Robertson et al., 1993; Treece and Fox, 1993; Wickins and Lee, 2002; Parnes et al., 2007). (3) Although sub-adult males mature spontaneously and spermatophores containing mature spermatozoa develop and regenerate after mating without the need for special treatment, dietary or otherwise (Ceballos-Vazquez et al., 2003a; Ogle, 1992; Parnes et al., 2004), captive male reproductive capacity is highly variable, and there is no known objective measure for evaluating it in any particular population (Ceballos-Vazquez et al., 2004; Rosas et al., 1993). (4) Mating success in captivity is poor, as shown by the fact that seemingly fully vitellogenic females are scooped out of the mating tank night after night and even several times on the same night and found not to be mated (Chamberlain et al., 1983; Leung-Trujillo and Lawrence, 1987; Alfaro and Lozano, 1993; Rosas et al., 1993; Misamore and Browdy, 1996; Pascual et al., 1998; Ceballos-Vazquez et al., 2004; Parnes et al., 2007).

In the last century two extensive monographs on shrimp reproduction and development were published, the first one being that of Heldt (1938, cited in Dall et al., 1990) and the second being Hudinaga (1942). King (1948) is now considered a classical work on the morphology and anatomy of the reproductive organs of a shrimp. Dall et al. (1990) included a comprehensive chapter on reproduction in their extensive review on the biology of the Penaeidae and Rothlisberg (1998) briefly reviewed some aspects of penaeid ecology relevant to aquaculture. Since then, many scientific and technological advances have been made and a substantial volume of data about shrimp reproductive biology has accumulated. This includes our own experience with *Litopenaeus vannamei*, characterizing its reproductive readiness from juvenility to adulthood and investigating vitellogenesis and spermatophore dynamics in relation to the molt-cycle in adults (Parnes et al., 2004, 2006, 2007; Raviv et al., 2006). In what follows we will mainly review the literature that has been published since Dall et al. (1990) and draw the readers' attention to several publications that are rarely, if ever, visited. The central aim of this review is to attempt to discern the reported biological differences between wild shrimps and those in captivity, as such differences may shed some light on the puzzling observations listed above.

FEMALE REPRODUCTION

Ovary and Oocyte Development

The most intriguing question regarding female reproduction is: why do females not start to mature spontaneously after reaching adulthood under captive conditions (as far as can be judged by their size and weight)? One can start by looking at the processes of ovary and oocyte maturation. The morphology of these processes has been described in at least 10 species of open- and closed-thelycum shrimp. Eyestalk-ablated and non-ablated wild females were examined, as well as pond-reared ones grown from wild-caught juveniles or those that were pond-reared over several generations (Browdy et al., 1990; Quinitio and Millamena, 1992; Courtney et al., 1995; Medina et al., 1996; Ayub and Ahmed, 2002; Ceballos-Vazquez et al., 2003b; Palacios et al., 2003; Peixoto et al., 2005). These processes were also described at the ultrastructural level in open- and closed-thelycum shrimp (Rankin and Davis, 1990; Carvalho et al., 1998a, b, 1999). The most conspicuous features of the mature oocyte, attracting the attention of all the authors cited above, are the cortical rods or crypts. These are large, club-shaped cortical specializations packed with highly specialized proteins in the form of feathery, bottle-brush structures that are expelled out of the egg upon contact with sea water and form a gelatinous layer around the egg (Dall et al., 1990; Clark et al., 1990; Clark and Griffin, 1993). This has been termed the "egg reaction" and the layer appears as a transparent corona surrounding the egg (Pongtippatee-Taweepreda et al., 2004). This feature, however, may be found in only some Penaeoidea genera, as a study on the shrimp *Aristaeomorpha foliacea* did not find them in mature oocytes (Kao et al., 1999).

It has been suggested that the layer functions as a barrier against polyspermy or to protect the egg from mechanical damage and the rods are known to contain carbohydrates and abundant protein that were characterized biochemically and molecularly (Khayat et al., 2001; Yamano et al., 2004; Kim et al., 2004, 2005). The gelatinous layer takes about a minute to completely disperse and disappear so its function may also be to slow down the eggs that are demersal. It is not known how fertilization takes place, as penaeid sperm, like all other decapods, are non-motile but, unlike other decapods, penaeid females are broad spawners and the eggs are quickly dispersed in the water by the female vigorously beating her pleopods (Clark and Griffin, 1993).

In a comparison between wild and captive females, spermatophores were found attached to the thelycum of cortical stage females only, independent of female origin (Palacios et al., 2003). However, a significant proportion of the cortical stage females did not have an attached spermatophore, which, in the authors' view, indicates poor mating success or spermatophore adherence (Palacios et al., 2003).

In a comparison of wild versus pond-reared *Melicertus kerathurus*, a closed-thelycum species, the ovaries of *non-ablated* pond-reared females did not contain oocytes with cortical rods, while *non-ablated* wild-caught females of similar size did (Medina et al., 1996). This was in spite of the fact that the pond-reared females started vitellogenesis even earlier in the season than the wild-caught ones. However, their ovaries never attained full maturity and subsequently regressed, while wild females were found to progress to full maturation. The authors suggested that light intensity may have caused this difference, as the photoperiod and salinity were the same for both populations (Medina et al., 1996).

As expected, in a comparison of wild versus pond-reared *ablated* females of the open-thelycum species *L. vannamei*, no differences were found in the ovary maturation process (Palacios et al., 1999b). A similar result was obtained in ablated females of the closed-thelycum species *Penaeus monodon* (Peixoto et al., 2005).

The oocytes of fully mature, wild, non-ablated *Fenneropenaeus indicus* and *Penaeus semisulcatus* contained cortical rods (Quinitio and Millamena, 1992; Browdy et al., 1990).

The importance of the gelatinous layer to captive breeding is greater than one may initially assume. The grave problem faced by an imminent spawner is that the eggs must separate from each other upon spawning in order for the egg reaction to proceed successfully. If a female is disturbed while spawning or it spawns under crammed conditions, it will descend to the bottom and continue to spawn but these eggs adhere to each other, forming clumps, and do not hatch (Hudinaga, 1942; Clark and Griffin, 1993; and pers. obs.). To our knowledge, the adherence phenomenon was never treated in detail but it appears to be the result of the egg reaction and, as commercial breeders are aware of it, efforts are made to avoid it. Nevertheless, a reasonable cause for the inhibition of female maturation might be the shallow depth of traditional maturation systems. A short time before spawning, the female rises from the bottom and swims upwards, and this is most probably a behavioral adaptation that enables, among other advantages, successful egg reaction at sea. Thus, the female might have the ability to sense depth and this character might be the one that is bypassed by the endocrine induction commonly used in captive breeding, namely, unilateral eyestalk ablation. This hypothesis will be further explored after the following discussion on the hormonal regulation of female shrimp reproduction.

Endocrine Control

Endocrine control of reproduction in male and female crustaceans is discussed in Chapters 6 and 7 of this book so it will only briefly be dealt with here with reference to penaeid shrimp. Huberman (2000) and Okumura (2004) are the most recent publications that specifically reviewed some of the work on the endocrine control and manipulation of shrimp reproduction. The former author pointed out that the general study of shrimp endocrinology is lagging behind the effort invested in the study of non-penaeid endocrine glands and hormones (Huberman, 2000). Subramoniam (1999) treated the subject at the level of "economically important crustaceans", but it is clear from his review that a major portion of the cited literature was done on shrimp and, indeed, they are used by this author as a model for decapod crustaceans in general. It must be emphasized, however, that penaeid shrimp are in fact a very small and peculiar group within the Decapoda and have unique reproductive physiology and behavior, so generalizations to other decapods and vice versa could be problematic.

In practice, almost all the research on endocrine control and manipulation in shrimp has been aimed at the females. This is probably the reason that the gonad-inhibiting hormone (GIH) is still sometimes referred to as vitellogenesis-inhibiting hormone, even though it is most probably the same substance in males and females (Huberman, 2000). The endocrine manipulation routinely used for the induction of vitellogenesis in female shrimp is the removal of one of the two X-organ-sinus-gland-complexes (XO-SG) by eyestalk ablation. The XO-SG complex is a neurosecretory system located in the eyestalk and the processes of carbohydrate metabolism, molting and gonad maturation were shown to be regulated, at least in part, by factors secreted from it (Bocking et al., 2002). The complex synthesizes and secretes a group of neuropeptides that was named the CHH/MIH/GIH family after the observed phenomena occurring in the above processes (the initials stand for crustacean hyperglycemic hormone, molt-inhibiting hormone and gonad-inhibiting hormone, respectively). The ablation is assumed to cause an immediate decrease of 50% in the amount of circulating XO-SG secreted hormones. This assumption was never investigated experimentally and the high degree of similarity between the CHH family members makes the specific detection of GIH or MIH very difficult (Bocking et al., 2002). Thus, even if there is an immediate decrease, one cannot investigate the compensatory responses evoked in the animal. The results of the induction vary tremendously between individuals, from precocious death to accelerated rate of vitellogenesis during 3-4 months. Nevertheless, when performed on large numbers of females, it gives predictable and repeatable results that clearly explain the common use of it, as is discussed in the following section. A reliable CHH immunoassay might have explained the vast differences in the response of individual females to eyestalk ablation. It has been shown that the removal of at least one of the two XO-SG complexes will bring about a precocious molt or gonadal maturation, depending on the molt cycle stage the animal is at. There is still much speculation and divergence of opinion about the process of gonad maturation in Crustacea and the simple model of Adiyodi (1985) is still widely accepted (Dall et al., 1990). However, more than twenty years and many research articles later, it now seems to be much more complex than previously thought. This scheme proposes that the actions of MIH and GIH are antagonistic and generally a sexually mature crustacean female (as in penaeid shrimps) can either mature its gonad or go through a somatic growth phase that involves molts. It was postulated by Adiyodi (1985) that, in the Decapoda, these two fundamental processes put a heavy energy burden on the female and in this respect they share a common feature, carbohydrate metabolism, which is controlled by CHH. Indeed, on the basis of experimental evidence, Bocking et al. (2002) concluded that it is not unlikely that, at least in penaeids, one of the members of the CHH family is the long-postulated GIH. Nevertheless, experimental work on different species unequivocally shows that the processes of molting and vitellogenesis in penaeids are not antagonistic in their energetic requirements but merely in their temporal occurrence, i.e., a penaeid female will always absorb its ovary before molt because of hormonal synchronization between the MIH and the GIH (Adiyodi, 1985; Raviv et al., 2006). Moreover, experience accumulated in maturation systems shows that the best female brooders perform several vitellogenic cycles within a single molt cycle and continue to molt and simultaneously gain weight for several months in a row, provided they are adequately fed.

Causes of Female Reproductive Arrest

The traditional hypotheses on the reasons for the penaeid female reproductive arrest in captivity and the treatments that were used by various authors in tackling this problem are clearly reflected in the literature. As will be shown, all of these hypotheses suggest that the majority of researchers believe that either an environmental input to the female sensory system or a dietary one or both are lacking. In contrast to non-penaeid decapods, which carry their embryos and thus have control over the place and time of their release, penaeids spawn their eggs into the water column, leaving their larvae to manage on their own. Thus, it is logical to assume that the female would not invest in reproduction if the conditions were not favorable for its offspring (e.g., there is a lack of planktonic food or other favorable ecological conditions). The input to the female sensory system, however, may be dietary, environmental or social, and many parameters that were suspected to affect female reproduction have been investigated over the years. Intensive research has created the scientific basis for the "standard protocol" currently in practice. Manipulations of many parameters were examined, including endocrine manipulations through eyestalk ablation and injections of various materials; dietary components such as natural food items, carbohydrates, proteins, fatty acids, carotenoids, vitamins and minerals; environmental conditions such as light intensity, photoperiod, water salinity and quality, temperature, maturation tank size and color; and social parameters such as animal density and sex ratio.

Endocrine Manipulations

Unilateral eyestalk ablation is the most commonly used manipulation for inducing ovary maturation in shrimp females since its description by Panouse in 1943 (cited by Adiyodi and Adiyodi, 1970; Subramoniam, 1999; Huberman, 2000). It is performed on many species (open and closed thelycum) in aquaculture operations worldwide and produces repeatable and predictable results (Bueno, 1990; Bray and Lawrence, 1990, 1998; Gendrop-Funes and Valenzuela-Espinoza, 1995; Hansford and Marsden, 1995; Aktas and Kumlu, 1999; Palacios et al., 1999a, c, 2003; Arcos et al., 2003a, b, 2004, 2005; Coman and Crocos, 2003; Hoang et al., 2003; Palacios and Racotta, 2003; Peixoto et al., 2003, 2004). Vitellogenesis caused by induction not only progresses faster but also does so in a different manner than in naturally occurring vitellogenesis, as shown with *Marsupenaeus japonicus* (Tsutsui et al., 2005). In that study, vitellogenin mRNA transcripts in the naturally maturing shrimp were maintained at low levels during the previtellogenic stage; thereafter, levels increased as vitellogenesis progressed but decreased during the latter stages of maturation in the hepatopancreas and ovary. In contrast, in eyestalk-ablated shrimp, changes in mRNA levels differed in both tissues; an obvious increment of mRNA levels was revealed in the ovary, whereas mRNA levels were negligible in the hepatopancreas. On that basis, the authors suggested that eyestalk ablation cannot be used to accurately simulate the natural process of vitellogenin gene expression during vitellogenesis (Tsutsui et al., 2005). Nevertheless, the end result in breeding is that viable larvae are produced and the effects of this procedure are common knowledge among shrimp researchers and breeders. Some of the induced females promptly die and about 30% never mature their ovaries (Table 1). From those females who do perform vitellogenesis and spawn, less than 20% contribute the majority (60%) of nauplii (Table 1).

Table 1 Summary of the literature regarding the biased distribution of captive penaeid female reproductive output.

Percentage of reproductively non-active females	Percentage of females that together yielded ≥ 50% of total production	Percentage of total reproductive output of broodstock	Species	Source
–	9.5	50	*L. vannamei*	(Wyban and Sweeney, 1991)
*24	–		*L. vannamei*	(Palacios and Racotta, 2003)
*48	19	57	*L. vannamei*	(Arcos et al., 2003b)
11.1	24 ablated	70	*L. vannamei*	(Palacios et al., 1999a)
57.9	18 unablated	64		
13	25 wild		*L. vannamei*	(Palacios et al., 1999a)
27	15 pond-reared			
*25	*23.5	65	*L. stylirostris*	(Bray and Lawrence, 1990)
30	*20	67.7	*P. monodon*	(Bray and Lawrence, 1998)
*44			*L. vannamei*	(Arcos et al., 2004)
10-40	–		*F. merguiensis*	(Hoang et al., 2003)
36.6	19.5	54.4	*L. vannamei*	(Parnes et al., 2007)
30.6 ± 13.8	**19.3 ± 4.9**	**61.9 ± 7.2**	Average for ablated females alone	

*Percentages marked with an asterisk were supplied by the sources cited. All other data were calculated from that presented in the sources cited.

It is interesting to note that such skewed distribution was reported in many unrelated fields such as economics, where it is known as "the 80/20 principle". This principle describes situations in which 80% of a firm's profits emanate from 20% of its customers, salespeople or products (Dubinsky and Hansen, 1982).

In their effort to find alternatives to eyestalk ablation, researchers have examined the effect of various chemical substances on penaeid ovaries by *in vivo* injections. The most recent attempts tried vertebrate steroids (Yano and Hoshino, 2006), serotonin (Vaca and Alfaro, 2000; Alfaro et al., 2004), and Opioid peptides (Reddy, 2000). Like previous studies, all these induction studies reported very limited success, if any, and in spite of the new knowledge about the CHH family, all the publications conclude, even if not expressly, that there is currently no alternative to eyestalk ablation for the induction of ovary maturation in captive female shrimp. A novel inductive avenue that suggests attractive possibilities is the use of RNAi techniques (Chan and Mak, 2006).

Diet

The most recent publication specifically reviewing the subject of broodstock diet is that of Wouters et al. (2001). However, it seems that the need and demand for a pelleted maturation diet to replace the expensive and cumbersome fresh-frozen natural diets still drive intensive research in this area (Wouters et al., 2002; Diaz and Fenucci, 2004). It is common knowledge that adult females may initiate vitellogenesis even without eyestalk ablation, especially if they receive a fresh-frozen maturation diet, but they will not proceed to full ovarian maturation. This was observed in our maturation system and was also reported by others (Medina et al., 1996). The reason that the currently used diet containing fresh-frozen animal items cannot be replaced might well be shrimp feeding behavior and not a dietary constraint. These organisms are well equipped with sensory organs and they are active hunters that can prey even on fast swimmers such as fish and squid (Dall et al., 1990). A review of the literature on the natural diet of about 40 shrimp species revealed many food items, but the ones that appear repeatedly are crustaceans, mollusks, polychaetes and foraminiferans (Dall et al., 1990). In captivity shrimp are known to prefer fresh-frozen and non-processed animal items such as polychaetes and chopped fish, squid, oysters, mussels and other shrimp (Ogle, 1991; Dall et al., 1990). Considering that eyestalk-ablated females mature and spawn healthy larva, it can be concluded that the currently used fresh-frozen maturation diet is sufficient.

Environment

As mentioned above, there is currently a widely used "standard" protocol for shrimp breeding based on many years of trial and error experimentation with various environmental factors (Wyban and Sweeney, 1991; Browdy, 1992, 1998; Robertson et al., 1993; Treece and Fox, 1993; Wickins and Lee, 2002). Consequently, very little research on environmental manipulations is still carried out without eyestalk ablation to induce maturation in females. The most recent of these studies examined the effects of changing day length, temperature and light intensity on *Fenneropenaeus merguiensis* and, although some success was achieved, the females were finally ablated to induce mass spawning (Hoang et al., 2003). All current systems attempt to mimic the tropical or subtropical conditions of photoperiod, light intensity, water temperature and salinity. In addition, they all try to maintain proper oxygen levels while maintaining low concentrations of nitrogenous compounds. These complex tasks are mostly overcome by using ocean water on a unidirectional flow-through basis. The reported rates for the percentage of ocean water exchange vary, but they may be as high as 100% to 500% per day (Bray and Lawrence, 1992). Currently all shrimp breeding is carried out along coastlines and such water replacement rates put a heavy burden on the environment, mainly because of the sludge in the effluent (Moss, 2002).

Two water parameters were never reported to have been dealt with: pressure and depth. The tanks of all penaeid maturation systems allow a water depth of usually 0.5-0.7 m, but sometimes less than that. Such depths are maintained to allow the easy retrieval of the females on a daily basis. As mentioned above, penaeid females rise from the bottom when spawning is imminent and swim in circles on the water surface, vigorously waving their pleopods so the eggs are scattered around in the water. At sea, spawning grounds are found off-shore in deeper waters and a female probably swims upwards more than 0.5 m. This behavior has its logic as the freshly laid eggs of species of *Penaeus* are extremely fragile and tend to stick to one another (Clark and Griffin, 1993). Eggs that do not stay free in the water column but instead pile up on the bottom will not hatch and will most probably be preyed upon (Hudinaga, 1942; also pers. obs.). Lindley (1997) has also pointed out that spawning penaeid shrimp will be greatly disadvantaged in shallow water. However, the hatching envelope can be seen after a few minutes (Pongtippatee-Taweepreda et al., 2004) and it is fully formed after about 30 min, at which time the egg is able to withstand aeration in the hatching tank. The fully formed eggs are non-adhesive (Pearson, 1939) and because of a very small perivitelline space they are demersal and so are rarely found in plankton sampling of the water column out at sea (Pearson, 1939; Dall et al., 1990). No studies have been published on sinking rates of penaeid eggs (Dall et al., 1990; Lindley, 1997), but from our measurements on several *L. vannamei* spawns, the sinking rate of eggs is about 0.04-0.1 cm $\times$ s^{-1}. These numbers correspond to other broad-spawning crustaceans that have demersal eggs whose rates were at 0.05-0.17 cm $\times$ s^{-1} (Lindley, 1997). All this suggests that in order for the eggs to hatch in clean water, not too far from the phytoplankton-rich zone, the females should be several meters above the bottom while spawning.

Data on natural spawning grounds of penaeid shrimp support the above argument and show that females with mature ovaries are usually not found in water shallower than 10-20 m (Pearson, 1939; Lindner and Anderson, 1956; Ohtomi et al., 2003). If the female is sensitive to water depth in the mating tank, it will not reproduce under the atmospheric pressure conditions that prevail in the currently used systems. As penaeids are known to undergo active migrations to spawning grounds (Dall et al., 1990), it can be assumed that these selected areas and the egg characteristics have coevolved to ensure that eggs are adapted to the prevailing conditions (Lindley, 1997). This discussion is not as theoretical as it may seem: Koyama et al. (2005) reported on successful larval hatching of deep-sea shrimp after decompression to atmospheric pressure using a decompression chamber. Thus, technically the process can be reversed and a desired depth can be simulated for animals in a compression chamber.

The abundance of larval food supply may affect the "willingness" of the females to reproduce, but current penaeid maturation systems do not support plankton inside the mating tanks. The females might sense in some way that the mating system is a poor environment in which to spawn their eggs but it is not clear how they are able to assess this. A pressure-sensing mechanism may be a solution and such mechanisms are already known in marine organisms; the first to be described was discovered in a crab (Fraser and McDonald, 1994).

Environmental Input to the XO-SG and Cuticular Sensors

Like any other endocrine system that regulates physiological processes, the XO-SG complex in penaeids must receive environmental input through some kind of sensory structure subsequently translated to a hormonal output. This seems to be a central endocrinological paradigm and several questions arise from it: what are the structures that receive the environmental signals? What exactly do they measure? How are these signals transduced to the XO-SG? Can they be manipulated?

From the events that follow XO-SG complex removal, it is reasonable to assume that the most relevant sensory structures are located in the eyestalk, with the most obvious one being the eye itself. However, the removal of one of the XO-SG complexes probably reduces the hormonal titer in the animal hemolymph as well as the environmental input to the animal. This is because the removal of the XO-SG also involves the removal of the eye. Thus, it is not possible to separate the effects of these two treatments. The retina is the most obvious sensory organ on the eyestalk but there exists another sensory organ on the eyestalk that until now was not treated by crustacean physiologists, namely the eyestalk sensory pores.

The Eyestalk Sensory Pores

The eyestalk sensory pore is a special structure referred to by various authors as the X-organ sensory pore, sensory papilla, or sensory tubercle complex (Carlisle and Passano, 1953; Carlisle and Knowles, 1959; Young, 1959; Adiyodi and Adiyodi, 1970; Bocking et al., 2002). These structures are located on the inner side of the median region of the eyestalk on what is known as the optic calathus (Young, 1959). They bear a chemoreceptive morphology but their physiological function is not known (Cooke and Sullivan, 1982; Dall et al., 1990; Chaigneau and Besse, 1994). As no proof of a sensory function is available, these structures are here termed the micropores areas, since it has been shown that the cuticle of these areas is perforated with dozens of micropores arranged in pairs (Chaigneau, 1973; Chaigneau and Laubier-Bonichon, 1980; Chaigneau and Besse, 1994) and that there are cells beneath the cuticle, bearing cilia arranged in doublets (Chaigneau, 1973). Figure 1 shows photographs taken in our laboratory of the micropores areas in *L. vannamei*. Some of the photographs are of intact organs and others are of freshly molted exuvia, clearly showing the pairs of 1μm pores scattered on the drop-shaped micropores area.

It can be argued that this complex structure is, at least in part, responsible for the environmental input received by the XO-SG complex and it could be easily manipulated without the need for eyestalk removal. Moreover, it seems that both structures, one on each eye, could be manipulated simultaneously while XO-SG removal in both eyes will, eventually, be lethal. What exactly is measured by these organs is open to speculation and future research.

MALE REPRODUCTION

Spermatophore Maturity and Mating Success

In contrast to females, in which all research attention has focused on the gonads, the most interesting part of the male reproductive system is the

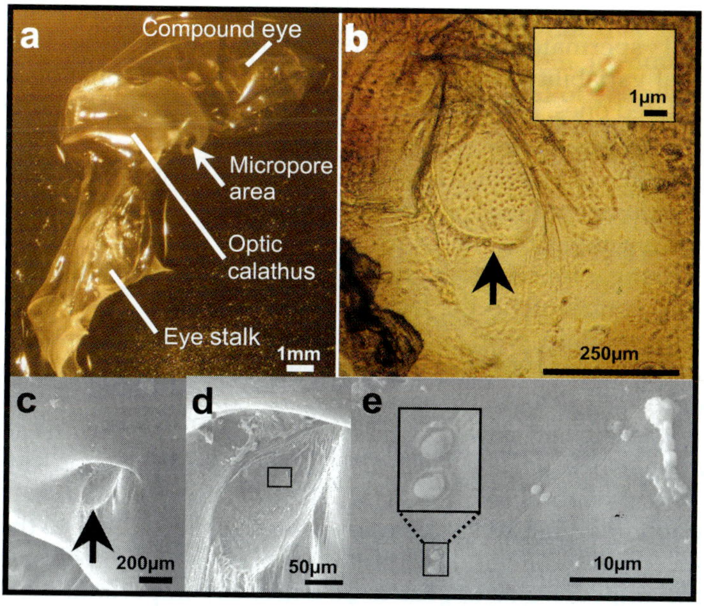

Fig. 1 Location and external morphology of the micropores areas in *Litopenaeus vannamei*. (a) The exuvia of the left eye. The micropores area is indicated by the white arrow. (b) An enlargement of the micropores area indicated in **a**. It is rouded and has a thin curicle and within it there is a drop-shaped element that contains several dozens of small pores arranged in pairs (the element is indicated with a black arrow). A pair of pores is shown in the inset. (c) The micropores area is indicated with a block arrow. (d) An enlargement of the micropores area shown in **c**. The framed area is enlarged in e. (e) Three pairs of pores from which the left one is framed and appears enlarged in the inset. Note that the pores seem to be covered with a lid. **a**. was taken through a dissecting mircroscope, **b** through a light microscope and **c**, **d** and **e** are scanning electron micrographs of an intact eye).

sperm duct. Small drops full of spermatozoa leave the testes and start their maturation journey through the convoluted sperm duct, where they are coated and molded into mature sperm transfer machines, the spermatophores (Subramoniam, 1993). In addition to the well-known studies of King (1948) and Malek and Bawab (1974a, b), there are several, more recent morphological descriptions of the process and the tissues involved (Ro et al., 1990; Bauer, 1991; Bauer and Cash, 1991; Chow et al., 1991a, b; Bauer and Min, 1993). These studies unequivocally show the high degree of complexity and variability of the sperm duct and spermatophores even between very close species. However, very little is known on the biochemical nature of the wrapping layers and only one study has begun to explore the subject (Malek and Bawab, 1971 see below, section on Melanization of the spermatophores). This lack of knowledge

in gonad size and sperm counts; however, none of the studies report a significant effect on larva production (Browdy, 1992). Since then, to the best of our knowledge, only three studies reported on the use of eyestalk ablation on males and the conclusions were the same (Alfaro and Lozano, 1993; Browdy et al., 1996). However, one of these studies also reported an increase in androgenic gland (AG) activity after bilateral eyestalk ablation (Kulkarni et al., 1984). This later study also examined the effect on the AG of injections of extracts from eyestalk, brain and thoracic ganglia (Kulkarni et al., 1984). Two studies reported on the effects of steroid hormones on male reproductive parameters. The first tested the effects of testosterone on AG and testis (Nagabhushanam and Kulkarni, 1981) and the second reported a significant improvement in spermatophore parameters after a single injection of 17α-methyltestosterone, while no effect was observed with 17α-hydroxyprogesterone (Alfaro, 1996).

In non-penaeid decapods, eyestalk ablation is known to cause hypertrophy of the AG and current research is aimed at isolating the androgenic hormone (Manor et al., 2006). However, bearing in mind the vast differences between the reproductive strategies of penaeid and non-penaeid decapods, the effect on penaeid breeding of having a commercially available AG hormone is not clear, as it is, for example, in the caridean prawn *Macrobrachium rosenbergii* (Sagi and Aflalo, 2005). In penaeid shrimp, sexual dimorphism is less noted but adult males are smaller than females. This stands in sharp contrast to non-penaeid species, where males are larger and sexual dimorphism is remarkable. Consequently, almost all the research on the AG is conducted on non-penaeid species. Only two studies examined the differentiation and development of the AG in the early life stages of a penaeid (Nakamura, 1992; Nakamura et al., 1992). In the non-penaeid Pleocyemata species in which the AG was studied, AG is known to cause the males to be larger than the females and it therefore seems strange that in penaeids, even though the males possess an AG, they are significantly smaller than the females. As the penaeids are considered the most ancient group in the Decapoda (Perez-Farfante and Kensley, 1997), it would be logical to ask what role is played by the AG in adult penaeid males.

Melanization of the Spermatophores

As mentioned already, in contrast to females, the sexual maturation of penaeid males occurs naturally in sea water as well as in saline, brackish water and there is no need for special dietary or environmental stimuli (Ogle, 1992; Ceballos-Vazquez et al., 2003a; Parnes et al., 2004). However, penaeid male reproductive performance is currently regarded as one of the major drawbacks in captive reproduction of these organisms.

Discussion of this problem can be divided into the time before ejaculation and the time after ejaculation. In relation to the time before ejaculation, two issues are examined: melanization of the spermatophores and the relationship between spermatophore development and the molt cycle.

In captive penaeid shrimps, such as *L. vannamei*, the brown pigment, melanin, accumulates in the spermatophores, and males with heavily melanized spermatophores are sexually impotent, since they cannot ejaculate (Leung-Trujillo and Lawrence, 1987; Talbot et al., 1989; Wyban and Sweeney, 1991; Alfaro and Lozano, 1993; Alfaro et al., 1993; Perez-Velazquez et al., 2001). Melanin specks may appear in the ampoule of sub-adult males even before the appearance of a spermatophore (Parnes et al., 2004). The phenomenon is characterized by the accumulation of melanin in the ripe spermatophores and an increase in the number of damaged sperm cells. Males with melanized spermatophores are sexually inadequate and cannot be used for reproduction purposes (Wyban and Sweeney, 1991). The progressive appearance of these symptoms was found to be directly related to the rise in water temperatures (Pascual et al., 1998; Perez-Velazquez et al., 2001). The fact that the spermatophores of sexually active males remain white is probably considered common knowledge and has been referred to in only three studies (Leung-Trujillo and Lawrence, 1991; Browdy et al., 1996; Parnes et al., 2006). Manually or electrically ejaculating the spermatophores can remove melanin debris from the ampoules, and many studies use these methods to keep males in good condition in the absence of females. Indeed, most of these studies reported that, over extended periods of time, spermatophores did not show deterioration (Leung-Trujillo and Lawrence, 1991; Alfaro, 1993; Alfaro and Lozano, 1993; Heitzmann and Diter, 1993; Pratoomchat et al., 1993; Alfaro, 1996; Pascual et al., 1998; Diaz et al., 2001; Ceballos-Vazquez et al., 2004). If the spermatophores are not evacuated (manually, electrically or through mating), they may become melanized in some individuals even within a few days and subsequently hardened enough to prevent ejaculation (Leung-Trujillo and Lawrence, 1987; Talbot et al., 1989; Alfaro et al., 1993; Perez-Velazquez et al., 2001). All these studies thus suggest that penaeid male readiness for the next copulation might actually be increased through the previous copulation.

Several studies have attempted to examine the physiological and biochemical basis of melanization. There is a general pattern reported by researchers in this field who worked with wild males introduced into artificial systems with no females. All the males had white spermatophores when brought into the systems, and most or all of them developed melanization of their spermatophores after several weeks or a few months (Chamberlain et al., 1983; Leung-Trujillo and Lawrence, 1987; Talbot et al., 1989; Alfaro et al., 1993; Pascual et al., 1998). There are no reports of melanization of the sperm duct in animals in the wild (King 1948; Lindner and Anderson 1956; DeLancey et al., 2005), although melanization of the cuticle and gills in wild shrimp has been observed (DeLancey, pers. comm.). Moreover, there is only one report of a single wild-caught *Litopenaeus* sp. specimen with melanized spermatophores (Chamberlain et al., 1983).

On a physiological level, the accumulated pigment in *L. vannamei* melanized spermatophores was shown to be melanin (Dougherty and Dougherty, 1989). The origin of spermatophores in *M. kerathurus* was traced with special reference to the chemical nature of constituent layers and their mode of hardening (Malek and Bawab, 1971, 1974a, b). It was found that the layers are of lipoprotein nature, the protein of which is rich in phenolic groups, owing to the presence of tyrosine. No free phenol could be detected but the enzyme phenoloxidase was demonstrated to be present, although the supporting data was never published (Malek and Bawab, 1971). On the basis of these findings, the authors suggested that the existing phenoloxidase could promote active oxidation resulting in the formation of quinones of which further polymerization results in melanin. Although the hardened layers failed to darken, they showed pronounced changes in affinity for stains and in isoelectric point. It was concluded that the spermatophore of *M. kerathurus* owes its hardness not to mere exposure to sea water but to a definite enzymatically catalyzed chemical transformation. The hardening of the spermatophores, or "phenolic tanning", was reported from several other decapod species (Subramoniam, 1993). According to our observations, melanization always started at the distal end of the spermatophores, next to the genital papillae that house the gonopores, environmentally the most exposed part. This suggests that there is a strong environmental basis for spermatophore melanization and that it is most probably an artifact of captive conditions that expose the males to long periods of time without receptive females.

In crustaceans, as in other arthropods, melanin synthesis is involved in the process of sclerotization and wound healing of the cuticle as well as in defense reactions against invading microorganisms entering the hemocoel. The enzyme involved in melanin formation is phenoloxidase (Sritunyalucksana and Soderhall, 2000). The enzymes of the phenoloxidase system (currently called prophenoloxidase or proPO) were found to be localized only in the hemocytes, the only tissue found to express the proPO mRNA.

Dougherty and Dougherty (1989) carried out a morphological and histological examination of the spermatophore melanization process in *L. vannamei*. They reported that no bacteria, fungi or hemocytes were detected in any sections of moderately pigmented black spermatophores. Nevertheless, they suggested that the possible role of phenoloxidase in these processes should be explored and, as no other data has been published since then, their suggestion is still valid.

Spermatophore Dynamics along the Molt Cycle

In many decapod crustacean species, such as shrimps, lobsters and crabs, female reproduction is synchronized with the molt cycle and is thus cyclic by definition (Adiyodi, 1985; Nelson, 1991). The nature of the association between molt cycles and reproduction in decapod males was until recently not known, although a few clues did exist, all from penaeid shrimp: manually ejaculated *L. vannamei* exhibited new pairs of spermatophores in their ejaculatory ducts (ampoules) only after molting (Heitzmann and Diter, 1993); a decline in spermatophore quality as the molt cycle progressed was observed in *F. indicus* (Muthuraman, 1997); and naturally mating *L. vannamei* and *L. setiferus* were shown to be carrying new pairs of spermatophores a few days after mating within the same molt cycle in which the mating had occurred (Leung-Trujillo and Lawrence, 1991). Some clues to the timed regulation of spermatophore behavior during storage within the male have also appeared in the literature. In *M. kerathurus*, a complete spermatophore is accommodated in the terminal ampoule, while another simultaneously forms in the medial vas deferens (Malek and Bawab, 1974a, b). Judging by the small variability in the dimensions of the medial vas deferens, Ro et al. (1990) concluded that the migration of a new spermatophore along the sperm duct does not occur at random but is precisely regulated.

Discovered recently in our laboratory is a molt-dependent mechanism by which old sperm is periodically removed from the reproductive system of male *L. vannamei* shrimp (Parnes et al., 2006). This programmed spermatophore degradation mechanism was elucidated by using the melanin specks mentioned above as color markers for male sexual activity. The mechanism is reported in Chapter 10 of this book, which deals with the more general issue of coordination of molting and reproduction in decapod crustaceans.

The Role of the Petasmata

When compared to pleocyemata species, penaeid shrimp generally seem to have a lower degree of sexual dimorphism. The most obvious trait is that adult males are significantly smaller than females (Dall et al., 1990). Morphologically, males have three structures not found in females: the genital openings located on the coxae of the fifth pair of walking legs, the petasmata and the masculine appendices (Perez-Farfante and Kensley, 1997). There is no direct evidence concerning the role of the latter two structures but, nevertheless, two studies have provided indirect but conclusive evidence of the essential role played by the petasma during mating. One of these studies was done on a sicyonid shrimp (Bauer, 1996) and the other, which is rarely cited, was done on several *Penaeus* species (Lin et al., 1991). In both studies the petasmata were ablated and the authors observed that no mating occurred in the maturation tank until the petasmata were fully regenerated.

SUMMARY

It is obvious that much is still to be learned on the reproduction of penaeid shrimp. It is our belief that male reproductive biology is relatively unexplored compared to that of females. The variation and complexity of spermatophore structure indicates strong sexual selection operating on these sperm transfer mechanisms. So far as females are concerned, the most enigmatic phenomenon is their "abstinence" from reproduction in captivity. There is still scope in both male and female shrimp reproductive biology for further basic biological observations to be made in captive and wild conditions.

References

Adiyodi, K.G., and R.G. Adiyodi. 1970. Endocrine control of reproduction in decapod crustacea. Biol. Rev., 45: 121-165.

Adiyodi, R.G. 1985. Reproduction and its control. *In:* The Biology of Crustacea, Vol. 9, Integument, Pigments, and Hormonal Processes. D.E. Bliss (ed.). Academic Press, New York, pp. 147-215.

Aktas, M., and M. Kumlu. 1999. Gonadal maturation and spawning of *Penaeus semisulcatus* (Penaeidae: Decapoda). Turkish J. Zool., 23: 61-66.

Alfaro, J. 1993. Reproductive quality evaluation of male Penaeus stylirostris from a grow-out pond. J. World Aquacult. Soc., 24: 6-11.

Alfaro, J. 1996. Effects of 17α-methyltestosterone and 17α-hydroxyprogesterone on the quality of white shrimp *Penaeus vannamei* spermatophores. J. World Aquacult. Soc., 27: 487-493.

Alfaro, J., and X. Lozano. 1993. Development and deterioration of spermatophores in pond-reared *Penaeus vannamei*. J. World Aquacult. Soc., 24: 522-529.

Alfaro, J., A.L. Lawrence and D. Lewis. 1993. Interaction of bacteria and male reproductive-system blackening disease of captive *Penaeus-setiferus*. Aquaculture, 117: 1-8.

Alfaro, J., G. Zuniga and J. Komen. 2004. Induction of ovarian maturation and spawning by combined treatment of serotonin and a dopamine antagonist, spiperone in *Litopenaeus stylirostris* and *Litopenaeus vannamei*. Aquaculture, 236: 511-522.

Arcos, G.F., A.M. Ibarra, C. Vazquez-Boucard, E. Palacios and I.S. Racotta. 2003a. Haemolymph metabolic variables in relation to eyestalk ablation and gonad development of Pacific white shrimp *Litopenaeus vannamei* Boone. Aquacult. Res., 34: 749-755.

Arcos, F.G., A.M. Ibarra, E. Palacios, C. Vazquez-Boucard and I.S. Racotta. 2003b. Feasible predictive criteria for reproductive performance of white shrimp *Litopenaeus vannamei*: egg quality and female physiological condition. Aquaculture, 228: 335-349.

Arcos, F.G., I.S. Racotta and A.M. Ibarra. 2004. Genetic parameter estimates for reproductive traits and egg composition in Pacific white shrimp *Penaeus (Litopenaeus) vannamei*. Aquacult. Res., 236: 151-165.

Arcos, G.F., E. Palacios, A.M. Ibarra and I.S. Racotta. 2005. Larval quality in relation to consecutive spawnings in white shrimp *Litopenaeus vannamei* Boone. Aquacult. Res., 36: 890-897.

Ayub, Z., and M. Ahmed. 2002. A description of the ovarian development stages of penaeid shrimps from the coast of Pakistan. Aquacult. Res., 33: 767-776.

Bauer, R.T. 1991. Sperm transfer and storage structures in penaeoid shrimps: a functional and phylogenetic perspective. *In:* Crustacean Sexual Biology. R.T. Bauer and J.W. Martin (eds.). Columbia University Press, New York, pp. 183-207.

Bauer, R.T. 1996. Role of the petasma and appendices masculinae during copulation and insemination in the penaeoid shrimp, *Sicyonia dorsalis* (Crustacea: Decapoda: Dendrobranchiata). Invert. Reprod. Dev., 29: 173-184.

Bauer, R.T., and C.E. Cash. 1991. Spermatophore structure and anatomy of the ejaculatory duct in *Penaeus setiferus, P. duorarum*, and *P. aztecus* (Crustacea: Decapoda): Homologies and functional significance. Trans. Amer. Microsc. Soc., 110: 144-162.

Bauer, R.T., and L.J. Min. 1993. Spermatophores and plug substance of the marine shrimp *Trachypenaeus similis* (Crustacea: Decapoda: Penaeidae): Formation in the male reproductive tract and disposition in the inseminated female. Biol. Bull., 185:174-185.

Bocking, D., H. Dircksen and R. Keller. 2002. The crustacean neuropeptides of the CHH/MIH/GIH family: structures and biological activities. *In:* The Crustacean Nervous System. K Wiese (ed.). Springer-Verlag, Heidelberg, pp. 84-97.

Bray, W.A., and A.L. Lawrence. 1990. Reproduction of eye-stalk ablated *Penaeus stylirostris* fed various levels of total dietary lipid. J. World Aquacult. Soc., 21: 41-52.

Bray, W.A., and A.L. Lawrence. 1992. Reproduction of *Penaeus* species in captivity. *In:* Marine Shrimp Culture: Principles and Practices. A.W. Fast and L.J. Lester (eds.). Elsevier, Amsterdam, pp. 93-170.

Bray, W.A., and A.L. Lawrence. 1998. Successful reproduction of *Penaeus monodon* following hypersaline culture. Aquaculture, 159: 275-282.

Browdy, C.L. 1992. A review of the reproductive biology of *Penaeus* species: perspectives on controlled shrimp maturation systems for high quality nauplii production. *In:* Proceedings of the Special Session on Shrimp Farming, 22-25 May 1992, Orlando, Florida. J. Wyban (ed.). World Aquaculture Society, Baton Rouge, Louisiana, USA, pp. 22-51.

Browdy, C.L. 1998. Recent developments in penaeid broodstock and seed production technologies: improving the outlook for superior captive stocks. Aquaculture, 164: 3-21.

Browdy, C.L., M. Fainzilber, M. Tom, Y. Loya and E. Lubzens. 1990. Vitellin synthesis in relation to oogenesis in in-vitro-incubated ovaries of *Penaeus semisulcatus* (Crustacea, Decapoda, Penaeidae). J. Exp. Zool., 255: 205-215.

Browdy, C.L., K. McGovern-Hopkins, A.D. Stokes, J.S. Hopkins and P.A. Sandifer. 1996. Factors affecting the reproductive performance of the Atlantic white shrimp, *Penaeus setiferus*, in conventional and unisex tank systems. J. Appl. Aquacult., 6: 11-25.

Bueno, S.L. de S. 1990. Maturation and spawning of the white shrimp *Penaeus schmitti* Burkenroad, 1936, under large scale rearing conditions. J. World Aquacult. Soc., 21: 170-179.

Carlisle, D.B., and F.B. Knowles. 1959. Endocrine Control in Crustaceans, Cambridge Monographs in Experimental Biology, Vol 10, M. Abercrombie, P.B. Medawar, G. Salt, M.M. Swann and V.B. Wigglesworth (eds.). Cambridge University Press, Cambridge.

Carlisle, D.B., and L.M. Passano. 1953. The X-organ of crustacea. Nature, 171: 1070-1071.

Carvalho, F., M. Sousa, E. Oliviera, J. Carvalheiro and L. Baldaia. 1998a. Ultrastructure of oogenesis in *Penaeus kerathurus* (Crustacea, Decapoda). I. Previtellogenic oocytes. J. Submicrosc. Cytol. Pathol., 30: 409-416.

Carvalho, F., M. Sousa, E. Oliviera, J. Carvalheiro and L. Baldaia. 1998b. Ultrastructure of oogenesis in *Penaeus kerathurus* (Crustacea, Decapoda). II. Vitellogenesis. J. Submicrosc. Cytol. Pathol., 30: 527-535.

Carvalho, F., M. Sousa, E. Oliviera, J. Carvalheiro and L. Baldaia. 1999. Ultrastructure of oogenesis in *Penaeus kerathurus* (Crustacea, Decapoda). III. Cortical vesicle formation. J. Submicrosc. Cytol. Pathol., 31: 57-63.

Ceballos-Vazquez, B.P., C. Rosas and I.S. Racotta. 2003a. Sperm quality in relation to age and weight of white shrimp *Litopenaeus vannamei*. Aquaculture, 228: 141-151.

Ceballos-Vazquez, B.P., I.S. Racotta and J.F. Elorduy-Garay. 2003b. Qualitative and quantitative analysis of the ovarian maturation process of *Penaeus vannamei* after a production cycle. Invert. Reprod. Dev., 43: 9-18.

Ceballos-Vazquez, B.P., B. Aparicio-Simon, E. Palacios and I.S. Racotta. 2004. Sperm quality over consecutive spermatophore regenerations in the Pacific white shrimp *Litopenaeus vannamei*. J. World Aquacult. Soc., 35:178-188.

Chaigneau, J. 1973. Fine structure of the sensory pore present in the eyestalk of Crustacea Natantia. Zeitschrift fur Zellforschung und Mikroskopische Anatomie, 145: 213-227.

Chaigneau, J., and C. Besse. 1994. The sensory tubercle on the eyestalk of the shrimp *Crangon crangon* (Linnaeus, 1758) (Decapoda, Caridea). Crustaceana, 66: 78-89.

Chaigneau, J., and A. Laubier-Bonichon. 1980. Particularites structurales et ultrastructurales du pore sensoriel de la crevette *Penaeus japonicus*. Arch. Zool. Exp. Gen., 121: 183-190.

Chamberlain, G.W., S.K. Johnson and D.H. Lewis. 1983. Swelling and melanization of the male reproductive system of captive adult penaeid shrimp. J. World Maricult. Soc., 14: 135-136.

Chan, S-M., and C.Y. Mak. 2006. Functional studies of an eyestalk molt inhibiting hormone (MIH-1) in black tiger shrimp *Penaeus monodon* by RNA interference. ICN conference abstract. Frontiers Neuroendocrinol., 27: 23–26.

Chow, S., M.M. Dougherty, W.J. Dougherty and P.A. Sandifer. 1991a. Spermatophore formation in the white shrimps *Penaeus setiferus* and *P. vannamei*. J. Crustacean Biol., 11: 201-216.

Chow, S., W.J. Dougherty and P.A. Sandifer. 1991b. Unusual testicular lobe system in the white shrimps *Penaeus setiferus* (Linnaeus, 1761) and *P. vannamei* Boone, 1931 (Decapoda, Penaeidae): A new character for Dendrobranchiata? Crustaceana, 60: 304-318.

Clark, W.H., Jr., and F.J. Griffin. 1993. Acquisition and manipulation of penaeoidean gametes. *In:* CRC Handbook of Mariculture, Vol. 1, Crustacean Aquaculture. J.P. McVey (ed.). CRC Press, Boca Raton, Florida, pp 133-151.

Clark, W.H., Jr., A.I. Yudin, J.W. Lynn, F.J. Griffin and M.C. Pillai. 1990. Jelly layer formation in Penaeoidean shrimp eggs. Biol. Bull., 178: 295-299.

Coman, G.J., and P.J. Crocos. 2003. Effect of age on the consecutive spawning of ablated *Penaeus semisulcatus* broodstock. Aquaculture, 219: 445-456.

Cooke, I.M., and R.E. Sullivan. 1982. Hormones and neurosecretion. *In:* The Biology of Crustacea. D.E. Bliss (ed.). Vol. 3, Neurobiology: Structure and Function. H.L. Atwood and D.C. Sandeman (eds.). Academic Press, New York and London, pp. 205-290.

Courtney, A.J., S.S. Montgomery, D.J. Die, N.L. Andrew, M.G. Cosgrove and C. Blount. 1995. Maturation in the female eastern king prawn *Penaeus plebejus* from coastal waters of eastern Australia, and considerations for quantifying egg production in penaeid prawns. Mar. Biol., 122: 547-556.

Dall, W., B.J. Hill, P.C. Rothlisberg and D.J. Sharples. 1990. The Biology of the Penaeidae. Adv. Mar. Biol., 27: 1-489.

DeLancey, L.B., J.E. Jenkins, M.B. Maddox, J.D. Whitaker and E.L. Wenner. 2005. Field observations on white shrimp, *Litopenaeus setiferus*, during spring spawning season in South Carolina, U.S.A., 1980-2003. J. Crustacean Biol., 25: 212-218.

Diaz, A.C., and J.L. Fenucci. 2004. Effect of prepared diet on the induction of precocious maturation in *Pleoticus muelleri* Bate (Crustacea, Penaeoide). Aquacult. Res., 35: 1166-1171.

Diaz, A.C., A.V. Fernandez Gimenez, N.S. Haran and J.L. Fenucci. 2001. Reproductive performance of male Argentine red shrimp *Pleoticus muelleri* Bate (Decapoda, Penaeoidea) in culture conditions. J. World Aquacult. Soc., 32: 236-242.

Dougherty, W.J., and M.M. Dougherty. 1989. Electron microscopical and histochemical observations on melanized sperm and spermatophores of pond-cultured shrimp, *Penaeus vannamei*. J. Invert. Pathol., 54: 331-343.

Dubinsky, A.J., and R.W. Hansen. 1982. Improving marketing productivity: the 80/20 principle revisited. Calif. Mgmt. Rev., 25: 96-105.

FAO. 2006. Fisheries Technical Paper. No. 500. State of World Aquaculture 2006. Food and Agriculture Organization of the United Nations, <http://www.fao.org>

Fraser, P.J., and A.G. McDonald. 1994. Crab hydrostatic pressure sensors. Nature, 371: 383-384.

Gendrop-Funes, V., and E. Valenzuela-Espinoza. 1995. Unilateral ablation of *Penaeus stylirostris* (Stimpson). Ciencias Marinas, 21: 401-413.

Hansford, S.W., and G.E. Marsden. 1995. Temporal variation in egg and larval productivity of eyestalk ablated spawners of the prawn *Penaeus monodon* from Cook Bay, Australia. J. World Aquacult. Soc., 26: 396-405.

Heitzmann, J.C., and A. Diter. 1993. Spermatophore formation in the white shrimp, *Penaeus vannamei* Boone 1931 — Dependence on the intermolt cycle. Aquaculture, 116: 91-98.

Hoang, T., S.Y. Lee, C.P. Keenan and G.E. Marsden. 2003. Improved reproductive readiness of pond-reared broodstock *Penaeus merguiensis* by environmental manipulation. Aquaculture, 221: 523-534.

Huberman, A. 2000. Shrimp endocrinology. A review. Aquaculture, 191: 191-208.

Hudinaga, M. 1942. Reproduction, development and rearing of *Penaeus japonicus* Bate. Japan. J. Zool., 10: 305-393.

Kao, H.-C., T.-Y. Cha and H.-P. Yu. 1999. Ovary development of the deep-water shrimp *Aristaeomorpha foliacea* (Risso, 1826) (Crustacea: Decapoda: Aristeidae) from Taiwan. Zool. Stud., 38: 373-378.

Khayat, M., P.J. Babin, B. Funkenstein, M. Sammar, H. Nagasawa, A. Tietz and E. Lubzens. 2001. Molecular characterization and high expression during oocyte development of a shrimp ovarian cortical rod protein homologous to insect intestinal peritrophins. Biol. Reprod., 64: 1090-1099.

Kim, Y.K., I. Kawazoe, N. Tsutsui, S. Jasmani, M.N. Wilder and K. Aida. 2004. Isolation and cDNA cloning of ovarian cortical rod protein in Kuruma prawn *Marsupenaeus japonicus* (Crustacea: Decapoda: Penaeidae). Zool. Sci., 21: 1109-1119.

Kim, Y.K., N. Tsutsui, I. Kawazoe, T. Okumura, T. Kaneko and K. Aida. 2005. Localization and developmental expression of mRNA for cortical rod protein in Kuruma prawn *Marsupenaeus japonicus*. Zool. Sci., 22: 675-680.

King, J.E. 1948. A study of the reproductive organs of the common marine shrimp, *Penaeus setiferus* (Linnaeus). Biol. Bull., 94: 244-262.

Koyama, S., T. Nagahama, N. Ootsu, T. Takayama, M. Horii, S. Konishi, T. Miwa, Y. Ishikawa and M. Aizawa. 2005. Survival of deep-sea shrimp (*Alvinocaris* sp.) during decompression and larval hatching at atmospheric pressure. Mar. Biotechnol., 7: 277-278.

Kulkarni, G.K., R. Nagabhushanam and P.K. Joshi. 1984. Neuroendocrine control of reproduction in the male penaeid prawn, *Parapenaeopsis hardwickii* (Miers) (Crustacea, Decapoda, Penaeidae). Hydrobiologia, 108: 281-289.

Leung-Trujillo, J.R., and A.L. Lawrence. 1987. Observations on the decline in sperm quality of *Penaeus setiferus* under laboratory conditions. Aquaculture, 65: 363-370.

Leung-Trujillo, J.R., and A.L. Lawrence. 1991. Spermatophore generation times in *Penaeus setiferus, P. vannamei, and P. stylirostris*. J. World Aquacult. Soc., 22: 244-251.

Lin, M.-N., Y.-Y. Ting, I. Hanyu, B.-S. Tzeng and C.-D. Li. 1991. Development and regeneration of petasma with reference to gonadal development in penaeid shrimp. J. Fish. Soc. Taiwan, 18: 145-154.

Lindley, J.A. 1997. Eggs and their incubation as factors in the ecology of planktonic crustaceans. J. Crustacean Biol., 17: 569-576.

Lindner, M.J., and W.W. Anderson. 1956. Growth, migrations, spawning and size distribution of shrimp *Penaeus setiferus*. Bull. Bur. Fish., 56: 555-645.

Malek, S.R.A., and F.M. Bawab. 1971. Tanning in the spermatophore of a crustacean (*Penaeus trisulcatus*). Experientia, 27: 1098.

Malek, S.R.A., and F.M. Bawab. 1974a. The formation of the spermatophore in *Penaeus kerathurus* (Forskal, 1775) (Decapoda, Penaeidae). I — the initial formation of a sperm mass. Crustaceana, 26: 273-285.

Malek, S.R.A., and F.M. Bawab. 1974b. The formation of the spermatophore in *Penaeus kerathurus* (Forskal, 1775) (Decapoda, Penaeidae). II — The deposition of the main layers of the body and of the wing. Crustaceana, 27: 73-83.

Manor, R., S. Weil, S. Oren, L. Glazer, E.D. Aflalo, T. Ventura, V. Chalifa-Caspi, M. Lapidot and A. Sagi. 2007. Insulin and gender: an insulin-like gene expressed exclusively in the androgenic gland of the male crayfish. Gen. Comp. Endocrinol., 150: 326-336.

Medina, A., Y. Vila, G. Mourente and A. Rodriguez. 1996. A comparative study of the ovarian development in wild and pond-reared shrimp, *Penaeus kerathurus* (Forskal, 1775). Aquaculture, 148: 63-75.

Misamore, M.J., and C.L. Browdy. 1996. Mating behavior in the white shrimps *Penaeus setiferus* and *P. vannamei*: A generalized model for mating in *Penaeus*. J. Crustacean Biol., 16: 61-70.

Moss, S.M. 2002. Marine shrimp farming in the Western hemisphere: past problems, present solutions, and future visions. Rev. Fish. Sci., 10: 601-620.

Muthuraman, A.L. 1997. Effect of ecdysial cycle on the semen and spermatophore production of the Indian white prawn (*Penaeus indicus* H.M. Edwards). Fish. Technol., 34: 11-14.

Nagabhushanam, R., and G.K. Kulkarni. 1981. Effect of exogenous testosterone on the androgenic gland and testis of a marine penaeid prawn, *Parapenaeopsis hardwickii* (Miers) (Crustacea, Decapoda, Penaeidae). Aquaculture, 23: 19-27.

Nakamura, K. 1992. Differentiation of genital organs and androgenic gland in the Kuruma prawn *Penaeus japonicus*. Mem. Fac. Fish. Kagoshima Univ., 41: 87-94.

Nakamura, K., N. Matsuzaki and K.-I. Yonekura. 1992. Organogenesis of genital organs and androgenic gland in the Kuruma prawn. Nippon Suisan Gakkaishi, 58: 2261-2267.

Nelson, K. (1991). Scheduling of reproduction in relation to molting and growth in malacostracan crustaceans. *In:* Crustacean Issues, Vol. 7, Crustacean Egg Production. A. Wenner and A. Kuris (eds.). Balkema Press, Rotterdam, pp. 77-113.

Ogle, J.T. 1991. Food preference of *Penaeus vannamei*. Gulf Res. Rep., 8: 291-294.

Ogle, J.T. 1992. A review of the current (1992) state of our knowledge concerning reproduction in open thelycum Penaeid shrimp with emphasis on *Penaeus vannamei*. Invert. Reprod. Dev., 22: 267-274.

Ohtomi, J., T. Tashiro, S. Atsuchi and N. Knhno. 2003. Comparison of spatiotemporal patterns in reproduction of the kuruma prawn *Marsupenaeus japonicus* between two regions having different geographic conditions in Kyushu, southern Japan. Fish. Sci., 69: 505-519.

Okumura, T., and K. Sakiyama. 2004. Hemolymph levels of vertebrate-type steroid hormones in female kuruma prawn *Marsupenaeus japonicus* (Crustacea: Decapoda: Penaeidae) during natural reproductive cycle and induced ovarian development by eyestalk ablation. Fish. Sci., 70: 372-380.

Okumura, T. 2004. Perspectives on hormonal manipulation of shrimp reproduction. Japan Agricult. Res. Q., 38: 49-54.

Okumura, T., H. Nikaido, K. Yoshida, M. Kotaniguchi, Y. Tsuno, Y. Seto and T. Watanabe. 2005. Changes in gonadal development, androgenic gland cell structure, and hemolymph vitellogenin levels during male phase and sex change in laboratory-maintained protandric shrimp, *Pandalus hypsinotus* (Crustacea: Caridea: Pandalidae). Mar. Biol., 148: 347-361.

Palacios, E., and I.S. Racotta. 2003. Effect of number of spawns on the resulting spawn quality of 1-year-old pond-reared *Penaeus vannamei* (Boone) broodstock. Aquacult. Res., 34: 427-435.

Palacios, E., D. Carreno, M.C. Rodriguez-Jaramillo and I.S. Racotta. 1999a. Effect of eyestalk ablation on maturation, larval performance, and biochemistry of white Pacific shrimp, *Penaeus vannamei*, broodstock. J. Appl. Aquacult., 9: 1-23.

Palacios, E., C. Rodriguez-Jaramillo and I.S. Racotta. 1999b. Comparison of ovary histology between different-sized wild and pond-reared shrimp *Litopenaeus vannamei* (= *Penaeus vannamei*). Invert. Reprod. Dev., 35: 251-259.

Palacios, E., I.S. Racotta and APSA. 1999c. Spawning frequency analysis of wild and pond-reared Pacific white shrimp *Penaeus vannamei* broodstock under large-scale hatchery conditions. J. World Aquacult. Soc., 30: 180-191.

Palacios, E., I.S. Racotta and M. Villalejo. 2003. Assessment of ovarian development and its relation to mating in wild and pond-reared *Litopenaeus vannamei* shrimp in a commercial hatchery. J. World Aquacult. Soc., 34: 466-477.

Parnes, S., E. Mills, C. Segall, R. Raviv, C. Davis and A. Sagi. 2004. Reproductive readiness of the shrimp *Litopenaeus vannamei* grown in a brackish water system. Aquaculture, 236: 593-606.

Parnes, S., S. Raviv, A. Shechter and A. Sagi. 2006. Males also have their time of the month! Cyclic disposal of old spermatophores, timed by the molt cycle, in a marine shrimp. J. Exp. Biol., 209: 4974-4983.

Parnes, S., S., Raviv, D. Azulay, and A.Sagi, 2007. Dynamics of reproduction in a captive shrimp broodstock: unequal contribution of the female shrimp and a hidden shortage in competent males. Invert. Reprod. Dev., 50: 21-29.

Pascual, C., E. Valera, C. Re-Regis, G. Gaxiola, A. Sanchez, L. Ramos, L.A. Soto and C., Rosas. 1998. Effect of water temperature on reproductive tract condition of *Penaeus setiferus* adult males. J. World Aquacult. Soc., 29: 477-484.

Pearson, J.C. 1939. The early life histories of some American Penaeidae, chiefly the commercial shrimp *Penaeus setiferus* (Linn.). Bull. Bur. Fish., 49: 1-73.

Peeters, L., and A. Diter. 1994. Effects of impregnation on maturation, spawning and ecdysis of female shrimp *Penaeus indicus*. J. Exp. Zool., 269: 522-530.

Peixoto, S., R.O. Cavalli and W. Wasielesky. 2003. The influence of water renewal rates on the reproductive and molting cycles of *Penaeus paulensis* in captivity. Brazil. Arch. Biol. Technol., 46: 281-286.

Peixoto, S., R.O. Cavalli, D. Krummenauer, W. Wasielesky and F. D'Incao. 2004. Influence of artificial insemination on the reproductive performance of *Farfantepenaeus paulensis* in conventional and unisex maturation systems. Aquaculture, 230: 197-204.

Peixoto, S., G. Coman, S. Arnold, P. Crocos and N. Preston. 2005. Histological examination of final oocyte maturation and atresia in wild and domesticated *Penaeus monodon* (Fabricius) broodstock. Aquacult. Res., 36: 666-673.

Perez-Farfante, I., and B. Kensley. 1997. Penaeoid and Sergestoid Shrimps and Prawns of the World. Memoires du Museum national d'Histoire naturelle, Paris, 233 pp.

Perez-Velazquez, M., W.A. Bray, A.L. Lawrence, D.M. Gatlin III and M.L. Gonzalez-Felix. 2001. Effect of temperature on sperm quality of captive *Litopenaeus vannamei* broodstock. Aquaculture, 198: 209-218.

Pongtippatee-Taweepreda, P., Chavadej, J., Plodpai, P., Pratoomchart, B., Sobhon, P., Weerachatyanukul, W. and Withyachumnarnkul, B. 2004. Egg activation in the black tiger shrimp *Penaeus monodon*. Aquaculture, 234: 183-198.

Pratoomchat, B., S. Piyatiratitivorakul, P. Menasveta and A.W. Fast. 1993. Sperm quality of pond-reared and wild-caught *Penaeus monodon* in Thailand. J. World Aquacult. Soc., 24: 530-540.

Quinitio, E.T., and O.M. Millamena. 1992. Ovarian changes and female-specific protein levels during sexual maturation of the white shrimp *Penaeus indicus*. Isr. J. Aquacult.-Bamidgeh, 44: 7-12.

Rankin, S.M., and R.W. Davis. 1990. Ultrastructure of oocytes of the shrimp, *Penaeus vannamei*: cortical specialization formation. Tiss. Cell, 2: 879-893.

Raviv, S., Parnes, S., Segall, C., Davis, C. and Sagi, A. 2006. The complete *Litopenaeus vannamei* (Crustacea: Decapoda) vitellogenin cDNA and its expression in endocrinologically induced sub-adult females. Gen. Comp. Endocrinol., 145: 39-50.

Reddy, P.S. 2000. Involvement of opioid peptides in the regulation of reproduction in the prawn *Penaeus indicus*. Naturwissenschaften, 87: 535-538.

Ro, S., P. Talbot, J. Leung-Trujillo and A.L. Lawrence. 1990. Structure and function of the vas-deferens in the shrimp *Penaeus-setiferus* – segments 1-3. J. Crustacean Biol., 10: 455-468.

Robertson, L., B. Bray, T. Samocha and A. Lawrence. 1993. Reproduction of penaeid shrimp: an operation guide. *In:* CRC Handbook of Mariculture: Crustacean Aquaculture. J.P. McVey (ed.). CRC Press, Boca Raton, Florida, pp. 107-132.

Ronquillo, J.D., T. Saisho and R.S. McKinley. 2006. Early developmental stages of the green tiger prawn, *Penaeus semisulcatus* de Haan (Crustacea, Decapoda, Penaeidae). Hydrobiologia, 560: 175-196.

Rosas, C., A. Sanchez, M.A.E. Chimal, G. Saldana, L. Ramos and L.A. Soto. 1993. The effect of electrical stimulation on spermatophore regeneration in white shrimp *Penaeus setiferus*. Aquat. Liv. Resour., 6: 139-144.

Rothlisberg, P.C. 1998. Aspects of penaeid biology and ecology of relevance to aquaculture. A review. Aquaculture, 164: 49-65.

Sagi, A., and E.D. Aflalo. 2005. The androgenic gland and monosex culture of freshwater prawn *Macrobrachium rosenbergii* (De Man): a biotechnological perspective. Aquacult. Res., 36: 231-237.

Skinner, D.M. 1985. Molting and regeneration. *In:* The Biology of Crustacea, Vol. 9, Integument, Pigments, and Hormonal Processes. D.E. Bliss (ed.). Academic Press, New York, pp. 43-146.

Sritunyalucksana, K., and K. Soderhall. 2000. The proPO and clotting system in crustaceans. Aquaculture, 191: 53-69.

Subramoniam, T. 1993. Spermatophores and sperm transfer in marine crustaceans. Adv. Mar. Biol., 29: 129-214.

Subramoniam, T. 1999. Endocrine regulation of egg production in economically important crustaceans. Curr. Sci., 76: 350-360.

Summavielle, T., P.R.R. Monteiro, M.A. Reis-Henriques and J. Coimbra. 2003. In vitro metabolism of steroid hormones by ovary and hepatopancreas of the crustacean Penaeid shrimp *Marsupenaeus japonicus*. Scientia Marina, 67: 299-306.

Talbot, P., D. Howard, J. Leung-Trujillo, T.W. Lee, W.Y. Li, H. Ro and A.L. Lawrence. 1989. Characterization of male reproductive-tract degenerative syndrome in captive penaeid shrimp (*Penaeus-setiferus*). Aquaculture, 78: 365-377.

Treece, G.D., and J.M. Fox. 1993. Design, Operation and Training Manual for an Intensive Culture Shrimp Hatchery. Texas A&M University Sea Grant College Program, Galveston, Texas.

Tsutsui, N., Y.K. Kim, S. Jasmani, T. Ohira, M.N. Wilder and K. Aida, 2005. The dynamics of vitellogenin gene expression differs between intact and eyestalk ablated kuruma prawn *Penaeus* (*Marsupenaeus*) *japonicus*. Fish. Sci., 71: 249-256.

Vaca, A.A., and J. Alfaro. 2000. Ovarian maturation and spawning in the white shrimp, *Penaeus vannamei*, by serotonin injection. Aquaculture, 182: 373-385.

Wouters, R., P. Lavens, J. Nieto and P. Sorgeloos. 2001. Penaeid shrimp broodstock nutrition: an updated review on research and development. Aquaculture, 202: 1-21.

Wouters, R., B. Zambrano, M. Espin, J. Calderon, P. Lavens and P. Sorgeloos. 2002. Experimental broodstock diets as partial fresh food substitutes in white shrimp *Litopenaeus vannamei* B. Aquacult. Nutr., 202: 249-256.

Yamano, K., G-F. Qiu and T. Unuma. 2004. Molecular cloning and ovarian expression profiles of thrombospondin, a major component of cortical rods in mature oocytes of penaeid shrimp, *Marsupenaeus japonicus*. Biol. Reprod., 70: 1670-1678.

Yano, I., and R. Hoshino. 2006. Effects of 17 β-estradiol on the vitellogenin synthesis and oocyte development in the ovary of Kuruma prawn (*Marsupenaeus japonicus*). Comp. Biochem. Physiol., A, 144: 18-23.

Wickins, J.F., and D.O.C. Lee. 2002. Crustacean Farming. Blackwell Science, Oxford, 446 pp.

Wyban, J.A., and J.N. Sweeney. 1991. Intensive Shrimp Production Technology. Argent Chemical Laboratories, Redmond, Washington, 174 pp.

Young, J.H. 1959. Morphology of the white shrimp, *Penaeus setiferus* (Linnaeus 1758). US Fish. Wildlife Serv., Fish. Bull., 59: 1-168.

12

Reproduction and Growth of Decapod Crustaceans in Relation to Aquaculture

G.G. Smith[1] and A.J. Ritar[2]

[1]Australian Institute of Marine Science, Marine Biotechnology, PMB No. 3 Townsville MC, Qld, 4810, Australia, Email: g.smith@aims.gov.au.

[2]Tasmanian Aquaculture and Fisheries Institute, Marine Research Laboratories, University of Tasmania, Nubeena Crescent, Taroona, Tasmania, 7053, Email: Arthur.Ritar@utas.edu.au

OVERVIEW OF THE MAIN CULTURED DECAPOD CRUSTACEANS

Marine shrimps (*Penaeid* or closely related spp.) and freshwater prawns (*Macrobrachium rosenbergii*) are the most widely cultured decapod crustaceans, with significant proportions of their production being based on hatchery-reared seed stock. These decapod crustaceans will be the focus of the chapter. The latest estimates of world shrimp production are in excess of 5.0 million MT, 25% being supplied through aquaculture (Josupeit, 2006). The species dominating marine aquaculture production are *Penaeus monodon, Litopenaeus vannamei, Fenneropenaeus chinensis, F. merguiensis, L. stylirostris, Marsupenaeus japonicus* and other *Metapenaeus* spp., produced predominantly in China, Thailand, Vietnam, Indonesia, India, and Ecuador. Dramatic increases are also evident in the freshwater aquaculture production of *M. rosenbergii*, which was static in the early to mid-1990s at 20,000 MT but increased to 280,000 MT by 2003; the species is produced primarily in China, Thailand, India and Vietnam (New, 2004). There are a number of other *Macrobrachium* species that are emerging as robust aquaculture candidates, including *M. nipponense* in China and *M. malcolmsonii* in India, Brazil and China (Nandeesha, 2003; Nair et al., 2005; New, 2005).

The establishment of a viable crustacean aquaculture sector based on marine shrimp has been possible only since the elucidation of the reproductive and larval rearing parameters of *M. japonicus* by Dr. Motosaku Hudinaga in 1935 (Hudinaga, 1935). It was another 30 years before commercial aquaculture facilities were established to enable the farming of penaeids, initially in Japan and in the western hemisphere, particularly in Galveston, Texas, where intensive larval rearing systems were devised based on high quality clear water systems. Parallel development occurred in the commercialization of an aquaculture industry based on *M. rosenbergii*, with Ling (1969a, b) describing their biology and Fujimura and Okamoto (1972) demonstrating their ability to be mass-cultured during studies on this species in Hawaii in the early 1970s.

Prior to the development of hatchery protocols for the production of crustacean seed stock, incidental culture had been conducted in Southeast Asia as a component of fish and rice farming for many hundreds of years. Juvenile crustaceans would be permitted to enter ponds as a component of the water supplied to these impoundments, feeding on aquatic vegetation, insect infauna and detritus, to be later harvested as a component of the primary cultivated crop. In many parts of Asia variations on this form of incidental cropping still occur today.

SHRIMP FARMING

Distribution

Shrimps are decapod crustaceans; they have numerous morphological features common to members of this grouping (Fig. 1).

They are distributed throughout the tropical and temperate regions of the world and are delineated into two specific morph types based on female genital structure, in particular that of the female thelycum. The thelycum is the external genital region of female penaeid. It is a modification of the posterior thoracic sternal plates and is located between the bases of the fifth walking legs (Bailey-Brock and Moss, 1992). Thelycum type is designated as being either "open" or "closed"; it serves primarily as the receptacle for storage of the sperm package

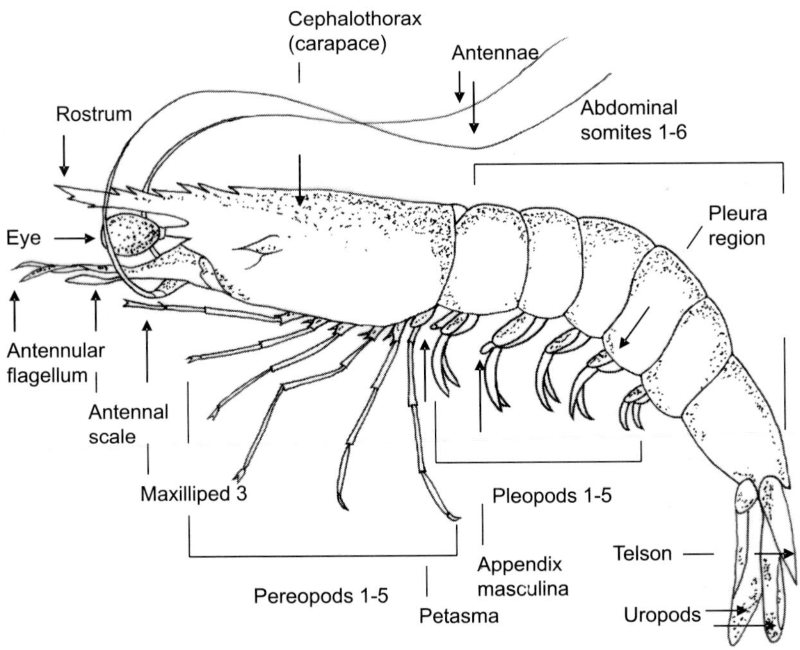

Fig. 1 The morphological features of a generalized marine shrimp; the petasma and appendix masculina are present only in males.

(spermatophore) delivered by the male during mating (Hudinaga, 1942; Primavera, 1988; Browdy, 1992). Open-thelycum species predominantly originate from the western hemisphere; both *L. vannamei* and *L. stylirostris* are native to the West Pacific coast of Latin America. Most other farmed shrimps possess a closed thelycum and are native to the East Asia-Pacific region.

In recent years, the traditional east-west separation of shrimp distribution and culture has been blurred, especially with large-scale introductions of *L. vannamei* into China and Taiwan in 1996 and into other shrimp-producing countries, namely the Philippines, Indonesia, Vietnam, Thailand, Malaysia and India, since 2000. Over this time, shrimp stocks moved relatively freely across borders, often without sanction, and generally without adequate quarantine procedures to ensure the exclusion of exotic diseases (Briggs et al., 2004). The reason for diversification into *L. vannamei* in the Asia-Pacific region rather than sole reliance on local shrimp (*F. chinensis* in China and *P. monodon* in many Asian countries) was related to the increased prevalence of disease in the local "native" species. In western countries, some progress had been made in producing Specific Pathogen Resistant stocks of *L. vannamei* and diagnostic tests to ascertain Specific Pathogen Free status. The use of a "new" species was seen as a way to circumvent disease problems; ironically, the same problems have plagued western producers in recent years. The spread of diseases associated with shrimp aquaculture has been well documented over the past two decades with undefined disease outbreaks occurring initially in Taiwan (1987) and subsequently in China (1993, White Spot Syndrome Virus or WSSV), Thailand, the Philippines, Indonesia (early 1990s, Yellow Head Virus or YHV and WSSV), Ecuador and Central America (mid-1990s, Tauro Syndrome Virus or TSV and from 1999 onwards WSSV) (Briggs et al., 2004). The shift to production of *L. vannamei* in China has not been without disease outbreaks but the sheer volume of production spread across a wide geographic region has masked localized crop failures. The profile of *L. vannamei* has increased dramatically in the Asia-Pacific region and now rivals production levels in traditional Latin American regions.

Reproduction

Sourcing Shrimp Larvae

A number of different strategies are employed by farmers to obtain penaeid larval seed stock. Extensive farming operations involve minimal farming input and allow the natural entry of larvae into rice fields, fishponds or impounded low-lying regions on the incoming tide. Juvenile shrimps are trapped at night and often at times of high abundance associated with the March-April spring tides in Vietnam and southern China. Water is exchanged and new seed added to these artificial catchments up to 10 times a month (Quynh, 1992). In such extensive farming operations, input costs (seed stock, supplementary feeding) are negligible, but returns are also variable. As an adjunct to this style of farming, impoundments may be stocked with juveniles from supplementary sources. They are often collected with scissor, seine, dip or trawl nets and transported to farms where they are cultured to harvest often without supplementary feed or fertilizer inputs (Wickens and Lee, 2002).

A second, and the predominant, method of obtaining larval seed stock is to catch mature, ready-to-spawn broodstock from the wild, or animals that are immature but capable of being matured in captivity. The practice of collecting ready-to-spawn broodstock was particularly common during the 1980s and 1990s. It still prevails, especially in areas where wild broodstock are abundant and infrastructure to operate continual maturation facilities is minimal. Females with fully developed ovaries will often undergo spontaneous spawning the night after capture stimulated by stress. This provides an immediate, although unreliable, source of larvae. To attain subsequent gonad maturation, freshly spawned broodstock or those with immature gonads must undergo conditioning in a maturation unit.

The third and increasingly important source of seed stock is domesticated broodstock, especially shrimps maintained for numerous generations (Racotta et al., 2003). By sourcing larvae from a closed life-cycle it is possible to undertake genetic selection to improve growth, disease resistance, feed conversion ratios, tolerance and adaptation to local conditions, while also quarantining breeding stock from disease outbreaks occurring in wild stock. Originally, one of the negative aspects of maintaining captive broodstock was their inability to mature spontaneously in maturation facilities. They often require eyestalk ablation to initiate maturation; this is a technique that has numerous adverse physiological side effects. However, a major benefit noted in some domesticated shrimps is the tendency for successive generations to undergo gonadal maturation and spawning without initiation by ablation. Some species are more conducive to natural maturation than others, with *L. vannamei* cultured for more than 17 generations in Venezuela no longer requiring ablation to stimulate spawning (Wickens and Lee, 2002), while *P. monodon* rarely spawn naturally even after many generations of culture. The source of broodstock for maturation units may initially be wild-caught spawners, but subsequent generations can be sourced from commercial growout ponds, in the case of species that mature readily during culture (*L. vannamei*), or from specialized culture facilities providing specific conditions to assist in the development of maturity (*P. monodon*).

Broodstock Development

The onset of maturity and optimal reproductive output in shrimp is often linked to species-specific size and age parameters. The suggested minimum size of mature *L. vannamei* broodstock for use in maturation facilities is 40 g for males and 45 g for females (Wyban and Sweeney, 1991). In large penaeid species such as *P. monodon*, males of 60 g and females > 90 g are preferred (Primavera, 1983; Hall et al., 2000). While the ideal broodstock size for hatchery production of larvae varies between species there is a general consensus that fecundity, both spawn size and frequency, is positively correlated with size (Hanford and Marsden, 1995; Palacios et al., 1998). It has been noted that there may be a negative effect of senescence on fecundity and larval viability in some older, large animals (Rothlisberg, 1998). Most species of shrimp do not live for more than 2 years; male and female *P. monodon* have maximum life-spans of 1.5 and 2 years, respectively (Dall et al., 1990).

Maturation Facilities

There are numerous environmental, physiological and nutritional parameters that must be met to facilitate maturation in penaeids. The development of maturation facilities that meet these needs has increased considerably during the past two decades, influenced by fluctuations in the availability and cost of wild spawners, the introduction and transfer of exotic diseases to shrimp-producing countries, and the desire to implement genetic stock improvements. The design and operation of maturation facilities are species and location specific; however, similarities in the life history of penaeids allow a number of general environmental parameters to be defined. In their natural environment, penaeids have a planktonic oceanic larval phase followed by movement on-shore as post-larvae, inhabiting coastal, estuarine and brackishwater environments as juveniles. As developmental maturity approaches, animals again move off-shore, undergoing the transition to adults in oceanic waters (Primavera, 1983). The provision of oceanic conditions is of fundamental importance to penaeid maturation success, and the way in which those parameters are attained will depend upon the infrastructure and technology available in specific locations.

Light Requirements. The quality, intensity and periodicity of the light supplied to broodstock in a maturation facility can have a large impact upon the ability of shrimp to successfully mature in captivity. Light quality should favour a blue-green wavelength and be of low intensity (12 μmol m^{-2} s^{-1}), mimicking deep oceanic water (Wurts and Stickney, 1984). A natural daylight period is often used when culturing native species; for translocated stock 12-16 h of light is effective for tropical species. A shorter day length of 8 h is suitable for subtropical and temperate species (Primavera, 1984). A 1 h sunrise and sunset fade-in and fade-out control is often used in systems using artificial lighting. The use of photoperiod phase shifting allows the scheduling of mating, egg collection and disinfection to be controlled and thus altered to occur at a time that is convenient for the hatchery operator (Browdy, 1992). Red light, a wavelength thought to be invisible to most crustaceans, may be used during the dark phase to observe animals while minimizing their disturbance.

Water Quality. The salinity in maturation facilities should be stable and, while maturation success has been reported at a range of salinities from 15 to 40‰, there is a consensus that oceanic salinities of 34-35‰ are optimal (Primavera, 1984; Wickens and Lee, 2002). Typical oceanic water also has high water clarity, stable pH (8.2), low dissolved organic content, and minimal numbers of bacterial and viral pathogens. Water exchange and filtration are two means of controlling the water quality in maturation systems. Generally flow-through systems are used with up to 400% water exchange per day. Where the water source is of variable quality, static or recirculation systems with daily exchange rates between 2.5 and 60% are used (Pimavera, 1984). Water quality may be adversely affected by high broodstock stocking density (which should be < 300-400 g/m^2), feeding (type, amount and frequency), and efficient removal of waste products from the tanks. When tanks are poorly managed, organic waste, protein, ammonia and nitrites accumulate and have a negative impact, particularly in static or recirculation systems. Optimal water temperature depends largely on the species cultured; tropical species prefer 26-32°C, while temperate and subtropical species require lower temperatures of 16-28°C (Primavera, 1984).

Physiological and Physical Factors. Physiological and physical factors such as a sex ratio of 1♂:1♀ and suitable tank design assist in the optimization of fertilization success in mixed-sex populations. Maturation tanks should be 0.5-0.8 m deep with a diameter and dimensions of ≥ 3.5 m in circular tanks, or, if rectangular tanks are used, long enough to provide uninterrupted swimming during mating. The sides of the tanks should be darkened to reduce ambient light levels while the provision of a light-coloured bottom assists in providing contrast when observing for shrimp gonad development. Alternatively, sand bottoms may be used, which are a requirement for burrowing shrimp species such as *M. japonicus*.

Nutrition. The ability of broodstock to successfully undergo maturation and produce healthy, competent larvae is heavily influenced by their nutritional status, in particular their ability to sequester sufficient nutrient stores from the hepatopancreas during vitellogenesis (production and storage of egg yolk) (Harrison, 1990). However, as maturation is often induced precociously in captivity by eyestalk ablation, dietary components need to play an immediate role in ensuring that vitellogenesis proceeds (Harrison, 1997). There are a number of lipid and vitamin components required to achieve optimum ovarian development (Alva et al., 1993a, b), fecundity and egg hatchability (Xu et al., 1992, 1994), including the essential fatty acids 20:4n6 (arachidonic acid), 20:5n3 (eicosapentaenoic acid) and 22:6n3 (docosahexaenoic acid) and vitamins A, E and C. The composition of vitellogenic yolk also includes proteins, free amino acids, triacylglycerides, cholesterol, carbohydrates, free sugars, nucleotides and nucleic acids. As not all requirements for vitellogenesis can be produced *de novo*, there are a number of natural dietary sources that have been successful in providing the necessary requirements (Harrison, 1997). Common maturation dietary components include squid, bivalves and marine polychaete worms. Supplementary feeds such as compound pellets and adult *Artemia* biomass (Naessens et al., 1997) may contain minerals, carbohydrate and protein essential for the metabolic functioning and ovarian development (Harrison, 1990, 1997), or may be specifically enriched with components such as carotenoids, lipids and vitamins (Smith et al., 2004). Feeds must be provided at a level to ensure broodstock are fed to satiation while minimizing excesses that would compromise water quality. Feeding is commenced at 10% of the wet weight of the broodstock biomass per day, divided into a number of portions (up to six) and distributed over the 24 hr period. The total daily feed amount and portion size is adjusted on the basis of consumption in individual tanks (Robertson et al., 1993; Wickens and Lee, 2002; Peixoto et al., 2005).

Reproductive Strategies

Male Maturity. Most commercial species of shrimp mature within 4-11 months of hatch. The onset of maturity is species specific and influenced by the timing of spawning (spring or autumn), local water temperatures and feed conditions (Primavera, 1988; Dall et al., 1990). Penaeid shrimps are dioecious and may be differentiated by their external reproductive organs. There are a number of external developmental changes that occur in adult males that are absent in juveniles, allowing them to be sexually differentiated. Mature males have the endopodites of the first and second pleopods fused to form the petasmata and appendix masculina respectively (Fig. 1). These specialized structures facilitate the transfer of the spermatophore (sperm package) from the two terminal ampoules located near the fifth pair of pereiopods to the thelycum of the female (Bailey-Brock and Moss, 1992). While the presence of advanced morphology does not guarantee sexual functionality, it is a requirement for mating. In most shrimp species, visual confirmation of male maturity can be obtained by noting the swelling and white colouration of the terminal ampoules. (Primavera, 1984; Bray and Lawrence, 1992). Biological functionality can be ascertained by removing the spermatophores from the terminal ampoules by squeezing them from the base to the tip. When they are expelled, the sperm can be examined microscopically for the presence of a developed spike on individual cells. The spike is used for sperm attachment during egg fertilization. Spermatophores are commonly removed using this procedure to obtain sperm for artificial insemination of mature females.

Mating Physiology. Many interspecies differences exist in mating behaviour in farmed shrimp. However, the major determinant of moulting and reproductive strategy is governed by female thelycum morphology. Mating occurs in open-thelycum species in the early evening and only during the hours prior to spawning; both sexes must have hard exoskeletons. In open-thelycum species (*L. vannamei, L. stylirostris*), there is no internal chamber to store spermatophores. Rather, the thelycum has a setose external region consisting of grooves, protuberances and ridges providing short-term attachment points for the spermatophore (Browdy, 1992). Open-thelycum females do not retain the spermatophore for subsequent moults, often mating several times during an intermoult period. (Yano et al., 1988). In closed-thelycum species (*P. monodon, F. chinensis, F. merguiensis* and *M. japonicus*), the thelycum is a flesh-covered chamber that is divided into two halves allowing individual deposition and storage of spermatophores delivered by the male during mating, which is only possible when the female is soft following a moult (Hudinaga, 1942; Browdy, 1992). Closed-thelycum species mate only once during a moult cycle and the female retains the spermatophore for use in subsequent spawning events.

The mating behaviour of shrimp species differs primarily in the courtship and placement of the spermatophore packages. Mating in the open-thelycum *L. vannamei* is typified by the males swimming behind and underneath the female for a short period prior to mating (10 min), their ventral sides brushing together in a brief embrace lasting only seconds, during which time the male spermatophores are transferred to the setose exterior of the thelycum (Misamore and Browdy, 1996). The orientation of the ventral surfaces is generally head to head although head-to-tail orientation has been observed (Yano, 1988). Mating behaviour is similar in open- and closed-thelycum species, except that in closed-thelycum species, the courtship and mating process is typically of a longer duration. The male often swims parallel to the female for a period before moving below her and positioning their ventral surfaces together with the female grasping the male carapace with her pereiopods. In the closed-thelycum *M. japonicus*, this embrace may last for 10 min before the spermatophores are transferred with the male squeezing the female. During courtship and mating in *P. monodon*, the embrace may last up to 3 h before the male wraps himself at right angles to the longitudinal axis of the female body and transfers the spermatophores to the thelycum (Primavera, 1979; Motoh, 1985; Dall et al., 1990).

Inducing Maturation

The development of husbandry protocols to induce maturation and spawning in captive shrimp has been examined for the last three decades.

Precocious maturation has been demonstrated in some species with successive generations of domestication, and this is especially true of *L. vannamei* and *L. stylirostris*. In other species such as *P. monodon*, the ability to mature in captivity without surgical intervention remains problematic, with only sporadic events reported. In animals that do mature unassisted in captivity, the problem of obtaining spawning synchrony over an extended time frame still exists. Synchronized spawning events are an important requirement for staged production of large numbers of shrimp.

Eyestalk Ablation. The discovery by Panouse (1943) that a hormone produced in the eyestalk was responsible for the inhibition of ovarian development in crustaceans was a significant factor in later efforts to domesticate shrimp species. Gonad-inhibiting hormone (GIH) is produced in the sinus gland X-organ complex of the eye and released into the bloodstream. It inhibits the production of gonad-stimulating hormones, a joint function of the brain and thoracic ganglia. Many in the shrimp industry in the mid-1970s adopted eyestalk ablation to overcome the inhibition of maturation and spawning by GIH in captive stock (Bray and Lawrence, 1992). Even today, eyestalk ablation is the only commonly used technique that provides predictable maturation and spawning results. The removal of the eyestalk effectively removes maturation inhibition. However, as the eyestalk is the source of numerous hormones including those required for moulting, sugar balance and metabolism, there are numerous adverse side effects (Browdy, 1992). Eyestalk ablation often results in spawner exhaustion, because ablated animals mature at a faster rate and have a reduced period between spawning events. This may affect their ability to sequester sufficient nutrients required for vitellogenesis and hence compromise fecundity and larval quality (Hansford and Marsden, 1995).

Unilateral eyestalk ablation (removal of one eye) is generally practised and provides sufficient inhibition to allow maturation to proceed without maximizing stresses associated with bilateral ablation (Wickens and Lee, 2002). There are a number of methods: cutting and cauterizing the eyestalk with hot scissors, tying the eyestalk with a ligand, or enuculation, slitting the eyeball and squeezing out the contents (Browdy, 1992). Eyestalk ablation is not practised in males, as it has not been shown to produce additional reproductive benefits (Browdy, 1992). In the decades since eyestalk ablation became commonplace, there has been a focus on research into shrimp endocrinology (Huberman, 2000). However, to date this has not produced viable alternatives.

Identifying Stages of Maturation. When maintaining captive broodstock, it is important to be able to identify the stage of ovarian development. In shrimp, the ovary extends from the cephalothorax region, predominantly behind the gut in a longitudinal direction finishing in the lower region of the abdomen (tail). It is a paired organ, although joined in the cephalothorax. Two oviducts connect the ovary with the external gonophores at the base of the third pair of pereiopods (Bray and Lawrence, 1992). The degree of maturation in shrimp can be observed through the tail section of shrimp, particularly in light-coloured species (*F. merguiensis*). In darker species (*P. monodon*), the shadow cast by the developing gonad is observed with the aid of torchlight. The developing gonad extends for much of the longitudinal portion of the female, with most of the tissue development occurring in the cephalothoracic region, often hidden by pigmentation and body tissue (Tan-Fermin and Pudadera, 1989). Numerous stages of gonad development can be recognized visually with similarities noted across the major farmed shrimp species (Fig. 2).

I. Undeveloped: the gonad consists of a thin clear vesicle present in the cephalothorax and extending the length of the tail with no discernible ovarian development.

II. Development begins: eggs undergo the first stage of meiotic division forming previtellogenic oocytes, which do not contain yolk. The ovary is faintly visible as a pale opaque or yellow (depending on the species) concertinaed structure that is centrally located in the tail section. When maturing females are at this stage of development, poor husbandry or adverse environmental conditions will halt maturation. If newly captured females have attained this stage of development, eyestalk ablation will induce them to spawn within 1 week.

III. Nearly mature: yolk is internalized within the oocytes in a process known as vitellogenesis and, as the yolk accumulates, the eggs begin to mature and take on a characteristic olive green to brown colour (may be yellow-orange-brown in some species). The colouration is due to pigments present in the yolk. At this stage the ovary is now clearly visible in the tail section of the female as it begins to take on a long concertina shape.

IV. Mature: eggs continue to develop and darken (brown-olive green) with a particularly wide section of the ovary visible immediately behind the carapace. This region is known as the saddle and the top view of the tail is dominated by the dark shadow cast down the length of the tail. Eggs are now mature and ready for ovulation into the oviduct for oviposition, release and mixing with sperm in the water column.

V. Spent: the ovary looks similar to Stage II ovary, with a watery appearance rather than an opaque colour. On some occasions, sections of the gonad are still visible in the tail post-spawning, which signifies incomplete spawning has occurred and the remaining oocytes will be reabsorbed. There are a number of factors that contribute to incomplete spawning, including partial blockages of the oviducts during spawning, low levels of the chemical trigger that induce ovulation (Hall et al., 2000) or incomplete maturation of oocytes infected with viral particles (Kuo and Lo, 1998). After each moult, the ovary is fully regressed and the process of maturation must occur from a primary state.

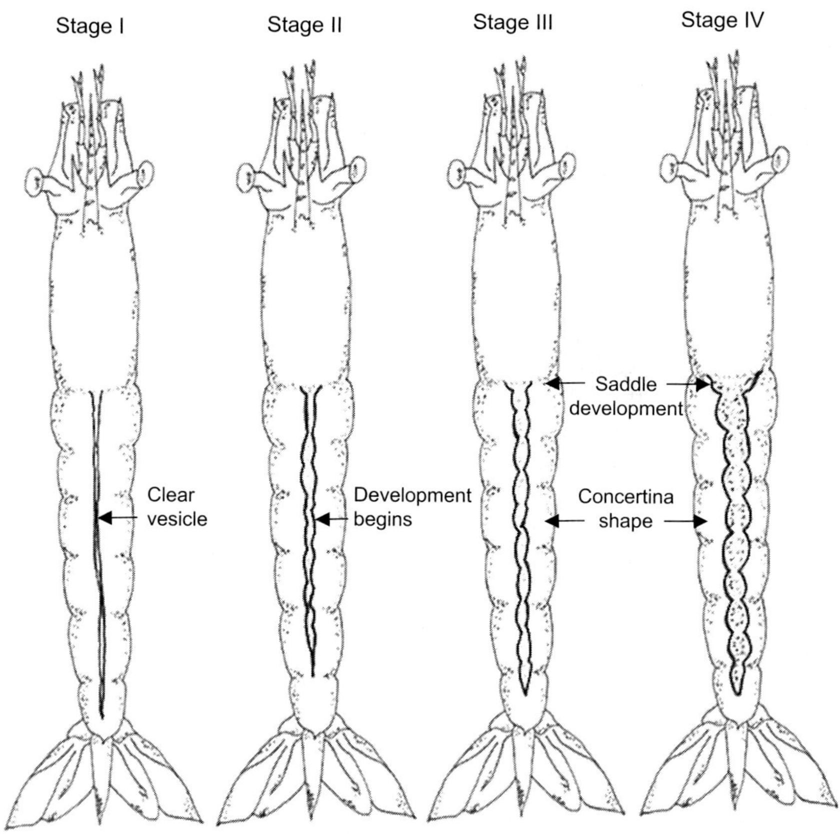

Fig. 2 The first four stages of ovarian development showing the progression of ovary maturation observed through the tail section of female shrimp. In light-coloured species development can be seen unassisted, in darker species the aid of a torch may be required to see the shadow cast by the developing ovary.

Hatchery Rearing

Spawning. Broodstock can be held in single-sex populations but are generally maintained as mixed populations. However, with the advent of selective breeding programmes, mixed populations may become the exception rather than the rule. Prior to spawning, open-thelycum species must be checked for mating and deposition of a spermatophore, and artificial insemination may be undertaken if they are unmated. Females are removed to separate individual spawning tanks late in the afternoon or early evening prior to spawning. This facilitates the collection of brood-specific information including number of eggs spawned, fertilization and hatch rates, individual vigour and health of the larvae, while minimizing the vertical and horizontal transfer of viral and bacterial diseases (Chanratchakool et al. 2005).

The volume of spawning tanks should be related to the expected fecundity of spawners with a maximum of 2,500 eggs l^{-1} and lower density used where possible (Primavera, 1983). Fecundity varies widely inter- and intra-species, with Motoh, (1981) reporting 250,000-800,000 eggs per spawn for *P. monodon* while Beard and Wickens (1980) obtained 19,000-460,000 in the same species. *Litopenaeus vannamei* commonly produces 60,000-200,000 eggs per spawn (Anon., 1983). There are a number of factors that contribute to fecundity, although female size appears to be the most dominant physiological factor, with large females producing more eggs (Bray and Lawrence, 1992). Spawning occurs in all shrimp species in the late evening or early hours of the morning, with simultaneous release of eggs from the ovipores and non-motile sperm from the thelycum accompanied by a rapid fanning action of the last three pereiopods. This ensures that eggs and sperm are mixed together and maximum fertilization can occur (Bray and Lawrence, 1992). The entire spawning process takes less than 5 min, and often only 1-3 min. The newly spawned fertilized eggs are neutrally buoyant and slowly sink towards the bottom, where they settle until hatch. To circumvent the vertical transfer of disease from broodstock to eggs in the hatchery, a second step of egg transfer and disinfection is often undertaken in the hours after spawning and during the early phase of egg incubation. There are numerous techniques for disinfecting eggs. Chanratchakool et al. (2005) reported that rinsing developing eggs for 5 min in filtered sea water, 30 sec in 100 ppm formalin, 1 min in 50 ppm povidone iodine and finally for 5 min in filtered sea water was effective. Recent studies on the tolerance of developing penaeid eggs to ozone (an effective bacterial and viral disinfectant) may also prove useful in the prevention of disease transfer (Sellars et al., 2005).

Larval Development. Shrimp eggs hatch a few hours after spawning with larval development continuing through 12 free-swimming planktonic instars completed with metamorphosis to a benthic post-larval stage over a 11-14 d period. During nauplius stages, larvae do not feed, surviving on endogenous yolk reserves. First feeding occurs in penaeids during the protozoeal stage where a range of fresh or preserved microalgae is provided (*Chaetoceros*, *Skelotenema* or *Thalassiorsira* spp.). There is a dietary preference for zooplankton (rotifers, *Artemia* nauplii) during the mysis stage shifting to *Artemia* nauplii and dry feeds for early post-larvae. During late post-larval and early juvenile stages, most shrimps consume prepared pellet feed, benthic invertebrate micro-organisms and other pond-based detritus.

Nursery Phase. There are a number of strategies used to culture shrimp larvae through the juvenile and adult stages of development until harvest. The commercial outcomes of survival and growth are highly dependent on the relevant management and husbandry inputs applied. For instance, when growout ponds are stocked with young larvae (PL10), the ability of animals to adapt to sub-lethal changes in pond conditions is minimal. In extensive and semi-intensive systems, mortality can be significant, owing to competition for food, high predation rates and limited ability of farmers to significantly alter water quality parameters through water exchange (Samocha et al., 1992; Wickens and Lee, 2002). By contrast, late PL or young juveniles readily acclimate to conditions that may be lethal to younger animals. Advanced PL or early juveniles can be sourced from specialized nursery facilities or farmers may choose to conduct an intermediate 15-50 d nursery phase themselves, producing 0.2-2 g animals. Survival can be enhanced and, equally important, predicted during growout operations when older, hardier shrimps have been stocked. In some instances, initial stocking rates have been reduced from 50 m^{-2} down to 7.5-15 m^{-2} in semi-intensive systems when older animals have been stocked, providing similar pond productivity at harvest (Wickens and Lee, 2002).

The advantage of including nursery systems as part of an integrated shrimp production system is the increased level of control in the provision of feed, temperature, water quality, exclusion of water and air-borne predators and improved ability to effectively monitor larval health. Nursery systems come in a variety of sizes and complexities; small pond enclosures (0.04-1.0 ha) may be stocked at densities up to 200 m^{-2} (Wickens and Lee, 2002), while very high density culture (1,200 PL m^{-2}) has been conducted in 0.1-0.4 ha temperature-controlled intensive greenhouse facilities (Wyban and Sweeney, 1991). Animals that have undergone a nursery phase prior to growout are highly regarded and provide a greater level of confidence when estimating future pond biomass.

Growout Phase. There are a number of factors that regulate the growth of shrimp in relation to aquaculture, including farming intensity, culture environment, stocking density and the provision of supplementary feeds. Shrimp aquaculture was founded on a history of extensive farming, with the incidental impoundment of shrimp in rice or salt farms. Today's extensive farms are a mixture of small family and state-run farms; pond sizes vary from < 3 ha up to 100 ha, are stocked at 0.2-5 m^{-2} and return yields of < 1 MT ha^{-1} without supplementary feeding (Wickens and Lee, 2002). Growth in extensive systems depends on shrimp being stocked at densities less than the natural carrying capacity of the aquatic system. Extensive systems may be fertilized with animal manures to enhance pond productivity through the establishment of a complex food web. Shrimps, being omnivorous, feed at different levels of this web, consuming a range of aquatic benthic invertebrates, insect larvae, macrophytes, algae and organic detritus to obtain their nutritional requirements for growth (Tacon and Akiyama, 1997).

The most common method of farming shrimp is classified as semi-intensive culture. Pond sizes in semi-intensive culture are generally smaller than in extensive systems at < 3 ha, stocked at densities of < 20 m^{-2} and returning yields of < 5 MT ha^{-1}. Ponds are stocked with hatchery-reared PL. Nursery culture is often carried out as an intermediate step to improve productivity prior to ongrowing. In these systems there is some control over water exchange, the provision of aeration and supplementary feeds of either farm-produced semi-moist diets or commercial extruded pellets (Wickens and Lee, 2002).

Intensive culture systems are a refinement of semi-intensive methods; stocking rates are higher and, therefore, monitoring and husbandry inputs such as water quality parameters and feed inputs must be more closely controlled to ensure that projected production levels can be attained. Varying the amount of water exchange provided can regulate water quality. The aim is to maintain a consistent low level of micro-algal growth in the pond, a healthy pond bottom free from excess organic material and stability in water quality parameters such as dissolved oxygen, pH, temperature and salinity. In addition to water exchange the use of mechanical aerators and paddlewheels assist in maintaining dissolved oxygen content, particularly before dawn, and the provision of water-stable extruded pellets to reduce the amount of organic waste present in the pond is integral to the success of intensive culture systems. Growout ponds are generally stocked with juveniles that have been hatchery reared before an intensive nursery phase. Intensive ponds are typically < 1 ha in size and stocked at densities of < 50 m^{-2}, yielding up to 15 MT ha^{-1} (Boyd and Musig, 1992; Wickens and Lee, 2002).

The fourth broad classification based on farming intensity is super-intensive culture and involves rearing shrimp at densities of up to 250 m^{-2} in 0.03-0.1 ha in lined and often covered ponds or raceways with yields as high as 50 MT ha^{-1}. The use of super-intensive systems is not widespread and is concentrated where infrastructure and labour costs are high, sophisticated technology is readily available and the end-product commands a premium market price, such as for live *M. japonicus* cultured in Japan. The risks of operating a super-intensive system are high; equipment failures, even for short periods of time, have the potential to result in mass mortalities.

MACROBRACHIUM FARMING

Distribution

Macrobrachium is a caridean prawn, naturally distributed throughout tropical and warm temperate regions of the world. There are currently 210 recognized species in this genus (Short, 2004), of which 49 are of commercial interest to either fisheries or aquaculture (Holthuis, 1980). There are a number of obvious taxonomic characteristics that distinguish caridean prawns from penaeid shrimps (Fig. 3). Caridean prawns have

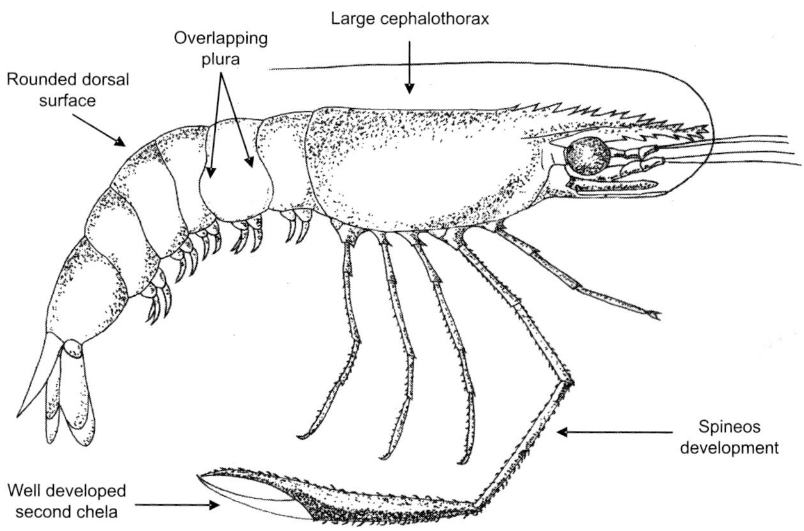

Fig. 3 Generalized drawing of a *Macrobrachium* spp. showing the prominent features that distinguish them from penaeid spp.

well-developed elongated second pereiopods with a strong chela (claw) used for food capture, mating and aggressive behaviour. Particularly in dominant males, they have a large cephalothorax, their abdomen has a round dorsal surface, and the second pleuron of the tail overlaps both the first and third pleura. In comparison, penaeid shrimps have small undeveloped second pereiopods, a dorsal abdominal ridge and sequential overlap of pleura segments from the first through to the fifth (New, 2002).

The major farmed species is *M. rosenbergii*. It has a natural distribution from northwestern India to Vietnam and southern China, the Philippines, Indonesia, New Guinea and northern Australia (Holthuis, 1980). Within this species there are thought to be numerous subspecies (Short, 2004), of which two are predominately recognized: the western *M. rosenbergii dacqueti* from India, Thailand, Malaysia, Sumatra, Java and Kalimantan and the eastern *M. rosenbergii rosenbergii* from the Philippines, Sulawesi, Irian Jaya, Papua New Guinea and northern Australia (New, 2002; Short, 2004). There appear to be a considerable number of phenotypic variations displayed across the natural geographic range of this species, for example, different head-to-tail size ratios and determinants of larval productivity including eggs size, fecundity and larval duration (Buranakanonda, 2002; Short, 2004). Such phenotypic diversity bodes well for selective enhancement of desirable aquaculture traits as well as improving current culture protocols through the determination of strain-specific rearing requirements (New, 2005).

Culture Status

Macrobrachium rosenbergii has a history of almost four decades of culture, initiated in Malaysia and Hawaii and spreading to almost every continent and many island states under a diverse range of conditions from extensive natural water impoundments in Vietnam to intensive systems in New Zealand and Russia using thermal wastewater from power stations. The growth of prawn farming has been increasing at a dramatic rate over the past decade with up to 30% annual boosts in production levels from 1999 to 2001. The latest conservative projections are at least 700,000 MT by 2010 and possibly 1,400,000 MT if production continues to increase at the current rate and all farmed *Macrobrachium* species are grouped together in the estimate (New, 2005). There are large numbers of *M. malcolmsonii*, native to Pakistan, India and Bangladesh (Holthuis, 1980), being farmed in Vietnam that are not specifically identified and therefore are excluded from world production figures, and *M. nipponense*, endemic to northern China, Japan and Taiwan, is thought to constitute a major proportion of the reported prawn production in China. Current production figures for these two species are unclear although they are expected to be significant and play an even greater role in the future aquaculture production of *Macrobrachium* species (New, 2005). The stimuli behind the dramatic increase in prawn production are a shift from extensive to semi-intensive production and an emphasis on environmentally sustainable production, often geographically removed from high value delicate coastal nursery grounds essential for the culture of marine species (New, 2002).

Morphology

Sexual Morphology

Macrobrachium rosenbergii is a dioecious species; male prawns have a large cephalothorax, the base of their walking legs are set close together and they have gonadopores covered by a clear membrane located at the base of the fifth set of walking legs. Obvious features of male prawns are the well-developed second set of pereiopods with large chela, and the modification of the second set of pleopods (appendix masculina) to assist in sperm transfer during copulation. Females have genital openings at the base of the third set of walking legs and a large gap between the last pair of walking legs. They have a wider abdomen shrouded by long pleura that forms a brood chamber during incubation, and their total body size is smaller than that of males of the same age (Ling, 1969a; D'Abramo and Brunson, 1996; New, 2002).

Male Morphology

There are considerable developmental differences between male prawns, often of the same age, with three distinct morph types generally recognized. Small males (SM), with a translucent body, possess undeveloped, slim, short second pereiopods and chela. The second group are strong orange claw males (SOC), sometimes called orange claw males (OC), having larger brown-olive bodies and long golden-orange claws with minor spineous development. The third group are the dominant blue claw males (BC), similar in body colour to OC males but larger and having extremely long blue claws (up to twice the body length) with heavy spineous development. There are intermediate stages of development between SM and SOC known as the weak orange claw (WOC) males and transforming orange claw (TOC) between the SOC and BC males (D'Abramo and Brunson, 1996; New, 2002).

The interaction between male prawns of different morph type is central to their hierarchical social structure; BC males are dominant over

OC males, which are dominant over SM. All male morph types are capable of participation in viable reproduction, although females prefer BC males. Interaction between BC males is generally restricted to posturing, with a BC and an SOC male usually engaging in fighting behaviour involving the use of claws, while BC males generally require only threat displays to deter SM (New, 2002). Transforming orange claw males are slow to progress to BC males in the presence of dominant BC males; BC male removal stimulates growth in SOC and TOC males, with the largest assuming the dominant hierarchical position and thus a reproductive advantage.

Reproduction in Macrobrachium

Mating and Egg Incubation

Female prawns become sexually mature 4-6 months post metamorphosis. Mating takes place only between intermoult males (hard) and post-moult females (soft). Prior to a pre-nuptial female moult, ovarian development can be observed in the dorsal region of the cephalothorax as a yellow-orange mass. Mature females possess abdominal pleura that are extended outwards to form a distinct brood chamber and the first four pairs of pleopods have ovigerous setae required for egg attachment (Ling, 1969a). Prior to mating, a male will engage in courtship display raising his body on extended legs, actively moving antennae and chelipeds for 10-20 min, and then moving forward to embrace the female with extended pereiopods. While holding the female, he cleans the ventral portion of her thoracic section for 10-15 min in readiness for sperm deposition. The female is turned upside down and the male brings his genital pores in contact with the cleaned surface between the fourth pair of walking legs and deposits a gelatinous sperm mass, with sperm deposition occurring within a few seconds.

Female *M. rosenbergii* extrude eggs 6-20 h after mating. The tail is extended forward towards the gonadopores located at the base of the third set of legs. The beating of the pleopods creates a current that moves eggs from the gonadopores over the sperm mass allowing them to be fertilized, and down into the brood chamber formed by the abdominal pleura (Ling, 1969a; D'Abramo and Brunson, 1996). The brood chamber is divided into separate regions delineated by the first four sets of pleopods. Eggs are attached to the pleopods by a thin membrane excreted from the ovigerous setae, with egg attachment commencing at the rear of the incubation chamber and filling sequentially towards the front (Ling, 1969a).

Newly extruded eggs are slightly elliptical in shape, measuring 0.6-0.7 mm longitudinally. They develop and remain attached to the pleopods for the 19-21 d of embryonic development at 26-28°C. The female constantly grooms the egg bundle using her first pair of thoracic legs to remove diseased, dying or dead eggs and aerates the remainder with pleopod movement. When eggs are first extruded they are bright yellow-orange fading to orange over a number of days. Two-thirds the way through the incubation period eggs fade to light brown and then to grey 2 d prior to hatch (Ling, 1969a; D'Abramo and Brunson, 1996).

Sourcing Broodstock

It is common practice to source berried females (ovigerous, carrying eggs externally) close to larval hatch in most tropical and subtropical regions where *M. rosenbergii* is cultured. The use of berried females negates the need for large-scale holding facilities and provides a high turnover of broodstock maintained often for < 1 wk in a facility. When prawns are obtained near the hatchery (< 1 h away) they may be transported in aerated live transport containers. This may be as basic as a 20 l bucket with an air stone for small numbers sourced from an adjacent pond. However, if large numbers of broodstock are to be moved they will require a specialized tank with baffles, hides and oxygen supply to reduce water movement and animal interaction while maximizing dissolved oxygen content during transport. Longer journeys of up to 24 h will require animals to be individually transported in double plastic bags in a 1:3 mix of water and oxygen; broodstock up to 25 g in size require 1-1.5 l of water. When transporting adult *M. rosenbergii,* the rostrum, telson and chelipeds should be covered with silicon tubing to prevent puncture of the transport bags.

Upon arrival at the hatchery, berried females should be disinfected (0.5 ppm copper sulphate) to reduce the incidence of fouling organisms (New, 2002) and sorted into stages of egg development. In areas where berried females are abundant females will be selected only in the latter stages of egg development to minimize holding requirements. Towards the end of egg development, berried females should be acclimated from their holding water, generally fresh water, to a salinity of 12-14‰ and pH 7 (Law et al., 2002), which improves hatch rates and optimizes larval viability (Cheng et al., 2003). Berried broodstock can acclimate to these conditions over a 48 h period without suffering noticeable side effects. Spent females are generally discarded after a 1-2 day egg hatch period, with new berried females sourced when required (New, 2002).

Moulting and egg production within groups of *M. rosenbergii* are asynchronous, making it difficult to produce significant numbers of larvae of similar age without using many females. In tropical regions, the timing of moulting and mating causes minimal problems because berried females are readily available for selection from natural or manmade impoundments, and hatchery requirements can be targeted towards a particular embryonic developmental stage. This is more difficult to achieve in temperate regions, where all broodstock are over-wintered, captive stock requiring dedicated holding and husbandry protocols.

Larval Rearing

Many hatcheries separate broodstock into hatching tanks 1-2 days prior to larvae hatching (grey egg mass). The tanks may be divided by a coarse mesh partition that will maintain broodstock on one side while allowing larvae to flow out into a holding tank. At hatch, larvae swim freely, albeit ventral side up and backwards. They are strongly photopositive, allowing light to be used to concentrate animals for collection and removal to culture tanks (New, 2002). The larval phase consists of 11 distinct morphological stages of development taking 15-40 d to achieve metamorphosis (D'Abramo and Brunson, 1996). Such a variation in time to metamorphosis indicates that there are a number of important physiological, nutritional and environmental factors that are yet to be fully understood in the larval culture of *M. rosenbergii*.

There are many husbandry methods used to culture larvae ranging from low density where 50-100 larvae l^{-1} are cultured in the same container for the duration of the larval phase to high density where up to 500 larvae l^{-1} are cultured for the first 10 days then thinned out to 50 larvae l^{-1} until metamorphosis. Water treatment protocols also exhibit a similar continuum, from green water systems where a background concentration of microalgae is maintained to stabilize water conditions to the use of mechanically filtered and disinfected water in recirculation systems (New, 2002; Wickens and Lee, 2002). The primary feed used to culture prawn larvae is newly hatched *Artemia*, which may be supplemented after 5-7 d of culture. There is no commercial formulated diet currently available. The composition of supplementary diets varies widely and is based on mixtures of fish or other seafood, eggs, vitamins and fish oil that are homogenized and partly cooked. Depending on the specific ingredients used in the prepared diet the mixture may need to be bound with a product such as alginate (New, 2002; Anon., 2003).

Sourcing Larvae

There are primarily three protocols to obtain *M. rosenbergii* larvae: collection of wild juveniles, larvae hatched from wild or pond-caught berried females, and larvae sourced from females mated and held in captivity. Juvenile seed stocks are collected from the wild using brushwood, shelter traps, stow and straw nets in tidal shallows and lower reaches of freshwater rivers (New, 2002). This form of farming is generally extensive in nature and is common in Bangladesh, India and Vietnam (New, 2005). As the value of farming prawns is being realized, extensive farming operations are giving way to semi-intensive methods, and farmers are moving to the use of hatchery-produced seed stock where available. The adoption of semi-intensive farming has resulted in increasing productivity and flow-on effects to improve family incomes (New, 1995; Nandeesha, 2003).

The production of hatchery seed stock is divided predominately on the basis of tropical and temperate regions. In tropical and to a certain extent subtropical regions, berried females can be sourced from local rivers if they are endemic to the region or removed selectively from growout ponds during partial or complete harvests. Berried females are available most of the year where the ambient water temperature does not fall below 25°C (New, 2002; Wickens and Lee, 2002). It has been the general practice in these regions to collect only berried females and not to hold mature males. It is likely that there will be trends towards establishing captive domesticated broodstock due to benefits that could be obtained from selective breeding programmes and perhaps gene transfer. Managers must be mindful of maintaining genetic diversity as declining pond productivity is often reported when stock, with often limited variation to begin with, is repeatedly targeted (New, 2002).

Obtaining seed stock in temperate regions of the world, notably the southern United States, requires that next year's broodstock be collected from ponds prior to the end of the growout season. During autumn the ambient water temperatures will drop below an optimal reproductive threshold of 25°C and in most temperate regions water temperatures will reach lethal levels during winter. Therefore, broodstock are over-wintered in heated indoor holding facilities, which provides an opportunity to commence hatchery production of larvae prior to ambient water temperatures stimulating natural breeding. Therefore, advanced juveniles are available to stock ponds when suitable ambient temperatures have been attained. An ideal sex ratio for captive broodstock is up to two BC males plus two to three OC males: 20 females. Mortality is generally high in prawn "maturation" facilities. Therefore, to service future larval requirements, a number of points must be considered: 50% mortality may be experienced during long-term holding (5-8 months), fecundity is low compared to shrimp, and < 10% of females may be carrying eggs at the same stage of development. Recommended stocking densities in broodstock maturation facilities are relatively low at 1 broodstock per 40 l, and negative social interaction can be reduced by the addition of vertical and horizontal substrates (D'Abramo and Brunson, 1996; Wickens and Lee, 2002; Anon., 2003).

The maximum number of eggs produced by prawns seldom exceeds 1000 eggs g^{-1} of body weight and is often only half this amount. Pond-reared females generally produce 10,000-30,000 eggs (D'Abramo and Brunson, 1996; New, 2002). As mentioned previously, females should be selected at the same egg development stage. Reducing variability in the timing of larval hatch ensures that larvae are of similar age and size in culture, allowing the scheduling of husbandry and feed protocols that provide optimal nutrition, minimize water pollution and prevent cannibalism. Following hatch and larval rearing, newly metamorphosed juveniles are cultured in nursery facilities under controlled conditions of temperature, water and feed quality for 1-2 months before they are transferred to outdoor ongrowing ponds during an abbreviated 4-7 month culture season (Tidwell et al., 2002a).

Macrobrachium Culture

Nursery Phase

Nursery culture is prevalent in temperate climates and provides the ability to stock animals at an advanced stage of development, particularly in regions with a short growout season due to low prevailing ambient water temperatures. However, this is not the only advantage of integrated nursery systems. Advanced juveniles are hardier than PL, being able to avoid predation from insect larvae and fish as well as being large enough to be graded, thus reducing the effect of heterogeneous growth of individual males suppressing population growth (New, 2002; Wickens and Lee, 2002; Anon., 2003). When prawn PL metamorphose they weigh approximately 0.01 g and can grow to 0.05-0.1 g after 1 month of nursery culture, and 0.3-0.8 g after a further month. Nursery culture can be conducted either indoors or outdoors. Indoor facilities do not generally exceed a tank size of 50 m^2 with water to 1 m depth, while outdoor nursery culture also may extend to the use of ponds up to 2,000 m^2. Stocking densities are often site and management specific; in general terms culture for 1 month with the addition of mesh substrate to increase tank surface area could be undertaken at densities of up to 2,000 m^{-2}. For longer periods of 2 months, stocking densities should be reduced to < 500 m^{-2} (New, 2002; Anon., 2003). However, there are many examples of greater and also lesser stocking densities being used successfully (Wickens and Lee, 2002).

Growout Phase

Macrobrachium species are cultured under a number of extensive and semi-intensive regimes. Extensive culture systems rely almost solely on wild-caught fry stocked in natural reservoirs, canals or rice fields at densities of 1-4 m^{-2}, where there is no supplementary feeding or fertilization and yields are often much less than 0.5 MT ha^{-1} yr^{-1} (Nandeesha, 2003). A recent study has shown that in rice and prawn dual cropping, the provision of fertilizer and commercial feed for the rice and prawns, respectively, can significantly increase both their yields (Giap et al., 2005). A variation on extensive farming is the polyculture of freshwater prawns with other species as the primary or secondary crop. The complementary species would ideally occupy an alternative ecological niche within the pond food web, thus not competing for food and complementing the culture of prawns. Of all polyculture candidates, tilapia, carp and milkfish appear to be the most likely species studied to date. When prawns were stocked at low density with tilapia, without extra feeding, fish productivity did not change, and prawns were harvested albeit at a smaller size than would have occurred under monoculture (Wickens and Lee, 2002). In some of these systems, the culture of prawns has become so successful that tilapia are no longer cultured and farmers now culture only prawns semi-intensively (New, 2002).

Semi-intensive culture of freshwater prawns is the most common farming practice. In these systems, PL or juveniles can be successfully stocked at rates from 4-20 m^{-2} and achieve yields up to 5.0 MT ha^{-1} yr^{-1}. Hatchery-produced PL or nursery-cultured juveniles are most commonly stocked because of the specific requirements for high numbers of seed stock of similar size and age. Semi-intensive prawn culture provides farmers with a number of challenges primarily in the management of male heterogeneous growth. There are a number of strategies used to minimize the effects of growth retardation by BC males. Grading juveniles on the basis of size and culturing them separately is effective. Juveniles should be separated in a 40:60 ratio of large to small animals and stocked in separate ponds, thus providing a maximum growth potential for animals in both ponds (Tidwell et al., 2005). Large juveniles are destined to become large adults. However, when small, slow-growing juveniles are cultured in the absence of larger animals they also have the capacity for a marked increase in growth rate (Wickens and Lee, 2002; Tidwell et al., 2005). The use of substrate within ponds is also an effective method of increasing surface area, providing shelter during moulting and reducing antagonistic behaviour likely to suppress the growth of prawns. Numerous products both natural and artificial can be used to provide substrate. Natural products such as branches, green vegetation and palm fronds can be effective substrates, although they are subject to decay and are difficult to remove during periodic harvesting procedures. The most popular artificial substrate, plastic barrier fencing, which can be suspended within ponds either vertically or horizontally, does not easily decompose and is easily removed prior to harvest (Tidwell et al., 2002a, b, 2005). Culturing monosex populations can also provide significant growth benefits, with males growing faster and females able to be cultured at a higher density to achieve increased biomass. It may also be viable to culture females in areas that have lower ambient temperatures (25°C instead of 28°C) as this has been shown to slow the onset of maturation allowing metabolic reserves to be partitioned towards somatic growth (Wickens and Lee, 2002; Tidwell et al., 2005). Manual sorting of juveniles into sexes is time consuming; therefore, the use of hormones to alter sex ratios (Sagi and Aflalo, 2005) or the production of functional neo-females (Baghel et al., 2004) may be future avenues of research that could provide monosex populations.

Macrobrachium farming is generally conducted as a batch culture operation. This ensures that all stock and predators are removed, eliminating the possibility of carryover of BC males and predators that would suppress growth in subsequent crops. The design and preparation of ponds is crucial to obtaining high yields. Prior to stocking a pond, a number of preparations should be made. The bottom should be completely drained, dried, cleaned, limed and fertilized, and inlet screens should be checked to ensure that they are able to exclude predators prior to refilling. During culture, water should be monitored for quality, aerated and exchanged as required. In areas where water supply is limited, additional water input may be used only to top up ponds to compensate for evaporation and seepage. When ponds have been filled, it is important to establish an algal bloom to increase pond productivity and the availability of feed for newly stocked animals. The maintenance of a diverse range of pond infauna through regular fertilizing somewhat reduces the requirement for artificial feeds in ponds stocked at low densities without negatively affecting prawn growth (Correia et al., 2003). Some of the most important husbandry practices undertaken during culture are management of feeds and antagonistic suppression of growth by BC males. Feeding can be difficult to manage in prawns; the use of feed trays is not effective because BC males exclude other males from accessing food. Therefore, the provision of feeds is based on estimates of pond biomass and consumption rates of different size-classes.

A number of methods and types of feeds are used to culture prawns. Traditional use of vegetable and animal matter, if carefully monitored, can provide both fertilization and feed. As prawns grow, the balance shifts predominately towards consumption rather than stimulating blooms of pond infauna. In farming systems based on the use of artificial diets, feeding is often commenced a few weeks after stocking juveniles. Animals should then be fed several times a day at a rate equal to 7% of the estimated pond biomass in animals up to 5 g in size, but reduced to approximately 3% for animals > 25 g (D'Abramo and Brunson, 1996; New, 2002). The depression of growth caused by BC males can be managed by ensuring animals of the same age and sizes are stocked in ponds. After they have been grown for 4-5 months, selective harvest and removal of the large individuals in the pond by seine nets will encourage subsequent growth in smaller individuals in the group. Selective harvest once a fortnight will allow relatively homogenous growth of the majority of individuals in the pond (D'Abramo and Brunson, 1996; New, 2002; Wickens and Lee, 2002).

GROWTH IN CRUSTACEANS

Moulting

Crustaceans have an exoskeleton incapable of peripheral expansion or growth once it has hardened during the post-moult interval. Moulting (or ecdysis) facilitates the phenomenon known as staged or discontinuous growth (Bray and Lawrence, 1992). For crustaceans to achieve an increase in size, they must shed their old smaller exoskeleton and expose a new soft exoskeleton capable of expansion. The process can be divided arbitrarily into five discrete sections:

A. Immediately postmoult the exoskeleton is soft and incapable of providing support.

B. Postmoult the exoskeleton is soft but able to support the animal.

C. Intermoult the exoskeleton has functional rigidity.

D. Premoult the new exoskeleton is produced and attachment to the old one deteriorates.

E. Ecdysis involves shedding the old exoskeleton and exposure of a new one.

The exoskeleton consists of a number of component parts layered over the epidermis and connective tissue of crustaceans and is composed of an inner endocuticle, a thicker calcified exocuticle, and an outer thin epicuticle (Dall et al., 1990). The involvement of the epidermis, sequence of formation and physical state of these layers is described in Table 1.

Table 1 Formation and physical nature of the exoskeleton layers during the ecdysis cycle. Ecdysis stages are designated as immediately postmoult (A), postmoult (B), intermoult (C), premoult (D) and ecdysis (E).

Stage	*Epidermis*	*1st Endocuticle*	*1st Exocuticle*	*1st Epicuticle*	*2nd Exocuticle*	*2nd Epicuticle*
A	Present	*----	Soft	Soft	*----	*----
B	Present	Secretion	Hardening	Hardening	*----	*----
C	Present	Hard	Hard	Hard	*----	*----
D	Present	Withdraw	Hard	Hard	Secretion	Secretion
E	Present	Separation	*----	*----	Soft	Soft

*---- signifies absence.

The expansion of the new exoskeleton during phase A of the ecdysis cycle is achieved by increasing the hemolymph volume through water influx across the epidermis, gills and gut regions. Many complex and energy-demanding biochemical processes occur post- and premoult involving the mobilization of hepatopancreas lipid reserves, the concentration and reabsorption of polysaccharides and minerals from the old endocuticle, ion scavenging, particularly calcium and bicarbonates in fresh water, and protein production to replace fluid uptake accumulated immediately postmoult (Dall et al., 1990; Chang, 1992). During phase D and just prior to phase E, the 1st endocuticle begins to separate from the 2nd epicuticle, the 1st endocuticle is partly reabsorbed, and there is an increased concentration of moulting hormone in the hemolymph (Chang, 1992).

Ecdysis lasts only a few minutes. The old exoskeleton opens dorsally adjacent to the thorax and abdomen allowing the animal to flick out. During this period crustaceans are vulnerable to cannibalization and they often minimize this risk by seeking shelter (Chang, 1992). During the hours and days postmoult, the exoskeleton hardens with no further size increases evident until after the next ecdysis. In crustaceans, the completion of ecdysis in larval stages may take hours, in PL it may take days, and in mature adults it may take several weeks to months or years. Growth rate is a function of the intermoult duration and the size of the moult increment; young crustaceans have short intermoult durations that increase with age (Bray and Lawrence, 1992). Therefore, a daily PL moulting pattern will expose vulnerable animals on a daily basis to changing environmental conditions compared to the infrequent exposure of older animals with longer intermoult durations. Environmental fluctuations that may adversely affect shrimp larvae include sudden changes in salinity, pH, dissolved oxygen, exposure to viral or bacterial pathogens and feed availability. It is important to maintain stable culture conditions for larvae.

Dietary Requirements

Feeding Behaviour

When shrimps and prawns are cultured in super-intensive, intensive and to a lesser extent semi-intensive systems the attractiveness, stability and composition of compound diets can profoundly affect dietary intake. Crustaceans are messy eaters. When they have located their food, they manipulate and move it with their pereiopods to their mouthparts (i.e., the maxillipeds, maxillae, mandible, paragnath and labrum), where food is often rejected prior to complete crushing, sorting, mastication and consumption (Garm et al., 2003). The process of feeding and mastication largely occurs externally in crustaceans; therefore, small food particles are often lost from the mouthparts and essential nutrients are leached, reducing the growth potential and water quality they are exposed to. The appropriately sized pellet (artificial diet) for crustaceans is generally small with a large ratio of surface area to volume, so it is susceptible to nutrient leaching if not consumed quickly. The primary determinant of how quickly a pellet is located and consumed is its attractiveness; if this is low, consumption will be slow and nutrient loss through leaching will be high.

Protein Requirements

Shrimps and prawns are considered omnivores, although within this classification there are extremes that affect the potential nutritional composition of the diet, particularly optimum protein levels. Protein is the most important macronutrient required for growth, especially during larval stages (Jones et al., 1997a, b), where the specific growth rate is high and large losses of protein may occur through the loss of exuviae at ecdysis. A number of crustaceans preferentially use protein as an energy source for normal metabolic functions (Guillaume, 1997). *Marsupenaeus japonicus* is primarily a carnivorous omnivore requiring a high dietary protein content of 42-55% to facilitate metabolism and growth (Koshio et al., 1993; Guillaume, 1997). Other crustaceans such as *L. vannamei* and *M. rosenbergii* can be cultured on a diet favouring herbivores with a protein inclusion of < 30%. Growth has been demonstrated in *M. rosenbergii* with dietary protein inclusion as low as 13% (D'Abramo and New, 2000). In animals that can survive on low protein diets, it is thought that almost all the protein is partitioned towards growth, rather than utilized for energy (Tacon and Akiyama, 1997; Guillaume, 1997). There are suggestions that even though *M. rosenbergii* survive adequately on a low protein diet, they still possess the ability to use protein as an energy source demonstrating better growth on a high protein, low lipid diet formulation (44% protein, 9% lipid) (Cavalli et al., 2001). More herbivorous species such as *M. rosenbergii* benefit from the inclusion of high fibre content in the diet (10%) increasing gut retention times and thus assisting in the absorption of nutrients (Gonzalez-Pena et al., 2002).

The requirement of protein for growth is not for protein *per se* but for the building blocks of protein, the amino acids (Guillaume, 1997). Ten amino acids have been identified as essential for crustacean growth: arginine, methionine, valine, threonine, isoleucine, leucine, lysine, histidine, phenylalanine and tryptophan (Kanazawa and Teshima, 1981). Compound feeds for crustaceans also include a range of lipids, carbohydrates, vitamins, minerals, carotenoids and attractants, with many species-specific requirements noted (Harrison, 1990, 1997). In general terms, the more intensive the culture system the more dependent the animals will be on the artificial feed supplied for all of their dietary requirements. Therefore, it is likely that the dietary requirements for shrimps will have to be more exacting than those for prawns, which generally are cultured less intensively because of their disposition towards hierarchical growth-limiting behaviour.

Acknowledgements

We thank Casey Smith for producing the line drawings used in this chapter.

References

Alva, V.R., A. Kanazawa, S.I. Teshima and S. Koshio. 1993a. Effect of dietary phospholipids and n-3 highly unsaturated fatty acids on ovarian development of Kuruma prawns. Bull. Jap. Soc. Sci. Fish., 59: 345-351.

Alva, V.R., A. Kanazawa, S.I. Teshima and S. Koshio. 1993b. Effect of dietary vitamins A,E, and C on the ovarian development of *Penaeus japonicus*. Bull. Jap. Soc. Sci. Fish., 59: 1235-1241.

Anon. 1983. Constitution of broodstock, maturation, spawning, and hatching systems for penaeid shrimps in the Centre Oceanologique Du Pacifique. *In*: CRC Handbook of Mariculture, Vol. 1, Crustacean Aquaculture. J.P. McVey (ed.). CRC Press, Boca Raton, Florida, pp. 105-127.

Anon. 2003. Culture of freshwater prawns in temperate climates: Management practices and economics. Bulletin 1138. Mississippi Agricultural and Forestry Experiment Station, 23 pp.

Baghel, D.S., Lakra, W.S., Satyanarayana Rao, G.P. 2004. Altered sex ratio in gaint freshwater prawn, *Macrobrachium rosenbergii* (de Man) using hormone bioencapsulated live *Artemia* feed. Aquacult Res., 35: 943-947.

Bailey-Brock, J.H., and S.M. Moss. 1992. Penaeid taxonomy, biology and zoogeography. *In:* Developments in Aquaculture and Fisheries Science, 23. Marine Shrimp Culture; Principles and Practices. A.W. Fast and L.J. Lester (eds.). Elsevier, Amsterdam, pp. 9-27.

Beard, T.W., and J.F. Wickens. 1980. Breeding of *Penaeus monodon* Fabricius in laboratory recirculation systems. Aquaculture, 20: 79-89.

Boyd, C.E., and Y. Musig. 1992. Shrimp pond effluents: Observations of the nature of the problem on commercial farms. *In*: Proceedings of the Special Session on Shrimp Farming. J. Wyban (ed.). pp. 195-197. World Aquaculture Society, Baton Rouge, Louisiana.

Bray, W.A., and A.L. Lawrence. 1992. Reproduction of Penaeus species in captivity. *In:* Developments in Aquaculture and Fisheries Science, 23. Marine Shrimp Culture; Principles and Practices. A.W. Fast and L.J. Lester (eds.). Elsevier, Amsterdam, pp. 93-170.

Briggs, M., S. Funge-Smith, R. Subasinghe and M. Phillips. 2004. Introductions and movement of *Penaeus vannamei* and *Penaeus stylirostris* in Asia and the Pacific. Food and Agriculture Organization of the United Nations, Regional office for Asia and the Pacific, Bangkok. RAP publication 2004/10: 88 pp.

Browdy, C. 1992. A review of the reproductive biology of *Penaeus* species: Perspectives on controlled shrimp maturation systems for high quality nauplii production. *In:* Proceedings of the Special Session on Shrimp Farming. J. Wyban (ed.). World Aquaculture Society, Baton Rouge, Louisiana, pp. 22-51.

Buranakanonda, A. 2002. Kasetsart University to improve giant freshwater prawn broodstock. Asian Aquaculture Magazine, May/June: 15-16.

Cavalli, R.O., P. Lavens and P. Sorgeloos. 2001. Reproductive performance of *Macrobrachium rosenbergii* females in captivity. J. World Aquacult. Soc., 32: 60-67.

Chang, E.S. 1992. Endocrinology. *In:* Developments in Aquaculture and Fisheries Science, 23. Marine Shrimp Culture; Principles and Practices. A.W. Fast and L.J. Lester (eds.). Elsevier, Amsterdam, pp. 53-91.

Chanratchakool, P., Corsin, F., Briggs, M. 2005. Better management practices (BMP) Manual for black tiger shrimp (*Penaeus monodon*) hatcheries in Viet Nam. http://library.enca.org/shrimp/publications/BMP_for_hatcheries_EN_pdf.

Cheng, W., C.-H. Liu, C.-H. Cheng and J.-C. Chen. 2003. Osmolality and ion balance in giant river prawn *Macrobrachium rosenbergii* subjected to changes in salinity: role of sex. Aquacult. Res., 34: 555-560.

Correia, E.S., J.A. Pereira, A.P. Silva, A. Horowitz and S. Horowitz. 2003. Growout of freshwater prawn *Macrobrachium rosenbergii* in fertilized ponds with reduced levels of formulated feed. J. World Aquacult. Soc., 34: 184-191.

D'Abramo, L.R., and M.W. Brunson. 1996. Biology and life history of freshwater prawns. United States Department of Agriculture, Cooperative State

Research, Education, and Extension Service. Southern Regional Aquaculture Center Publication 483.

D'Abramo, L.R., and M.B. New. 2000. Nutrition, feeds and feeding. *In:* Freshwater Prawn Culture. M.B. New and W.C. Valenti (eds.). Blackwell Science, Oxford.

Dall, W., B.J. Hill, P.C. Rothlesberg and D.J. Sharples. 1990. The biology of the Penaeidae. *In:* Advances in Marine Biology, Vol. 27. J.H.S. Blaxter and A.J. Southward (eds.). Academic Press, London.

Fujimura, T., and H. Okamoto. 1972. Notes on progress made in developing a mass culturing technique for *Macrobrachium rosenbergii* in Hawaii. *In:* Coastal Aquaculture in the Indo-Pacific Region. T.V.R. Pillay (ed.). Fishing News Ltd., West Byfleet, pp. 313-327.

Garm, A., E. Hallberg and J.T. Høeg. 2003. Role of maxilla 2 and its setae during feeding in the shrimp *Palaemon adspersus* (Crustacea: Decapoda). Biol. Bull., 204: 126-137.

Giap, D.H., Y. Yi and C.K. Lin. 2005. Effects of different fertilization and feeding regimes on the production of integrated farming of rice and prawn *Macrobrachium rosenbergii* (De Man). Aquacult. Res., 36: 292-299.

Gonzalez-Pena, M.C., A.J. Anderson, D.M. Smith and G.S. Moreira. 2002. Effect of dietary cellulose on digestion in the prawn *Macrobrachium rosenbergii*. Aquaculture, 211: 291-303.

Guillaume, J. 1997. Protein and amino acids. *In:* Crustacean Nutrition: Advances in World Aquaculture, Vol. 6. L. D'Ambro, D.E. Conklin and D.M. Akiyama (eds.). World Aquaculture Society Ltd., Baton Rouge, Louisiana, pp. 26-50.

Hall, M.R., N. Young and M. Kenway. 2000. Manual for the determination of egg fertility in *Penaeus monodon*. http://www.aims.gov.au/pages/research/mdef/mdef-00.html

Hansford, S.W., and G.E. Marsden. 1995. Temporal variation in egg and larval productivity of eyestalk ablated spawners of the prawn *Penaeus monodon* from Cook Bay, Australia. J. World Aquacult. Soc., 26: 396-405.

Harrison, K.E. 1990. The role of nutrition in maturation, reproduction and embryonic development of decapod crustaceans: a review. J. Shellfish Res., 9: 1-28.

Harrison, K.E. 1997. Broodstock nutrition and maturation diets. *In:* Crustacean Nutrition: Advances in World Aquaculture, Vol. 6. L. D'Ambro, D.E. Conklin and D.M. Akiyama (eds.). World Aquaculture Society Ltd., Baton Rouge, Louisiana, pp. 390-408.

Holthuis, L.B. 1980. FAO species catalogue, Vol. 1. Shrimps and prawns of the world. An annotated catalogue of species of interest to fisheries. FAO Fisheries Synopsis 1, 1-261.

Huberman, A. 2000. Shrimp endocrinology. A review. Aquaculture, 191: 191-208.

Hudinaga, M. 1935. Studies on the development of *Penaeus japonicus* (Bate). Report. Hayatomo Fish. Inst., 1(1).

Hudinaga, M. 1942. Reproduction development and rearing of *Penaeus japonicus* Bate. Jap. J. Zool., 10: 305-393.

Jones, D.A., M. Kumlu, L. Le Vay and D.J. Fletcher. 1997a. The digestive physiology of herbivorous, omnivorous and carnivorous crustacean larvae: a review. Aquaculture, 155(1-4): 285-295.

Jones, D.A., A.B. Yule and D.L. Holland. 1997b. Larval nutrition. *In:* Crustacean Nutrition: Advances in World Aquaculture, Vol. 6. K. Akiyama, D.E. Conklin and L. D'Abramo (eds.). World Aquaculture Society, Baton Rouge, Louisiana, pp. 353-389.

Josupeit, H. 2006. Aquaculture production and markets. FAO Globefish. www.globefish.org/

Kanazawa, A., and S. Teshima. 1981. Essential amino acids of the prawn. Bull. Jap. Soc. Sci. Fish., 47: 1375-1377.

Koshio, S., S.I. Teshima, A. Kanazawa and T. Watanabe. 1993. The effect of dietary protein content on growth digestion efficiency and nitrogen excretion of juvenile kuruma prawns, *Penaeus japonicus*. Aquaculture, 113: 101-114.

Kuo, G.-H., and C.-F. Lo. 1998. Practical strategies for combating shrimp White Spot Syndrome Virus. Joint Taiwan Aquaculture and Fisheries Management Resources and Management Forum. 2-8 November 1998. Taiwan Fisheries Research Institute, Keelung, Taiwan.

Law, A.T., Y.H. Wong and A.B. Abol-Munafi. 2002. Effect of hydrogen ion on *Macrobrachium rosenbergii* egg hatchability in brackish water. Aquaculture, 214: 247-251.

Ling, S.W. 1969a. The general biology and development of *Macrobrachium rosenbergii*. FAO Fish. Rep., 57: 589-606.

Ling, S.W. 1969b. Methods of rearing and culturing *Macrobrachium rosenbergii*. FAO Fish. Rep., 57: 607-619.

Misamore, M.J., and C.L. Browdy. 1996. Mating behavior in the blue shrimps *Penaeus setiferus* and *P. vannamei*: a generalized model for mating in *Penaeus*. J. Crustacean Biol., 16: 61-70.

Motoh, H. 1981. Fisheries Biology of *Penaeus monodon*. Technical Report No. 7. Aquaculture Dept, Southeast Asian Dept. Centre, Tigban, Iloilo, Philippines.

Motoh, H. 1985. Biology and ecology of *Penaeus monodon*. *In:* Proceeding of the First International Conference on the Culture of Penaeid Prawns/Shrimp. Y. Taki, J.H. Primavera and J.A. Llobrera (eds.). Aquaculture Department Southeast Asian Fisheries Development Center, Iloilo City, The Philippines, pp. 27-36.

Naessens, E., P. Lavens, L. Gomez, C.L. Browdy, K. McGoven-Hopkins, A.W. Spencer, D. Kawahigashi and P. Sorgeloos. 1997. Maturation performance of *Penaeus vannamei* co-fed *Artemia* biomass preparations. Aquaculture, 155: 87-101.

Nair, C.M., M.B. New, M.N. Kutty, P.B. Mather and D.D. Nambudiri. 2005. Editorial Freshwater Prawns 2003 – special issue on the international symposium on freshwater prawns. Aquacult. Res., 36: 209.

Nandeesha, M.C. 2003. Commercialization of giant freshwater prawn culture in India by farmers. Aquacult. Asia, 8: 29-33.

New, M.B. 1995. The status of freshwater prawn farming: a review. Aquacult. Res., 26:1-54.

New, M.B. 2002. Farming freshwater prawns. A manual for the culture of the giant river prawn (*Macrobrachium rosenbergii*). FAO Fisheries Technical Paper 428, Food and Agriculture Organization of the United Nations.

New, M.B. 2004. Cultured Aquatic Species Information Programme – *Macrobrachium rosenbergii*. Cultured Aquatic Species Fact Sheets. FAO Inland Water Resources and Aquaculture Service. http://www.fao.org/figis/servlet/static?dom=culturespecies&xml=Macrobrachium_rosenbergii. Updated Thu Jan 26 10:59:22 CET 2006.

New, M.B. 2005. Freshwater prawn farming: global status, recent research and a glance at the future. Aquacult. Res., 36: 210-230.

Palacios, E., A.M. Ibarra, J.L. Ramírez, G. Portillo and I.S. Racotta. 1998. Biochemical composition of egg and nauplii in White Pacific Shrimp, *Penaeus vannamei*, in relation to the physiological condition of spawners in a commercial hatchery. Aquacult. Res., 29: 183-189.

Panouse, J.B. 1943. Influence de l'ablation du pedoncle oculaire sur la croissance de l'ovaire chez la crevette *Leander serratus*. C.R. Acad. Sci. Paris, 217: 553-555.

Peixoto, S., R.O. Cavalli and W. Wasielesky. 2005. Recent developments on broodstock maturation and reproduction of *Farfantepenaeus paulensis*. Brazil. Arch. Biol. Technol., 48: 997-1006.

Primavera, J.H. 1979. Notes on the courtship and mating behaviour in *Penaeus monodon* Fabricius (Decapoda: Natantia). Crustaceana, 37: 287-292.

Primavera, J.H. 1983. Broodstock of Sugpo, *Penaeus monodon* Fabricus. Southeast Asian Fisheries Development Center, Extension Manual 7. Aquaculture Department SEAFDEC, Iloilo, The Philippines, 26 pp.

Primavera, J.H. 1984. A review of maturation and reproduction in closed-thelycum Penaeids. *In:* Proceedings of the First International Conference on the Culture of Penaeid Prawns/Shrimp. Y. Taki, J.H. Primavera and J.A. Llobrera (eds.). Aquaculture Department Southeast Asian Fisheries Development Center, Iloilo City, The Philippines, pp. 47-64.

Primavera, J.H. 1988. Biology and Culture of *Penaeus monodon*. Extension Manual No. 7, 3rd ed. Aquaculture Department, Southeast Asian Fisheries Development Center. Tigbauan, Iloilo, The Philippines.

Quynh, V.D. 1992. Shrimp culture industry in Vietnam. *In:* Marine Shrimp Culture: Principles and Practices. A.W. Fast and L.J. Lester (eds.). Elsevier Science Publishers, New York. B.V., New York pp. 729-756.

Racotta, I.S., E. Palacios and A.M. Ibarra. 2003. Shrimp larval quality in relation to broodstock condition. Aquaculture, 227: 107-130.

Robertson, L., B. Bray, T. Samocha and A. Lawrence. 1993. Reproduction of penaeid shrimp, an operations guide. *In:* CRC Handbook of Mariculture, 2nd ed., Vol. 1, Crustacean Aquaculture. J.P. McVey (ed.). CRC Press, Inc., Boca Raton, Florida, pp. 107-132.

Rothlisberg, P.C. 1998. Aspects of penaeids biology and ecology of relevance to aquaculture: a review. Aquaculture, 164: 49-65.

Sagi, A., and E.D. Aflalo. 2005. The androgenic gland and monosex culture of freshwater prawn *Macrobrachium rosenbergii* (De Man): a biotechnological perspective. Aquacult. Res., 36: 231-237.

Samocha, T.M., A.L. Lawrence and W.A. Bray. 1992. Design and operation of an intensive raceway system for penaeids. *In:* CRC Handbook of Mariculture, 2nd ed., Vol. 1, Crustacean Aquaculture. J.P. McVey (ed.). CRC Press, Inc., Boca Raton, Florida, pp. 173-210.

Sellars, M.J., G.J. Comon and D.T. Morehead. 2005. Tolerance of *Penaeus (Marsupenaeus) japonicus* embryos to ozone disinfection. Aquaculture, 245: 111-119.

Short, J.W. 2004. A revision of Australian river prawns, *Macrobrachium* (Crustacea: Decapoda: Palaemonidae). Hydrobiologia, 525: 1-100.

Smith, G.G., A.J. Ritar and M.R. Brown. 2004. Uptake and metabolism of a particulate form of ascorbic acid by *Artemia* nauplii and juveniles. Aquacult. Nutr., 10: 1-8.

Tacon, A.G., and D.M. Akiyama. 1997. Feed ingredients. *In:* Crustacean Nutrition: Advances in World Aquaculture, Vol. 6. L. D'Ambro, D.E. Conklin and D.M. Akiyama (eds.). World Aquaculture Society Ltd., Baton Rouge, Louisiana, pp. 411-472.

Tan-Fermin, J.D., and R.A. Pudadera. 1989. Ovarian maturation stages of the wild giant tiger prawn, *Penaeus monodon* Fabricus. Aquaculture, 77: 229-242.

Tidwell, J.H., S. Coyle, R.M. Durborow, S. Dasgupta, W.A. Wurts, F. Wynne, L.A. Bright and A. Van Arnum. 2002a. Kentucky State University Prawn Production Manual. Kentucky State University Aquaculture Program, 40 pp.

Tidwell, J.H., S.D. Coyle, A. Van Arnum and C. Weibel. 2002b. Effects of substrate amount and orientation on production and population structure of freshwater prawns *Macrobrachium rosenbergii* in ponds. J. World Aquacult. Soc., 33: 63-69.

Tidwell, J.H., L.R. D'Abramo, S.D. Coyle and D. Yasharian. 2005. Overview of recent research and development in temperate culture of the freshwater prawn (*Macrobrachium rosenbergii* De Man) in the South Central United States. Aquacult. Res., 36: 264-277.

Wickens, J.F., and D. O'C. Lee. 2002. Crustacean Farming: Ranching and Culture, 2nd ed. Blackwell Science, Mead, Oxford, 446 pp.

Wurts, W.A., and R.R. Stickney. 1984. An hypothesis on the light requirements for spawning Penaeid shrimp, with emphasis on *Penaeus setiferus*. Aquaculture, 41: 93-98.

Wyban, J.A., and J.N. Sweeney. 1991. Intensive Shrimp Production Technology, The Oceanic Insititute Shrimp Manual. The Oceanic Institute, Honolulu, Hawaii, 158 pp.

Xu, X.I., W.J. Ji, J.D. Castell and R.K. O'Dor. 1992. Influence of dietary lipid sources on fecundity, egg hatchability and fatty acid composition of Chinese prawn (*Penaeus chinensis*) broodstock. Aquaculture, 119: 359-370.

Xu, X.I., W.J. Ji, J.D. Castell and R.K. O'Dor. 1994. Effect of dietary lipid sources on fecundity, egg hatchability and fatty acid composition of Chinese prawn (*Penaeus chinensis*) broodstock. Mar. Fish. Res., 13: 13-19.

Yano, I., R.A. Kanna, R.N. Oyama and J.A. Wyban. 1988. Mating behavior in the penaeid shrimp *Penaeus vannamei*. Mar. Biol., 97: 171-175.

13

The Crustacean *Nephrops norvegicus*: Growth and Reproductive Behaviour

C.J. Smith and K.N. Papadopoulou
Institute of Marine Biological Resources, Hellenic Centre for Marine Research (HCMR), P.O. Box 2214, Heraklion 71003, Crete, Greece. E-mail: csmith@her.hcmr.gr

THE SPECIES

Nephrops norvegicus (Linnaeus, 1758) (Fig. 1), known as Norway lobster, scampi, langoustine and Dublin Bay prawn, is a malacostracan decapod crustacean within the Family Nephropidae, which also includes clawed lobsters of the genus *Homarus*. The Norway lobster is the only species within the genus *Nephrops*, so hereafter it is referred to by genus alone. There are a number of closely related and similar-looking species within the genus *Metanephrops*, for example, the New Zealand lobster *Metanephrops challengeri* (Balss, 1914), which lives in the Indo-West Pacific region (Holthuis, 1991). *Nephrops* is found throughout the continental shelf and upper slope of the northeastern Atlantic and the Mediterranean, from Iceland and Norway in the north to Morocco and Greece in the South (Farmer, 1975; Fischer et al., 1987; Holthuis, 1991; Abello et al., 2002). It is absent from the Baltic Sea, the Black Sea and the Levantine Sea. Its occurrence is restricted to certain types of sediment with a component of silt and clay, ranging from muddy gravel through muddy sand and sandy mud to mud. Although *Nephrops* density is related to sediment

Fig. 1 *Nephrops norvegicus*.

grain size, this relationship is not linear (Bell et al., 2006). In the Mediterranean it is commonly found on muddy bottoms at 400 m with a range of 200-700 m depth, although in some areas characterized by a wide continental shelf such as the Adriatic Sea or close to the Ebro delta in the western Mediterranean *Nephrops* is found in depths of less than 100 m (Abelló et al., 2002). In the Mediterranean the upper depth distribution appears to be dependent on temperature, requiring waters that do not exceed 16°C. In the northeastern Atlantic and northern and northwestern Iberian Atlantic, *Nephrops* is found in depths ranging from 90 to 600 m (Fariña and González Herraiz, 2003).

LIFESTYLE

Nephrops is a non-migratory burrowing species with juveniles and adults living within the sediment in often complex burrow structures. Burrows range from simple unbranched tunnels to complex multi-branched systems with many openings (Fig. 2). Most commonly burrows have two or three openings. The most complex burrows are those in which the burrows of juveniles interconnect with that of an adult. Underwater video, diving observations, burrow-casting, and field and laboratory experimentation have been used to study its burrows and lifestyle, including aspects of burrow construction and maintenance, emergence patterns and foraging behaviour (Rice and Chapman, 1971; Chapman and Rice, 1971; Marrs et al., 1996; Smith and Papadopoulou, 2003; Smith et al., 2003; Aguzzi et al., 2004a).

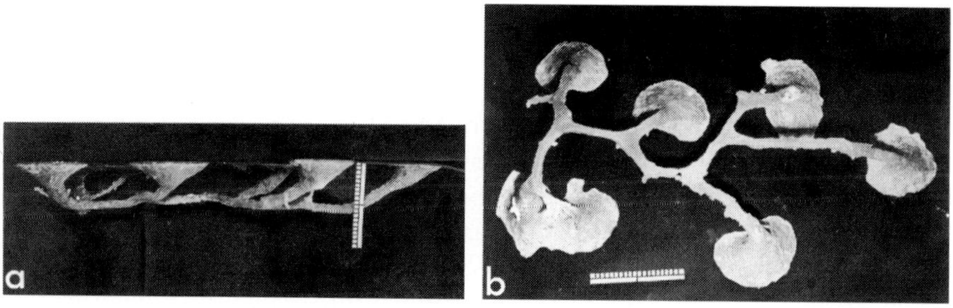

Fig. 2 Resin cast of large complex *Nephrops* burrow from Camas Nathais in the west coast of Scotland. Scale bar shows 1 cm divisions (source: Smith, 1988).

Emergence is probably regulated by both endogenous activity rhythms (Atkinson and Naylor, 1976) and exogenous mechanisms such as light and availability of prey species (Aguzzi et al., 2003; Bell et al., 2006). Emergence varies daily with ambient light level, in relation to the lunar tidal cycle and seasonally with light intensity and spectral quality (number of photons per unity of time and wavelength range), photoperiod duration, reproductive state (females only, with reduced foraging of ovigerous females during incubation) and moulting (both sexes, with different frequency and/or at different times) (Aguzzi et al., 2003; Bell et al., 2006). Diurnal activity patterns are clearly depth-related. In shallow waters (< 30 m) peak emergence occurs during nighttime, at intermediate depths at dusk and dawn and at slope depths during daytime. Diurnal and seasonal rhythms affect catchability and result in different exploitation patterns and stock assessments for males and females (Fariña and González Herraiz, 2003; Aguzzi et al., 2004b).

COMMERCIAL IMPORTANCE

The *Nephrops* resource has a high social and economic relevance with more than 30 individual stocks regularly assessed by International Council for the Exploration of the Sea (ICES) in the northeastern Atlantic (Graham and Ferro, 2004). Advice for the Mediterranean comes through the General Fisheries Commission for the Mediterranean (GFCM). The advice and assessments of ICES and GFCM are considered at the EU level, through the work of the Scientific, Technical and Economic Committee for Fisheries (STECF). *Nephrops* is of high commercial importance in both the northeastern Atlantic and the Mediterranean (Sardà, 1998; Abelló et al., 2002; Fariña and González Herraiz, 2003; Aguzzi et al., 2004a; Catchpole et al., 2006). In Europe, more has probably been written about *Nephrops* than about any other decapod crustacean (Sardà, 1998).

The importance of *Nephrops* varies considerably between fisheries (Graham and Ferro, 2004). In 2005 the total European reported landings (source: FAO) were 60,500 t, with the UK accounting for 50% of EU *Nephrops* landings and Italy for 80% of Mediterranean *Nephrops* landings. In some areas, *Nephrops* is the main component, with other species considered as bycatch, the opposite being true elsewhere. The current EU regulation framework in *Nephrops* fisheries differs between regions and especially so between the Mediterranean and the rest of the EU fishing areas. Differences include variations in minimum landing sizes (e.g., 20 mm carapace length (CL) in the Mediterranean, 25 mm in the North Sea), minimum mesh sizes (e.g., 40 mm in the Mediterranean, 70-120 mm elsewhere in Europe), bycatch limitations and use of selectivity devices to moderate bycatch (e.g., square mesh codends, grids).

The most widespread fishing method is otter trawling, with *Nephrops* usually being caught along with gadoids such as whiting (*Merlangius merlangus*) in the North Sea and hake (*Merluccius merluccius*) and rose shrimp (*Parapenaeus longirostris*) in the Mediterranean (Sardà, 1998; Bergmann et al., 2002; Catchpole et al., 2006) (Fig. 3). In parts of Scotland, the Skaggerak and Kattegat in Sweden, and more recently in Greece and Italy, *Nephrops* fisheries use baited pots, traps or creels (Fig. 3) (Eggert and Ulmestrand, 1999; Papadopoulou et al., 2006). Although the tonnage caught is much less than in the trawl fisheries, the value of creel-caught *Nephrops* is high, particularly in the case of animals marketed live. In some localized small-scale artisanal Mediterranean gillnet fisheries, a configuration is used with a reduced number of floats on the net and bait is attached to it; this rigging allows the net to lie on the seabed while the bait attracts *Nephrops* that become entrapped in the mesh.

Fig. 3 Mixed trawl catch from deep waters and *Nephrops* trap catch.

REPRODUCTIVE SYSTEM

The reproduction of *Nephrops* has been studied throughout its geographical range using both field and experiment approaches (Eriksson, 1970; Farmer, 1975; Figueiredo et al., 1982; Sardà, 1995; Tuck et al., 1997a; Orsi-Relini et al., 1998; Fariña et al., 1999; Briggs et al., 2002; McQuaid et al., 2006).The reproductive system of *Nephrops* is basically similar to that of other macruran decapods (Farmer, 1974a). Both male and female systems were studied in detail in the 1960s and 1970s and more recent work focused on studies of fecundity and maturity. Farmer (1974a) gives extensive descriptions of the reproductive systems of male and female *Nephrops*, the development of spermatozoa and oocytes, and copulation behaviour.

The Male System

A brief presentation of the male system is given here based on Farmer's (1974a) work. The paired testis is H-shaped with two anterior arms extending around the foregut to the regions just behind the cephalic ganglion, and the two posterior arms extending backwards into the abdominal segments, where they lie between the dorso-lateral abdominal muscles along the midgut. The paired vasa deferentia (each composed of three sections) arise on the outside of the posterior arms of the testis and open to the coxopodites of the fifth pereiopods. The genital apertures are normally closed by a thin membrane, which becomes ruptured when the spermatophores are released. The spermatophores are produced in the secretory region of the vas deferens and are stored in the sphincter muscle and ejaculatory regions until copulation takes place. The spermatophore is composed of a single convoluted strand of spermatozoa enclosed in a thick membrane (Farmer, 1974a). Drawings and photos of histological sections are provided by Farmer (1974a) and McQuaid et al. (2006).

The Female System

A brief presentation of the female system is given here based on Farmer's (1974a) work. The ovary is H-shaped, and the anterior and posterior arms project into the cephalic region and the abdominal segments, respectively. An oviduct arises at approximately the mid-point on each side of the ovary just posterior to the connection between the two halves. These open on the coxopodites of the third pereiopods. The genital apertures are covered by thin membranes that are ruptured at egg-laying. The thelycum, an invagination of the exoskeleton, is a cavity on the underside of the thoracic segments composed of the fused sternites, which accommodates the spermatophore after copulation (Farmer, 1974a). The inseminated females carry the spermatophore through ovarian maturation until egg-laying. The exact mechanism for fertilization is still unknown. Farmer (1974a) suggests that fertilization must occur internally, through the temporary formation of tubules from the thelycum to the oviduct. Drawings and photos of histological sections are provided by Farmer (1974a) and McQuaid et al. (2006).

COPULATION

Copulation behaviour was studied in the early 1970s in animals kept in captivity in aquaria. Farmer (1974a) details this behaviour along with photographic records. The male lobster searches for the female, digs in the substratum as part of a cleaning activity, strokes the female with his antennae, approaches from the rear, straddles and turns the female on her back, and adjusts her position until the thelycum is opposite the first pair of pleopods, which together with the second pleopods are lowered in the copulatory position. The tips of the first pair of pleopods enter the thelycum and a spermatophore is forced into the thelycum by sliding the appendices masculinae (on the second pair of pleopods) along the grooves on the inside of the first pleopods. Penetration lasts no more than 2 seconds. The pair is separated immediately after penetration and the male usually shows no further interest in the female.

MATING

Copulation occurs between mature males and sexually mature females that have recently moulted (within a few days during the "soft" phase). Placement of a male in a tank from which a newly moulted female has been removed elicited an immediate response, suggesting the production of a pheromone by the moulting female (Farmer, 1974a). Mating in mature females occurs after larvae have hatched from the previous batch of eggs carried on the pleopods, this being followed by a moult preceding copulation. Females are cryptic when soft but then become a significant part of the fishery during the time they are intensively feeding before egg-laying and returning to their burrows during the egg-bearing period. Latitudinal variation has been seen in the occurrence and duration of these processes, with environmental temperature probably being the main determining factor (Sardà, 1995; Bell et al., 2006).

Multiple mating and multiple paternity in wild *Nephrops* stocks off the west coast of Portugal were recently shown to occur via DNA analyses (Streiff et al., 2004). The causes involved in the process (i.e., multiple mating during the short soft phase and, although traditionally considered impossible, intermoult mating or sperm storage over breeding seasons) are currently unknown. Multiple paternity has been suspected on the basis of captivity behavioural studies (e.g., the European lobster *Homarus gammarus*) and/or observed earlier in other lobster and decapod populations (e.g., the American lobster *H. americanus*).

SPERMATOGENESIS

Several studies have been done on spermatogenesis in *Nephrops*. These include the work of Farmer (1974a) and Mc Quaid et al. (2006) in the Irish Sea, Mouat (2002) in the Clyde Sea, west coast of Scotland and northern North Sea and Figueiredo and Barraca (1963) in Portuguese waters. These studies suggest that spermatogenesis occurs throughout the year and spermatophores are carried in the vasa deferentia at all times. From recent studies in the Clyde Sea, it would appear that male *Nephrops* are not able to increase production in the testes and therefore do not show a peak in reproductive output to coincide with the main period of female moulting (Mouat, 2002). The presence of spermatophores in the vas deferens of males and in the thelycum of females has been used as an indicator of sexual maturity (physiological maturity). Previous studies (Farmer, 1974a) reported male *Nephrops* smaller than 18 mm CL may not be able to copulate successfully: although having fully developed spermatozoa in the tubules of the testis, they did not have spermatophores in the vasa deferentia. Recently, physiologically mature male *Nephrops* as small as 15 mm CL have been recorded (McQuaid et al., 2006). However, there is no evidence to suggest that physiologically mature small males take part in the reproductive output of the population (Mouat, 2002).

OVARY AND OOCYTES

In contrast to the paucity of work carried out on males, many studies have been carried out on the reproductive biology of females, including research on the ovary maturation cycle, and on ovary and oocyte staging (Figueiredo and Barraca, 1963; Fontaine and Warluzel, 1959; Dunthorn, 1967; Thomas, 1964; Farmer, 1974a; Chapman, 1980; Orsi-Relini et al., 1998; Fariña et al., 1999; Smith et al., 2001). Ovary staging (with some variations between studies) is routinely based on ovary colour (as visible through the back of the carapace). The validity of this method is based on work by Farmer (1974a), with subsequent refinements (see Chapman, 1980). White (in non-preserved specimens) stage 1 ovaries are found in juveniles. Stage 2 ovaries are cream in colour, denoting early maturation.

Stage 3 ovaries are pale green, this being visible through the cephalothorax, and stage 4 ovaries are dark green, clearly visible through the exoskeleton and extending into the abdomen to near the point of extrusion on to the pleopods. A further category may be recognized in some animals that represents resorbing ovaries (Bailey, 1984). Ovigerous (berried) females are occasionally staged on the basis of egg development. This may be done on the basis of colour alone (from dark green to brownish) or may take account of embryonic development stages (e.g., presence of yolk, appendages, eye pigments, segmentation), the number of stages recognized varying between studies (Fontaine and Warluzel, 1959; Dunthorn, 1967; Fariña et al., 1999; Smith et al., 2001).

SEASONAL CYCLES

The seasonality of the reproduction of *Nephrops* is well studied across its geographic range (Farmer, 1974a; Chapman, 1980; Orsi-Relini and Relini, 1989; Sardà, 1995; Orsi-Relini et al., 1998; Fariña et al., 1999; Smith et al., 2001; Bell et al., 2006). In the Mediterranean, differences in the timing of ovarian maturation and laying of the eggs are attributed to depth (shelf/slope habitats) and latitude (north/south areas). A shallower depth and lower latitude causes slightly earlier onset of reproductive events (Orsi-Relini et al., 1998). The duration of embryonic development, i.e., the incubation period, also varies with latitude, increasing towards the north (Chapman, 1980; Sardà, 1995; Fariña et al., 1999; Bell et al., 2006). Spawning (egg laying on the pleopods) in populations with an annual reproductive cycle (e.g., in the Mediterranean, Fig. 4, around the Iberian peninsula, the Irish Sea, the Bay of Biscay and most of the North Sea) occurs in late summer to early autumn, while eggs hatch predominately in winter in the Mediterranean (e.g., Ligurian Sea) and in the spring in the Atlantic (e.g., Bay of Biscay) (Orsi-Relini and Relini, 1989; Orsi-Relini et al., 1998; Fariña et al., 1999; Briggs et al., 2002; dos Santos and Peliz, 2005; Bell et al., 2006). Mediterranean larvae have significantly higher wet mass, dry mass and protein values, which may represent adaptive traits allowing for early post-hatching survival under food-limited conditions in an oligotrophic environment (Rotllant et al., 2004). Spawning is annual in the Mediterranean and the south (although with varying duration), with biennial periodicity seen in the most northern stocks around Iceland and the Faeroe Islands (Sardà, 1995; Fariña et al., 1999; Bell et al., 2006).

FECUNDITY

A large number of studies have been carried out on aspects of fecundity, most of which focus on the northern areas of *Nephrops* distribution, but

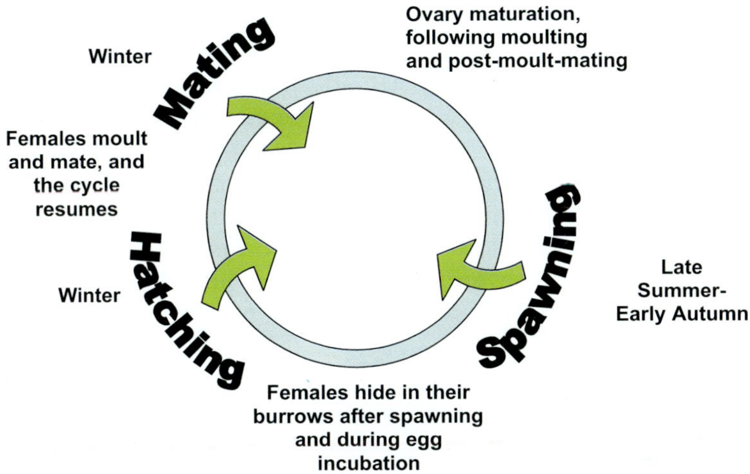

Fig. 4 Schematic of the annual reproductive cycle of *Nephrops* in the Mediterranean.

with a number of Mediterranean studies (Orsi-Relini et al., 1998; Fariña et al., 1999; Tuck et al., 2000; Smith et al., 2001; Briggs et al., 2002). Fecundity can be estimated in a number of ways. Potential fecundity is estimated from the number of oocytes in the mature ovary (Thomas, 1964; Tuck et al., 2000). Actual, realized and effective fecundity is based on counts of the number of eggs on the pleopods at the time of capture, fertilization and hatching, respectively (Tuck et al., 2000; Farmer 1974a; Figueiredo et al., 1982). The difference between effective fecundity and realized fecundity is the egg loss during the incubation period. Potential fecundity has been rarely studied (Tuck et al., 2000; Smith et al., 2001, 2007).

Potential fecundity is exponentially related to female body size and increases from 600 in females of 25 mm CL to 3,200-4,800 oocytes in females of 45 mm CL (Bell et al., 2006). Effective fecundity is always lower than potential fecundity. The initial number of eggs on the pleopods is reduced by progressive egg loss during the course of development, which will depend on a number of factors, including failure of the eggs to adhere to the pleopods, losses of unfertilized eggs, mechanical abrasion of egg clusters and the duration of the incubation period (Farmer, 1974a; Smith, 1987). Egg loss has been studied by both sequential monthly sampling over the incubation period and experimental captivity studies (Mori et al., 1998; Briggs et al., 2002; Fariña et al., 1999). Reported values range between 18 and 68% (reviewed in Fariña et al., 1999 and Briggs et al., 2002), with available data implying fewer losses in the north despite the fact that the incubation period lasts longer in the northern latitudes (Fariña et al., 1999).

Large differences have been recorded in fecundity (embryos hatched) and sexual maturity (size at onset of sexual maturity, SOM) both between and within Mediterranean and Atlantic *Nephrops* populations. Even between different parts of the same population, differences in fecundity can be as high as 33% (Tuck et al., 2000). In general, fecundity decreases with latitude in the northeastern Atlantic from western Ireland to south of Portugal, with the Mediterranean populations showing lower values than those in the Atlantic although, considering the periodicity of spawning, the annual egg production per female may be similar between areas and may even increase towards the south (Fariña et al., 1999).

Geographic variation in fecundity is thought to reflect growth-related differences (e.g., in oocyte and ovary volume) and to be linked with SOM, density-dependent processes and environmental characteristics (Tuck et al., 2000; Fariña et al., 1999). Significant spatial differences associated with ovary maturation, possibly related to growth and food availability, are also seen in the biochemical composition of ovary and hepatopancreas (Tuck et al., 1997b). Reproduction and gonadal maturation, which have large associated energy costs (i.e., increased biosynthetic work to support the lecithotrophic strategy of the embryos and pre-feeding larval stages), may be influenced by or even synchronized with seasonal feeding or food availability (Rosa and Nunes, 2002).

SIZE AT ONSET OF SEXUAL MATURITY

The concept of size at onset of sexual maturity is important for understanding the reproductive strategy and fitness of a species as a determinant of reproductive output (McQuaid et al., 2006). Spatial variations in SOM have important implications for stock management with technical measures (e.g., mesh sizes) based on area-specific values to avoid capture of immature females (Briggs, 1988; Tuck et al., 2000).

Female SOM has been studied (Briggs, 1988; Redant, 1994; Bianchini et al., 1998; Orsi-Relini et al., 1998; Tuck et al., 2000; Smith et al., 2001; McQuaid et al., 2006) by measurement of the smallest ovigerous female (Farmer, 1974b) and by estimation of 50% maturity from ovary examination (Bailey, 1984). In the absence of an easy-to-apply macroscopic external index of male maturity (although recent work shows that the appendix masculina is a good microscopic index; McQuaid et al., 2006), very few studies have investigated male maturity and even fewer have done so on the basis of histological examinations and the presence of spermatophores in the vasa deferentia (physiological maturity). Similarly, very few studies have investigated female maturity on the basis of presence of spermatophores in the thelycum of females. With allometric changes taking place at maturity both male and female maturity can be investigated from morphometric data (functional maturity) (Farmer, 1974b). Changes in CL in relation to cutter and crusher propodal length and width as well as appendix masculina length have been used to study male maturity and the relationship between CL and abdominal width for the study of female maturity (Tuck et al., 2000; Smith et al., 2001; McQuaid et al., 2006).

The age at which SOM occurs varies with sex and latitude (see Sardà, 1995; Chapman, 1980; Bell et al., 2006). Estimates of SOM are higher for males than for females, for example, 21-34 mm CL for females and 29-46 mm CL for males in the Clyde Sea (Tuck et al., 2000). The size at 50% maturity (L_{50}) for Mediterranean female *Nephrops* ranges from 30 to 36 mm CL with a mean of 31.9 mm, which is clearly higher than that (25.6 mm) found in ICES areas in Atlantic stocks (Orsi-Relini et al., 1998). The corresponding age at 50% maturity, however, is roughly the same (3-4 years) in all areas (Bell et al., 2006). Concurrent growth and male maturity studies from the Clyde Sea suggest that age at the onset of maturity ranges between 4 and 4.5 years for the faster-growing *Nephrops* populations and between 5 and 5.5 years for slower-growing populations (Tuck et al., 2000). Registered differences in size and age at maturity are thought to be related to different somatic growth rates (Bailey and Chapman, 1983; Orsi-Relini et al., 1998; Tuck et al., 2000). Morphometric differences between populations may result from or be linked with density-dependent processes, food availability, environmental parameters and differing states of exploitation (Sardà et al., 1998; Tuck et al., 1997a; Parslow-Williams et al., 2001; Maynou and Sardà, 1997).

GROWTH

As with other crustaceans, growth is a non-continuous process, with moults separated by intermoult periods. Moulting is known to be more frequent in juveniles and decreases with increasing size (Bailey and Chapman, 1983). Adult females moult once a year usually in winter, early spring (in the Mediterranean) or spring (in the Atlantic) after hatching their eggs (Gramitto, 1998). Adult males may moult throughout the year when small, but later less frequently. Populations often show one male moulting peak in late summer-autumn and a second one co-occurring with that of females. Growth is a function of two components, moult increment and moult frequency. Growth of juveniles is isometric, becoming allometric following the pubertal moult such that positive allometry is seen in cheliped growth in males and abdomen width in females (see Farmer, 1974b and section on SOM, above).

Existing techniques for the estimation of growth in *Nephrops* are based on the analysis of length-frequency distributions (from port sampling or sampling at sea), mark-recapture data from tagging, and direct observations on animals kept in captivity in aquaria. Published growth studies (Tuck et al., 1997a; González-Gurriarán et al., 1998; Mytilineou et al., 1998; Smith et al., 2001; Castro et al., 2003) recorded a number of similarities and differences in growth parameters within and between Atlantic and Mediterranean sites with, for example, the Alboran Sea male component of the *Nephrops* population having a higher growth rate than that of the Catalan Sea. However, differences seen may have important site-specific attributes related to environmental factors (such as sediment type, water circulation and temperature) and biological factors (such as density and food availability) as well as fishing pressure. The variability seen in growth rates between *Nephrops* stocks is likely the combined effect of several of these driving forces (Bell et al., 2006).

Acknowledgements

The authors would like to thank Prof. R.J.A Atkinson (UMBSM, UK) and Dr. I. Tuck (NIWA, New Zealand) for helpful comments on an earlier draft of this paper.

References

Abelló P., A. Abella, A. Adamidou, S. Jukic-Peladic, P. Majorano and M.T. Spedicato. 2002. Geographical patterns in abundance and population structure of *Nephrops norvegicus and Parapenaeus longirostris* (Crustacea: Decapoda) along the European Mediterranean coasts. Sci. Mar., Suppl. 2: 125-141.

Aguzzi, J., F. Sardà, P. Abello, J.B. Company and G. Rotllant. 2003. Diel and seasonal patterns of *Nephrops norvegicus* (Decapoda: Nephropidae) catchability in the western Mediterranean. Mar. Ecol. Prog. Ser., 258: 201-211.

Aguzzi, J., A. Bozzano and F. Sardà. 2004a. First observations on *Nephrops norvegicus* (L.) burrow densities on the continental shelf off the Catalan coast (Western Mediterranean). Crustaceana, 77(3): 299-310.

Aguzzi, J., F. Sardà and R. Allué. 2004b. Seasonal dynamics in *Nephrops norvegicus* (Decapoda: Nephropidae) catches off the Catalan coasts (Western Mediterranean). Fish. Res., 69: 293-300.

Atkinson, R.J.A., and E. Naylor. 1976. An endogenous activity rhythm and the rhythmicity of catches of *Nephrops norvegicus*. J. Exp. Biol. Ecol., 25: 95-108.

Bailey, N. 1984. Some aspects of reproduction in *Nephrops*. ICES CM 1984/K: 33.

Bailey, N., and C.J. Chapman. 1983. A comparison of density, length composition and growth of two *Nephrops* populations off the West coast of Scotland. ICES CM 1983/K: 42.

Bell, M.C., F. Redant and I. Tuck. 2006. *Nephrops* species. Chapter 13. *In:* Lobsters: Biology, Management, Aquaculture and Fisheries. B.F. Phillips (ed.). Blackwell Publ., Oxford, pp. 412-462.

Bergmann, M., S.K. Wieczorek, P.G. Moore and R.J.A. Atkinson. 2002. Discard composition of the *Nephrops* fishery in the Clyde Sea area, Scotland. Fish. Res., 57, 169-183.

Bianchini, M.L., L. di Stefano and S. Ragonese. 1998. Size and age at onset of sexual maturity of female Norway lobster *Nephrops norvegicus* L. (Crustacea: Nephropodidae) in the Strait of Sicily (Central Mediterranean Sea). Scientia Mar., 62(1-2): 151-159.

Briggs, R.P. 1988. A preliminary analysis of maturity data for Northeast Irish Sea *Nephrops*. ICES CM 1988/K: 21.

Briggs, R.P., M.J. Armstrong, M. Dickey-Collas, M. Allen, N. Mc-Quaid and J. Whitmore. 2002. The application of fecundity estimates to determine the spawning stock biomass of Irish Sea *Nephrops norvegicus* (L.) using the annual larval production method. ICES J. Mar. Sci., 59(1): 109-119.

Castro, M., P. Encarnação and P. Henriques. 2003. Increment at molt for the Norway lobster (*Nephrops norvegicus*) from the south coast of Portugal. ICES J. Mar. Sci., 60: 1159-1164.

Catchpole, T.L., C.L.J. Frid and T.S. Gray. 2006. Resolving the discard problem—A case study of the English *Nephrops* fishery. Mar. Policy 30: 821-831.

Chapman, C.J. 1980. Ecology of juvenile and adult *Nephrops*. *In:* The Biology and Management of Lobsters, Vol. II. S. Cobb and B. Phillips (eds.). Academic Press, New York, pp. 143-178.

Chapman, C.J., and A.L. Rice. 1971. Some direct observations on the ecology and behaviour of the Norway lobster *Nephrops norvegicus* (L.). Mar. Biol., 10: 321-329.

dos Santos, A., and A. Peliz. 2005. The occurrence of Norway lobster (*Nephrops norvegicus*) larvae off the Portuguese coast. J. Mar. Biol. Assoc. UK, 85: 937-941.

Dunthorn, A.A. 1967. Some observations on the behaviour and development of the Norway lobster. ICES CM 1967/K: 5.

Eggert, H., and M. Ulmestrand. 1999. A bioeconomic analysis of the Swedish fishery for Norway lobster (*Nephrops norvegicus*). Mar. Res. Econ., 14: 225-244.

Eriksson, H. 1970. On the breeding cycle and fecundity of the Norway lobster at Southwest Iceland. ICES CM 1970/K: 6.

Fariña, A.C., and I. González Herraiz. 2003. Trends in catch-per-unit-effort, stock biomass and recruitment in the North and Northwest Iberian Atlantic *Nephrops* stocks. Fish. Res., 65(1-3): 351-360.

Fariña, A.C., J. Freire and E. González-Gurriarán. 1999. Fecundity of the Norway lobster (*Nephrops norvegicus* (L.)) in Galicia (NW Spain) and a review of geographical patterns. Ophelia, 50(3): 177-189.

Farmer, A.S.D. 1974a. Reproduction in *Nephrops norvegicus* (Decapoda: Nephropodidae). J. Zool., Lond., 174: 161-183.

Farmer, A.S.D. 1974b. Relative growth in *Nephrops norvegicus* (L.) (Decapoda: Nephropodidae). J. Nat. Hist., 8: 605-620.

Farmer, A.S.D. 1975. Synopsis of biological data of Norway lobster, *Nephrops norvegicus* (Linnaeus, 1758). FAO Fish. Syn., 112: 1-97.

Figueiredo, M.J., and I.F. Barraca. 1963. Contribução para o conhecimento da pesca e da biologia do lagostim (*Nephrops norvegicus L.*) na costa portuguesa. Notas Estudos de l'Instituto de Biologia Marina de Lisboa, 28: 1-44.

Figueiredo, M.J., O. Margo and M.G. Franco. 1982. The fecundity of *Nephrops norvegicus* (L.) in the Portuguese waters. ICES CM 1982/K: 29.

Fischer, W., M. Schneider and M.L. Bauchot. 1987. Fiches FAO d'identification des espèces. Méditerranée et Mer Noire. Zône de Pêche 37, Vol. 1, Végétaux et invertébrés. FAO, Rome, 760 pp.

Fontaine, B., and N. Warluzel. 1959. Biologie de la langoustine du Golfe de Gascogne (*Nephrops norvegicus L.*). Rev. Trav. Inst. Pêches marit., 33: 223-246.

González-Gurriarán, E., J. Freire, A.C. Fariña and A. Fernández. 1998. Growth at molt and intermoult period in the Norway lobster *Nephrops norvegicus* from Galician waters. ICES J. Mar. Sci., 55: 924-940.

Graham, N., and R.S.T. Ferro (eds.). 2004. The *Nephrops* fisheries of the Northeast Atlantic and Mediterranean—A review and assessment of fishing gear design. Report prepared by the Ad hoc Group of Fishing Technology and Fish Behaviour and the Working Group on Fishing Technology and Fish Behaviour, ICES Cooperative Research Report No. 270, 38 pp.

Gramitto, M.E. 1998. Molt pattern identification through gastrolith examination on *Nephrops norvegicus* (L.) in the Mediterranean Sea. Sci. Mar., 62 (suppl. 1): 17-23.

Holthuis, L.B. 1991. Marine Lobsters of the World. FAO Fisheries Synopsis No. 125, Vol. 13, 292 pp.

Marrs, S.J., R.J.A. Atkinson, C.J. Smith and J.M. Hills. 1996. Calibration of the towed underwater TV technique for use in stock assessment of *Nephrops norvegicus*. EC DGXIV Final Report, Study Project 94/069, 155 pp.

Maynou, F., and F. Sardà. 1997. *Nephrops norvegicus* population and morphometrical characteristics in relation to substrate heterogeneity. Fish. Res., 30: 139-149.

McQuaid, N., R.P. Briggs and D. Roberts. 2006. Estimation of the size of onset of sexual maturity in *Nephrops norvegicus* (L.), Fish. Res., 81(1): 26-36.

Mori, M., F. Biagi and S. Ranieri. 1998. Fecundity and egg loss during incubation in Norway lobster (*Nephrops norvegicus*) in the North Tyrrhenian Sea. J. Nat. Hist., 32: 1641-1650.

Mouat, B. 2002. Reproductive dynamics of the male Norway lobster, *Nephrops norvegicus* (L.). PhD Thesis, University of Glasgow, 312 pp.

Mytilineou, C., M. Castro, P. Gancho and A. Fortouni. 1998. Growth studies on Norway lobster, *Nephrops norvegicus* (L.) in different areas of Mediterranean Sea and adjacent Atlantic. Sci. Mar., 62 (suppl. 1): 43-60.

Orsi-Relini, L., and G. Relini. 1989. Reproduction of *Nephrops norvegicus* L. in isothermal Mediterranean waters. *In:* Reproduction, Genetics and

Distributions of Marine Organisms, 23th EMBS. S. Ryland and P. Tyler (eds.). Olsen and Olsen, Fredensborg, pp. 153-160.

Orsi-Relini, L., A. Zamboni, F. Fiorentino and D. Massi. 1998. Reproductive patterns in Norway lobster *Nephrops norvegicus* (L.), (Crustacea Decapoda: Nephropodidae) of different Mediterranean areas. Sci. Mar., 62 (suppl. 1): 25-41.

Papadopoulou, K.N., C.J. Smith, L.A. Lioudakis and K. Skarvelis. 2006. Trap fishing, an alternative *Nephrops* fishery in the Aegean Sea. ICES Conference on Fishing Technology in the 21st century, Boston, USA, ICES BOSO6 2, 127.

Parslow-Williams, P.J., R.J. Atkinson and A.C. Taylor. 2001. Nucleic acids as indicators of nutritional condition in the Norway lobster *Nephrops norvegicus*. Mar. Ecol. Prog. Ser., 211: 235-243.

Redant, F. 1994. Sexual maturity of female Norway lobster, *Nephrops norvegicus*, in the central North Sea. ICES CM 1994/K: 43.

Rice, A.L., and C.J. Chapman. 1971. Observations on the burrows and burrowing behaviour of two mud-dwelling decapod crustaceans, *Nephrops norvegicus* and *Goneplax rhomboides*. Mar. Biol., 10: 330-342.

Rosa, R., and M.L. Nunes. 2002. Biochemical changes during the reproductive cycle of the deep-sea decapod *Nephrops norvegicus* on the south coast of Portugal. Mar. Biol., 141: 1001-1009.

Rotllant, G., K. Anger, M. Durfort and F. Sardà. 2004. Elemental and biochemical composition of *Nephrops norvegicus* (Linnaeus 1758) larvae from the Mediterranean and Irish Seas. Helgoland Mar. Res., 58(3) 3: 206-210.

Sardà, F. 1995. A review (1967-1990) of some aspects of the life history of *Nephrops norvegicus*. ICES Mar. Sci. Symp., 199: 78-88.

Sardà, F. 1998. *Nephrops norvegicus*: comparative biology and fishery in the Mediterranean Sea. Introduction, conclusions and recommendations. Sci. Mar., 62 (suppl. 1): 3-15.

Sardà, F., J. Lleornart and J.E. Cartes. 1998. An analysis of the populations dynamics of *Nephrops norvegicus* (L.) in the Mediterranean Sea. Sci. Mar., 62 (suppl. 1): 135-143.

Smith, C.J. 1988. Effects of megafaunal/macrofaunal burrowing interactions on benthic community structure. PhD Thesis, University Marine Biological Station Millport and University of Glasgow, 265 pp.

Smith, C.J., and K.N. Papadopoulou. 2003. Burrow density and stock size fluctuations of *Nephrops norvegicus* in a semi-enclosed bay. ICES J. Mar. Sci., 60: 798-805.

Smith, C.J., K.N. Papadopoulou, A. Kallianiotis, P. Vidoris, C.J. Chapman and D. Vafidis. 2001. Growth and natural mortality of *Nephrops norvegicus*, with an introduction and evaluation of creeling in Mediterranean waters. European Commission Final Report DG XIV Study Project 96/013, 195 pp.

Smith, C.J., S.J. Marrs, R.J.A. Atkinson, K.N. Papadopoulou and J.M. Hills. 2003. Underwater television for fisheries-independent stock assessment of *Nephrops norvegicus* from the Aegean (eastern Mediterranean) Sea. Mar. Ecol. Prog. Ser., 256: 161-170.

Smith, C.J., K.N. Papadopoulou, C.J. Chapman, B. Catalano, A. Kallianiotis and P. Vidoris. 2007. Fecundity and size at onset of sexual maturity of *Nephrops norvegicus* from two sites in the Aegean, submitted.

Smith, R.S.M. 1987. The biology of larval and juvenile *Nephrops norvegicus* (L.) in the Fyrth of Clyde. University Marine Biological Station Millport and University of Glasgow, 265 pp.

Streiff, R., S. Mira, M. Castro and M.L. Cancela. 2004. Multiple paternity in Norway lobster (*Nephrops norvegicus* L.) assessed with microsatellite markers. Mar. Biotechnol., 6: 60-66.

Thomas, H.J. 1964. The spawning and fecundity of the Norway lobster (*Nephrops norvegicus* (L.)) around the Scottish Coast. J. Cons. Int. Explor. Mer., 29: 221-229.

Tuck, I.D., C.J. Chapman, and R.J.A. Atkinson. 1997a. Population biology of the Norway lobster, *Nephrops norvegicus* (L.) in the Firth of Clyde, Scotland. I: Growth and density. ICES J. Mar. Sci., 54: 123-135.

Tuck, I.D., A.C. Taylor, R.J.A. Atkinson, M.E. Gramitto and C.J. Smith. 1997b. Biochemical composition of *Nephrops norvegicus*: changes associated with ovary maturation. Mar. Biol., 129: 505-511.

Tuck, I.D., R.J.A. Atkinson and C.J. Chapman. 2000. Population biology of the Norway lobster, *Nephrops norvegicus* (L.) in the Firth of Clyde, Scotland. II: fecundity and size at onset of sexual maturity. ICES J. Mar. Sci., 57: 1227-1239.

14

Implementation of Failure Mode and Effects Analysis (FMEA), Cause and Effect Diagram, HACCP and ISO 22000 to the Reproductive Cycle and Growth of Crustaceans in Cultured Conditions

I.S. Arvanitoyannis

Associate Professor, University of Thessaly, School of Agricultural Sciences, Department of Ichthyology and Aquatic Environment, Fytoko Str., Nea Ionia Magnsias, 38446 Volos, Greece. E-mail: parmenion@uth.gr

INTRODUCTION

Crustaceans are a dominant and successful group represented by a high number of species exhibiting a great array of life styles and occupying quite dissimilar habitats. This diversity is a result of their life patterns and reproductive strategies. Research on crustacean reproductive biology is required in order to maintain natural and fishing stocks. Reproductive cycles involve a series of events in a population (Reigada and Negreiros-Fransozo, 2000).

Fisheries, including aquaculture, provide a vital source of food, employment, recreation, trade and economic well-being for people throughout the world, both present and future generations, and should therefore be conducted in a responsible manner. This Code sets out principles and international standards of behavior for responsible practices with a view to ensuring the effective conservation, management and development of living aquatic resources, with due respect for the ecosystem and biodiversity (http://www.fao.org/DOCREP/005/v9878e/v9878e00.htm). Crustaceans have been farmed all over the world for decades. Ninety-nine percent of the crustaceans that aquaculture farmers are interested in are in the decapods. The majority of these are caridean shrimps, but some are also penaeid shrimps, crabs, lobsters and crayfish. As in most cultured species the growth rate of the organisms cultured is important. The faster the growth rate, the sooner the farmer sees a return on his investment. Another important issue is the final size that the organism reaches: the larger the animal, the greater the price that can be charged. Choosing a species that has fast growth and a large size leads to the greatest profit (http://www.tracc.org/my/Borneocoast/aquaculture/crustacean_aquaculture.htm).

Shrimp is intensively cultured, which can lead to overcrowding and stress. When the shrimp is overly stressed, bacterial disease sets in, causing more stress; then viral disease strikes, wiping out the whole stock if not treated early enough. Shrimps are never exported live because of the high risk of importing a disease. Shrimp farming involves spawning, growing larvae and post-larvae, and then farming in ponds or in the wild. Spawning is quite hard to achieve and is generally done by cutting off the eyes of the organism so that the gonad-inhibiting hormone is reduced. It is common for mature, berried females to be caught in the wild and then placed in tanks so that the eggs can be collected. The larvae are fed large amounts of microalgae and larvae depending on their larval stage. This food can be replaced with artificial feed, and a large amount of research is focused on finding the best combination of vitamins and minerals to provide the best larval growth (http://www.tracc.org/my/Borneocoast/aquaculture/crustacean_aquaculture.htm).

Fisheries management should promote maintenance of the quality, diversity and availability of fishery resources in sufficient quantities for present and future generations in the framework of food security, poverty alleviation and sustainable development. Management measures should ensure the conservation of not only target species but also species belonging to the same ecosystem or associated with or dependent on the target species (http://www.fao.org/DOCREP/005/v9878e/v9878e00.htm).

Hazard Analysis and Critical Control Points (HACCP) is a systematic preventive approach to food safety that addresses physical, chemical and biological hazards as a means of prevention rather than finished product inspection. It is used in the food industry to identify potential food safety hazards so that key actions, known as Critical Control Points (CCPs) can be taken to reduce or eliminate the risk of those hazards. The system is used at all stages of food production and preparation processes. Today HACCP is applied to industries other than food, such as cosmetics and pharmaceuticals (http://en.wikipedia.org/wiki/HACCP). Risk assessment of food products has been interwoven with risk analysis system and HACCP, the determination of CCPs, and the undertaking of actions against every recognized hazard. The HACCP system also refers to physical, chemical and microbiological hazards occurring in raw materials or processes during food production (Mortimore and Wallace, 1995). This method, which in effect seeks to phase out unsafe practices, differs from traditional "produce and test" quality assurance methods, which are less successful and inappropriate for highly perishable foods. HACCP is based on seven established principles: (1) conduct a hazard analysis, (2) identify CCPs, (3) establish critical limits for each CCP, (4) establish CCP monitoring requirements, (5) establish corrective actions, (6) establish record-keeping procedures, and (7) establish procedures for verifying that the HACCP system is working as intended (http://en.wikipedia.org/wiki/HACCP).

Failure mode and effects analysis (FMEA) is a method (first developed for systems engineering) that examines potential failures in products or processes. It may be used to evaluate risk management priorities for mitigating known threat-vulnerabilities. FMEA helps select remedial actions that reduce cumulative impacts of life-cycle consequences (risks) from a systems failure (fault) (http://en.wikipedia.org/wiki/Failure_mode_and_effects_analysis). The FMEA methodology was developed and implemented for the first time in 1949 by the United States Army. In the 1970s, thanks to its strength and validity, its application field extended first to the aerospace and automotive industries, then to general manufacturing (SVRP, 1997). The application of FMEA methodology and the integration of its results in the HACCP system already implemented in the company applying the FMEA empowered the study and analysis of both the aspects of food quality. FMEA is a systematic process meant for reliability analysis (James, 1998). It improves operational performance of the production cycles and reduces their overall risk level. This task is achieved by preventing the system potential failures that have been identified through the preliminary analysis and the collection of plant historical data (Sachs, 1993). The FMEA technique considers each item that constitutes the total system. On the basis of best expert opinion and historical information for similar items, an analysis is carried out of all the ways that each component or subsystem might fail to fulfil its intended function (James, 1998). Each part of the system is described, and the consequences of its failure are listed. In most formal systems, the consequences are then evaluated by three criteria and associated risk indices: severity of the worst potential outcome due to the failure in terms of safety and system functionality (severity), relative probability that the failure will occur (occurrence) and probability that the failure mode will be detected and/or corrected by the applicable controls installed on the production lines (detection) (Stamatis, 1995). Each index ranges from 1 (lowest risk) to 10 (highest risk). The overall risk of each failure is called Risk Priority Number (RPN), which is the product of severity, occurrence, and detection rankings: RPN = S × O × D. The RPN (ranging from 1 to 1000) is used to prioritize all potential failures to decide on actions that reduce the risk, usually by reducing likelihood of occurrence and improving controls for detecting the failure (http://en.wikipedia.org/wiki/Failure_mode_and_effects_analysis).

In September 2005 the International Organization for Standardization announced the publication of ISO 22000. The standard was published in response to 15 countries, among them Denmark, the United States, and the United Kingdom, interested in the creation and realization of a standard to ensure safe food supply chains worldwide (http://certification.bureauveritas.com/webapp/servlet/RequestHandler?mode=PT&pageID=32154&nextpage=siteFrameset.jsp). It is designed for all types of organizations within the food chain, including feed producers, primary producers, food manufactures, transport and storage operators, and subcontractors to retail and food service outlets. It is the first time that a certifiable standard providing an internationally harmonized framework for safety and quality requirements has been available for the whole food chain (http://www.lrqausa.com/4-1.cfm?id=1029). ISO 22000 specifies the requirements for a food safety management system in the food chain where an organization needs to demonstrate its ability to control food safety hazards in order to provide consistently safe end-products that meet both the requirements agreed with the customer and those applicable to food safety regulations (http://www.bsi-global.com/Food_Management/FoodSafety/bseniso22000.xalter). ISO 22000 has been designed to have such flexibility that it enables a tailor-made approach to food safety depending on which segment of the food chain a company is involved in. Bearing in mind that the standards and procedures required for high risk areas in one food sector may not be appropriate in another, it does not take a one-size-fits-all approach, nor does it take a checklist approach. Central to its procedure is that all companies will follow the local laws and those of the market they are exporting to, as well as take heed of client needs (http://certification.bureauveritas.com/webapp/servlet/RequestHandler?mode=PT&pageID=32154&nextpage=siteFrameset.jsp). The standard has three parts: requirements for good manufacturing, requirements for HACCP according to HACCP principles of the Codex Alimentarius, and requirements for a management system. The requirements for good manufacturing practices are not listed in the standard but are referenced. It will not have a detailed list of requirements for good practices since practically it is impossible to make a list that covers all requirements for all organizations and all situations. However, ISO 22000 will require the implementation of good practices and expects organizations to define the practices that are appropriate to them (http://www.lrqausa.com/4-1.cfm?id=1029). The standard combines generally recognized key elements to ensure food safety along the food chain including interactive communication, system management and control of food safety hazards through prerequisite programs and HACCP plans and continual improvement and updating of the management system. The standard will further clarify the concept of prerequisite programs. These are divided into two subcategories: (1) infrastructure and maintenance programs and (2) operational prerequisite programs (http://www.bsi-global.com/Food_Management/FoodSafety/bseniso22000.xalter).

The aim of this research is to look into the factors affecting both the reproductive cycle of crustaceans and to assess the risk of all parameters involved per stage by applying FMEA, Cause & Effect Diagram, HACCP and ISO 22000.

The implementation of HACCP in the reproductive cycle of crustaceans is given in Table 1. Since no Prerequisite Programs (PP) are considered to be in place, all stages are considered to be CCPs in agreement with the tree diagram for CCP determination (FAO, 1994). However, the implementation of ISO 22000, which presupposes the application of Good Manufacturing (GHP) leads to considerable loss of CCPs (only 5 in Table 2) compared to 12 CCPs in Table 1. Figure 1 also shows the identification of five CCPs in the flow diagram. Judging from the completion data they form, an FMEA table, where hazardous parameters are assessed and RPN is calculated, can be seen in Table 3.

Table 1 Application of HACCP in reproductive cycle of Crustaceans.

Reproductive cycle	*CCP*	*Hazard*	*Preventive action*	*Corrective action*	*Responsibility*	*Records*
Broodstock catch by fishing vessel	1	Physical	Improved cultural techniques	Reduction of salinity	Fishing vessel manager	Sanitation and water analysis
1st selection	2	Microbiological (disease-free worms)	Microbiological analysis	Rejection of batch	Fishing vessel manager	Disease control
Transport to hatchery	3	Physical and chemical	Hygiene control	Rejection of batch	Van driver	Water analysis
2nd selection	4	Microbiological (*E. coli*, *Vibrio parahaemolyticus*)	Microbiological analysis Pathogen control	Rejection of batch	Quality control (QC)	Pathogens analysis
Maturation	5	Chemical (presence of hormones)	Temperature control Water control (pH, salinity, etc.)	Rejection of batch	QC	Chemical analysis
Spawning	6	Cross contamination due to inadequate cleaning Pathogen growth due to temperature abuse	Temperature control Appropriate place Control breeding programs	Rejection of batch	QC	Water analysis
Hatching	7	Chemical (heavy metals, pesticides)	Temperature control Control of environmental and cultural conditions	Rejection of batch	QC	Temperature
Larval rearing	8	Microbiological (fungi, bacteria)	Good hygiene UV radiation	Antibiotic treatment to control bacteria contamination	QC	Temperature and water analysis

Contd.

Contd.

Juvenile nursery	9	Chemical (heavy metals, pesticides, antibiotics)	Optimal temperature pH control Ammonia/nitrogen Control Salinity control	Body weight increase Cleaning	QC	Water analysis
Juvenile growout	10	Chemical	Optimal temperature pH Ammonia/nitrogen Salinity control	Body weight increase Cleaning	QC	Water analysis
Harvest	11	Microdiological and phyoical	Less transport between tanks	Recall of batch	QC	Personnel sanitation
Processing	12	Microbiological due to presence of pathogens	Personnel hygiene control Removal of affected crustacean	Recall of batch	QC	Temperature during freezing, microbiological control and sanitation

Table 2 Application of ISO 22000 in reproductive cycle of Crustaceans.

Reproductive cycle	*CCP*	*Hazard*	*Preventive action*	*Corrective action*	*Responsibility*	*Records*
Broodstock catch by fishing vessel	PP	Physical	Improved cultural techniques	Reduction of salinity	Fishing vessel manager	Sanitation and water analysis
1st selection	1	Microbiological (disease-free worms)	Microbiological analysis	Rejection of batch	Fishing vessel manager	Disease control
Transport to hatchery	PP	Physical and chemical	Hygiene control	Rejection of batch	Van driver	Water analysis
2nd selection	2	Microbiological (*E. coli*, *Vibrio parahaemolyticus*)	Microbiological analysis Pathogen control	Rejection of batch	QC	Pathogens analysis
Maturation	PP	Chemical (presence of hormones)	Temperature control Water control (pH, salinity, etc.)	Rejection of batch	QC	Chemical analysis
Spawning	3	Cross contamination due to inadequate cleaning Pathogen growth due to temperature abuse	Temperature control Appropriate place Control breeding programs	Rejection of batch	QC	Water analysis
Hatching	PP	Chemical (heavy metals, pesticides)	Temperature control Control of environmental and cultural conditions	Rejection of batch	QC	Temperature

Contd.

Contd.

Larval rearing	4	Microbiological (fungi, bacteria)	Good hygiene UV radiation	Antibiotic treatment to control bacteria contamination	QC	Temperature and water analysis
Juvenile nursery	PP	Chemical (heavy metals, pesticides, antibiotics)	Optimal temperature pH control Ammonia/nitrogen control Salinity control	Body weight increase Cleaning	QC	Water analysis
Juvenile growout	PP	Chemical	Optimal temperature pH Ammonia/nitrogen Salinity control	Body weight increase Cleaning	QC	Water analysis
Harvest	5	Microbiological and physical	Less transport between tanks	Recall of batch	QC	Personnel sanitation
Processing	PP	Microbiological due to presence of pathogens	Personnel hygiene control Removal of affected crustacean	Recall of batch	QC	Temperature during freezing, microbiological control and sanitation

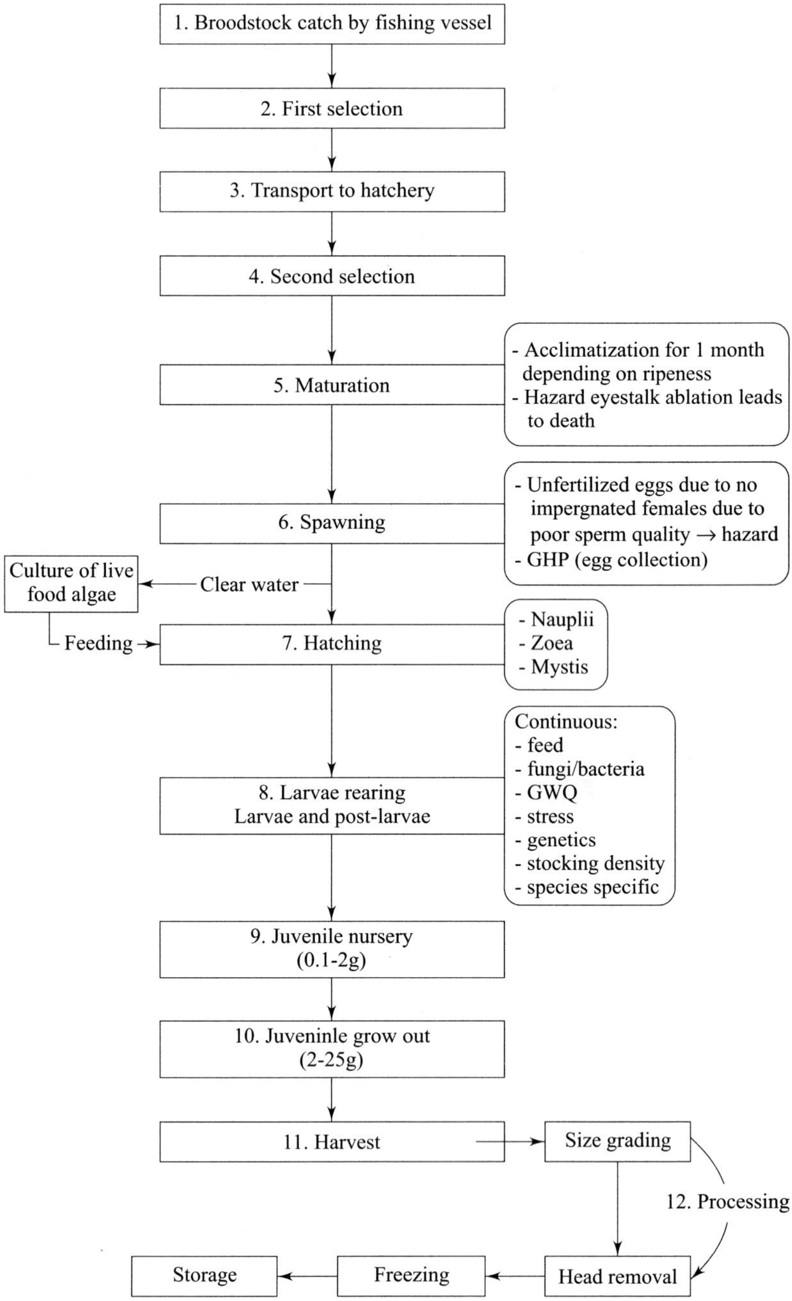

Fig. 1 Flow diagram of reproductive cycle of Crustaceans.

Table 3 Failure Mode and Effects Analysis for cultural conditions.

Reproductive stage	*Hazards*	*Causes*	*S*	*O*	*D*	*RPN*	*Corrective actions*	*S*	*O*	*D*	*RPN*
Maturation	Ineffective fish feeding	GM feed/ ingredient	9	7	8	504*	Change of supplier	9	3	3	81
		Inappropriate composition	8	6	7	336*	Change of supplier	8	3	2	48
		Antibiotics (prohibited, higher concentration than permitted)	9	7	7	441*	Change of supplier	9	3	3	81
		Improper feeding method	9	7	3	189*	Training of personnel involved	9	5	2	90
		Improper feeding timetable	8	6	5	240*	Training of personnel involved	8	4	2	64
Spawning and hatching	Low growth rate	Low spawning and hatching tank	9	8	8	576*	Change of tank (wider)	9	3	3	81
		Inappropriate shape of tank	8	7	7	392*	Change of tank (wider)	8	3	2	48
		Inappropriate tank material	8	6	7	336*	Stainless steel	8	3	2	48
Larval rearing	Growth of pathogens	Improper sanitation	9	9	7	567*	Frequent cleaning	9	5	2	90
		Low water exchange rate	8	8	6	384*	Continuous water exchange	8	4	3	96
	Low growth rate	Low pH	8	6	6	288*	pH adjustment	8	3	2	48
		High salinity	8	6	5	240*	Salt concentration control	8	4	2	72
		Low aeration	8	8	7	448*	Control of dissolved oxygen in water	8	4	2	64

Contd.

Contd.

Juvenile nursery and growout	Low growth rate	Unstable feeding program	8	7	7	392*	Improvement of feeding schedule (usually twice a day)	8	3	3	72
		Low nutrient concentration	9	5	6	270*	Control of ammonia/ nitrogen concentration in water	9	2	2	36
		Improper feeding timetable	8	6	5	240*	Training of personnel involved	8	4	2	64

S, severity; O, occurrence; D, detection; RPN, risk priority number.

CASE STUDY PRESENTATION: REPRODUCTIVE CYCLE OF CRUSTACEANS

The reproductive cycle of Crustaceans including their processing is shown in Fig. 2.

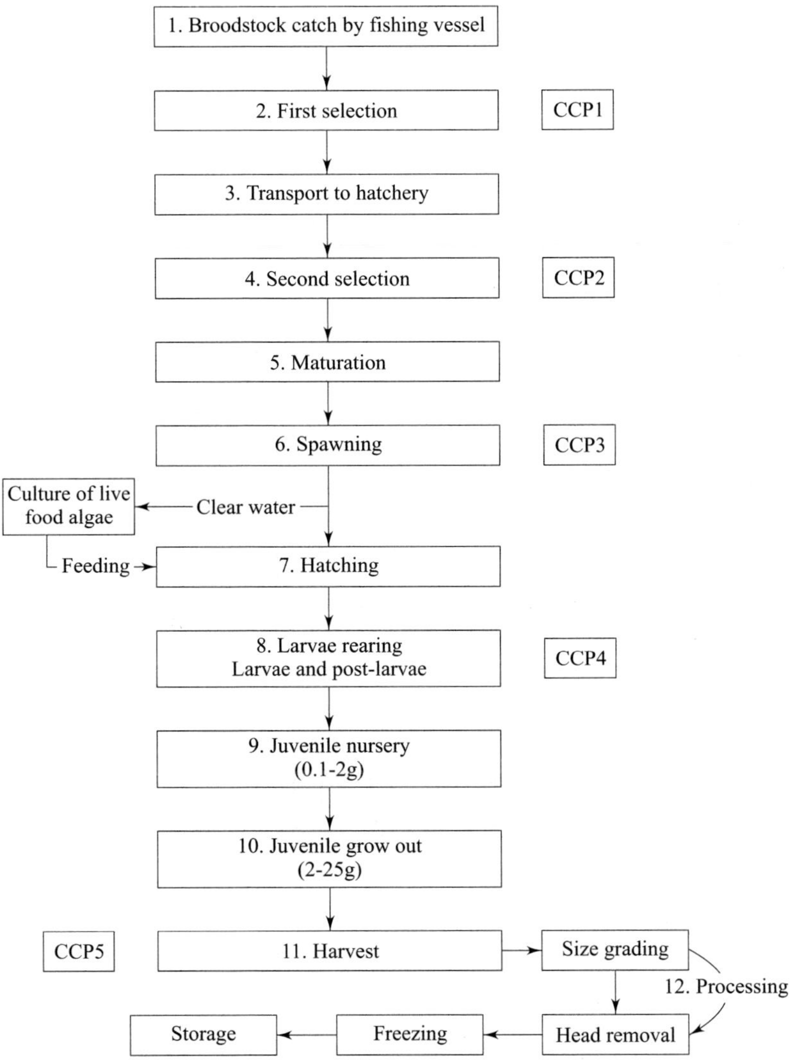

Fig. 2 Identification of CCPs in the flow diagram of reproductive cycle of Crustaceans.

Broodstock Catch by Fishing Vessel

The following handling practices should be continuously controlled: good water quality, low salinity and low turbidity.

A few researchers and commercial hatcheries in Venezuela, Ecuador, Australia and Mexico managed to select families in a breeding program and obtain egg development, mating and spawning of captive penaeids without ablation. Temperature and photoperiod manipulation alone have not produced sustainable commercial operations without ablation. As a rule, hatcheries have not been able to run a long-lasting, profitable, and highly productive commercial operation without ablation. The non-ablation approach is the desired route for hatcheries in the future but presently only makes up a very small percentage (Treece, 1999). Prevention and control of disease in hatcheries can make a tremendous difference to overall mortality levels and help to improve final product quality (http://www.scotmas-food.com/booklets/booklet-Fish Production & Processing.pdf).

First Selection

It is of great importance for the very first selection to ensure that the broodstock is free of disease and worms (visual observation) and there are no epizootic organisms.

Transport to Hatchery

Apart from the handling practices (worm and parasite growth) stress can be caused by poor control of temperature, salinity, pH (~7) and turbidity and overcrowding, and the result is low spawning.

Bacterial infection can often adversely affect the quality of eggs during transport to the hatchery. Left untreated, high bacterial counts can adversely affect hatchability of the eggs and pose the risk of transferring infection to the hatchery from the egg source (http://www.scotmas-food.com/booklets/booklet-Fish Production & Processing.pdf).

Second Selection

The second selection involves removal of dead fish, addition of clear water, and continuous control of physical and chemical characteristics of water.

Maturation

The maturation stage can be divided into two parts: **(1)** Ripe females are transferred into the tanks with males (natural way). Males deposit spermatophores and the females spawn a few hours later. **(2)** The broodstocks (genitors of optimal age and size and healthy) are constituted from adults (females and males) caught in the wild (selection). Sometimes eyestalk ablation is applied so that the gonad-inhibiting hormone is reduced and vitellogenesis is accelerated and the animals reach sexual maturity more rapidly.

The above-mentioned method occasionally results in a low survival rate but it is species specific. Other parameters also affecting the survival rate are feed pellets and fresh feed quality. Statistics show that 70% of females in the broodstock cycle successfully extrude eggs. Furthermore, there is a natural attrition rate (egg loss) of 36%.

It is well known that eyestalk ablation induces ovarian development and oviposition. This is because the source (X-organ-sinus gland complex) of the vitellogenesis-inhibiting hormone (VIH) is removed by the ablation. It was revealed that bilateral eyestalk ablation induced ovarian development and oviposition in lobster, indicating that VIH also occurs in this species. However, VIH has not yet been identified in any shrimps. The regulatory mechanism of VIH is partially understood. It was clarified in an *in vitro* study that purified eyestalk peptides, including VIH, inhibited protein-synthesis activity of ovaries, so VIH may act on the vitellogenin synthesis sites directly (Okumura, 2004). Females were cultured under natural temperature and high temperature (21°C) conditions to examine ovarian development and oviposition from autumn to the beginning of breeding season in the following spring. Ovaries developed because of vitellogenesis of oocytes and their developmental state did not change during the overwintering period. In spring, some ovaries reached prematuration and maturation stages and oviposition began under natural temperature conditions. Females reared in tanks at high temperature regimes oviposited earlier than those reared in tanks at natural temperature. Therefore, it is inferred that temperature and photoperiod are important environmental factors controlling ovarian development and oviposition (Katsuyuki et al., 2004).

The achievement of functional sexual maturity can be noticed by macroscopic observations of gonad size and color. In some species, gametes are produced rather continuously and the sequential offspring production is interrupted only during the incubation period. On certain occasions, consecutive egg batches are laid within a single intermolt period. Otherwise, egg production is separated by molting in some cases

(Reigada and Negreiros-Fransozo, 2000). Crustacean reproduction is regulated by a complex chain of hormonal interactions in which the crustacean hyperglycaemic hormones A and B and the gonad-inhibiting hormones play a primary role. These neurohormones are produced in the same neuroendocrine cells situated in the eyestalks (De Kleijn et al., 1998). The female crustacean is sexually receptive only for a few days after molting, while her shell is still soft and flexible. At that time the eggs are released from the openings of the oviduct at the bases of the third pair of walking legs. As they pass toward the swimmerets, sperm is released to fertilize them. The eggs are then carried on the swimmerets until the following year. In the past it was assumed that after hatching the eggs the female would molt and then mate again to recommence a two-year reproductive cycle. However, it seems that very large females are able to conserve enough sperm from a single mating to fertilize two or even three broods of eggs. They may spawn several times without an intervening molt (http://www.mi.mun.ca/mi-net/fishdeve/lobster.htm).

The onset of sexual maturity of female lobsters can be established by observations of the ovarian condition, either color or weight, and staging of cement glands but cannot be detected by the morphometry of their abdomens. To study the reproductive cycle of females, molt stage, ovarian development, and egg spawning were monitored by dissections in the laboratory and by tagging studies in the field. The majority (80%) of small mature females in the GSL had a typical two-year reproductive cycle with molting (with copulation) and spawning in alternating years. A small percentage (5%) of small mature females could also skip molting or spawning for a year. Temperature data suggested that the length of the female reproductive cycle, and possibility of molting and spawning in the same year, were related to the number of degree-days in a particular season (http://www.bioone.org/perlserv/?request=get-abstract&issn=0278-0372&volume=022&issue=04&page=0762).

Another deviation from the normal two-year reproductive cycle is the case of females that molt, mate and extrude eggs all in one season. This phenomenon seems to be limited to individuals reproducing for the first time in warmer water habitats. Each female produces enormous numbers of eggs. The size at which the average female is mature and capable of producing eggs to replenish the population is very important in the regulation of the fisheries harvest, because if females are not given the chance to contribute to the reproduction of the species, the lobster population will obviously suffer a decline (http://www.mi.mun.ca/mi-net/fishdeve/lobster.htm). For most species of decapod crustacean, reproduction is thought to be regulated by various hormones. The complex interactions between several neuroendocrines and endocrine organs that play a key role in control of gonad development and secondary sexual characteristics have been identified in males as well as females. The two antagonistic neurohormones regulating crustacean gonad maturation were identified. Several neurotransmitters have been demonstrated to affect the release of reproductive hormone in crustaceans, for instance, dopamine and serotonin. Some research has been carried out on improvement of aquaculture production via hormonal manipulation, which could be done in many species of fish (Prasert et al., 2005).

Spawning

Nowadays, the volume of spawning tanks is 150 l; they are cylindroconical with a perforated plastic plate, thus preventing females from entering the conical bottom and eating the eggs. Constant filtered water flow is required in conjunction with overflow passes through a concentrator provided with a mesh that retains eggs. After spawning, the tanks are drained and the eggs are collected in the concentrator. The eggs are then poured into a bucket (10 l) and samples are taken for counting (unfertilized eggs). It was found that the number of eggs greatly depends on the weight of the female. The percentage of normal eggs (normally fertilized with a fecundation membrane present of well-developed nauplii) varies from 60 to 100% depending on the species and on what they eat in the maturation process. They need to eat pellets and fresh food. Unfertilized spawning is frequently due to lack of impregnated females due to poor sperm quality.

The capture of wild females with mature ovaries for immediate spawning in captivity, known as "sourcing", was the only method known and practiced for inducing penaeid females to spawn in captivity until the early 1970s. As discussed earlier, shrimp eyestalk ablation was not used in commercial shrimp maturation until the early 1970s. The female reproductive system consists of paired ovaries, paired oviducts and a single thelycum; the first two are internal and the last is an external organ. The male genital system consists of spermatophore and spermatozoa. The spermatozoa are non-motile and have been described as resembling a golf ball on a tee. The number of spermatozoa is directly related to the size of the male (Treece, 1999).

From the spawning tank, a sample of eggs is counted to determine the number of eggs spawned per female. In normal condition, fertilized eggs hatched within 12-15 hours. The hatching rate is measured by assessing the number of hatched nauplii. They are then reared directly in the large tank up to the 25th post-larval day. On the other hand, if the number of nauplii is less than 0.5 million, they are stocked in 2.5 t indoor tanks at a density of 100-150 larvae per liter. The larvae are reared either to the third mysis stage or one-day-old post-larvae. They are then transferred to the outdoor 40 t nursery tanks for further nursing. Aeration is stopped, nauplii are allowed to swim toward the surface and then collected in a beaker and the water is drained through a filter net for total harvest. The nauplii are then transferred to a jerrycan, plastic jar or plastic bag (http://www.fao.org/docrep/field/003/AC232E/AC232E10.htm).

Hatching

Hatching tanks are of different dimension and size from spawning tanks and the direction of water flow is also different. The fertilized eggs have to be properly placed for hatching purposes. Once nauplii hatch, they swim to the surface and are borne away by the overflow into the sieve. The temperature in the hatching tanks should be high. Depending on the species they hatch either in the morning or the night after spawning or in the day after spawning, so the duration of hatching varies considerably. About 60% of the eggs are fertilized depending on the species and 70% of them produce nauplii. In other species, all fertilized eggs hatch. In still others, the number of hatched nauplii is lower than the number of normally fertilized eggs after spawning. Hatching rate % = total number of nauplii/total number of eggs x 100%.

There are four hatching stages: (1) nauplii (day 0-2), (2) zoea (day 2-7), (3) mysis (day 8-11), and (4) larvae (restocking in sea). The larva stage itself consists of three parts: (a) nymph (post-larvae) on day 12-15 (PL12), (b) juveniles in nursery (0.1-2 g), and (c) juveniles in growout cage (2-25 g).

Fertilized eggs of most marine crustaceans are attached with a funiculus to the abdominal appendages of the female and are ventilated by the female during embryonic development. When the development is complete, the outer egg membrane breaks, and the larvae hatch. These larvae are released into water with a special fanning behavior of the female's abdomen. Larval release is generally a short-lived event, and the timing is often correlated with such environmental periodicities as day-night, tidal, or lunar cycles. Such a short and precisely timed event implies that the timing of larval emergence from the egg capsule must be synchronized within each batch of embryos (Masayuki, 1992). The fertilized egg of crustaceans contains large quantities of maternally derived ecdysteroids and corresponding polar metabolites, which decline during nauplii development. These are presumed to be important in nauplii development. However, during metanaupliar development, dramatic increases in ecdysteroid levels are seen, which are believed to originate via *de novo* ecdysteroid synthesis by the developing Y-organ. The embryonic Y-organ has been observed only in palaemonid shrimps during the period just prior to eye development (Chung and Webster, 2004). In decapod crustacean and other aquatic invertebrate larvae, rates of survival, development and growth, as well as overall physiological fitness are affected by numerous extrinsic and intrinsic factors. Among the principal environmental variables, temperature, salinity, quality and quantity of nutrition, viral, bacterial and parasitic infections, as well as several chemical parameters of water quality and technical details of the rearing methods may influence the success of larval development in artificial cultures. The strength of those extrinsic effects greatly depends also on interactions with intrinsic factors such as the larval stage, the stage of the molting cycle and genetic factors and on late effects of previous conditions affecting maternal or embryonic quality (Anger, 2005).

The ease with which the crustaceans can be cultured depends on fecundity, larval life span, larval densities and the survival rate of the organism being farmed. The perfect organism for culture would have a high fecundity, short larval life, an ability to be kept at high density and a good survival rate (http://www.tracc.org/my/Borneocoast/aquaculture/crustacean_aquaculture.htm). At the time of egg hatching, pheromones are released from the eggs. These pheromones induce a stereotypic larval release behavior in which the female vigorously pumps her abdomen. This action breaks open the unhatched eggs and results in synchronized release of larvae. Another study suggested the pheromone was composed of a group of small peptides and identified active peptides. The larval release behavior is evoked by di- and tripeptides, which have a neutral amino acid at the amino terminus and a basic amino acid at the carboxy terminus. The most effective peptides also have hydrophobic functionality at the amino terminus. The two most active peptides tested were leucylarginine and glycyl-glycyl-arginine (Forward et al., 1987).

Once the eggs have hatched, the water is fertilized to stimulate the growth of diatoms. Predetermined amounts of fertilizer and sea water are added each day to the tank until the larval shrimps reach the last mysis stage. Because it is too costly and time consuming to separate the crushed shell from the meat, the shell eventually covers the pond bottom, resulting in a substrate that hampers the burrowing of the shrimps. Thus, ponds must be drained or dredged periodically to remove the shell debris (http://www.lib.noaa.gov/japan/aquaculture/report1/mock2.html). The same operational procedure for rearing of mysis stage used in the small-tank system is employed in the outdoor nursery tanks. After the nursery tank is filled with water to its full rearing capacity, approximately 30% of the water is changed daily. Early post-larval stages are fed with brine shrimp nauplii at a rate of 100-200 per post-larva per day (http://www.fao.org/docrep/field/003/AC232E/AC232E10.htm).

Larval Rearing (Larvae and Post-larvae)

Larval rearing tanks are of different sizes and cylindronical water renewal aeration is provided by a central aerator. Temperature greatly depends on the species, salinity is 35 ppt, pH is 8.2, the water is filtered and the oxygen is high. Air stones are also included. Larvae feed on algae (phytoplankton), which are produced in another tank (algae need oxygen water flow and light to be cultured in long plastic tubes. After that, they feed on artemia nauplii (zooplankton). Artemia is produced in another tank (they are cysts that hatch in well-aerated tanks). Larvae are frequently treated for fungal contamination. Hazards consists of water turbidity, inadequate agitation, and high stocking density. Furthermore, a small amount of antibiotic treatment is added to control bacterial contamination.

The control schedule to be adopted morning and afternoon is as follows: (1) stocking density, (2) fatality survival rate, (3) feed, (4) larvae counting in a sample (density), (5) color observation, (6) behavior (active, passive), (7) appearance of larvae (size), (8) larval stage determination, (9) ecdysis determination, (10) weak and dead larvae determination, (11) fungi and necrosis determination, (12) feeding counts of algae (amount of feed consumed) and (13) feeding counts of artemia nauplii.

Survival between nauplius stage and post-larvae stage varies from 65 to 80% and for some species is 45%. The main failures are related with fungal attacks, bacterial necrosis, badly shaped nauplii and nutrition problems. The major problems due to which larvae do not feed at stage Zoea 1 are the following: (1) in larvae with black stomach, a black plug in the stomach obstructs the digestive tract and (2) in larvae with a grey stomach the algae are not ingested. In both cases therapy with proper antibiotics is required.

After hatching, the newly hatched nauplii are stocked at a density of 100-150 nauplii per liter in 2.5 t larval rearing tanks with fresh filtered sea water filled up to 3/4 of tank capacity. No feed is required at the nauplii stage since the nauplius still utilizes its yolk as food. However, diatoms are inoculated immediately after stocking to ensure availability of feed when the nauplii molt into the protozoa stage. Immediately after stocking, diatom starters are inoculated to ensure bloom of the desirable species. Technical-grade fertilizers can be directly used to enhance algal growth (http://www.fao.org/docrep/field/003/AC232E/AC232E10.htm). Fish excrete ammonia through their gills. The ammonia is waste nitrogen produced by the fish's liver. The ammonia is toxic to most fish in concentrations in excess of 0.5 mg/l. As well as increasing stress levels in the fish, these conditions promote the build-up of slimy and filamentous algae, which can reduce available oxygen levels even further (http://www.scotmas-food.com/booklets/booklet-Fish Production & Processing.pdf).

Although larval rearing techniques are primarily the same today as they were 10 years ago, research in shrimp culture has expanded because of three important factors: (1) the rising demand and costs for fresh food items to be fed to the shrimp, (2) the rising wages of employees and (3) disease problems encountered (http://www.lib.noaa.gov/japan/aquaculture/report1/mock2.html).

Nursery (Juveniles 0.1-2 g)

Nursery tanks or ponds may have natural feed as well; juveniles consume pellets (dry feed). Optimization of the following parameters is required: temperature, pH, ammonia/nitrogen, salinity, dissolved oxygen in water, stocking density and feeding time. Daily feeding as a percentage of body weight is initially 25% for post-larvae and is reduced to 2.5% by the time they reach market size. High growth rate in conjunction with high survival rate is the target of this stage. Any uneaten feed and faecal material should be thoroughly and meticulously siphoned out.

Grow-out Tanks (Juveniles 2-25 g) (Similar to Stage 9)

This series begins at the end of the juvenile phase and includes gonad development, gamete differentiation and growth, reproductive behavior related to molt stage, spermatophore transfer during mating in males, egg development, extrusion and incubation in females. On the other hand, some species present a restricted breeding season. This length is dependent on certain environmental factors, temperature being one of the most influential factors (Sastry, 1983).

Nursery or intermediate culture involves transfer of the spats to the open sea and rearing them until they attain 7 mm shell height. Prior to their transfer, the rearing water temperature should be lowered by 1-2°C every day to approximate the temperature of the sea. This is important for increasing their survival. Each transfer leads to better survival and better growth rate. Because of the differences in the size of spats, condition of the sea area, culture materials and management, high mortality occurs after the spats are transferred to the open sea (http://library.enaca.org/NACA-Publications/MarinecultureWorkshop/SpecialReview-The%20Status%20of%20Marineculture%20in%20North%20China.pdf).

Alevins and juveniles are often transported short distances from the hatchery to growing cages just off-shore. Such transport is short enough for ammonia build-up not to be considered a problem, but the high stocking densities involved mean that a biocide application is necessary to prevent unnecessary transmission of infection between fish (http://www.scotmas-food.com/booklets/booklet-Fish Production & Processing.pdf).

Juveniles are manually moved from settlement plates to nursery culture plates at approximate size of 3-7 mm. During early juvenile stages, an artificial diet is exclusively used as food supply. Several commercial diets, from various diet producers, are available to aquaculturists. The major ingredients of artificial diets are dried kelp powder and fish meat powder. Fresh kelp is the principal food for older juveniles. The quality of artificial diet plays an important role in post-settlement survival. Rearing juvenile sea cucumbers may take as long as 6 months if the rearing conditions are not favorable. As the juveniles grow, the water quality and dissolved oxygen must be maintained at the optimal level. It becomes necessary to increase aeration and water exchange rates. The oxygen level has to be maintained above 5 mg/l. It is also important to use formulated feed that can be digested and absorbed easily. The juveniles are placed in temperature-controlled boxes for transportation. They should not be fed for 1-2 days prior to this operation. The temperature should be maintained below 18°C. The shrimp post-larvae are generally transported in oxygen-filled plastic bags with a sufficient quantity of sea water (http://library.enaca.org/NACA-Publications/MarinecultureWorkshop/SpecialReview-The%20Status%20of%20Marineculture%20in%20North%20China.pdf).

Harvest

Losses during the rearing period from juvenile to market size can be up to 25%. The harvesting, handling, processing and distribution of fish and fishery products should be carried out in a manner that will maintain the nutritional value, quality and safety of the products, reduce waste and minimize potential negative impacts on the environment (http://www.fao.org/DOCREP/005/v9878e/v9878e00.htm). The quest for high quality fish products begins at the harvesting stage. Maintenance of hygienic conditions in fish holds, crates and market areas is required by law (Directive EU 92/48/EEC, effective 1/1/1993) (http://www.scotmas-food.com/booklets/booklet-Fish Production & Processing.pdf).

Most of the harvesting sites are located far from urban settlements and sources of industrial pollution. Rich grounds are also found near urban and industrial centers but are likely to be affected by contaminants and waste products from these centers. The nature and distance of markets and availability of ice and transport facilities determine the type of product processed at the harvest site. In most cases supply and demand are attributed to the marketability of the products and the availability of raw material at a particular time. Harvesting is generally carried out manually by fisherfolk at low tide. The bivalves gathered are then collected into boats and ferried ashore when the tide moves in (http://www.fao.org/docrep/field/003/AB710E/AB710E19.htm).

Processing

The main steps of processing are size grading, heading, freezing and storage. Sometimes there is a thermal treatment (for canning).

Fish processing facilities typically employ a combination of primary and secondary wastewater treatments, depending on the degree to which organic materials are collected separately or mixed into the effluent stream. Waste water from fish processing facilities typically has a very high organic and nitrogen load that can be effectively treated in aerobic or anaerobic systems, including lagoons. However, care should be taken to reduce odors, including those from extended biological and/or chemical phosphorus removal. Fish processing comprises two main activities: processing the main product and processing the by-products. Mollusk and crustacean processing activities contain fewer stages and concentrate on washing, cooking, cooling, processing, and packaging (http://www.ifc.org/ifcext/policyreview.nsf/AttachmentsByTitle/EHS Draft Fish/$FILE/IFC+DRAFT+-+Fish+Processing August+2006.pdf).

Water is one of the most important raw materials used in the processing system, and it must of be of high quality to prevent contamination of the food product. Water is bound to be a bacteria carrier. Without proper treatment, the bacteria will grow in water storage tanks and pipeline to infect the water passing through (http://www.scotmas-food.com/booklets/booklet-Fish Production & Processing.pdf).

Raw materials are received at the fish processing facility from a commercial fishing vessel or fish farm. For some fish species, gutting, cleaning, and head removal can take place at sea on board fishing vessels to maintain optimal quality. This is often the case for white fish with a low oil content, which are then kept on ice or frozen until they arrive at the processing facility. Fatty fish fillets may have an oil content of up to 30% and are usually not gutted until they arrive at the processing facility. When fish are processed at sea, the offal is discarded into the sea. Although this reduces the amount of offal produced at land-based fish processing facilities, if taken ashore, the offal could be turned into a potentially valuable by-product (http://www.ifc.org/ifcext/policyreview.nsf/AttachmentsByTitle/EHS_Draft_Fish/$FILE/IFC+DRAFT+-+Fish+Processing_August+2006.pdf).

Optimum packing conditions for the transport of hatchery-reared and wild grouper larvae were investigated under simulated condition or actual air transport. Simulation of transport motion was carried out through the use of an electric orbit shaker to identify the best packing conditions for the transport of grouper larvae at various ages. The increase in total ammonia level was dependent on temperature, packing density and size of larvae. High packing density and temperature (28°C) resulted in enhanced ammonia level and mortality rates during transport (http://cat.inist.fr/?Modele=afficheN&cpsidt=14646451).

The most important aspects of stages of production can be classified as environmental parameters and cultural conditions. Environmental parameters are optimal temperature, optimal salinity, pH, ammonia/nitrogen, aeration, oxygen, water exchange rate and sensitivity to light or photoperiod (which also determines ovarian development). Cultural conditions are species specific and include size of spawning and hatching tanks (conical for agitation purposes), stocking density, acclimatization, feed used, feeding time, survival rate (mortality 20%), losses due to transport between tanks (5%), cleaning, water treatments and rearing method (community culture method, monoculture method or separate tank method).

RESULTS AND DISCUSSION

The implementation of Cause and Effect to low growth rate, low reproduction rate and high death rate is shown in Figs. 3, 4 and 5, respectively. In Fig. 3, the low growth rate (at first level) is attributed to inappropriate use of gonad hormones, improper larval rearing, improper feeding time, inadequate control of pH, inadequate control of pellets and fresh feed, too rapid dissolution of pellets, and incomplete staff training. At the second level in Fig. 3, the factors are lack of sexual maturity, larvae of different sizes, lack of knowledge in personnel, lack of equipment, bad management and high cost. Factors at the third level of Fig. 3 are fewer

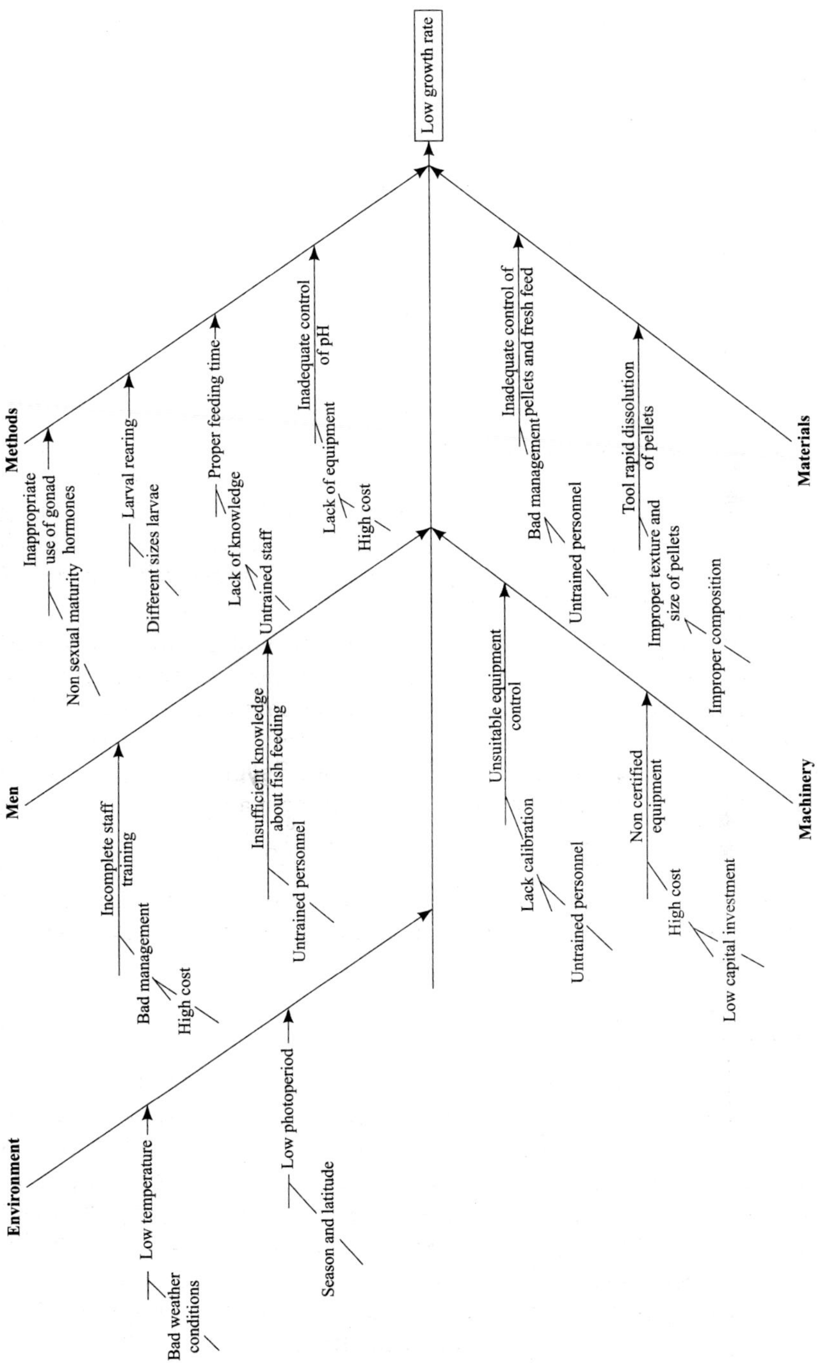

Fig. 3 Application of the Cause-and-Effect diagram (Ishikawa diagram) to low growth rate.

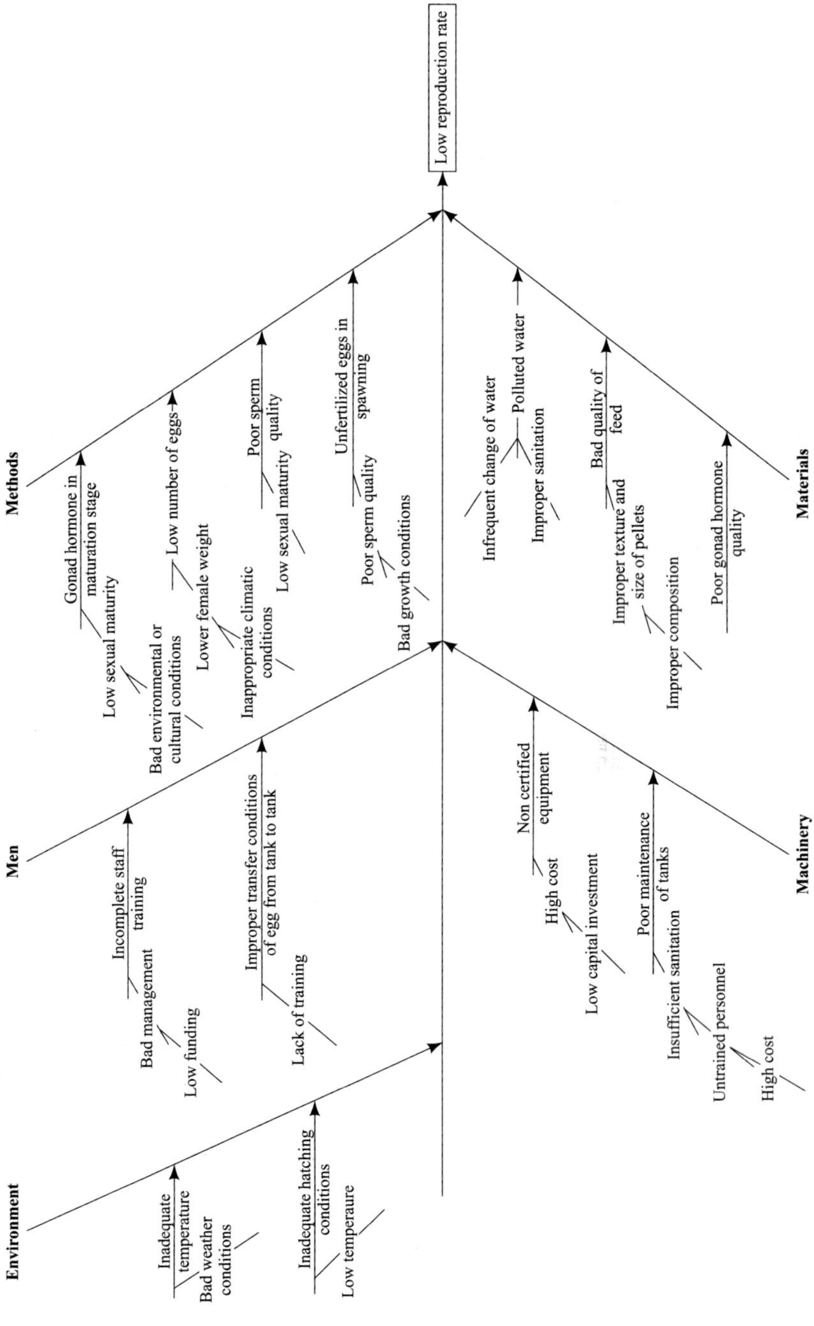

Fig. 4 Application of the Cause-and-Effect diagram (Ishikawa diagram) to low reproduction rate.

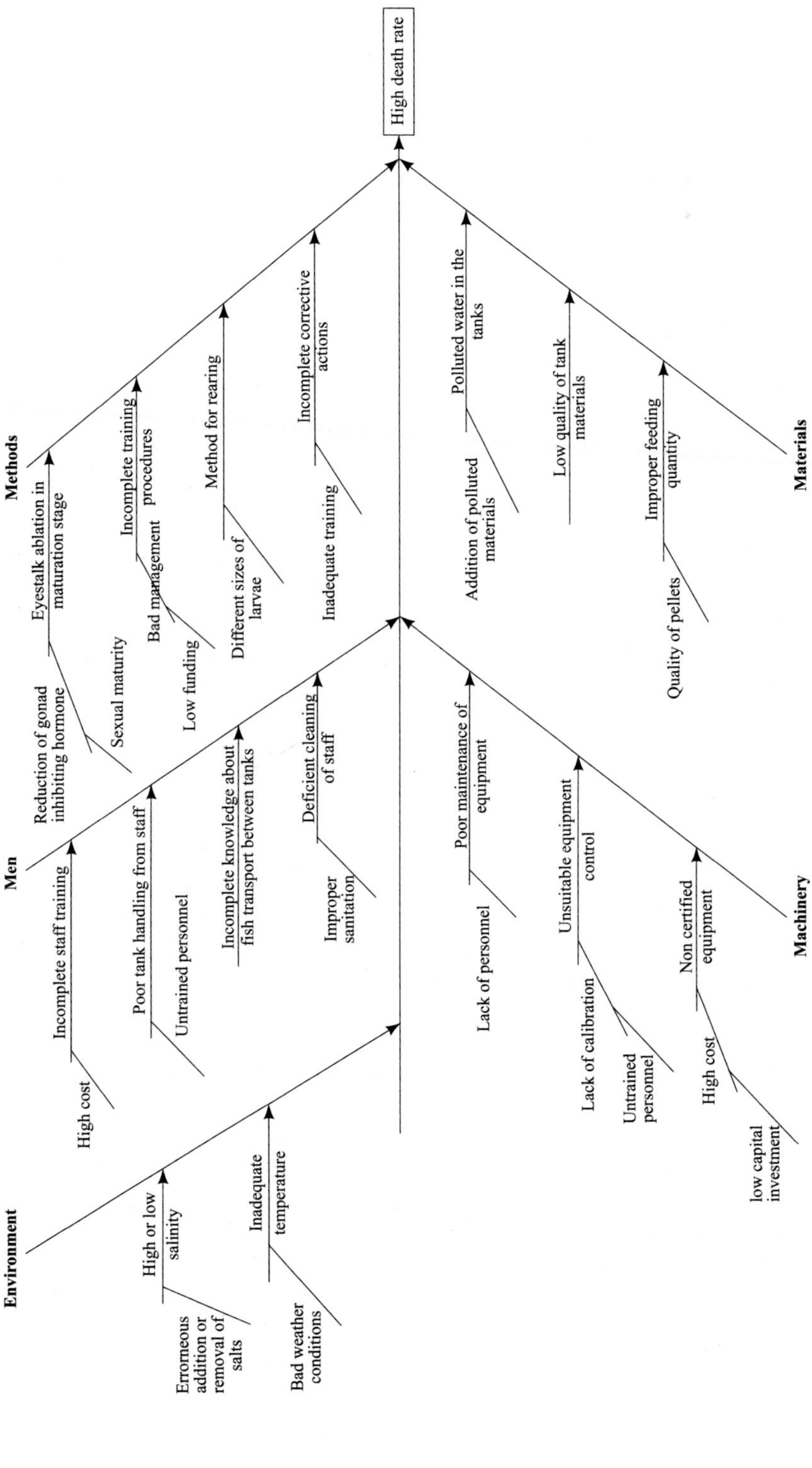

Fig. 5 Application of the Cause-and-Effect diagram (Ishikawa diagram) to high death rate.

than at the second level: untrained personnel, low capital investment and improper composition of pellets. In Fig. 4, the low reproduction rate is imputed to gonad hormones in maturation stage, low number of eggs, poor sperm quality, unfertilized eggs in spawning stage, polluted water, bad quality of feed, incomplete staff training, inadequate temperature and inadequate hatching conditions. At the second level in Fig. 4, the factors are low sexual maturity, low female weight, poor sperm quality, improper sanitation, improper texture and size of pellets, high cost and bad weather conditions. At the third level, the factors are focused on bad environmental and cultural conditions, high cost, untrained personnel and improper composition of pellets. In Fig. 5, the high death rate (at first level) is related to eyestalk ablation in maturation stage, incomplete training procedures, incomplete corrective actions, polluted water in the tanks, low quality of tank materials, improper feeding quantity, poor maintenance of equipment, poor tank handling from staff and inadequate temperature. At second level, the factors are limited to untrained personnel, bad management, improper sanitation and high cost. Finally, at third level, there are only two factors: untrained personnel and low capital investment.

Generally speaking the Cause and Effect analysis at second and particularly at third level reveals a convergence in terms of causes. For example, the common causes are focused on untrained personnel, bad management and high cost.

In Pareto diagram 1 (Fig. 6), an analysis is shown of causes for low growth rate occurring in terms of frequency in various aquaculture plants. The causes of low growth rate are summarized in order of decreasing frequency from low spawning to low salinity. The 82.84% cumulative frequency stands for the first four causes. Therefore, according to the Pareto theory, if corrective actions were taken for these four factors, 82.84% of the problems would be resolved. Figure 7 shows the Pareto analysis after the undertaking of corrective actions. The frequency of the first two factors is significantly reduced because of corrective actions. In Pareto diagram 2 (Fig. 8), an analysis is shown of causes of low reproduction rate occurring in terms of frequency in various aquaculture plants. The causes of low reproduction rate are given in increasing frequency, from unstable feeding program to inappropriate shape of tank. The 79.09% of cumulative frequency stands for the first two causes. Therefore, according to the Pareto theory, if corrective actions were taken for these two factors, 79.09% of the problems would be solved. Figure 9 shows the Pareto analysis after corrective actions. The frequency of the first two factors is reduced and the factor of improper feeding method is at a low level. Finally, in Pareto diagram 3 (Fig. 10), an analysis is shown of causes for high death rate in terms of frequency in

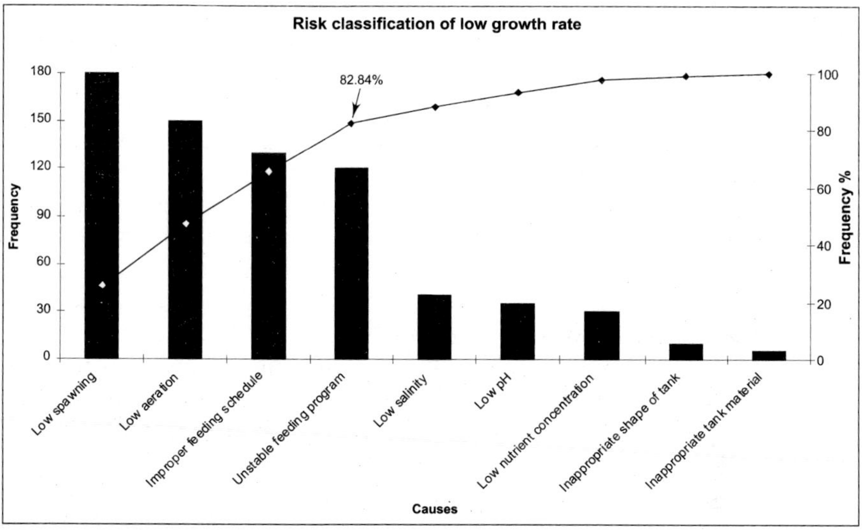

Fig. 6 Pareto diagram: causes of low growth rate risk classification prior to corrective actions.

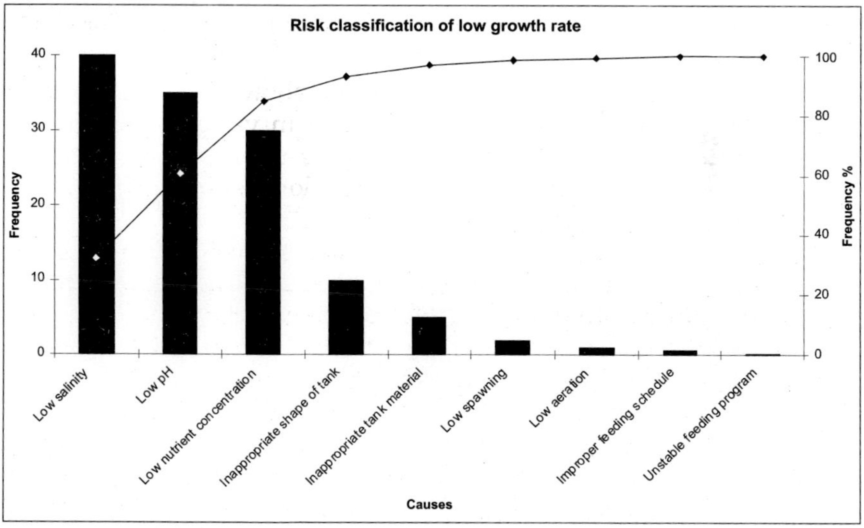

Fig. 7 Pareto diagram: causes of low growth rate risk classification after undertaking corrective actions.

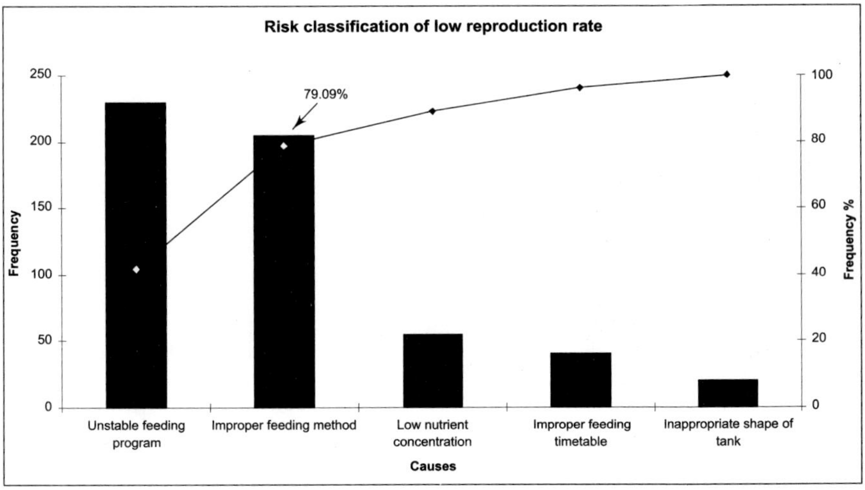

Fig. 8 Pareto diagram: causes of low reproduction rate risk classification prior to corrective actions.

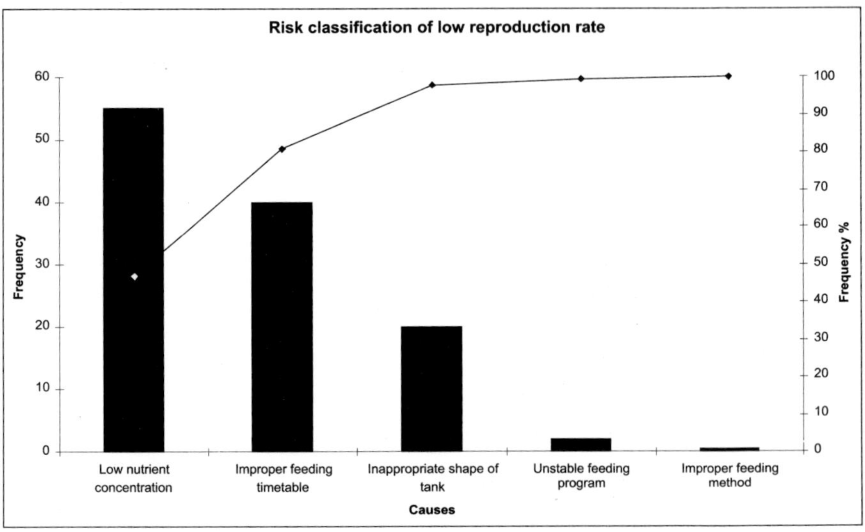

Fig. 9 Pareto diagram: causes of low reproduction rate risk classification after undertaking corrective actions.

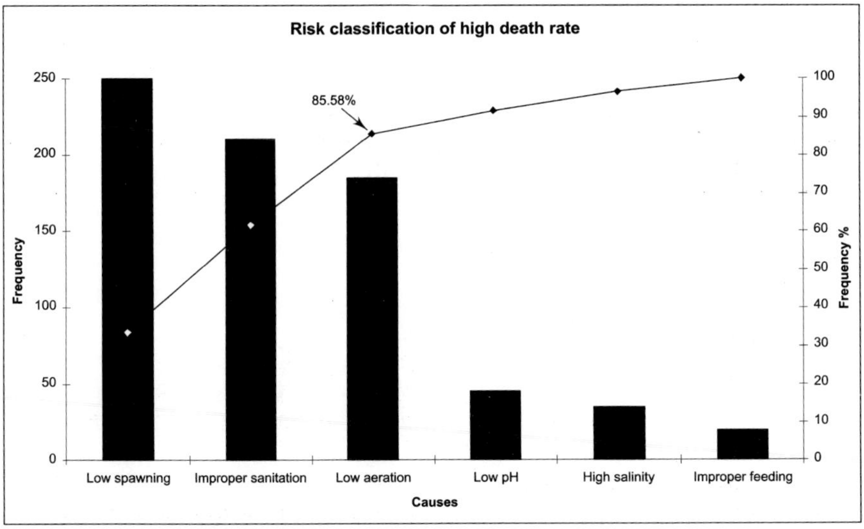

Fig. 10 Pareto diagram: causes of high death rate risk classification prior to corrective actions.

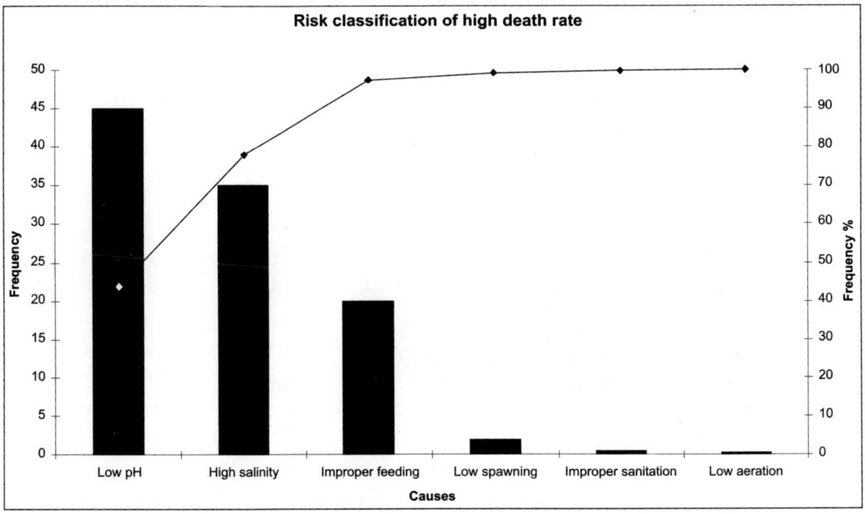

Fig. 11 Pareto diagram: causes of high death rate risk classification after undertaking corrective actions.

various aquaculture plants. The causes of high death rate are recapitulated in increasing frequency from low spawning to improper feeding. The 85.58% of cumulative frequency stands for the first three causes. Therefore, according to the Pareto theory, if corrective actions were taken for these three factors, 85.58% of the problems would be resolved. Figure 11 shows the Pareto analysis after corrective actions. The frequency of the first three factors is reduced and especially the factors of improper sanitation and low spawning are of very low level.

References

Anger, K. 2005. Key factors in the biology of decapod crustacean larvae: effect assessment and implications for cultivation, ePIC (electronic Publication Information Center, http://web.awi-bremerhaven.de/Publications/Ang2005b_abstract.html accessed 20 Oct. 2006).

Chang, Y., and J. Chen. 2006. The status of marine culture in North China. (http://library.enaca.org/NACA-Publications/MarinecultureWorkshop/SpecialReview-The%20Status%20of%20Marineculture%20in%20North%20China.pdf).

Chung, J.S., and S.G. Webster. 2004. Expression and release patterns of neuropeptides during embryonic development and hatching of the green shore crab, Carcinus maemas. Published by The Company of Biologists, Development, 131: 4751-4761.

De Kleijn, D.P.V., K.P.V. Janssen, S.L. Waddy, R. Hegeman, W.Y. Lai, G.J.M. Martens and F.V. Herp. 1998. Expression of the Crustacean hyperglycaemic hormones and the gonad inhibiting hormone during the reproductive cycle of the female American lobster *Homarus americanus*. J. Endocrinol., 156: 291-298.

FAO. 1994. Marine Fisheries and the Law of the Sea: A Decade of Change. FAO Fisheries Circular No. 853, p. 35.

Forward, R.B., D. Rittschof and M.C. De Vries. 1987. Peptide pheromones synchronize crustacean egg hatching and larval release. Chem. Senses, 12: 491-498.

James, E.B. 1998. Risk Analysis: Two Tools You Can Use to Assure Product Safety and Reliability, [online], Available at: http://www.1stnclass.com/risk_analysis.htm [Accessed 02 February 2005]

Katsuyuki, H., I. Hideyuki, A. Nobuhiko and F. Kyohei. 2004. Ovarian development and induced oviposition of the overwriting swimming crab *Purtunus trituberculatus* reared in the laboratory. Fish. Sci., 70.

Kumamoto, H., and E. Henley. 1996. Probabilistic Risk Assessment and Management for Engineers and Scientists, 2nd ed. IEEE Press, Piscataway, New Jersey, USA.

Masayuki, S. 1992. Control of hatching in an estuarine terrestrial crab I. Hatching of embryos detached from the female and emergence of mature larvae. Biol. Bull., 183: 401-408.

Mock, C.R. 2005. Crustacean Culture, Contribution No. 343 from the National Marine Fisheries Service, Biological Laboratory, Galveston, Texas (http://www.lib.noaa.gov/japan/aquaculture/report1/mock2.html).

Mortimore, S., and C. Wallace. 1995. HACCP: A Practical Approach. Chapman & Hall, London, Glasgow, pp. 25-47.

Okumura, T. 2004. Perspectives on hormonal maniputation of shrimp reproduction. JARQ, 38(1).

Prasert, M., S. Prasert, D. Praneet, W. Kanokphan, S. Supatra, S. Anchalee and W. Boonsirm. 2005. Possible mechanism of serotonin induces ovarian maturation in giant freshwater prawn broodstock, *Macrobrachium rosenbergii*, de man. 31st Congress on Science and Technology of Thailand at Suranaree University of Technology, 18-20 October 2005.

Reigada, A.L.D., and M.L. Negreiros-Fransozo. 2000. Reproductive cycle of *Hepatus pudibundus* in Ubatuba, sp, Brazil. Revista Brasileira de Biologia, Sao Carlos, 60(3): 29-36.

Sachs, N. 1993. Failure analysis of mechanical components. Maint. Technol. Mag., Sept. : 8-16.

Sastry, A.N. 1983. Ecological aspects of reproduction. *In:* The Biology of Crustacea: Environment adaptations, Vol. 8. W.B. Vernberg (ed.). Academic Press, New York, pp. 179-270.

Stamatis, D.H. 1995. Failure Mode and Effects Analysis: FMEA from Theory to Execution. ASQ Quality Press, New York.

Surface Vehicle Recommended Practice (SVRP). 1997. Draft proposal from the SAE J1739 Main Working Committee, April.

Treece, G.D. 1999. Shrimp maturation and spawning, UJNR Tech. Rep. No. 28.

Electronic Sources

http://www.tracc.org/my/Borneocoast/aquaculture/crustacean aquaculture.htm (accessed 15 Oct. 2006)

http://www.fao.org/DOCREP/005/v9878e/v9878e00.htm (accessed 15 Oct. 2006)

http://www.scotmas-food.com/booklets/booklet-Fish Production & Processing.pdf (accessed 18 Oct. 2006)

http://www.mi.mun.ca/mi-net/fishdeve/lobster.htm (accessed 20 Oct. 2006)

http://www.bioone.org/perlserv/?request=get-abstract&issn=0278-0372&volume=022&issue=04&page=0762 (accessed 18 Oct. 2006)

http://www.fao.org/docrep/field/003/AC232E/AC232E10.htm (accessed 15 Oct. 2006)

http://www.tracc.org/my/Borneocoast/aquaculture/crustacean aquaculture.htm (accessed 18 Oct. 2006)

http://www.scotmas-food.com/booklets/booklet-Fish Production & Processing.pdf (accessed 20 Oct. 2006)

http://www.fao.org/docrep/field/003/AB710E/AB710E19.htm (accessed 20 Oct. 2006) http://www.ifc.org/ifcext/policyreview.nsf/AttachmentsByTitle/EHS_Draft_Fish/$FILE/IFC+DRAFT+-+Fish+Processing_August+2006.pdf (accessed 15 Oct. 2006)

http://cat.inist.fr/?Modele=afficheN&cpsidt=14646451 (accessed 15 Oct. 2006)

http://en.wikipedia.org/wiki/HACCP (accessed 18 Oct. 2006) http://en.wikipedia.org/wiki/Failure_mode_and_effects_analysis (accessed 15 Oct. 2006)

http://certification.bureauveritas.com/webapp/servlet/RequestHandler?mode=PT&pageID=32154&nextpage=siteFrameset.jsp (accessed 20 Oct. 2006)

http://www.lrqausa.com/4-1.cfm?id=1029 (accessed 18 Oct. 2006)

http://www.bsi-global.com/Food_Management/FoodSafety/bseniso22000.xalter (accessed 15 Oct. 2006)

Index